APOLLONIA ON MY MIND

sidestonepress

APOLLONIA ON MY MIND

The memoir of a paraplegic ocean scientist

NICHOLAS C. FLEMMING

Published by Sidestone Press, Leiden
www.sidestone.com

Lay-out & cover design: Sidestone Press
Photograph cover: Howard Rosenstein

Series Editor: Miranda Richardson
Copyediting: Bob Holtzman
Proofreading: Alva McSherry
Indexing: Nicola King

ISBN 978-94-6426-032-8 (softcover)
ISBN 978-94-6426-033-5 (hardcover)
ISBN 978-94-6426-034-2 (PDF e-book)

The book is dedicated to Jay, Kirsten, Peter, Matan, and Lucio.

Closed and sacred, filled with a weightless fire,
Fragment of earth offered up to the light,
This place is pleasing, overspread by flames,
Composed of gold, dark trees and stone, where so
Much marble trembles on so many shadows,
The faithful sea asleep across my tombs!

From *The Cemetery by the Sea*, by Paul Valéry, translated by Nathaniel Rudavsky-Brody.

Contents

Preface

This book is a big surprise for me. I never expected to live past the age of 30. I had a natural inclination to take calculated physical risks with adventurous escapades. I had also been trained in the Royal Marines and Royal Air Force Volunteer Reserve to practice potentially dangerous activities such as oxygen diving and flying light aircraft with proper professional care. Nevertheless, someone who repeatedly takes calculated risks, and further embellishes life with a few impulsive uncalculated ones, is likely to make a mistake sooner or later. On this assumption, I concluded that I might as well enjoy my twenties according to my nature, taking the rest as a bonus if I survived.

Now, to settle down at the age of 80 to write a memoir requires optimism: at the very least, the assumption that one will live past the age of 81. So far, so good.

Viewed close-up, my life has been a confused muddle during any particular day or week. Viewed in retrospect, from a distance, there are more or less consistent themes that run on, or disappear for a few years and then recur, decade after decade: diving exploration in the sea, underwater ancient cities, sea levels and climate change, global ocean forecasting, ocean exploitation and economics, submerged prehistoric archaeology. But there were many false trails that led nowhere and were abandoned.

In this book, I have described eleven aspects or episodes of my life. By adopting a separate theme as the driving motive for each chapter, I can show events following logically one from another, even if some temporal jumps are left empty in that chapter. Some topics inevitably occur, or are referred to, in several different chapters. The techniques and practice of diving under the sea with breathing equipment recur several times. Only one chapter is focused on the practice and safety of diving itself as a technical exercise in the conduct of scientific research. That includes the maintenance of safe training standards, avoidance of fatalities, safety of employees, and risk management. In the other chapters, diving is treated as a tool or service.

Some chapter themes run right through my adult life, so that the discussion of that topic concludes with the situation as it is today. Other themes are short-lived and terminate at an earlier date. I have replicated attitudes and experiences as they were felt in the past, so far as possible without hindsight. What were the priorities and dangers then, how did we evaluate them then, and why did we do things then in the way that we did? Sometimes I explain what has happened since, or how the problem would be tackled now, but in most cases, the effect of 30 or 50 years of change will be obvious to

In: Flemming, N.C., 2021, *Apollonia on my Mind. The memoir of a paraplegic ocean scientist*, 11-18. Leiden: Sidestone Press.

the reader without comment. The irony of what we thought and did then, even if it was obviously wrong from a modern perspective, does not need emphasis.

I have burrowed through dusty boxes of diaries, field reports, and my early scientific and academic or technical publications, sometimes consulting the web to ensure accuracy so far as possible. If a minor error would make no difference, I have relied on my memory and personal documents where available. Errors that remain are of course mine. There is a short bibliography suggesting further reading and websites for each chapter at the end of the book.

Some scientists claim that they are or were the first to discover something, or observe it, or define a grand new principle, or perfect a new piece of research equipment that made a great discovery possible. Yet one of the greatest scientists of all time, Isaac Newton, said, "If I have seen further (than others) it is by standing on the shoulders of giants". Newton was an egocentric and cantankerous man, but it is still a good saying and a worthy principle. I respect scientists who give credit to their forerunners.

Growth of knowledge in the present era is an exponential curve. In the early 1960s, when I started my research on past sea level change using the evidence of coastal archaeological buildings and underwater caves, I was conscious of earlier observers extending back in time more than 100 years. People had seen buildings under the sea in the 18th century, and I enjoyed such obscure discoveries narrated in long-dead journals. For example, Anders Celsius and Carl Linnaeus, two of the greatest scientists of the 18th century, walked together along the north Baltic coast of Sweden and noted that the land had been exposed by a recent retreat of the Gulf of Bothnia. They wrote a field report in Latin. Celsius and Linnaeus are of course well remembered and well honoured for amazing achievements, but not for their early observations on sea level.

Once you are aware of a known but unexplained anomaly in nature, you can look at the facts to figure out what they mean, approaching the problem in a new way. You gather many new observations more accurately, analyse the data more thoroughly, capture more facts with new techniques or equipment, establish new principles using new analysis. I felt that I was picking up the torch of research from generations before me to advance knowledge by a significant and useful increment. Decades later, many of my precursors in the 19th and 20th centuries have been forgotten, and I am treated by some as a pioneer. It is sometimes not possible to say that you originated an idea, but by taking action you become the first person to achieve a factual discovery or to confirm or reject that idea.

I enjoy finding the earliest and most original sources of the ideas that I need to build on, giving those people the credit for their research. It makes me feel that I am part of an intellectual river of creation, adding ideas and thoughts in a cumulative flood to solve bigger and bigger problems.

* * *

Many of the events in this book, or at least the results from them, are published in official reports, government policy statements, UN committee reports, academic research papers by myself or collectively with others in learned journals, and edited books or conference proceedings. There is no need here to go into exhausting technical details, and if you do want to check those, please see the bibliography. Except for Chapter 3 on my childhood, the topic of every chapter has already been the subject of dozens, perhaps hundreds of technical publications. In this book I describe my experience of these events.

I enjoyed an unusual childhood, and have tried to bring that enjoyment to the next generation. My wife, Jay Kleinberg, and our children, Kirsten and Peter, feature as active participants in several chapters. The upbringing of children in a busy household with two workaholic parents and many international commitments required Jay and I to adapt our routines. I have included some of the stories of the children's early days, including diving expeditions where they accompanied us. Family life, with all its unpredictable dramas, scares and joys, has been an enlightening and sobering experience.

My purpose now is to describe what life was like while doing the work, the people I met, the sensations I experienced, the risks and calculations when things could have gone wrong and sometimes did go wrong. In short, to re-live many of the things, and the fun or distress, that the official reports always leave out. A scientific report tends to include only the experiments or measurements that worked and turned out well. The author may explain that so many days or months were lost due to bad weather or equipment failure but then glosses over the problems to focus on the success. In this book I will be a bit more explicit about problems, why decisions were taken, what were the alternatives, and what happened when things went wrong. One can tell a second story, a more human and richer story of people, events, and motives, that ultimately supports the sterilized "official" version. It is also sometimes quite funny: quirks of character, chance meetings, or random events often determined what happened, started a positive chain reaction, or killed a good idea stone dead.

* * *

The alert or prudent reader who scans this book before purchase will notice that in some pictures I am using a wheelchair. I broke my back just below the ribcage in a car accident in Yugoslavia in July 1969. Neither in my memory when awake, nor in dreams when asleep, do I distinguish my life before and after the accident, and I therefore make no distinction in narrating chapters which describe the strands of activity in which I have been employed and enjoyed myself, whether they are before or after 1969. My academic CV and my list of scientific and technical publications show no discontinuity or change of pace at 1969-70. In my daily life after 1969, there were very few work tasks that I declined to do because of disability. In fact, I cannot think of any. My employers knew that I was available to work anywhere in the world and, so far as I know, they never hesitated to ask me to do a job or accept a responsibility because of my disability.

I have grouped all the aspects of disability into Chapter 7, which I have subtitled "A Statistical Sample of One". My reason for this is that I cannot and do not preach. I have no theories or political view of disability. I don't know the priorities of the various disability lobbies or charities, but I am sure they are justified. They are based on facts and figures that I do not possess. I am only describing the experience of one disabled person with one particular handicap, acquired at a particular age, and treated in a particular hospital. Everyone else's experience is different. I cannot generalize, and I do not attempt to do so. In Chapter 7 I do, however, explain some of the tricks of the trade as to survival in a wheelchair for a paraplegic.

* * *

In the world of diving, whether as a sport or a profession, accidents and fatalities unfortunately do occur. Sport diving has a low accident rate that is consistently similar to that of a range of other outdoor adventure sports such as mountaineering, sailing, riding, or skydiving. When divers are at work as defined in law, there are additional responsibilities and risks, and the accident rate varies between different types of employment, professions, industries, and self-employment. As chair of the UK Scientific Diving Supervisory Committee from 1978 to 1987, and then as vice chair until 1997, I was responsible for diving safety for British scientific divers in government laboratories and universities, and to a certain extent, for standards internationally. I took that responsibility seriously.

Diving by British scientists while at work has an unblemished safety record for the past 60 years. There has not been a diving fatality for a scientist at work covered by the diving regulations during all that time, out of a community varying from 200 to 400 professional scientists in employed work and some hundreds of students occasionally diving in official projects at universities. Nevertheless, the way we dived in the 1950s is not the way we dive now, and those readers who are experienced modern divers may be shocked to learn what we did. I do not think that we were reckless or negligent, and we were fully aware of the risks, both medical-physiological and from traumatological accidents, taking the precautions that were deemed necessary at the time. The history of scientific diving and the evolution of dive safety management is reviewed in Chapter 8. I suggest a simple comparison. The first commercial airline I flew on was Olympic Airways in 1957, on a flight from Corfu to Athens, and we flew in a converted military aircraft with the main passenger door open, as there was no air-conditioning. When I started flying regularly on international flights in 1965, the carry-on hand luggage was stowed in open racks above the passengers' heads, with no restraints. After a few substantial but survivable crashes, the accident-analysis experts realized that many injuries were caused by luggage crashing down on the passengers and hurtling forward while the passengers were fixed in their seats. Nowadays, all carry-on luggage is enclosed in sealed overhead lockers, or stowed securely under seats.

Similarly with the management of safe diving. When numbers of divers were very small, the less likely combinations of risk that trigger an accident only occurred and coincided at long intervals, so the anticipation of, and prevention of, that risk was only factored into training later, as experience and statistics accumulated. These considerations then become very important when assessing risk management for large numbers of people. Unlikely events, and unlikely combinations of events, are much more probable as the numbers at risk increase. In complete contrast, that argument does not excuse or justify some of the more risky things that I did as an individual, which can only be explained by youthful hormones.

* * *

Every writer of autobiography or memoir is forced to confront the deceptive nature of memory. Modern brain research emphasizes the errors, psychological bias, and confusions of memory still further. In the simplest terms, there is my original experience, stimulus, or sensation, which is uncatchable in its entirety; then the mental record that I made of the events at the time; the conscious self-perception of the experience, and the short-term memory as I recalled it and sometimes wrote it down within a few hours or days. A decade or 50 years later, some memories have disintegrated, others have grown dusty, and some have been taken out, polished, used, and put back into my mind so many times that they are no longer much like the original. I can only write my recollections as they are now, refreshed by the records written down at different dates.

In addition to the quirks and errors of memory, there are different types of written record of what happened, such as diaries, expenses claims, tax returns, and scientific papers. Over many years, I have often smiled to myself at the incongruity as I fill in expense claims or tax deductions for expenses as I remember a meal with a pleasant companion or a ferry ride, and think how I really experienced and enjoyed that journey. I recall the two Turkish military police with submachine guns who arrested me in that restaurant, and it just appears as item "X", cost "Y", in a table or spreadsheet which, for the purpose of expenses, is a complete description. Similarly, a month of work on the coast of Turkey or Israel or the Aleutian Islands may appear in one form as a scientific paper of five or ten pages in an academic journal, or in another as a memory of weeks of sunlight, swimming, new friends, good food, or ice and brown bears, and wildly unexpected cultural discoveries.

As a scientist, traveller, bureaucrat, explorer, and plain ordinary human being, I have kept many different kinds of records: a daybook of events, meetings, and decisions; official minutes of meetings and committees where I was secretary or chair; action lists of things to do; field reports of data gathered and errors in observations; maps resulting from surveys and measurements; calculations and drafts when writing scientific papers prior to publication; the published academic papers when they appear in print, or more recently, on the web; diaries that I sometimes kept in parallel with fieldwork records; and the occasional published work which was not purely of an academic nature. Although I have moved house many times since I graduated from university in 1960, and experienced a divorce which meant abandoning many of my possessions in 1977, I have somehow accumulated a ridiculous quantity of papers, maps, documents, notebooks, press cuttings, and diaries filling many dozens of file boxes in the conservatory of my present house.

As I searched through boxes of papers, I became progressively astonished by the detail in which I had recorded events, scenery, meals, impressions, and even dialogue, as well as by some of my own past actions and opinions. While reading and sorting these files, there were times when I felt guilty, as if I were snooping or spying on the private diaries of another person. I normally consider that I have a fairly good memory, but reading in detail the way I lived and thought 50-60 years ago has been quite a shock. The sheer detail that I wrote down is, to me, overwhelming, and I only did that for a few active weeks in each year, and only in a few years. Prior to beginning this memoir, I had not consulted or read these documents since the year or so in which they were written, by which time the factual scientific data had been extracted and distilled into an academic publication. After that, the notebooks and boxes gathered dust as they were packed and unpacked in successive offices and house moves. Now perhaps they have served their purpose. My field notebooks, original maps and technical photographs are now available in the special collections of the Hartley Library at the University of Southampton.

* * *

I grew up in the great era of illustrated weekly news magazines, with the adage that a picture is worth a thousand words. Photography has immense value in environmental or archaeological fieldwork, but I am not a talented photographer, and instinctively I try to measure things rather than take a picture. The data to be presented in a scientific paper must be quantitative, and this implies accurate measurements which can, in some circumstances, be supported by pictures. For this reason, I try to travel with a photographer, or include a photographer in an expedition team, in addition to assuming that many team members will take their own casual pictures. Jay is an excellent photographer and took many of the pictures in this book. From the perspective of obtaining linear dimensions, a photograph postpones the critical moment when you have to extract the precise numbers from a picture that may have been distorted. Many experts have, over the years, produced excellent ways of solving those problems and extracting accurate measurements, but that in itself is difficult under water when the visibility is limited. In recent decades, improvements in multibeam sonar imaging, and robotic digital photogrammetry, in which thousands of electronic images can be integrated into a precisely-scaled, continuous 3-D picture, have revolutionized the speed and scope of multi-image recording and made quantitative data more accessible from images. But I do not regret my cautious decision for most of the years covered by this book to measure features directly in the field if possible, with photography as a back-up.

Photography had and has many wider valuable uses. It confirms that the scientist is not lying, and was actually at the place doing the work that is reported in a quantitative way. For a geo-scientist, photo-documentation records the state of the feature at that date, which may be valuable for researchers who revisit the site later. I have thousands of pictures of coastal archaeological ruins from dozens of countries for this reason alone. Finally, the pictures can be used for publicity, lectures, fundraising, and entertainment.

* * *

I hope to show in these stories my appreciation for the pleasure and help that many people have given me. I must first thank Jay, who has accompanied me on many voyages of exploration, and our two children, Kirsten and Peter. They are all creative, hardworking people and good companions. Jay has her own professional life as a historian and has carried the extra and unusual burden of providing the back-up support for a paraplegic husband. I make a cheerful stab at being independent, but the risks I have taken when travelling alone were only acceptable because Jay provided the safety net and default support in case and when things went wrong. She has been unendingly patient with my own work obsessions and the time taken to write this book. I thank my siblings: my two sisters, Sarah and Mandy, and my late brother, John, whom I remember with a mixture of happiness and sadness. For several years, my family have been encouraging me to write a memoir, but I prevaricated and always found new research that I felt needed doing and thought to be more important.

This book would not have been written without the consistent chivvying, prompting, and analysis of ideas provided by my long correspondence and occasional meetings with Professor John Hale, Director of Liberal Studies, University of Louisville. He has acted as friend, intellectual inspiration, and agent provocateur. I met John at an archaeological site in southern Greece in 2010, and he explained that he taught his archaeology students that the mapping of the underwater Greek city of Apollonia, near Benghazi, Libya, which I had conducted in 1958-59, was the first fully scientific project in underwater archaeology. He wanted to know how and why I had done that at the age of 21. I explained the training in underwater survey methods that I had received during my military service in the Royal Marines and told him also that I had returned to examine Apollonia in 2003, and that the city was seriously decayed and damaged in the intervening 44 years. This dialogue produced the structural symmetry of a memoir, opening and closing with the birth and death of Apollonia (Chapters 1 and 11).

I have worked with many hundreds of professional colleagues and associates in dozens of organizations in many countries, and a fair sample of them feature in these pages.

John Woods appears in several chapters. We first met as undergraduates when he was at Imperial College London and I was at Cambridge. Our mutual interest in diving, underwater research, oceanography, the commercial and social benefits of the sea, and the creation of a Global Ocean Observing System, resulted in our careers criss-crossing frequently over several decades. His research achievements in global oceanography are remarkable, and his role in establishing the British National Oceanography Centre in Southampton was a truly immense achievement. John was at times my boss and has always encouraged and promoted my research. I am very grateful for his support and friendship.

Yehuda Melamed at the Haifa Hyperbaric Centre was a naval officer when we first met in 1974 when, together, we managed a scuba diving course for disabled men. We became close friends afterwards, enjoying many sociable occasions, and Yehuda has saved my life more than once in the decades of our acquaintance.

Before presenting a list of the many people to whom I owe gratitude and appreciation, I want to acknowledge those colleagues, pioneers, and explorers who, in the early days of diving, died in the sea: Conrad Limbaugh, Peter Small, Peter Gill, Rob Palmer, and Keith Muckelroy. These men, in very different ways, pushed forward the techniques of diving with scuba apparatus, and advanced scientific and archaeological discovery. Their deaths were a sad loss to the diving community, and to me. Their loss influenced the way in which I developed policies for diving safety, described in Chapter 8.

For many different kinds of help and collaboration, I thank, in alphabetic order: Abdul Hamid Abussaid, George Bass, Geoff Bailey, Paul Bennett, Debbie Bliss, Lucy Blue, Mike Davis, Sir George Deacon, Catherine Giangrande, Catherine Goetschel, Harry Gould, Sir Ludwig "Poppa" Guttmann, Jon Henderson, Bill Hemmings, Bill Hill, Vane Ivanovic, Solvey Koetschau, Elisha Linder, Sally Marine, Sir John Mason, Colin McLeod, Jean-Pierre Misson, Keith Muckelroy, Iouri Oliounine, Ken Oliver, Avner Raban, Howard Rosenstein, Dimitris Sakellariou, Angi Stefanon, Pat Troy, Claudio Vita Finzi, Sir Arthur Weller, Stewart Willis, Nick Wood, Alexander Yankov, and Nikos Yialouris. As I write this list, I am saddened to note how many are no longer with us. I have not indicated those who have died before me. They are alive in my mind.

I thank the Honor Frost Foundation for supporting and endorsing the publication of this book, and for their appointment of Miranda Richardson and Bob Holtzman as editors. I must thank them both for achieving a transformation that a scientist finds difficult to explain. Editing, which suggests correcting the use of commas and semicolons, does not begin to describe the subtle changes of style, compressions, deletions, additions, and clarifications that they achieved. Karsten Wentink and Eric van den Bandt at Sidestone Press managed a busy schedule of checking graphics and page layout to produce a handsome final volume with efficient computer support. I thank them both.

Finally, I thank all the boatmen, ship's captains, masters, crew, bosuns, fisher-folk and deckhands who provided and operated craft of every size, shape, and type over 60 years of diving and underwater research. Their skill and care kept me and my companions safe from the perils of the sea.

1

Apollonia on My Mind

Sea spray drifts over me and I shiver in spite of the African sun. Clad in swimming trunks, I crouch on the ridge of a high wall of solid rock from which I stare down into the sea, and try to understand the mind of a Greek king over 2500 years ago. Another wave rises out of the Libyan Sea, curling over to crash in clouds of foam. I am thinking.

Why was this island off the coast of Libya sculpted this way over one hundred human generations ago? I must watch and think. Be patient. A human mind created this carved shape for a purpose.

A wave crest rears and plunges with a roar into the lagoon as the engulfed air explodes upwards into jets of blue-green with glinting flecks and dark shadows that break into white spray. The cloud of salt spume drifts against the cliff face of the island.

The blue-black water sucks back again from the ridge that separates the lagoon from the open sea. I watch the straight smooth rock, exposed and slimy in the brilliant sunlight. Watch the water rise and plunge again, falling, falling, and the spray rises, but the island behind the ridge on which I am perched stays dry. Is this what he intended 2500 years ago? To protect the island with its precious buildings that were full of weapons and stores and, above all, to protect the war galleys?

When the ancient city of Apollonia was being built in about 600 BC, a weather-beaten battle-hardened man, probably a Phoenician sailor-turned-engineer, told a Greek king that if his slaves cut the rocks just so, and just so, his city would last for ever and defy Poseidon and all the storm-gods of the sea (Fig. 1.1).

The king ordered slaves, sponge-divers, and murex-purple-shell-divers to excavate the rocks as planned. They toiled and struggled in the surf day after day while they worked underwater holding their breaths for several minutes on each dive. The gang-masters exhorted them without mercy. How many divers drowned in the task? How did they cut solid rock more than 2 m (6 ft) deep in the water using only bronze tools and abrasive powder? They succeeded in cutting a trough 70 m (231 ft) long and several metres wide, deep into the bedrock, leaving a low residual rock barrier on the seaward side, and creating a vertical artificial cliff face on the island. The water-filled trough ran along the whole of the outer face of the island.

The result was a carved island-fortress protected from storms by this artificial wave-trap, creating a work of technical art and cunning as inspired in its way as any statue of marble. On the sheltered side of the island, away from the open sea, facing the harbour and the city centre, were ten slipways for storing the king's lightweight fighting galleys

In: Flemming, N.C., 2021, *Apollonia on my Mind. The memoir of a paraplegic ocean scientist*, 19-60. Leiden: Sidestone Press.

Figure 1.1. Much of the ancient city of Apollonia, Cyrenaica, lies submerged in the bay shown here between the theatre and the rocky islands, 250 m off the modern shore.

Figure 1.2. West Island at Apollonia where the author sat on the top of the rock wall which appears dark at the back of the island. The tops of the slipways are easily visible, as is the network of low walls under the sea in shallow water close to the beach. The gap between the submerged block forts and the channel to the eastern harbour is on the right margin of the picture.

when they were hauled out of the sea to prevent weed growing on the hulls. That way they stayed light, smooth and fast. They were only launched for military conflicts or at times of tension for training. This feat of harbour engineering was the Phoenician's masterpiece. His was the art of turning the destructive power of nature against itself to produce a safe haven for human activities in a fortified coastal city (Fig. 1.2). He had forced the natural physics of the waves to destroy their own forward momentum. They dissipated their energy within the artificial lagoon. The war galleys were safe. Perhaps the king paid him well for his work.

Unfortunately, after a thousand years, nature played an unexpected trick. Earthquakes wrecked the city by dragging the harbour and the islands down into the sea. Then the biggest waves overtopped the rubble breakwaters and cruised across the wave-trap so that they struck too fiercely on the rock walls. The wealth and grandeur of a commercial port city were destroyed in minutes. Multi-storey buildings cracked as masonry fell into the sea. Sailors and dock workers, slaves and merchants, scrambled for the high ground on the islands, or fled to the towers of the city wall that did not collapse. Broken houses and warehouses, fallen columns and roof tiles, roads, statues of marble, coins and pottery are now beneath the sea in the 25 hectares (62 acres) between the islands and the mainland. The abandoned remnants are exposed to the force of storms. This rock wall where I sit makes the last partial defence against the waves. Plate tectonics shows that Africa is moving slowly towards Europe, crushing the Mediterranean Sea, and that process has already crumpled the earth's crust, taking the city of Apollonia down nearly three metres (10 ft) deep. The calculations of the Phoenician engineering genius have come to nothing.

Figure 1.3. Martin Minns (left) and the author working on the ruins of Apollonia. Note my Formica drawing board hanging on a string.

Figure 1.4. Looking down from the cliff top on the West Island at Apollonia, Libya, into the artificial wave-trap cut by divers and engineers about 2500 years ago. The ridge of rock just visible beyond the white water has been created by cutting the trough between it and the cliff. The waves break in the trough, and do not attack the cliff. The town has sunk about 2 m into the sea, so originally the wall of the wave-trap would have risen 1.5 m above still water level.

Behind me, between the island and the Libyan shore of Africa, divers were swimming on programmed tracks, placing their survey poles on the corners of the drowned buildings of the lost Greek city. For the past three weeks our team had been diving every day, searching wide-eyed, peering through the haze that cloaked the ruins of Apollonia (Fig. 1.3). On shore, Nick Wood, the survey architect, was hunched over the plane table and telescopic alidade plotting the divers' positions. Day after day, blasted and bleached by salt and sun, we created a map of the city under the sea, street by street, wall by wall, quays and docks. We did that in Libya back in August 1958.

We made the first-ever scientifically accurate map of an underwater city anywhere in the world. The divers each day swam over collapsed buildings as we sought to interpret the strange shapes of the ruins and to understand the motives of the people who lived there.

Sixty years later, I can zoom down with Google Earth to focus on the yellow-grey rocky island off the coast east of Benghazi, finding the wall where my 21-year-old body crouched while my eyes burned and burned into the waves, searching for the truth. But the modern aerial photographs show that something has changed, because that great rock wall has gone. Simply vanished. It was solid sandstone, 2 m thick, 5 m high, and 20 m long (7 x 17 x 66 ft) linking two high massifs of rock. It effectively joined together the two halves of the island at a constant height, so as to protect the city of Apollonia from the storm waves that the north wind brings in across the Libyan Sea. It was the second line of defence against the sea, after the wave-trap, and a necessary part of the military defence of the city.

Every surface of that island had been quarried, cut, carved, sculpted, shaped for a human purpose, as surely as Michelangelo released a human body from its block of Carrara marble. The residual barrier of cut rock just offshore from the cliff and extending parallel to the seaward side of the island (Fig. 1.4) forced the waves to break and plunge into the pool, from which the water escaped sideways. The precious warships, food stores, sails, ropes, spare masts, shipwrights' tools, barrels of pitch, sheets of copper and lead, oars, weapons and the garrison of troops were safe and dry. The island was once the focus of marine and military trades, with a constant noise of workshops, the smell of freshly cut wood or molten lead, the creaking of rigging, ropes, and oars, the clang of hammers. Now the defences are shattered. When the city subsided deep into the water, the waves began to strike directly against the island and against the rock wall, until late in the 20th century a roaring storm cracked the wall and tore the solid stone from the bedrock, sending huge slabs tumbling. Hundreds of tons of rock crashed into the torrent. Smashed chunks of rock were driven like pebbles by the colossal waves. The island broke in half so that jagged stone blocks as big as cars rolled into the city and crunched to a halt on the slipways where the Greek war galleys had once been safe. Now the ocean breakers surge over the wave-trap, re-gathering their momentum so that the white-crested surf streams across the broken island into the former city centre. *Sic transit Apollonia.*

* * *

I first heard of Apollonia while I was a member of a diving club in London in the summer of 1957. The British Sub Aqua Club London Branch held its training sessions once a week in the evening at the Marshall Street Baths, near Holborn. I joined the club after finishing my National Service with the Royal Marines in February 1957, and I converted my service diving training, breathing pure oxygen on closed circuit, to civilian air-breathing equipment quite quickly. Once I had become familiar with compressed-air sports diving, I took a diving holiday in Greece (Chapter 4) and started thinking of how I could use intelligent observations under water as a means of scientific research when I would start at Cambridge University the following October.

The BSAC-Club members were a lively and idiosyncratic bunch, amongst whom I soon noticed an energetic young woman who seemed different from the others. She was already an excellent diver. Her name was Natalka Czartoryska, although she had anglicized it as Natalie Pater, and four years later, in 1961, I married her. As the divers scattered to pubs and apartments after an evening's pool training, Natalka suggested that we have coffee with a friend of hers, Judy Jordan. Coffee and coffee-bars were still a post-war social novelty.

We sat round the low coffee table gossiping about the diving world and the latest films from Jacques-Yves Cousteau and Hans Hass. The film *The Silent World* had been a sensation in cinemas that year, with images of divers carrying flaming magnesium torches under the sea. Diving was a fabled frontier that had the magic to take you out of the grey routine of post-war London. There was an irrational buzz with no limits to the imagination. Technology solved problems. Judy's husband, John, was a professional cameraman for film productions who specialized in aerial photography from helicopters and light aircraft. I asked him about his most recent work.

John described his contract for the British Army in Libya where he had been filming from a helicopter at a bathing beach used by the Army for rest and recreation. It was a village called Marsa Susa, east of Benghazi. He mentioned that you could see ruins on the seafloor, with more ruins on the beach, including many standing columns. I quizzed him for more details, but he could not provide any information, apart from the name of the ancient city, Apollonia, and the postal address of the officer who had arranged the contract.

The next day I posted a letter to Captain D. Forrow, Royal Engineers, at British Forces Post Office Tobruk. The reply ten days later sparked my imagination. Forrow confirmed that there were walls and square stones to be seen over a wide area underwater. He even included a very rough sketch of the places between the mainland and the two islands where most of the drowned buildings seemed to lie. My mind teemed with questions and doubts.

Was it possible that the story was pure imagination, a fantasy, and the supposed ruins were just debris, fallen stones and natural rocks? What did the supposed ruins look like under water? Were they really square? Why were they under the sea? Could a diver draw them or measure them accurately? Could I enlist a team of divers to make a map of the underwater city? That would take time. How would such a project raise funding? Would sculptures or tombs have survived the inundation? The Atlantis legend distorted rational planning, and I knew that accurate surveying under water would not be easy or romantic.

At the start of the Michaelmas Term at Cambridge University in October 1957, I found two other freshmen with diving experience, Bill Hemmings and David Fagan, and together we started a new diving club which we called Cambridge University Underwater Exploration Group (CUUEG), which still thrives. I toured the college notice-boards sticking up typed Gestetner-duplicated announcements inviting members for the new club. As the winds from the fens attacked the trees, and leaves accumulated in soggy drifts in the streets, the black-gowned undergraduates scurried between lectures clutching piles of books. I persevered with recruiting would-be divers. There was so little time. I was studying mathematics, physics, chemistry, and metallurgy in the Natural Sciences Tripos (the broad category of science exams) but my extra-mural activities distracted from lectures and course-work which I often started after midnight. The colleges were quite tolerant of a wide range of talents in those days, and academic dedication was not essential. The so-called "well-rounded individual" was still accepted. Some of my contemporaries at Pembroke College were very well-rounded indeed.

Richard Adrian, a don in Corpus Christi College, helped us to write a constitution for the club, since the university rules required a don to supervise the finances. When we had elected the Club Chair and Secretary, we were ready to start training our recruits. Richard was a flamboyant, elegant and cultured junior don, driving around Cambridge in a gorgeous pre-war Bentley Lagonda. His rooms were decorated with sophisticated abstract paintings in muted colours. He was a cellular physiologist whose later career was extremely distinguished, and he became a Fellow of the Royal Society, Master of Pembroke College, and Vice Chancellor of the University of Cambridge. During that Michaelmas term, at a Club committee meeting in Richard's rooms in Corpus, I proposed a diving expedition to the underwater city of Apollonia. From my experience in the Royal Marines, combined with Bill's diving experience in the Royal Engineers, I was sure that we could devise a method of mapping an area as big as a whole city under water. Bill was a keen photographer who developed his own photographic negatives, had already experimented with a simple underwater camera, and was teaching himself to develop colour negatives. His skill in the sea would be a great help.

The treatment of underwater archaeology on feature films and television documentaries at that time stressed the melodramatic discovery of Roman and Greek shipwrecks in the Mediterranean. There were frequent long upward-tracking shots of divers rising in plumes of bubbles towards a bleached-out sunlit surface while clutching a huge teardrop-shaped amphora. This cliché did not help us, as there was seldom any reference to serious collection of measurements, documentary photography, or mapping. Divers had recovered statues and amphorae from several wrecks between 1900 and 1956, but the goal was always to find objects of artistic value. Plotting the positions of finds within the wreck was approximate, with the simplest measurements by tape when required. In some shipwrecks divers had made measurements of the surviving hull timbers before raising them, but a whole underwater town of tens of acres would need a different strategy. Nobody had ever attempted to make a precise archaeological map of the seabed on this scale. I would have to apply a mixture of science with my military experience to find a way to do it. I also became fascinated by the concept of the ancient port and its operations. For two years I had worked with the sea in the military, and then on the Rhine in Germany, and I wondered how the same activities were carried out 2000 years ago. How were sailing and oared ships docked or loaded and unloaded by hands alone?

During the Michaelmas term, Martin Minns, a fellow undergraduate at Pembroke College, and I did the background archaeological reading, after which I wrote to potential sponsors, in the style that had proved so successful for the Cambridge Explorers and Travellers Club. Adventurous travel was very much in the news. It was part of the post-war explosion of ideas and impulsive experimentation. Martin was the grandson of the distinguished late Professor of Archaeology Sir Ellis Minns and, although he was reading for a degree in English, he was naturally a keen amateur archaeologist. He was also an excellent swimmer. Many potential sponsors thought that the project was too ambitious for undergraduates. Eventually the Royal Geographical Society, the British School at Athens, the Department of Classical Archaeology, and the *Sunday Times* agreed to support us with grants of equipment and money. Various companies supplied stores, or reduced their prices for us, including Kodak film, and Bell & Howell Ltd who made movie cameras. The De La Rue Company provided numerous sheets of Formica, a new plastic surface for kitchen counters, which we had found was ideal for writing on underwater, while ICI provided a crate of newly invented plastic bags of various sizes as an experiment to find out if they were useful. They asked us to report back on the ways that we used the transparent bags, as they were not sure how to market them.

I wrote to Richard Goodchild, the British archaeologist who supervised antiquities in Libya, which the British had freed from Italian occupation during the war, and he suggested that we could find cheap accommodation and store sheds for our gear at the Dolphin Hotel in Marsa Susa, close to the ruins of Apollonia. The British School at Athens agreed to transfer an air compressor that they had been using in Crete for our use in Apollonia. A few weeks later Goodchild was in Cambridge to give a lecture to the Department of Classical Archaeology. This chance was almost too good to be true.

Professor Jocelyn Toynbee invited me to attend the lecture where I would have the opportunity to meet the speaker at the sherry party afterwards. Goodchild was a large cheerful man who immediately confirmed my idea that there was plenty of original research to do under the sea at Apollonia. He gave me contacts in the Libyan Embassy in London and the British Consulate in Benghazi, and assured me that he could grant a permit for our work under water.

* * *

Greek explorers from the volcanic island of Thera, otherwise known as Santorini, landed on the coast of Cyrenaica, in the eastern part of Libya, in the 7th century BC. There is evidence from Minoan pottery in Cyrenaica, and Egyptian pottery in the Aegean, that regular contacts across the eastern Mediterranean already existed in the earlier Bronze Age, but settlement and foundation of a permanent Greek colony only occurred later. Herodotus narrates the account of the expedition that founded Cyrene in 631 BC, a story that is confirmed by a 4th-century inscription found at Cyrene. Over-crowding of the island of Thera provided the main motive for the expedition, which was steered towards its eventual destination by the usual cryptic advice of the Delphic oracle.

After several abortive attempts at settlement by the Greeks over a period of about six years, local Libyans led them to the position of Cyrene, and encouraged them to settle there. From the Greek point of view the high rainfall in the mountains of the Jebel Akhdar was an advantage. Cyrene grew and prospered, acquiring ruling status in the five cities of the Pentapolis during the Hellenistic period, and was absorbed under Roman colonial rule in 96 BC. Apollonia was the port of Cyrene. After a turbulent history of revolts, wars, and periods of improved prosperity, Cyrene declined in the 3rd century AD. After AD 395 Cyrene was ruled from Constantinople, within the eastern half of the Roman Empire that we know as Byzantium. Earthquakes are common in Cyrenaica, and a large one in AD 365 destroyed many cities in Crete and Libya, including Cyrene, and caused a tsunami in Alexandria. After steadily declining security and attacks from barbarians from the south, the cities of the Pentapolis in Cyrenaica fell to the Arab invaders from the east in about AD 642.

The prosperity of Apollonia rose and fell with that of Cyrene for many centuries, but it continued to grow in the Byzantine period when Cyrene was in decline. It even acquired the status of capital of the Libyan Pentapolis in the middle of the 5th century AD, but in spite of the city's substantial fortifications, the Byzantine governor abandoned Apollonia when it was attacked by the Arabs.

Cyrene is in a beautiful location at a height of 520 m (1700 ft) in the mountains, with distant views down a succession of terraces to the coast 5 km (3 mi) away. Since the wealth and luxury of the city, and the grandeur of its temples, depended upon trade and agriculture, especially exports of the mysterious drug silphium, Cyrene needed a secure port on the coast. Mediterranean trade by ship at this time included grain, olive oil, honey, wine, raw glass, perfumes, garum (fish paste), henna, leather, almonds and other nuts, dyed cloths, ingots of unprocessed metals, timber, bronze tools, faience plaques, scarabs, saffron, ivory, and occasionally jewellery and precious metals such as gold or silver. Apollonia was a natural focus for south-north trade across the Mediterranean, and was also the safest port between Alexandria to the east and Leptis Magna and Carthage to the west on the North African coast. Its prosperity benefitted from this coastwise trade, including products from the Middle East, and possibly even pearls from the Red Sea.

Each element of trade used containers, usually jars or amphorae of various sizes and shapes. Thousands of fragments of these containers exist amongst all archaeological remains in the Mediterranean, and divers have found many shipwrecks stacked with hundreds or even thousands of unbroken amphorae. The management of trade and cargoes for ships requires specialist buildings and restricted areas in a port where products of high value await embarkation or dispersal inland. Record keeping was essential. Bureaucracy goes back a long way: written or inscribed tallies and records of production, trade, deposits, loans, and debts existed on clay tablets for many centuries, perhaps millennia, before anyone started to write down poetry, moral thoughts, or legends such as the glorious *Epic of Gilgamesh.* A busy harbour needs official, locked store rooms, bonded warehouses and armed guards even in a time of peace. The risk of theft and looting or avoidance of taxes was ever-present, as it is today. Guards were always alert for possible threats to peaceful trade. The slipways protecting the warships were probably roofed with timber, and the risk of fire or deliberate arson was a military security threat. For all these reasons the buildings around a port would have proportions and shapes that were uncommon in inland cities. Above all they were large, strong, defensible, and probably embellished with symbols of their importance. Mapping these unique structures was the challenge we faced as we planned our expedition during the winter 1957-58.

Research in the university library and the British Museum Library revealed that the first attempt at making a map of the underwater structures at Apollonia dated back to a book published in 1827, in which Captain F. W. Beechey of the Royal Navy published a very passable sketch map with three of Apollonia's most prominent submerged buildings (Fig. 1.5). The stimulus for mapping the Mediterranean coastline and shallow seas in

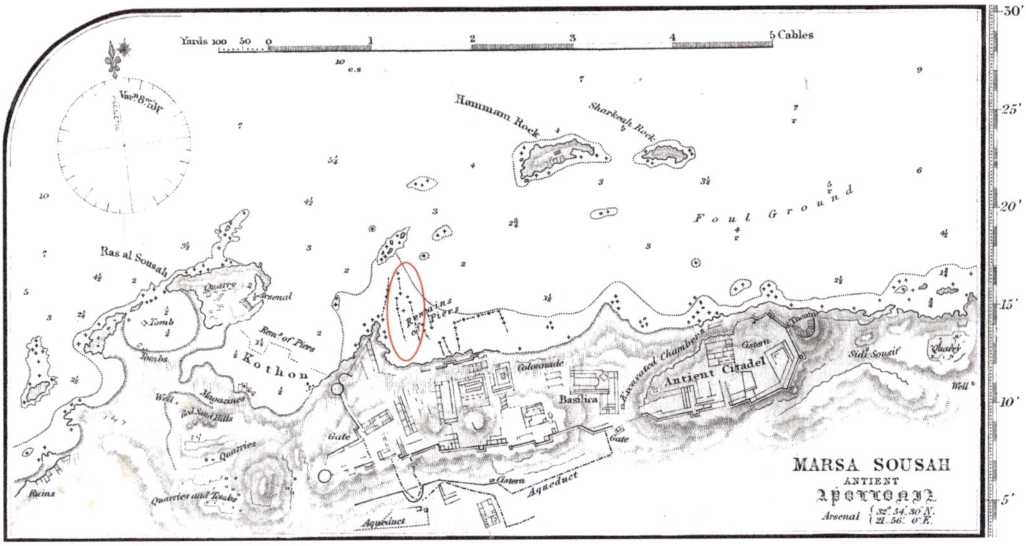

Figure 1.5. In 1827 the Royal Navy Captain Beechey surveyed the coast of what is now Libya, and made this map of Apollonia, showing some walls in the water. The building shaped like a two-pronged fork, circled in red, is quite accurate. Most of the buildings under the sea were not discovered or mapped at this date. This map is copied from chart 241, dated 1861.

the early 19th century was Britain's ongoing naval rivalry with France. Royal Navy captains had a classical education, and many of these surveys produced supplementary volumes describing the antiquities found on the shore, with careful correlation between the ancient historians and the surveyors' modern findings. In the true spirit of Enlightenment science, the same captains noted that coastal port structures from the ancient world were sometimes uplifted high above the water, or sometimes deep beneath the waves. Sailors looked down through glass-bottomed buckets, a trick used by fishermen, to observe and sketch buildings on the seabed.

My military diving experience provided the skills needed to measure and accurately map such a large area of buildings under the sea. We also needed to link the measurements of each submerged building to a baseline on shore and then join that consistently to a matching survey on land. But the most obvious survey tool, a theodolite that measures angles numerically, would be expensive, difficult to maintain in a rough environment, and prone to damage.

Nowadays, in 2020, with GPS and laser rangefinders, it is hard to imagine the difficulties we faced. Nick Wood, studying architecture, proposed that we used a more robust instrument called a telescopic alidade instead of a theodolite. He suggested a straight baseline along the shore marked with poles at intervals of 30 m (100 ft) measured by surveyor's tape. At each point we could position a survey tripod table, and sight through the alidade that rested on the drawing paper, giving the exact bearing relative to the baseline to a diver holding a pole on the surface of the sea over the wall to be fixed. The telescope of the alidade stood on a metal base with a straight edge that could be used to draw a line directly onto the map exactly parallel to the telescope itself. If this was done from two stations on the baseline, with the baseline already drawn on the otherwise blank map, any position out to sea could be fixed. We would need to repeat this routine for each point which would be marked by holding a coloured pole over the designated structure on the seabed. In practice we would fix ten or more points from one station, and then repeat all ten from a second baseline station, as this would be more efficient. With corners and junctions of walls fixed on the map, divers could then fill in the details of wall thickness, block sizes, and orientations of cut blocks, writing notes on Formica sheets. They could also sketch in details such as fragments of pottery, fallen columns, cuttings in the rocks, and lead dowels. When Nick Wood transferred these sketches to paper, the drowned city would live again.

Ranging poles marked in coloured 20 cm (8 inch) bands would provide rigid measuring rods under water for distances of a few metres, while a standard wooden or metal carpenter's rule would measure blocks with an accuracy of a single centimetre. In 1958 there was no such thing as a plastic tape with which we could measure distances underwater, while a normal surveyor's canvas tape would stretch and shrink in the water, and probably rot after a few days. Martin was brimming with ideas for fancy gadgets, and we adapted equipment, such as the chest-mounted reels of cord I had used in the Royal Marines Special Boat Service for beach profile surveys. Martin and I then constructed a chest-mounted reel, supported by a harness of canvas straps, holding 50 m (164 ft) of rust-proof copper wire used for the aerials of battle tanks marked with coded tapes inserted every metre. It looked clumsy, but it worked perfectly.

Photography and filming were a fundamental part of our plan. Our underwater still camera, loaned by the British School at Athens, was a Leica 35 mm with a 50 mm lens, which was a rather narrow angle on land, and gave an even narrower field of view when you take into account the refraction of light under water.

Bell and Howell had given us generous terms with the 16 mm spring-wound movie camera that held 30 m (100 ft) reels of film. With this I could shoot several minutes of action at a single loading of the camera, shooting at 24 frames per second. The 15 mm wide-angle lens gave broad images under water, and reduced the need to adjust focus between shots since it would be in focus from close-up to almost infinity. Colour-correcting red Wratten filters helped to recover some of the colour which is lost in the blue of the underwater world. The only 16 mm colour film available was Kodachrome ASA10, which meant that the lens aperture would have to be wide open most of the time to collect enough light under water. But because of the short focal length, the pictures would still be in focus because of the good depth of field. This was the best photographic documentary system we had under water. It is very difficult to obtain a wide view picture in shallow water, and it was not until ten years later that Nikonos and Calypso Phot produced a wide-angle underwater lens for use with 35 mm still cameras.

The university engineering laboratory helped by designing an underwater case for the Bell and Howell camera. At that date the idea of filming underwater was still wildly exotic, so my request caused considerable surprise. Professor Baker could see that the challenge would be valuable to some of his staff, and they offered to make the case for us at cost. The cadmium-plated case they designed had the camera mounted securely with screw-clips onto a thick circular steel base, with the controls coming through the base in penetrations sealed with rubber O-rings. The camera could be easily lifted off the base-plate for adjustment or loading film reels.

The other half of the case was like a large saucepan that fitted over the camera, and had a circular glass-plate porthole in the side that lined up with the lens. The "saucepan" was bolted onto the base-plate by a dozen hefty screws with wing-nuts, and was sealed again by a large O-ring (Fig. 1.6). Everything was easy to handle and looked strong.

Figure 1.6. The underwater movie camera case, made in the engineering laboratory at Cambridge, disassembled to show the internal arrangement. The external view-finder is designed to match the wide-angle lens that is shown in the picture. The scale bar is in cm.

We took the camera in triumph to the Leys School swimming pool to test a roll of film. It worked perfectly, but the engineers had been so clever in making the "saucepan" a tight fit around the camera that they had neglected to calculate the effect on its buoyancy. It was so heavy that it was impossible to swim with. We postponed a solution to this problem until the end of the academic year in June.

* * *

During the early months of 1958 our potential expedition team members learned to understand and respect each other's skills. Those who were new to scuba diving developed confidence in their own ability under water, and trusted their buddy divers. It is essential when diving that you can rely on your co-diver. By the spring we had a team of ten. In the Easter vacation the CUUEG members spent a week diving in the sea around Portland, on the British south coast, where we experienced open sea conditions and some deep diving to 30-50 m (100-165 ft).

We tried to recruit an undergraduate archaeologist to the team, but failed, in spite of strong encouragement from Professor Toynbee. In that absence, Nick Wood, our architect draftsman, had the most archaeological experience, with Martin's added dedication as a strong amateur. Further members included myself, Richard Adrian, Bill Hemmings who would specialize in still photography and biological sampling, and Richard Windred who was studying chemistry and was interested in the psychology of diving; then Dick Longbottom and David Fagan both studying medicine; Richard Everington reading English, and Natalka Czartoryska who worked part time at the Natural History Museum in London. With four Richards in the team, we had to rely on abbreviations and nicknames. We needed time at the beginning of the summer vacation to test equipment. A departure for Libya at the end of July would enable us to dive with the best weather and wind conditions on the Libyan coast.

During the spring, Martin and I continued the background archaeological investigations, and I bought charts of the Libyan coast from J.D. Potters, at 142 The Minories, just behind the Tower of London. Much-lamented, Potters, which had probably sold charts to Joseph Conrad, closed in 2002 after centuries of service to the merchant mariners of the world.

Back home in Chiswick after the end of the Easter Term in June, preparing for departure in six weeks' time, I finally tackled the problem of the heavy movie camera. Richard Everington and I obtained permission to scuba-dive in the public pool at Chiswick in the early mornings before it was officially open so that we could test the camera itself, and the range of measuring and surveying gadgets, drawing boards, and tools that we had devised. Some worked, some didn't. The fresh morning stillness in the vast pools prompted clear thinking. The camera and case weighed more than 4.53 kg (10 lb) when under water, which was fine while you were standing on the bottom of the pool, but you could not swim with it. We shot some reels of colour film to test exposures and the filters while I tried to think of a buoyant device that could compensate for the weight.

The only lightweight strong sphere that I could think of was a metal ball-cock float like the one in a lavatory cistern. I calculated that it would have to be about 35.6 cm (14 inches) in diameter to give the lift required: more than twice the diameter and nearly ten times the volume of the usual ones in home lavatories. I hoped that factories and big buildings used cisterns with ball-cocks of large size. I went up to the Chiswick High Road to visit the biggest plumber's shop.

"You want a copper ball-cock 'ow big?" gasped the man at the counter.

"Er, about 14 inches diameter," I repeated nervously.

"Nothing like that 'ere," he said. "Perhaps you'd get it at Johnsons in Westminster Bridge Road. Big firm, they are."

I took the Underground to Westminster. At Johnsons I had a similar exchange.

"The only people that could supply that would be the big wholesale plumber at 120 King's Road in Chelsea," said the salesman after recovering from the shock. "Really big industrial company. That's your best bet."

Arriving in the King's Road I searched for number 120, and when I found it I saw the proud sign in huge golden letters across the pale-blue shop-front: "Thomas Crapper and Co. Sanitary Appliances by Appointment to Royalty". It was logical.

"I need a fourteen-inch copper ball-cock, please," I said to the salesman, feeling more confident this time.

"Fourteen and a half inches is the nearest standard, Sir."

"That's fine,"

The salesman returned from a distant store shelf and dumped a large gleaming copper sphere on the counter.

"That'll be 17 shillings and sixpence," he said sharply. He had the professional's contempt for a non-plumber.

I left happily with my prize.

At the Chiswick swimming pool, I suspended the camera below the copper sphere and found that it was now a couple of pounds too light. A few bits of lead tied below the sphere made a perfect balance, and the whole array had the unexpected and delightful advantage that the camera was now held tautly between the buoyancy of the float and the weight below it, so that it was always steady in the water. It did not wobble or twist when I was swimming with it. The buoyant float acted like an inverted tripod for steadiness. I painted the slogan "Cambridge Apollonia Expedition" in bright yellow on the ball-cock (Fig. 1.3).

***

The journey to Apollonia by way of Benghazi required driving across Europe to Sicily, where we would board the ship *Città di Livorno*, crossing to Benghazi in three days at sea, followed by a further day for the drive to Apollonia. Sandy May, a Canadian friend, had lent us his Volkswagen van for the drive from Dieppe to Syracuse in southern Sicily. The whole trek to Apollonia would take 11 days. Nick Wood, Richard Everington, and Bill Hemmings departed early for excursions in Europe, and we arranged rendezvous to pick them up on the way south. In the days before mobile phones a series of rendezvous like this was risky, and if they failed the expedition would have to proceed with people missing. David Fagan, Dick Longbottom, and Natalka Czartoryska would travel to Syracuse by train a week after the main party and take the next ship to Benghazi.

On Tuesday, 29th July, I drove the bundles and crates of gear to Newhaven in a hired van, leaving them in customs bond at the docks. Two days later I met Adrian, Martin, and Richard Windred at Victoria station. The crossing to Dieppe was uneventful and Sandy was full of cheer and welcome as we loaded the gear into his van. There was the conventional hysterical screaming row with the French customs and the bank-bond officials who inspected my papers and doubted that Barclay's Bank had the credit-worthiness to compensate for the risk of our selling all the gear in France. We left Dieppe, driving off in high spirits, and at midnight we said goodbye to Sandy in Paris. We drove all night, and through the next day, down that big road in the middle of France that is like driving through wine labels. Near Fréjus, we parked in a quiet spot by the sea at sunset, cooked a meal on the Calor gas stove, and slept on the ground in our sleeping bags.

As we headed south through France and along the coast through the Alpes Maritimes the roads became narrower and more tortuous. The heavily burdened Volkswagen was slow, and we took turns driving, making for long days on the road while the non-drivers chatted or slept amidst the baggage. We became addicted to that hypnotic Mediterranean blue that was so novel to us then.

After the Italian border at Ventimiglia, we spent the third night on a hill above La Spezia, from which we watched a spectacular display of fireworks launched from barges in the bay. The reflections of the soaring rockets gave us a two-for-one. Next morning, we drove on quickly. I parked the bus at 0945 where we had agreed to meet Nick Wood by the leaning tower of Pisa in the main square. We climbed the tower. By the time we came down Nick was waiting for us, suntanned and cheerful, looking very Italian. To escape the heat of the city we drove to the sea nearby for a swim, driving back to check the tower at every hour, but Richard Everington and his Lambretta scooter were missing. We left without him.

Because of our late departure from Pisa, we had to drive all night, arriving in Rome at two in the morning. Barefoot and wearing shorts and singlets, we wandered around the great architectural monuments in deep shadows with the silence broken intermittently by brief lectures from Nick. I remember standing beside an isolated huge marble foot that must have been part of a statue to make Ozymandias jealous. (The colossal foot, accompanied by its anatomical twin, is now exhibited in the courtyard of the Capitoline Museum. It was a fragment of a gigantic seated statue of Emperor Constantine). We drove on all night, trundling the next morning into the outskirts of Naples. Mist hung over the bay concealing Vesuvius, Sorrento, and the islands.

We drove through heavy traffic to the seafront on Viale Anton Dohrn where the Stazione Zoologica, one of the oldest marine stations in the world, occupies the centre of an extensive scrubby tropical garden. Bill Hemmings welcomed us at midday, beaming with the successful results of his studies of Mediterranean species. We loaded his kit bag into the long-suffering Volkswagen, spent the afternoon in Pompeii, and continued south under a clear sky with Vesuvius rampant. I slept for the next part of the trip, waking in some befuddlement to find the bus stopped at dusk in a field near some Greek temples. There were two perfect temples close by, and a third a bit further back parallel to the road.

With no training in archaeology, I was surprised and baffled.

"Nic, we have to visit the Greek temples of Paestum," explained Adrian. Nick Wood agreed. So we piled out of the bus, vaulted a fence, and trooped across the grass. The red sun was sinking already into the sea and the last rays shone through the ribs of the black temples like the vision of the Ancient Mariner.

There were no houses nearby, no tourists, and no traffic on the road. There was almost complete silence, although the sea was just a mile distant. We climbed onto the ruined platform or crepidoma of the Temple of Hera, the most complete of the temples, and admired the extreme bulging curves of the fluted Doric columns. Every column was still intact and in place. Nick Wood explained that they had been built about 500 BC. The architraves and entablature were almost complete, although there were no decorative sculptures. I could see no other ancient buildings or ruins of a town, although parts of the ancient city do survive. The three exquisite temples dominated the fields in neglected grandeur.

It was the beginning of my love affair with Magna Graeca, Great Greece, that continued for many years. In this part of southern Italy there were so many prosperous Greek cities in the centuries before Roman power absorbed them, that the whole region was known as Great Greece. Constantine Cavafy wrote a bitter-sweet poem about Poseidonia, the original Greek name for Paestum, in which he imagines the Greek citizens of the once-cultured and sophisticated city bewailing their humiliation by the barbarians of Rome. As Rome triumphed, so the Greeks continued clumsily celebrating their own half-forgotten ancient festivals. The cities such as Sybaris, Croton, Poseidonia, Tarentum and Syracuse produced a wealth of music, mathematics, geometry, physics, sculpture, architecture, trade, and engineering that paralleled the achievements of mainland Greece and the Aegean. Anecdotes about the outrageously decadent luxury of Sybaris were a staple of comic Greek story-telling, and the town gives us the modern word "sybarite".

With every mile that we travelled south of Paestum the road became steeper and more twisted with hairpin bends and narrow bridges over dry gorges. Sometimes the narrow road clung to the cliff above the sea, and sometimes it plunged inland skirting behind a particularly resistant mountain peak. We rolled at sunset into the ferry port of Reggio di Calabria where the dock-hands weighed the Volkswagen and could not believe that it was twice the weight they expected, even without the six passengers who stood by and watched. As the incoming ferry unloaded its cargo we saw two heavily barred railway coaches carrying imprisoned bandits being shunted off the ship into a siding on the mainland. We crossed on the ferry from Reggio to Messina, spending the next night in the hills of Sicily. We had been advised by the ferry captain to beware of bandits, and, since we were not sure that the train-load we had seen was a clean sweep, we slept with spear-guns and diving knives close to hand.

* * *

The last miles of southeastern Sicily took us through sun-baked mountains and coast roads, round the curvaceous cone of Mount Etna, through the industrial port towns of Catania and Augusta, and finally into the magnificent ancient city of Syracuse, famed as the colony that defeated Athens and precipitated its final collapse as a Mediterranean power. The main port is the Porto Grande where fleets of hundreds of Athenian triremes and galleys had been trapped and destroyed in desperate battles at close quarters. We found the modern port jetty, where the well-worn passenger-freighter *Città di Livorno* was berthed, tied up to bollards on the dockside. I negotiated with a nearby garage-owner to leave the Volkswagen in a lock-up shed while my companions unloaded our gear and stacked it beside the gang-plank

I heard shouting, laughter, and rapid tooting of a motor horn. Richard Everington had put-putted into the dockyard at full speed in a cloud of dust, exhausted, begrimed, with smoke belching from the over-heating scooter, and a wide grin above his wind-blown beard. We parked the scooter with the Volkswagen and settled down to a yarn-swapping evening meal before embarking at eight o'clock. Richard had missed the last rendezvous at Pisa by an hour and had followed us down the whole length of Italy a few miles behind us.

The *Città di Livorno* was a cargo vessel that carried a few passengers, but the food was good. On the planked deck we acclimatized to the sun, while we checked our stores documentation, and revised the survey procedures, plans, and logistics. The ship stopped for an afternoon in Grand Harbour, Valletta, Malta, and we arrived in Benghazi on the morning of the third day, with a strong onshore wind making a choppy sea. The breakers smashing over the harbour wall, the yellowish haze over the dull-coloured city, and the apparent lack of life combined to give a down-trodden and uninspiring impression. Things changed quickly. A gang of red-hatted Arabs and Africans gathered on the quay as we came alongside where they unloaded our equipment onto a donkey-cart. We trekked across the dusty cobbles to the customs office, in which a smart African officer chalked our baggage with customs code marks and then we were ushered out through the wire gates into the town. We loaded everything onto a hired minibus and refreshed ourselves at a hotel, before driving eastwards for Apollonia with an Egyptian driver.

Libya at that time was ruled by King Idris and the country was beginning to modernize, while the British Army provided security. During the Second World War the territory of Cyrenaica changed hands several times, and derelict army vehicles and thousands of munitions still littered the countryside and the shore. By 1958 the main streets of Benghazi were lined with large European buildings, many built by the Italians. The roads were wide and filling with cars as we drove away from the port. Between the road and the sea, there was a large grove of palm trees with a shanty town of corrugated-iron and cardboard shacks packed around the trunks of the trees.

We drove through the foothills, across numerous dry waddis spanned by military girder bridges, and up into the Jebel Akhdar, the Green Mountains. In those days the modern village next to ancient Cyrene, Shahat, had only a single street that was shaded by whispering eucalyptus trees where it twisted between the low white shops. Surrounding the village, dotted amongst the trees, were the magnificent ruins of temples, columns, ancient market places, baths, stone sports tracks, and theatres, sprawling over graded terraces and slanting down the mountainside. The scale was almost unimaginable. Beyond these were the scattered remains of the suburbs of the once-great city, half-buried in sand and overgrown with trees and thorns.

The road from Cyrene to Apollonia descended the Jebel in a series of hairpin bends bordered on either side by hundreds of Greek and Roman tombs in the form of miniature houses carved into the cliff. As we rounded a curve in the hills the view of the coast emerged below us with a chain of headlands and islands slanting away from the shore to the east, sheltering little bays and harbours. The late afternoon sun was behind us, and the sun-gilded islands were cast into heavily moulded relief of brightness and black shadow against a hazy sea. Modern Apollonia, the village of Marsa Susa, was a cluster of neat houses just to the west of the ancient port, and the poorest members of the community still lived amidst the ruins of the ancient city that have remained on dry land.

The street into the village opened into a beautiful little square where palm trees and all manner of greenery flourished, with occasional bursts of red flowers. The Dolphin Hotel, which had been converted from the original Italian post office, was run by Captain Spencer, a retired British Army officer, who briskly indicated where to stow gear and showed us the various bedrooms, which would need to be shared in twos and threes. Natalka would have a small room for herself when the second party arrived. Gratefully, we washed and showered and unloaded the equipment. Captain Spencer was astonished at the quantity of cylinders, tripods, cameras, the inflatable boat, and boxes and bags that piled up in the shed. He advised us to employ a *gaffir* or caretaker to watch over it.

After a British-style meal, I wandered out towards the ruins to size up the lie of the land. The sun had set long ago, but there was a bright moon and I easily followed the path into the ruined city. Though the wind had dropped after a recent storm, the rollers still came in with a constant rumble of noise, and the hulks of the islands were riding a sea of phosphorescent white. I walked along the bank above the beach where I saw how the masonry walls broke off into the sand and surf. On my right were the silhouetted columns of the Eastern Christian basilica, two rows of smooth marble, spaced on a mosaic floor. I stopped walking. I looked out again at the lines of breakers sweeping in, over the submerged city, and felt the sorrow of loss. So much had gone already and would never be retrieved. A helpless sense of the destructive power of earth and sea sobered my adventurous anticipation. What could we rescue from that desolation in just four weeks?

I returned slowly to the Dolphin Hotel.

* * *

Preparation for the first day in the hot climate required us to think both of the time to do the most work that would avoid the heat of the day, and how to manage our own health and diet. Early in the morning we tidied and checked the stores before breakfast. Before starting work in the sea, we rubbed olive oil on our bodies, and donned T-shirts to wear even while swimming. We planned to eat plenty of salt with meals, while always drinking plenty of water. The temperature at midday was 32° C (90° F). Although nobody knew then about the risks of skin cancer, and acquiring an oak-like tan was thought to be both healthy and virile, we had enough sense to know that burning lobster-red was not a good idea.

From the work-sheds and store rooms that we used opposite the hotel we could see ruins close by on the extreme west of Apollonia, outside the city proper. Standing in the shallow sea, there were the skeletal remains of a tomb carved out of the solid rock. On the beach on the opposite side of the village the waves had split open a massive circular tower, and shattered blocks projected from the remaining segment overhanging the water. The land surrounding the village was almost bare and sparsely cultivated. Where the 300-ton cargo ships of Imperial Rome once docked to load cargo after cargo of wheat, the peasant farmers now scraped a living from soil which they no longer irrigated.

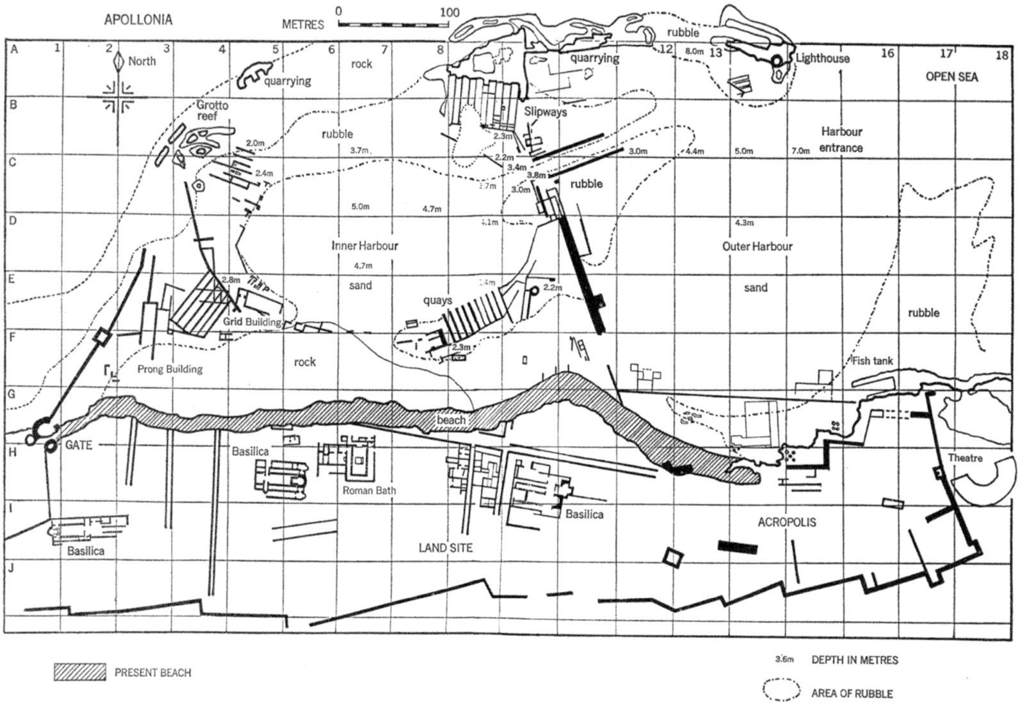

Figure 1.7. The complete map of the underwater city made by divers in 1958 and 1959. The hatched strip is the modern beach, so everything south of that is on land, and everything north of it is under the sea, except for the islands.

After a hefty breakfast prepared by Mrs Spencer, Nick Wood and Dr Adrian went out to find the best position for the survey baseline, while Richard Windred and I started the air compressor. We checked the purity of the air and charged two diving cylinders. The others were experimenting with swimming over the ruins for the first time using masks, fins, and snorkels, and carrying plastic sketching-boards. Richard and I soon followed them. We walked through the broken circular tower which was the western city gate, from which we could see the external city wall running inland about 200 m (650 ft). It then turned to continue its serrated path more or less parallel to the shore as far as the acropolis and theatre at the eastern limit. The wall was part of the great Hellenistic defence system, but most of the buildings visible within it were Byzantine Christian. The structures in the sea might turn out to be of any date between 635 BC and AD 600.

We climbed down the rubble slope, donning our masks as we scrambled over the fallen blocks, then waded into the water over the submerged foundations of the tower. I swam out, gazing with fascination down into gloom where I saw the foundations of a broad stone wall overgrown with a short fringe of green and brown algae, more like moss than seaweed. The stones were laid lengthwise across the thickness of the wall, known by archaeologists as "headers", and they rose only one or two courses above the surrounding rubble. Upper parts of the building had collapsed around the lower walls, protecting them from further erosion. I thought that many small objects must be buried. I curved out over the wall into the deep water to the west, and saw a steep slope of rubble plunging into the blue; on the inner side I found more low walls, almost covered by the tumbled masses of square blocks. This was the western limit of the city. There were defences here, both against the sea and against attack from warships.

I reached a low, square building 100 m (325 ft) from the shore, which had three complete walls, and in one place the masonry rose to within inches of the surface. The long swell waves swirled and boiled over the stonework, driving clouds of bubbles down into the clear water. A fat mérou, a slow-moving, rock-dwelling grouper, lay motionless in one corner, gazing from beneath a fallen block. I swam slowly round the building, drawing its outline on the Formica plastic board, first the wall which joined it to the shore, and then other walls leading off from it towards the unknown. I followed the faint traces of a thin wall eastwards into the city until it met a cross-wall. I sketched the junction and swam left. The morning passed with the constant hiss and slap of water over my ears and the ever-expanding vision of buildings emerging from the blue before my eyes.

By midday I had outlined a fork-shaped building with two prongs towards the shore, which we later called Prong Building (Fig. 1.7 grid square E.3), and immediately east of this, a complex tangle of walls overlapping one corner of a series of long parallel structures which we called the Grid Building (Fig. 1.7, E.4). I swam back to the beach, and over lunch we put our drawing boards together to compare notes. Some areas of ruins were so confused that it would require several days of careful tracking and sketching to sort them out before we could start the main survey with accurate measurements.

The sea was calming down, although the residual swell from the windy day before was still stirring up some flecks of sand amongst broken fragments of weed. The largest and best-preserved buildings were in the shelter of the islands, where the other snorkellers had found walls ending in solid masonry towers reaching almost to the surface, causing the swell to pile up and break into surf over the ashlar blocks. We gave more names to the biggest buildings which reflected how they were found, who found them, or some token of their construction and position. Amongst them were "Bill's House", "Tyre House", "West Island Slipways", "Five Silos", and "Nine Quays". The reef to the west of the city, "Grotto Reef", where we could see the entrance to a submerged tunnel, was exposed to incoming ocean waves that were very dangerous, even in the relatively calm weather. We were not able to swim near it. Also, the water there was so full of bubbles that it was difficult to see much, and sharp rocks would suddenly loom up in front of you.

This preliminary sketching occupied most of the team for the first week. Each sector of the town had to be drawn with an unambiguous set of shapes and corners, disregarding the true lengths and distances, but sufficiently accurately that anyone with the sketch could identify each corner or junction for later precise fixing. With a diver in the water having a similar copy of the sketch map as the surveyor with the alidade table on shore, we could make fixes of agreed points in logical sequence without mistakes. Nick Wood had unpacked his drawing board, T-square, rolls of linen-backed drawing paper, and drawing instruments in readiness. During this first week Dr Adrian found that Mrs Spencer had been running an informal medical clinic for the villagers, with the result that his professional skills were soon engaged every morning working together with her.

In four days we had familiarized ourselves with much of the city and we had set up the baseline on shore. The line was 810 m long (2657 ft) and marked every 30 m (100 ft). It started by the Main Gate in the west and finished by the Greek Theatre in the east. We were ready to start the accurate survey, while sketching still continued in the far corners of the underwater town. Work progressed well, but our discoveries showed that the scale and architectural complexity confronting us were daunting.

Underwater archaeological mapping has made great advances since 1958. Where the seabed structures are large masonry blocks or concrete buildings with rectangular corners and plane faces, the use of high frequency multibeam sonar can produce reliable preliminary outlines to be checked by divers. Where the walls of buildings have collapsed into surrounding rubble, or where uncut stones have been used like dry-stone walls on land, the human eye of the diver is necessary from the start. Even with modern Total Station laser rangefinders reflecting off poles equipped with reflecting prisms and automatic data logging, somebody has to make the preliminary sketch map so as to define the recorded points, and this phase of mapping underwater cities continues to be laborious. The only system that might speed things up and enable faster surveys is the use of robotic automatic vehicles that simultaneously take overlapping stereo-photographs while recording position by GPS (Chapter 6).

* * *

Every morning for the next few days, after collecting our kit, we walked through the scorching heat of the village to the beach, dressed only in swimming trunks and T-shirts. Bill Hemmings and Richard Everington took aqualungs and underwater cameras, Nick took the alidade and plane table, while the rest of us carried swimming-gear, Formica plastic sheets, ranging poles and tripods. This procession caused much amusement, and Arab children who called everyone "Johnny" always accompanied us. They often helped us carry things. Every morning we passed the same donkeys, camels, and chickens scratching and rolling in the thick red dust. On arrival at the day's survey base, Nick

Figure 1.8. Nick Wood working with the telescopic alidade on the tripod, positioned on the West Island. The yellow arrows permit the tripod to be put in exactly the same place on different days. Nick is viewing a diver in the sea holding a coloured ranging pole, like the one in the background.

set up the plane table while Martin, Richard Windred, and I swam out to sea with the ranging poles. Using duplicate copies of the preliminary drawings, drawn on Formica, it was easy for the swimmers to position themselves over 20 or 30 points in turn, while Nick took the bearings on the same points and signalled when he was ready for us to move (Fig. 1.8). We took two or three hours to fix 30 points in this way, and when Nick wanted to cool off in the water, somebody would take his turn at the plane table. The children taught us to count in Arabic and how to say a few polite phrases, but we never got very far, finding it easier to talk to the older students who often visited us, and who usually spoke excellent English. They were very interested in our work as we explained the principles of triangulation to them, which they easily understood.

Sometimes I would sit on the rocks of the West Island, or on the beach, redrawing my plans or transferring them to paper, and gaze out over the water to where the tiny figures of swimmers cruised back and forth, hesitating and turning with slight splashes as they followed the elusive walls flickering in the refracted sunlight below them. Each day the snorkel swimmers swam several miles as they stationed themselves over one key point after another. When we landed on the island we beached the dinghy on the trireme slipways cut by the Greek colonists 2500 years earlier. Just east of the slipways (Fig. 1.7 B.9) was the entrance into the harbour basin, guarded by the two solid towers which we called the Block Forts (Fig. 1.7, C.10).

The towers of the Block Forts rose massively from the sandy floor and broke off just at the sea's surface, so that in the slightest wind or swell the waves broke white with foam (Fig. 1.9). South of the Block Forts was a maze of walls that was proving most intractable. We swam over the area in a collegiate shoal, trying to identify every stone, checking and cross-checking, and discussing our different drawings. A colossal wall nearly 10 m (30 ft) thick stretched from the shore near the Great Basilica (Fig. 1.7, H.10) to the southern Block Fort, near which were numerous small rooms, one almost complete and with a door space. This was the only area of the city where kelp grew thickly, and rank brown fronds of laminaria concealed the details. The massive wall separated the inner harbour basin from the bay to the east, forming a section of the military defensive walls of the city. Swimming westwards, we followed the curved

Figure 1.9. Southern Block Tower, guarding the channel between the two harbour basins, showing the stepped-in courses of ashlar masonry (Photo Carlo Beltrame).

wall that led round the inner harbour to the nine parallel buildings which we dubbed the Nine Quays (Fig. 1.7, E.9), and the foundations of a round tower (Fig. 1.7, E.10). By repeated observations, discussion, and redrawing, we eventually eliminated most of the ambiguities in the original sketches.

As the survey progressed and the map grew larger, we discovered more rock cuttings that looked like slipways by the East Island. We also found that the galley slipways on the West Island had buildings right on top of them, which was very odd; and below the end of the slip was a large wall that would have prevented the entry of any ships. The city had been rebuilt over time so that we were seeing stratigraphic superimposition of buildings from different periods, one on top of another. There were also the foundations of a house on top of the Grid Building in the southwest corner of the harbour; and overall there seemed to be two generations of buildings, one defining a circumference about 20 m (65 ft) nearer the centre of the harbour than the other. At some date the harbour basin had contracted, or the sea level had dropped. We found fragments of amphorae everywhere, and several whole pots and jars embedded in the black earth below the sand. Miraculously, after a storm had shifted the sandbanks on the seabed, we found the arm of a large marble statue which we raised and gave to Richard Goodchild. The discovery confirmed that in its heyday Apollonia had been decorated with fine sculptures. We searched for hours for the rest of the body, but had no luck.

Every evening after dinner we discussed the progress of the work. Nick Wood confirmed that the accuracy of positions measured from the baseline to buildings in the sea was consistently of the order of 20 cm (8 inches); although on one occasion he found an error in the day's work, and we had to repeat it. After a week we planned to extend our activities. Bill had shot several rolls of black and white film which he had developed successfully, so that he was ready to start shooting colour under water. He had also been collecting biological samples which he stored in the transparent plastic bags. Nick Wood had moved his architect's drawing kit to the peace of a beach hut, where he had plotted all the survey points to date. He suggested that we should start measuring wall thicknesses and block sizes the next day. The time was also right to start filming movie, using the West Island as our base.

Next morning, with a stack of aqualungs, tripods, boards, poles, and towels piled into the inflatable boat, Nick perched on the top and rowed cautiously across the calm sea to the island. His loyal following of Arab children stood wide-eyed and bereft on the beach for some time, then ran away to find other amusement. Dave, Dick, Martin, and Richard Windred swam after the boat and helped Nick with the surveying, while Bill and Natalka collected zoological specimens, and Richard Everington and I prepared the underwater ciné-camera. Adrian was by now committed to helping Mrs Spencer running the surgery for the local population, and could not be with us in the mornings. By eleven, when we gathered on the hot, rocky slopes of the West Island, the sun was burning down on the dead calm sea. We planned to shoot a film series of the different stages of surveying around the Block Fort towers, and we discussed the exact sequence of shots as we fitted our aqualungs. I plunged into the water first, and briefly tested the motor of the underwater movie camera. Some wrasse swam towards me, their mosaic colours shimmering in the green light, and I slowly brought the camera into position and shot a few feet of film. The divers were on their way, and the fun was soon fast and furious. There were six divers in the water in front of the camera, with Bill gliding around the outskirts of the group taking still pictures. On either side of us the flanking towers rose up to the surface, and the floor of our stage was carpeted with the building stones quarried from the islands by the Greeks and Romans. While we were swimming in this mysterious and peaceful setting, we felt periodically the sharp clang and pressure-shock of underwater explosions. Arab divers were blasting for scrap on a wreck just along the coast. We hoped that they would not come too close.

One morning I was finning luxuriously over a curious area of superimposed buildings dating from different periods, when I saw a fisherman's rowing boat about 50 m away, with dark figures waving to me. I could not make out whether I was meant to approach or keep away. But as I guessed that they were bombing fish, I thought I would rather be in their boat when the big bang came than anywhere in the water. I swam towards them and climbed into the stern of the boat. A tall, handsome, one-legged African stood amidships, handling the oars, with his stump-leg propped on the thwart. In the bows sat two Arabs, gazing into the water. The first wore only trunks and was still wet and shiny. He gripped a glass-bottomed bucket with his good hand, while his other hand, which had three fingers missing, was curled painfully on the side of the boat. His friend held a homemade bomb, constructed out of sardine-tins filled with the explosive from shells, and he too had the fingers of one hand missing. The whole crew seemed very happy, and the big African smiled at me as he turned the boat this way and that, but nobody talked as they waited for the fish. There was a long, slow swell coming into the bay, on which the boat rose and fell silently, rocked occasionally by the gentle creaking movement of the oars. They could find no fish, so I thanked them for considering my safety, and they smiled again as I dropped back into the water.

* * *

The slipways were amongst the most perfectly preserved features of the whole city, since they were in the lee of the West Island. You could lie in the shallow water gazing down the parallel sloping walls and imagine the slender bow of a galley nosing onto the central runner of the slip after a long voyage. Then the shouting and heaving as the vessel was hauled, dripping and weedy, into the heat of the sun, before being pulled up under the wooden roof, to be scraped and re-caulked. Now the slips were completely submerged, and littered upon them were fragments of columns, rubble, and three-inch shells from the wrecked ammunition ship nearby. At the top end of the slipways was an excavated work space or storage area carved out of the island, with rock-cut doors leading into hollowed-out caves, from one of which steps led up onto the top of the island. There had probably been a watchtower up there. This would have given advance warning of attack from the sea. The fortification of the West Island confirmed both the military and commercial importance of Apollonia.

We had anticipated that parts of the city might be covered in deep sand so that we would need a powered device to excavate down to the building stones. Martin and I had constructed a small airlift consisting of 3 m (10 ft) of sections of rigid pipe that slotted together, with an air inlet welded into the side near the lower end. Injecting high-pressure air at this point would create an upward flow of water mixed with bubbles that dragged in more water, and sucked up the sand. An aqualung cylinder strapped onto the lower end of the pipe provided the air supply. We took this out in the dinghy and lowered it to the east of the Nine Quays to test it. I forced the bottom end of the pipe into the sand before turning on the high-pressure air. Plumes of muddy sand shot out at the top, as I dug a small pit. Soon the pipe had cut through the sand layer and was resting on packed mud and stones. The machine worked well, but after further exploration, we realized that the ruins projected above the sand almost everywhere, so we did not use it often.

As the map grew, it began to reveal how the city reached its present state. The West Island slipways were Greek, similar to those at Piraeus, being just wide enough to hold triremes. Thus they were probably built in the earliest period of the city's history, in the 5th century, and should indicate the level of the sea surface at that time. If boats are to be pulled up a slip, the water must be deep enough at the bottom for the keels to rest on the slip, and it must be long enough for the boats to be dragged high and dry. From the dimensions and slope the water level had risen by just over 2 m (6 ft).

The slips were 6 m (20 ft) wide and 40 m (131 ft) long, with the top just awash and the foot submerged by 2.8 m (9 ft 2 in), making a gradient of 4 degrees (Fig. 1.10). The centre of each slip was depressed relative to the sides, so that it had a shallow V-shaped cross-section, with thick walls dividing each slip from its neighbours. These walls, which were cut out of the solid rock, were 75 cm (2 ft 6 in) thick, and stepped irregularly as if to support a masonry wall of rectangular blocks above, or perhaps wooden pillars and a roof. No blocks were found in position, and it is possible that the large patch of rubble overlying slips three to five contained the remains of some fallen walls, or that they were removed deliberately before the later buildings overlying slips six to nine were built. Polybius, in his description of the sack of Carthage, mentions slipways with storage chambers above them and a column at the foot of each dividing wall. The slips at Piraeus seem to have been separated by rows of columns, but the columns at Apollonia all lay on slips six to eight, suggesting that they were relics of the overlying buildings rather than the slips themselves.

Slips three, four, five, seven, and eight had central runners, each about 5 cm (2 in) high, varying from 30 cm (1 ft) wide at the top to just over 1 m (3 ft 4 in) at the bottom (Fig. 1.10). Slipway two had a more complex runner with a slot in it, presumably to take a timber beam to provide a safe skidding surface. Later work on slipways at other ports, and literary study, show that the stone cut slipways were probably the foundation for wooden skids with supporting frames to prevent damage to the fragile hulls.

Even with the lower water level, there would still be a depth of a few metres in the gap between the West Island and the Grotto Reef. Since the prevailing wind is from the north-northwest, so that there is often a heavy sea from that direction, the harbour would have been almost useless unless that gap had been blocked up. This had been done at other Greek harbours, and it seems very likely that there was a dyke or mole connecting the reef and the island, especially as we found large quantities of rubble in this area. The need for the early construction of this dyke was also dictated by the large number of people frequently working on the West Island: soldiers, guards, watchmen, shipwrights, and the crews of the galleys would need to go back and forth easily to the main city on foot. The dyke connecting the Grotto Reef and the West Island also formed part of the defensive perimeter.

To account for the buildings over and beyond the foot of the slipways, and for the inner ring of dockside buildings, it seems that the level of the sea surface dropped at

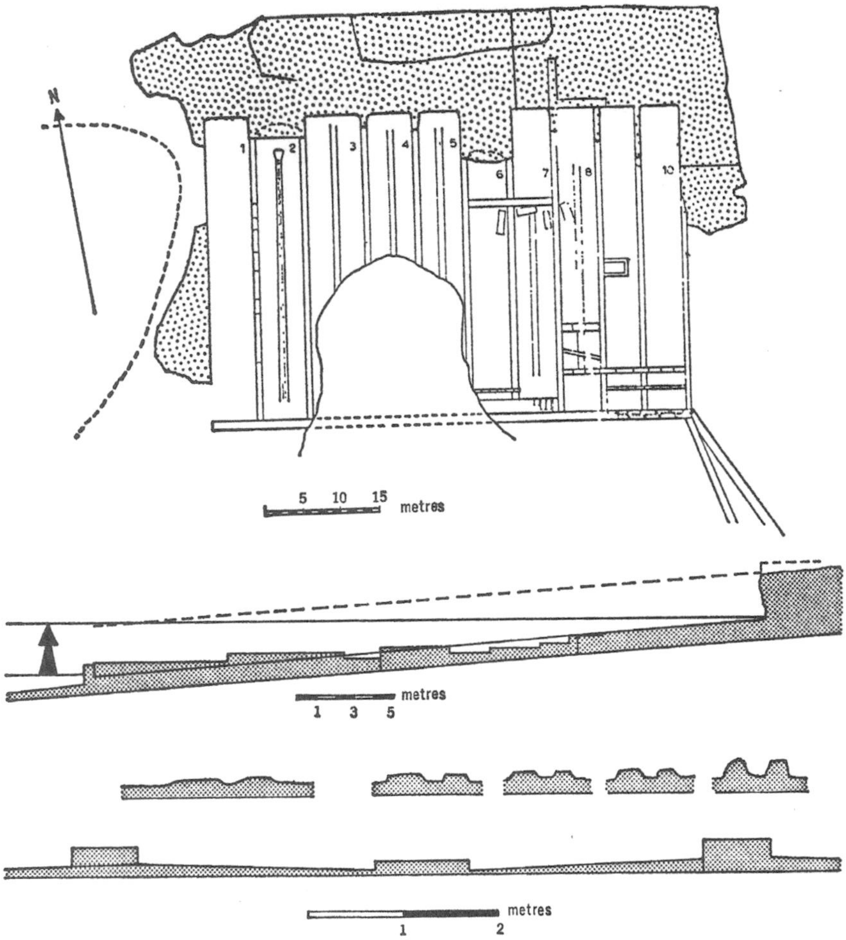

Figure 1.10. The ten slipways at Apollonia probably had wooden skids fitted into the central runner to prevent damage to the galleys when they were hauled out of the sea, and the whole complex may have been roofed in. Sand and stones cover slipways 3-4-5, and walls of a later date have been constructed over the slipways, when they were disused. The lower drawings show the slipways in profile and cross-section.

some time, or that the land uplifted. It is likely that the lowering of the water occurred, or became serious, during the Roman period. A sequence of earthquakes could account for first an uplift followed by a major subsidence. In later years I spent a great deal of time studying the interaction of earthquakes and global sea level change (see Chapter 6), but in 1958, sitting right beside the puzzling ruins, we did our best to decipher the cryptic clues with no established science to guide us.

* * *

We soon became acquainted with a team of Arab divers who were spending a season at Apollonia. They lived on the modern concrete jetty beneath a canvas shelter supported by oil drums. They had cushions to sit on and a radio for entertainment, and they would have been very happy had it not been for the bends, or decompression sickness. The wrecked ammunition ship that they were cutting up for scrap lay at a depth of 24 m (80 ft). They used old-fashioned standard helmet gear, with a large copper helmet and baggy leather suit provided with compressed air by a long hose from a pump on the surface. After dives of many hours, they were in the habit of surfacing without stops. As a result, one man was groaning in agony on a heap of rugs while his companions comforted him by singing, waiting for him to get better before they continued their work. We tried to explain that the pains would not occur if they dived for a shorter time or stopped on the way up. They would hear none of it.

"It is the will of Allah," said the afflicted one. "All divers die young. My father died at 27. I will soon be dead. It is the will of Allah."

Meanwhile, during Mrs Spencer's morning surgery for the village, Dr Adrian's medical treatment of the wife of the captain of the Arab divers had been so successful that the grateful skipper offered to take us out in his boat any time we wished. I asked if we could use his boat to dive for a short while in the open sea to a depth of 60 m (200 ft). The Hydrographic Office charts, combined with the Arab's local knowledge, enabled us to anchor a mile offshore at the correct depth, where I organized a sequence of buddy diving pairs supported by standby divers who had not dived. I wanted to see what the seafloor was like at this depth, checking for rock outcrops. The charts showed a flat gravel seabed, but one could not expect detail in such a remote area.

The sea was so unbelievably clear as we plunged "into the blue" that there was almost a feeling of vertigo when you looked down as if floating in the air. We descended rapidly down a vertical shot-line. The colder water at depth shocked our skin as we finned downwards, and gradually the darkness of the seabed loomed below. It came into focus as ribbed undulating gravel, with a few fish looking startled by our arrival. Huge elephant ear sponges were dotted about like gooseberry bushes, and I cut some chunks and stuffed them into a plastic bag. We had no wetsuits and, at this depth, where the sun could not warm the water, we felt the cold badly. I scooped a sample of the gravel into another bag, and after a few more minutes I signalled to ascend, coasting slowly upwards, conscious of the need to be careful, to save air for decompression stops. We used the British Sub Aqua Club decompression table, which demanded a series of progressively longer stops, starting at 20 ft (6 m), and then 10 ft (3 m), before reaching the surface.

We hovered at the 20 ft stop where we rested, hanging on the shot-line which was secured to a buoy on the surface. Since stops are usually boring, I gazed around in all directions, trying to find something to pass the time. The water at this level seemed to be cloaked in a translucent film, like a horizontal sheet of gauze that stretched away to infinity. The water above and below the film was pure and clear. Sunlight filtering down from the surface was reflected back from the film like headlights of a car bouncing back from fog, although the film was so thin that we had not noticed it at all during our descent. It was only visible when you were very close to it and you looked at it at a very oblique angle, almost parallel to the film itself. Even more puzzling was the fact that the film undulated with long slow waves a few inches in height that travelled at a crawling, creeping velocity, like a movie being shown in very slow motion. I was familiar with the idea of a thermocline, an interface where warmer water lies on top of cooler water with a sharp temperature change at the boundary, but I never expected to see one in the flesh. The discontinuity in density caused by the change of temperature had created a catchment on which falling dust and plankton had come to rest, making the layer visible. This visibility also revealed the internal waves that travelled slowly on the interface. That decompression stop was not boring. I had made my first observation of physical oceanography. The Arab divers watched with incredulity and mirth as we held onto the anchor rope, just flapping our fins slowly, doing nothing, easily visible from the surface. Finally, we climbed back up the diving ladder and, after other buddy-pairs had dived, returned to shore, where we thanked the captain of the boat for his time and help.

In the next two days we surveyed and fixed the location of all the corners of the Nine Quays (Fig. 1.7, E.9), confirming that there was an unbroken chain of scattered buildings from the quays to the Grid Building, thus completing the circuit of the southern border of the ancient harbour. The quays are not closely similar to any structures in other harbours, either ancient or modern, but can only have been used for the berthing of slender ships, either civil or military. The 6 m width of the slipways on the West Island would have been sufficient for vessels with a beam of 4-5 m (13-16 ft), but the spacing of the quays is only 3.5 m (12 ft). Whether this is the maximum beam of the largest vessel, or whether only multiple smaller vessels were berthed at the quays, is not certain. The dock spaces between the quays are 25 m (82 ft) long, and if the ships were this length, they would have had a length-to-beam ratio of about 7:1, which is high for a cargo boat, but very likely for a fast boat built more to be rowed than to carry a large sail. From the rough rule that a stable rowing boat draws one-third to one-quarter of its beam, these boats would have drawn less than 1 m (3 ft). The function of the quays has been much disputed in subsequent years, but I stand by the proposal here, which seems the best explanation. This point is discussed more fully in Chapter 11.

We continued the underwater survey of Apollonia by mapping the walled channel which led into the harbour from the east. Straight walls, spaced 14 m (46 ft) apart, stretched 65 m (213 ft) northeast from the Block Forts. The sides of the channel contained fallen rubble, which sloped down to a sandy floor. The channel was partly silted up, but was 4 m (13 ft) deep at the inner end, sloping to 6 m (20 ft) deep at the outer. The top four courses of the Block Forts were each inset from the course below by 10-15 cm (4-6 in), so that the buildings tapered slightly (Figs 1.11, 1.12). The tops of the Block Forts were completely flat, with no suggestion of a superstructure or peripheral walls. Maybe they supported wooden towers. The small square building just south of the southern Block Fort had a door in its eastern wall opening onto the area of rubble (Fig. 1.7, C11), where one or two foundations were visible.

Sitting on the sandy floor of the channel and looking up, you could imagine triremes with bronze rams at the bow passing slowly overhead, blotting out the sunlight. They might be carrying soldiers returning from a patrol where they had searched for pirates. Armed guards on either side of the channel lowered the heavy chains between the block forts that closed it to unauthorized vessels, and the triremes rowed over to the slipways.

For the first time we were able to approach the Grotto Reef and fix its limits (Fig. 1.7, B.4). We left one man on the island to guard the equipment during lunch, where he was surrounded by Arab boys throwing unexploded shells about in trials of strength. The afternoon was exceptionally calm. Four of us swam out to the reef accompanied by the dinghy, after setting up the surveying alidade on the West Island from which Nick surveyed the prominent corners of the reef.

We found a long boomerang-shaped lobe of bedrock curving round the outside edge of the city to form its northwest corner. The outside face of the rock was cut into vertical planes,

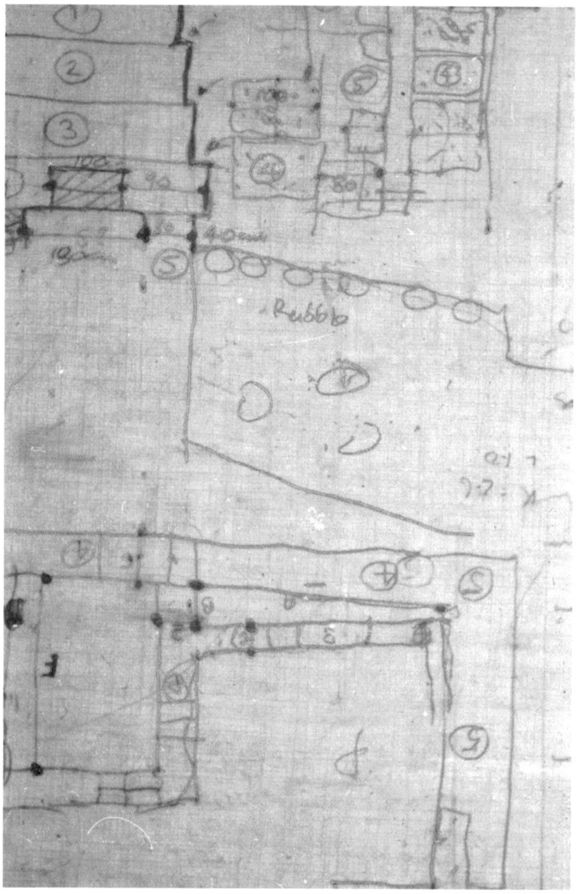

Figure 1.11. A drawing made under water of the southern Block Fort showing the stepped-in courses of ashlar masonry. This gives an idea of the level of detail that was recorded in 1958-59 for the main well-preserved buildings. The architect's rendition of this is shown in Figure 1.12.

Figure 1.12. Nick Wood's accurate survey drawing based on observations by divers working round the southern Block Fort, and photographs. The two fort bases were on the south and north sides of the channel between the two harbour basins. The channel could have been closed with a chain for defence.

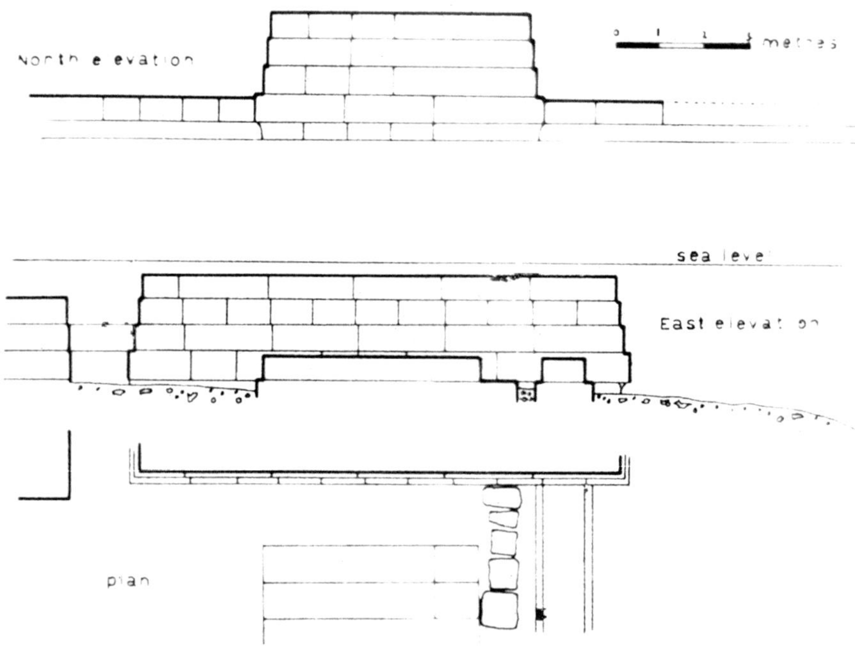

and the upper surface broke the water at a few points. Outside the main block there was a second series of planar rock walls parallel to the first, separated from them by a channel 10 m (33 ft) wide. It was like a recessed road with a solid rock fortification protecting it on the seaward side. The most extraordinary feature was a tunnel cut right through the main block, some 17 m (56 ft) long, 2 m (7 ft) high, 1 m (3 ft) wide, and with a perfectly flat floor and flat roof. The rock roof was just below the surface of the water. This tunnel was one of the most beautiful and impressive remains of the city. At the inner entrance facing into the ancient harbour, there was a great overhanging slab of rock projecting like a dark porch, with the waves rippling gently on top of it. Inside, the straight walls and flat floor were menacingly dark, but in two places the roof had cracked, admitting silver wedges of sunlight that illuminated the drifting plankton like dust in a curtained room. Pale-pink diaphanous fish clustered against the ceiling, trailing transparent fins. I swam slowly through the darkness until, at the outer end, the tunnel opened into a small courtyard or chamber, from which a wide doorway led beneath a slender horizontal beam carved from the solid rock into the channel or roadway beyond (Fig. 1.13).

Swimming under that roof of the tunnel where the soldiers of the city had once marched through to stand guard on the outer defences was the moment when I felt most powerfully the haunting presence of the dead. The sense of identity with the original users of the shortcut through the rocks onto the sea wall was visceral. We measured the internal dimensions of the passage with the 2 m (7 ft) rule, and the outside with the 50 m (164 ft) wire. Apart from these crude measurements, this exquisite structure has

Figure 1.13. This blurred image is a still taken from the 16 mm movie shot at 24 frames per second. It is the only existing record of the rock-cut arch at the outer end of the tunnel through the Grotto Reef. The diver has swum through the roofed tunnel that ends in the dark shadow on the right-hand side of the picture. This has all been destroyed by storms or fish-bombing since 1959.

not been drawn or measured in detail, and the only record of it is the 16 mm movie. I copied the film onto DVD in the 1990s, and in 2020 I had the original Kodachrome film digitised at high resolution direct onto solid state hard disc memory. In 1958 there was no time to measure the tunnel precisely, draw the roadway, or photograph the beautiful arch at the outer end: nobody anticipated that it might be almost impossible to dive at Apollonia for decades to come. French divers who photographed the reef in the 1980s found that much of the roof had been smashed by the waves, or perhaps by the fish bombers. Professeur Claude Sintes, to whom I sent a copy of the DVD, said that most of the tunnel or corridor and the arch had collapsed.

* * *

During our last days at Apollonia the weather was bad with the north wind first bringing waves roaring over the Grotto Reef, before swinging round to the south so that the sky was red with dust. The divers worked singly or in pairs measuring block sizes or patches of masonry that had proved too small or isolated to be surveyed from the shore baseline. These were fixed in position by trilaterating measurements using the 50 m (164 ft) wire from known fixed points. Many buildings had collapsed so completely that only further excavation could possibly reveal their former shapes. On the last full day of diving and snorkelling, Martin found that a rectangular excavation under the sea just close to the acropolis was a segmented tank. It was almost certainly a fish preservation tank, or *piscina loculata*. This discovery was exciting, but also frustrating, since there was no

Figure 1.14. Martin Minns found rectangular submerged tanks at the east side of Apollonia on the last few days that we were there in 1958. The tank was divided into sections by internal walls, and we concluded that it must have been a fish-tank, or piscina loculata used by the Romans for keeping different species of fish in different compartments (Photo Carlo Beltrame).

time to measure it accurately (Fig. 1.14). The Romans had perfected ways of keeping sea-fish living in protected pools, but we knew nothing of the details.

We were exhausted by the work, the heat, and countless abrasions and small cuts on our hands, feet, and legs that did not heal because of the constant exposure to salt water. These sometimes became infected, in spite of the application of mercurochrome – an old fisherman's standby – or, in extreme cases, magnesium sulphate paste, which were the only treatments available in our medical kit.

We had avoided siestas during the four weeks at Apollonia, but tried to use the hottest hours of the day to do work in the stores, repairing equipment, filling air cylinders, drafting the maps and drawings, writing reports, or developing films, minimizing the risk of sunburn. We developed minor attacks of dysentery or "gippy tummy", but luckily nobody was badly affected. Dr Adrian had departed to visit friends in Tripoli a week before the end of the work at Apollonia, so that we no longer had his expert advice. David Fagan had brought some new sulpha drugs to test for a pharmaceutical firm, but we found that they made people drowsy and lethargic, which was dangerous if you were snorkelling or diving. Instead we relied on kaolin clay or charcoal to control the dysentery. An occasional visitor during our work was Dr Wolfgang Berka, who, in his capacity as Doctor to King Idris, drove a pale-blue Mercedes which unfortunately had a big dent in it where it had been kicked by a camel. Wolfgang suggested that the infected cut on my foot would benefit from an injection of penicillin, which he administered over a drink of pastis at the Dolphin Hotel.

The mapping of Apollonia was so clearly unfinished. It dawned on me with frustration and sorrow that I had been naïve to even think of finishing such a task in four weeks. In probability it would never be finished. Brigadier Martin arranged for some army trucks that had been conducting exercises nearby to transport us back to Benghazi, where we were entertained with a refreshing and light-hearted reception party at the British Consulate. With heroic feelings of achievement somewhat restored we boarded the *Città di Livorno*, returned to Syracuse, reclaimed the Volkswagen minibus, and drove back across Europe

As soon as we were back at Cambridge, Nick Wood, Martin Minns, and I set about thanking sponsors, recording details from numerous notebooks, documenting, and tidying the records and maps (Fig. 1.15), while Bill developed and printed the still photographs and sorted his biological specimens. The 16 mm film had turned out quite well, requiring me to spend many hours squinting at a small illuminated editing screen, winding the film through by hand, and cutting it into short lengths, before sticking the good segments together again in a logical order to make a presentable story. Editing was an absorbing skill, but the smell of the adhesive cement was overpowering.

One result of our mapping was to show the horizontal erosion of the coastal ruins at Apollonia since the survey by Beechey 130 years earlier. Comparing Figures 1.5 and 1.7, you can see that the circular western gate tower was securely on dry land in the 1820s, with a solid triangle of land stretching about 100 m (325 ft) to seaward along the western city wall, up to the square tower in the city wall near Prong Building. The sea had eroded away this area by the time we arrived in 1958, although the foundations of the broad outer city wall had survived on the seabed. This maximum erosion on the western side of the city confirmed the damaging effects of the northwest wind and storms. Other parts of the beach and shallow ruins are better protected by the refraction effects of waves obstructed by the island barriers on the seaward edge of the city. The term "Foul

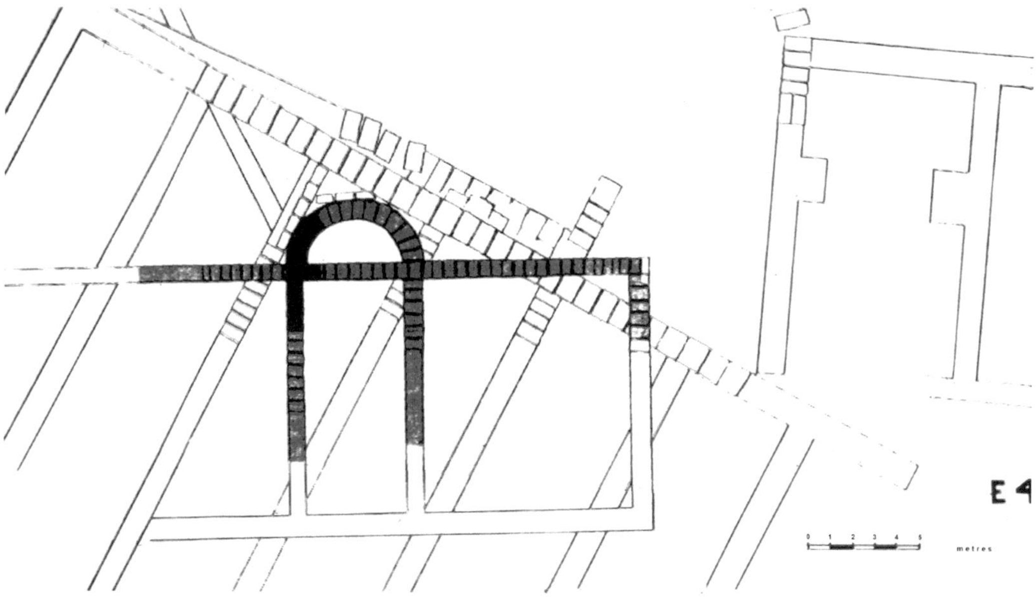

Figure 1.15. One of the most complex pieces of stratigraphic evidence at Apollonia. There are four layers of superimposed buildings of different dates, culminating in the small apsidal building using smaller ashlar blocks at the top. The shading of the walls of the apsidal building indicate one course of masonry, two, or three and more courses intact. The successive phases of use of the buildings suggests that the diameter of the inner harbour contracted at one stage.

Ground" is applied to most of the underwater city area, this being the navy term for a seabed that is likely to snag your anchor and make it difficult to recover.

During that autumn I tried to establish whether the submergence of Apollonia was due to local earthquakes or a global rise of sea level in the past 2500 years. I thought that this would take a few days. (Sixty years later we are still working on the details of an accurate answer to that question. See Chapter 6). I visited the National Institute of Oceanography at Wormley, Surrey, where, after a day with the experts, they provided the classic answer that "It could be a bit of both". I learned that global sea level could change due to the heating and expansion of the sea itself, or to melting or freezing of the polar continental ice sheets. The knowledge that sea level really could change over the whole planet started a train of thought that motivated me for the rest of my life.

The Director of the Royal Geographical Society, Laurence Kirwan, invited me to give a lecture on our team's work at Apollonia, to be presented in the spring of 1959. I gave the lecture, illustrated with slides and a short excerpt of the underwater film, which provoked many eager questions afterwards. Sir George Deacon, Director of the National Institute of Oceanography (NIO), came up to the lectern and said very quietly, "If you ever need a job, please apply to me." I was informed later that I was the youngest person ever to give a public evening lecture at the RGS, and so far as I know, this is still the case.

I wrote an article that was published in the *Geographical Magazine*, followed by short articles for the house magazines of several of the sponsoring companies. After that I settled down to study the Natural Sciences Tripos at Cambridge with renewed energy.

* * *

During the winter we worked over the results of the expedition, finding, to our irritation, that the deductions we made from our map would need further fieldwork to confirm. Was there a lighthouse on the East Island? Was the rectangular tank really a *piscina loculata*? Was there a further breakwater enclosing the eastern harbour basin? What were the accurate depths in the centres of the harbour basins? What was the relationship between the great port and the neighbouring small ancient towns along the coast? Did they all trade independently with Rome, or was produce transported coastwise to be concentrated at Apollonia, for transfer into larger cargo ships?

But there was no plan to return to Libya.

A few weeks before leaving for a project in Sicily in July 1959, I found the opportunity to arrange a second visit to North Africa as a result of the repeated generosity of the Royal Geographical Society, the British Academy and *The Geographical Magazine* Trust Fund.

Our diving team that left Sicily for Libya at the end of August 1959 was six people, including some old-hands from Apollonia 1958, and some excellent new members. The core team was myself, Martin Minns, and Natalka Czartoryska; the new divers were Martin's younger brother Jonny, John Dick from Cambridge University, and a wonderfully energetic Australian named Hugh Edwards. Although I have not met Hugh again since 1959, I have followed his career in Australia, noting that he became a successful writer who was awarded several literary prizes. Hugh was an expert on Australian shipwrecks, but he enjoyed the novelty of an underwater Greek city, which he has mentioned several times in his later writings. He had a sharp eye for detail and was an invaluable member of the diving team. Machines loved Jonny. The most battered outboard or rusty compressor would purr to life instantly when he touched it.

On the morning of September 3rd, the twin domes of the Italian Catholic Cathedral of Benghazi crystallized out of the early mist on the horizon, and we carried our gear onto the deck of the *Città di Livorno* for unloading. There was a small compressor, an outboard motor, four aqualungs, two underwater cameras (one still, one ciné), plus a mass of personal equipment and odds and ends. We had no boat of our own, but one of the sailors on board assured us that his friend Nino of the fishing vessel *Freccia Nero*, who would be in Benghazi, would be delighted to lend us a dinghy, so I gave him a large bottle of Marsala wine and hoped that this would help him to find Nino.

We cleared customs quickly. By chance we met Richard Goodchild in the street, and he arranged for us to stay in an unused school building in Marsa Susa. The main party set off in a bus with all the equipment, while I stayed in Benghazi to collect the dinghy from Nino's fishing boat in the harbour the next day. I contacted the Royal Army Service Corps who told me that they had a patrol of two Land Rovers and a 3-ton truck going east, on which they would take the dinghy and one or two people. That evening, as the air cooled from the heat of the day, I sat at a café in the square calculating the odds on getting any useful work done in only ten days at Apollonia so late in the season. We were committed to make a map of Phycus, an unsurveyed town west of Apollonia, to confirm the positions of the submerged rubble moles around Apollonia, and to do a plan of Martin's fish tanks, all in so short a time with only a slight chance of good weather.

I accompanied the army lorry next morning to the dockyard, where Nino had some hands ready to load up the dinghy. We drove out of the docks, met the Land Rovers at the Turkish Fort in the foothills of the Jebel Akhdar, and drove on to Barce and Cyrene. We arrived at Apollonia in the early evening, where, to the west, the ranges of hills were blurred by the glare of the setting sun.

We started work early next morning. Martin was continuing the detailed measurements of the submerged *piscina loculata*, his all-absorbing task. The *piscina loculata* is a fish storage pool divided into compartments connected by channels controlled with sluice gates, each compartment containing different varieties of fish (Fig. 1.16). The technology is described in the Roman literature of Columella and Vitruvius. (Archaeologists have documented and measured many ancient fish tanks on the Adriatic coast of Italy since 1960, and in a few in other parts of the Mediterranean). Continuous circulation of fresh seawater is assured by synchronizing the opening and shutting of the main sluice gates with the small Mediterranean tide, so that water flows in and out, keeping the tank fresh (Fig. 1.17). The Apollonia tank was cut out of the solid rock at the foot of the acropolis, so that, diving from the rock platform at the water's edge, you plunged straight down to the bottom of the pool, where you could see the dividing walls projecting from the sand and rubble. The water was about 3 m (10 ft) deep here, and so clear and still that Martin could work for many hours without an aqualung, measuring and writing on a board the minute details of the tank's construction.

Hugh went to the extreme eastern end of the city, where he was searching for remains of a sea wall or breakwater, while John and Jonny took the dinghy to make depth measurements all over the ruins and the central harbour basin. One great surprise was the discovery that some of the huge ashlar blocks of the southern Block Fort had been slightly dislodged so that the fresh smooth surfaces with no weed on them projected free. What could have caused them to move? Was it storms, or possibly fish bombers? The implications were worrying. How long would Apollonia last if this was typical damage in one winter?

The time had come to find and explore the ruins of Phycus, the next known Greek town to the west. The Army provided two Jeeps for Natalka and me to make the journey along the coastal dirt track. We drove westwards until we reached Ras Aamer at mid-morning, where we found a team of Arab divers at work on the remains of a wartime wreck, which they had started to cut up the year before. The tangled mass of rusty steel lay against the rocks in 9 m (30 ft) of water with the swell waves breaking slowly over it, disappearing into the gaping holes in the deck to reappear as great plumes of spray. The divers used standard Siebe-Gorman helmet gear, a hand pump and oxy-acetylene cutters. At this depth there was very little risk of bends.

Shortly before midday we reached a stretch of the coast where a line of small dome-shaped hills bordered the sea, clustered around a deep-blue inlet that opened to the northeast. Strange rectangular shadows betrayed the overhangs of rock-cut tombs. We came to a stop on the crest of a hill and looked across the ashlar walls and bits of column which scattered the slopes and projected from the sand. The rocky shore was honeycombed with the hollows of ancient tunnels and tombs.

We climbed out of our seats, took pencils and pads, and started to draw rough plans of the hill and buildings on it. The shore itself was the most important and here we found a submerged shelf fringing the headland, only about 60 cm (2 ft) underwater, with depressions about 3 m (10 ft) deep incised in it, connected to the open sea by narrow channels. The rock headland was about 9 m (30 ft) high with a series of broad straight-

Figure 1.16. Diver examining the small channels connecting the separate tanks of the piscina loculata found by Martin Minns at the eastern end of the submerged city. Fresh seawater could flow through the tanks, driven by the small Mediterranean tide, and the flow was regulated by perforated stone grilles and sluice gates (Photo Carlo Beltrame).

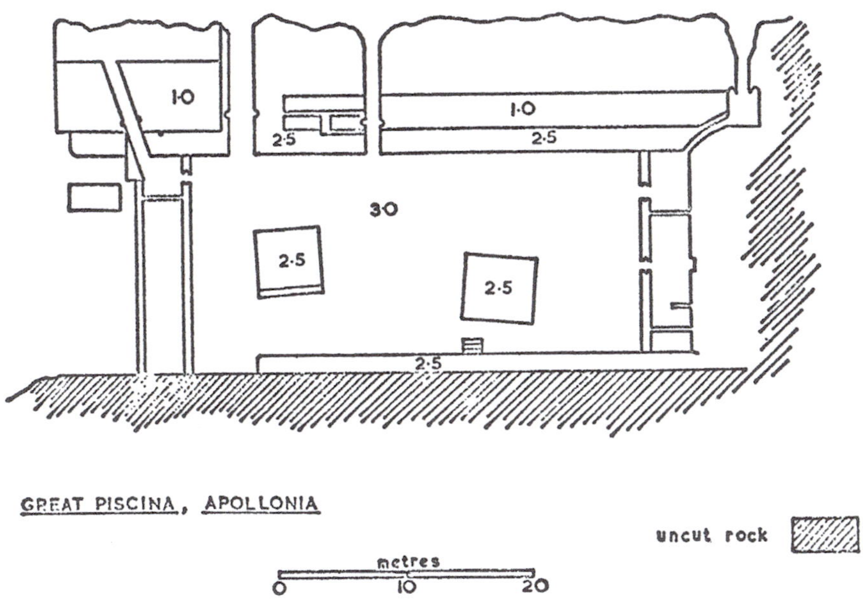

Figure 1.17. Map of the Roman piscina loculata. Note the separate containers for fish of different species and the channels connecting the tanks to the open water. The numbers show the depth in metres. The relative sea level change must have been a bit more than 2.5 m, so that the walkway around the tank was dry.

Figure 1.18. Phycus is an ancient town 30 km west of Apollonia that had never been mapped or excavated beyond preliminary sketches. A network of rock-cut chambers and submerged roadways are incised into the cliffs, flooded by seawater. The ancient town has sunk into the sea by about 1.0-1.5 m.

sided cuts through it, with the flat floor almost like a roadway, but submerged beneath the rippling wavelets coming in from the sea. From these "roadways", rock-cut doors and windows led into tomb chambers, each with alcoves and niches for monuments or votive relics (Fig. 1.18). It is unlikely that the carved "roadways" were just quarries. You don't quarry 10 m down into the rock when you can cut blocks at sea level just nearby. The "roadways" have vertical sides, with openings into chambers with rectangular windows, which all had a purpose, though it is now difficult to understand.

Close to the sea, in places where the rock rose only a few feet above the water, we found evidence of successive cutting of channels, tunnels, silos, and storage tanks, since one cutting would sometimes slice straight through another. Many of these chambers looked as if they had been partly filled with concrete, but closer investigation showed that it was a natural formation of blown sand and chemical salts from the sea, cemented together by sunlight and evaporation. The network of tanks probably formed a storage depot for oil, freshwater, grain, fish, oysters, crayfish, and so on for supplying the ships which called at Phycus in the Byzantine period in the 4th and 5th centuries AD, but many of the features that we found still defy explanation. The produce stored here may have been for transport to Apollonia before assembling in larger cargoes. We swam round the headland, tracing the seaward limits of the walls of the tanks, collected a few pieces of amphorae, and left in the early evening. We drove through the decaying village of El Haniyah, then climbed the hairpin bends to the summit of the Jebel and returned to Apollonia by the mountain road.

Phycus was submerged by about 1.0-1.5 m (3-5 ft), so the earthquakes that had dragged Apollonia beneath the sea may have shattered towns for many miles along the coast of Cyrenaica. Or had the sea level risen also? (See Chapters 6 and 11 for the answers).

* * *

We were outrageously lucky. For the next few days, a gentle east wind blew. It was so soft that it did not stir the water to cloudiness, but was sufficient to cause a slow movement of sand from the eastern limits of the ruins of Apollonia onto the part that we had mapped the year before. The result was that many ruins were revealed in the area of the city that we knew least about, and we worked there furiously for fear that the conditions would change. We discovered a continuous broad band of rubble projecting from the sandy bottom in a depth of 9 m (30 ft) leading out to sea from the acropolis almost to the East Island. At the outermost end there were many enormous rectangular blocks of stone littered about the bottom, with smaller blocks of white marble between them. Knowing that the Greeks were quite capable of building a sea wall in this depth, we concluded that there had been a mole here enclosing a second harbour, and that there had been a guard tower at the end of it. Storm waves had progressively flattened the piled rubble of the mole to leave only the present broad scatter on the seabed.

Although the gap between West and East Island was deep enough for a channel, there was so much rubble lying at the bottom that it seemed that this too had, at some date, been blocked up, so that the real entrance to the port was between the East Island and the newly discovered Eastern Mole. Hugh was certain that the gap between the islands had been blocked and was less than 6 m (20 ft) deep. He was so sure of this that he volunteered to be tied by 20 ft of rope to a large rock and dropped over the side without an aqualung. Luckily we checked the depth first and found it to be 8 m (25 ft), but clearly it had not been used as the port entrance channel.

While we were in England, Sir Mortimer Wheeler, a famous archaeologist of the time, had examined our 1958 map, and had suggested that we search the East Island for the remains of a lighthouse. The sides of the island were all artificially quarried, cut plumb vertical, so that it was very difficult to land. Martin climbed the cliff with a rope which he used to haul up the cameras and surveying instruments from the boat. On the top we found a wide circular foundation partly of curved blocks, part cut from the solid rock, which projected from the cliff like a bay-window on the south side: the lighthouse had been here. As we stood on the lighthouse foundations looking shorewards towards the acropolis, we could see in the depths of water the dark shadow which marked the end of the Eastern Mole. We had discovered the true entrance to the eastern port of Apollonia. This was where the cargo ships entered to load up with hundreds of tons of grain to feed Rome. The narrow channel between the block forts then led further into the inner basin with the slipways.

There were only two days to go. We had achieved all we had set out to do except for finding ruins on the island chain immediately to the west of Apollonia. And still the weather held. Hugh and John spent a morning swimming westward from island to island down the coast, while the sun rose vertically into the sky, but they had to return, defeated, across the scorched red earth. To compensate for this unlucky start to the day, the afternoon provided a succession of surprises. Most of the divers were working on last-minute measurements, photography, and gap-filling, but Hugh was nowhere to be seen. By mid-afternoon we finished filming and swam back to the shore, where we found that Hugh had just emerged from the water. He was sitting on a large square stone, looking very pleased with himself.

"What have you been doing?" I asked abruptly. I was irritated because I felt that he had been slacking. But I knew that he must have found something exciting, so I tried to conceal my feelings. He jumped up from his seat, revealing it to be a wedge-shaped chunk of carved sandstone with three large holes cut in it through the narrower end.

"Just look at this, boys!" he shouted. "The finest Greek anchor you ever saw. It's better than any we found in Sicily. An absolute beaut."

"Where on earth did you find that?" I asked.

"I was just cruising along the little square building near the Nine Quays when I saw the corner of a stone sticking out of the sand. It had a hole drilled in it, so I guessed what it was. It was less than 10 ft down so I went and got it out. It was a honey. The trouble was it was so goddam heavy that I couldn't swim with it and had to walk along the bottom, dump it, come up for a breath, and then go down and get it again. I'm shattered."

The journey of 100 yards had taken him quarter of an hour before he was able to sit down on the anchor, exhausted but happy.

Figure 1.19. Hugh Edwards found this ancient Greek stone anchor lying on the bottom of the harbour at Apollonia and, since he had no scuba gear with him, carried it onto the shore by walking along the bottom holding his breath.

"Too bad," I said. "Shattered or not, you're going to do it all over again for the cameras. It'll make the funniest picture for years."

We whistled for the dinghy, and Jonny rowed it over. I was getting into the boat when Jonny prodded me and spoke in a hurt voice.

"Haven't you noticed anything?" he asked. I was puzzled. "This," he said, and triumphantly held up a slab of lead with a slot in it. "A Roman ballast weight. The plumb-line landed slap on it. I was looking to see that the line was straight, and there it was. Too easy." This was a wonderful find, but Jonny was cheated of his glory by the excitement over Hugh's anchor as we stowed everything into the boat. Sure enough, Hugh, staggering along the sandy bottom, hugging the great lump of stone and half falling over his fins, was a splendid scene for the film, but he found it too tiring to be funny (Fig. 1.19).

In the evening, after measuring some foundations which had just been uncovered by the sand in the surf zone, Natalka and I returned to the house to find the others already sitting down to start dinner. I was sitting down myself when Martin got up, walked to the corner of the room and said dramatically, "Hugh has done it again. Meet our new friend." He pulled a coloured towel away from a chair, and there was the gleaming white marble head and torso of a smiling faun. I was too astonished to say anything, so Martin provided the details. Hugh had been helping him finish measurements in the

Figure 1.20. Hugh Edwards was diving in the Roman fish-tank when he found this small marble statue of a faun. It was more than half-buried in the sand and gives an indication of the decorative surroundings that originally embellished the fish-tank.

fish-tank when he had glimpsed the head of the statue half concealed by the sand under the edge of a large boulder as he swam across with a measuring rod. Together they winnowed away the sand and raised the statue to the surface (Fig. 1.20). I began to see that the ancient fish-tank had been an ornamented pool with statues, possibly shaded by vines and awnings. It had been a place to relax in pleasure as much as a store of fresh food. Watching fish hover and glide is always soothing. The prosperity of Apollonia had enabled some people to live well.

We drank to the success of Hugh's wonderful eyesight, feeling very pleased with ourselves. But our luck had not run out yet.

A friend of ours, the Arab school teacher, came in to say that he had found something that might interest us, and he asked for a pencil and paper. We provided him with a large sheet of paper, on which he drew a map of Apollonia and the islands nearby to the west, explaining that he and some friends had been swimming around one island where they had found ruins which disappeared down into the water on all sides. He pointed awkwardly, as if he was afraid that we might not be interested. I assured him that this was marvellous news, that we were extremely grateful to him. The next day we were due to leave, and a truck driver from the village had offered to take us back to Benghazi if we were prepared to travel overnight, sleeping on top of his cargo beside our dinghy. First thing in the morning we drove with an Army Jeep along to the school teacher's new island where Martin took a camera in an underwater case, drawing boards, pencil, and ranging poles and swam over to it, while the rest of us returned to base to pack. In the afternoon we re-enacted the discovery of the marble faun statue for the movie camera.

The island that Martin sketched and photographed contained several large tanks cut deep into the rock, and there were numerous fallen curved blocks that must have formed an arched or vaulted roof over a huge storage tank. The roof had collapsed into the tank, and many of the blocks were set in the congealed natural cement that we had seen at Phycus. The carvings, cuttings, tanks, and silos descended into the water on the flanks of the island. At the time when the sea level was lower, when the port of Apollonia was thriving, the island would have been the tip of a rocky promontory joined to the mainland, but there was no sign of a settlement or ruined habitations. This was a depot purely for storing produce and supplies in the short term for transfer to coastwise shipping. From there agricultural products were transported to Apollonia for the great sea-crossing voyages to Rome or Athens. It is easy to imagine a few small ships lying in the shelter of the headland while their crews carried the amphorae or sacks into the holds.

At 11 o'clock that night we said goodbye to our friends who had helped us so much, and the lorry in low gear roared up the steep hill to Cyrene on the Jebel. We arrived in Benghazi at dawn, where we returned the dinghy to Nino's fishing boat and boarded the *Città di Livorno* once more.

* * *

In 1959 we had been lucky with the sea state and had discovered extraordinary things, but there was so much still to do, so many features of the ancient city that remained inexplicable. We had begun to reveal the decoration and luxury that beautified the working façade of the city, but it would take years of diving excavation to do justice to such a vast and unique complex. There were few clues as to how the great port really functioned, and there were many buildings that had no parallel in land-locked cities ... or indeed in other known ancient ports. How were they used? Fallen shards, debris, coins, tools, glass, lead, bones, statues, anything buried amongst the stones and tiles would help to reveal their purpose. How was the annual transport of thousands of tons of grain to Rome managed? How were the huge grain ships loaded with their cargo? What were the phases of building and abandonment as the sea level changed and the half-ruined city evolved over the centuries? There was nowhere else to go for the answers which all lay under the sea. We scattered at Syracuse and took our different ways back to England.

The marble faun had been given to Richard Goodchild as a valuable antiquity for curation in the museum at Cyrene, but he said we could keep the stone anchor, so I had slung a rope through the end hole and somehow between us we carried it like a suitcase across Europe. Lillywhites, the famous sports store on Piccadilly Circus, exhibited our finds and maps from Apollonia in its street-front display windows, but I had a memorable experience afterwards trying to travel on the Underground train carrying the heavy stone anchor on its loop of rope. I was told by a porter that it was too dangerous for other passengers. Rather surprised and disgruntled, I trudged with it to the next station, and boarded the train there.

In addition to the main map that Nick Wood had drawn, he had also prepared larger scale drawings of the multi-layer stratigraphy of the Apse building over the Grid Building (Fig. 1.15), and the staggered layers of ashlar masonry of the southern Block Fort (Fig. 1.12). These were probably the first accurate drawings of ancient underwater masonry anywhere. We took it all as routine work as the project morphed into the stages of drafting reports, writing lectures, and thanking sponsors. But the achievement is worth noting, both for the accuracy of the measurements by the divers, and the careful draftsmanship of Nick Wood.

The *Sunday Times* ran an illustrated feature on Apollonia, while I published academic articles and reports. Bernard Eaton, the editor of the British Sub Aqua Club magazine, *Triton*, ran a feature that emphasized the diving itself, to encourage amateur archaeologists. I gave lectures and showed the 16 mm film at meetings of the club and at an international diving conference in Barcelona in 1960. In the world of marine archaeology, I was known for a year or two as the man who had mapped Apollonia in the first scientifically controlled mapping of an entire underwater city.

The techniques and systematic wall-to-wall thoroughness that we had applied were innovative and ahead of their time. Every corner of a building was either triangulated from the baseline on shore, or trilaterated from two known points on the seafloor. Furthermore, every member of the team dived. Underwater archaeological projects before this date had employed naval or salvage divers to work on the seabed, supervised by an archaeologist on the surface. Trained minds did not observe the seabed directly. Unknown to me at the time, Professor Nino Lamboglia, in Italy, had instructed sailors to apply a grid of fixed tapes, 10 x 12 m, with 1 m spacing (33 x 40 ft with 3 ft spacing) over the shipwreck at Spargi in 1958, the year we started at Apollonia. Lamboglia described his innovation as "... to initiate this new experiment in the history of underwater archaeology ... and to test the co-operation between archaeologists and free divers." He still saw the grid of tapes as a way for non-academic divers to report their observations to the archaeologist on the surface.

Within a few years, shipwreck archaeology, by a process of evolution and diffusion, adopted as standard the use of fixed rigid grids laid over the wreck throughout the survey and excavation on the seafloor, fixing every artefact and key point surveyed into position in three dimensions. Gerhard Kapitän developed the techniques in Sicily. Peter Throckmorton, Honor Frost, and George Bass carried out meticulous surveys of shipwrecks on the coast of Turkey, followed by excavation, while other groups in Germany, France, Italy, and Israel applied rigorous survey techniques to submerged ruins. A team from Imperial College London, led by John Woods, mapped a deepwater wreck off Xlendi Reef, Gozo, Malta, in 1961, at a depth of 60 m (200 ft), from which they raised many amphorae under controlled conditions. A team from the USA led by Bob Marx surveyed and excavated parts of Port Royal, Jamaica, in the late 1960s.

An accurate map, whether of a wreck or a city, is the basis for the next stage of interpretation and selective excavation. I expected that having demonstrated the extraordinarily good preservation of the great city of Apollonia, showing the unique complexity of the unusual structures round the harbour basins, professional archaeologists would seize any occasion to continue the work for years to come. We had created the opportunity for original archaeological research by other people. I was happy for them to take the lead. But in Britain there was no professional reaction, and the initiative passed to French divers decades later (Chapter 11).

I also had to obtain my degree in Natural Sciences. I loved the sense of historical time passed and passing, the visceral contact with cultural roots, and I savoured the indulgence of desolation and decay in remote places, but my professional and intellectual focus became the change of sea level, the changes of climate, and the occurrence or absence of earthquakes. The intense memories of those days at Apollonia faded into the background, although a key piece of logic had stuck in my mind. I realized that there were hundreds of ancient ports around the Mediterranean coast, every one of them potentially containing enough remnant structure to tell me about the changes of sea level during its history. It seemed possible that mapping hundreds of sites would reveal for the first time how earthquakes had been distributed around the Mediterranean coast for thousands of years, how much the ground had moved up or down, and whether the global sea level had changed in the past 2000 years. My life progressed in chaotic and unpredictable zigzags from one chance opportunity to the next. Life was exciting and producing new sensations and experiences all the time, so that the mapping of Apollonia, which had started in my first year as an undergraduate, came to seem almost a juvenile escapade.

2

For Queen and Country Under the Sea

"Red on! Stand in the door!"
"Green on! Go!"
"Go!"
"Go!"

The Royal Marine Sergeant dispatcher tapped my shoulder and I jumped as number three, the last man in the stick.

I bounced on the slipstream of the departing aircraft, dropped into the still air below, and heard the satisfying "Phwupp" as the canopy opened and took my weight on the parachute rigging lines. The static cord attached to the aircraft had pulled the canopy cover off, and I was drifting sweetly above the sea. Swaying gently like thistledown.

The sun shone on the white horses dancing in the sea below. There was a gentle breeze. I was happy. I whistled pop-tunes of the time, like *See the Pyramids Along the Nile*, the *Tennessee Waltz*, and *Beside the Zuidr Zee*. A circling seagull looked at me with a malevolent and resentful glare. It kept flying around me with a nasty look in its eye as I drifted down. Seagulls never look happy.

I pulled the seat-strap forward under my rump and sat four-square in the parachute harness.

For landing in the sea you had to pull the seat-strap forward, loosen the shoulder straps, and plan ahead to drop out of the parachute harness well above the water, since it would be fatal to hit the sea with the parachute rigging lines and dozens of square yards of silk on top of you. Drowning would be quick. So there I was with the seat-strap under my backside, undoing the safety buckle and shoulder straps, and pushing the reserve 'chute to one side. I was as free as that pretty *Lady on the Swing* by Fragonard. Not a care in the world. At least you can't break your legs when landing on water. There were rumours in the base that in the previous year a marine had fallen out of his harness at a height of 400 ft and been killed on hitting the water, so I made sure not to lean forward. I whistled with happiness and the seagull glared at me.

"It was my twentieth year to heaven". I was paid to do things I enjoyed. A young man cannot ask for much more. It was April 1956. We were the democratic good guys. The Russian Communists with thousands of tanks were a threat to Europe. Life was simple.

I looked around me at the descending parachutes of the marines who had jumped before me in the previous sticks, usually of three or five men per stick. Away to the south was the sand-spit of Poole Sandbanks, now a haunt of millionaire footballers and Russian oligarchs, but then just a sparse sandy promontory dotted with small houses and one or

In: Flemming, N.C., 2021, *Apollonia on my Mind. The memoir of a paraplegic ocean scientist*, 61-100. Leiden: Sidestone Press.

two grander villas. Behind me was Brownsea Island covered in trees, and behind that the wetlands, creeks, and marshes of the Arne Peninsula fringed by the Wareham estuary. Everything shone with colour and glinting highlights in the spring sunshine.

The Royal Marines Amphibious Squadron used battered blue-grey landing craft left over from D-Day to pick up the parachutists as they fell into the choppy water. I scanned the sea below and saw some men already being picked up, crawling onto the bow-ramps while the bow-men dragged their soggy parachute canopies out of the sea with boat-hooks. Then I started to worry. There was a landing craft heading towards the place I was expecting to hit the water, but each time I looked at the craft the bearing was exactly the same. We had been dropped from 1000 ft (305 m) to give us relative beginners the chance to get organized during the descent, so I had plenty of time to work out what was happening. If two moving objects hold on convergent courses while the bearing between them remains constant, they will collide. It's a basic rule of navigation, even in three dimensions. That meant that unless the coxswain of the craft changed course or hove-to and stopped, I was going to skewer myself on a jagged rusty chunk of metal with vertical edges and the parachute would probably land on top of the men already on board. Bifurcation seemed an unpleasant possibility. I tried to side-slip by pulling down the rigging lines on one side of the canopy, but could not get a secure purchase as I was no longer strapped in tightly by the harness.

I was wearing a full Navy-pattern dry diving suit (Fig. 2.1) so I could easily drop into the sea from a height of 3-6 m (10-20 ft), or even more if necessary to avoid an accident.

Figure 2.1. Navy diving suits and oxygen breathing apparatus, as used by the Royal Marines SBS in the late 1940s. It has not proved possible to find pictures from the 1950s showing the breathing sets of that date. By the 1950s the soda-lime canisters, which look square in this picture, were cylindrical and in the centre of the oxygen rebreathing bag. The suits and fins look similar to those still in use in 1956, but we had bigger face-masks (Courtesy RM SBS Archive).

I watched carefully as the coxswain maintained the same course and speed. No change. Collision was inevitable. I slid forward off the seat-strap still 9 m (30 ft) above the sea and hung by my hands gripping the harness, ready to let go.

At the last second the bow-man saw the danger and turned to shout aft to the coxswain that he must stop and go astern. But he was too late. I had already dropped from the harness and crashed into the sea from a height of 4 m (15 ft). I bobbed up immediately because of the buoyancy of the drysuit. I could see what happened next.

My parachute billowed forward in the wind and the heavy reserve 'chute and harness straps and buckles struck the bow-man in his back. He pitched into the landing craft where the rigging lines and canopy collapsed onto the crew and the parachutists packed in already.

I can only imagine the chaos that followed. I was warm and dry, bobbing peacefully in the waves, watching the landing craft lying motionless for a long time. I could hear nothing because of my drysuit hood and the noise of the water slopping over my ears. There must have been a desperate struggle going on under the mass of fabric and tangled nylon cords.

Eventually the Marine corporal emerged on the lowered bow-ramp to extend his boat-hook as I scrambled out of the sea onto the cold ribbed-steel surface.

"What kept you, Sir?" he said. "Playing with the mermaids, were we?"

"Have a cuppa kai Sir!" called another marine. Kai is Royal Navy slang for a concentrated hot chocolate drink. It was supplied in solid blocks that needed a chipping-hammer to break it up before mixing with boiling water.

"Did you do some sunbathing?" asked another.

"Welcome aboard!" cried another laughing.

They knew I had done the only thing possible to avoid a potentially dangerous accident that would have got everyone into trouble, so there was no criticism in their joshing. Maybe they had already vented their opinions to the coxswain while disentangling themselves from the parachute. My reputation as "Young Sir" was unblemished. The kai tasted delicious and we were soon motoring back to the base at Hamworthy in Poole Harbour at a good clip, with the waves smashing rhythmically under the flat-bottomed hull.

* * *

In the 1950s young men still had to undergo a two-year compulsory period of military training, called National Service, unless they were medically unfit, or were engaged in some work of national importance from which they could not be spared. I wanted to use this opportunity to join a unit which would provide me with the maximum of adventure and new skills, which is how I found myself being trained in the Special Boat Service and became involved in the events described above.

I enjoyed my National Service, subsequently joining the University Air Squadron where I learned to fly military training aircraft in the RAF Volunteer Reserve while at Cambridge University. I am not attracted to the military life as a career, and I am not a warmonger. But conducting diplomacy with belligerent and expansionary military opponents will only bring peace if we possess a reserve of disciplined force under civil control, and if there is the knowledge that if pushed beyond an uncertain point, that force will be used. This policy in various guises maintained the stalemate of the Cold War, avoiding nuclear bombing for more than 40 years, in spite of the obvious outbreak of regional wars at times. My respect for the military grew early in my life because of what I learned from my father.

My father volunteered to join the Army straight from school in 1916. He served in the Light Infantry, the "Green-jackets", where he was shot, patched up once, sent back into the front line, wounded again, and was in hospital still when the war ended. He could reminisce calmly about his life in the trenches without reticence or undue stress, even having a few anecdotes about trench life. While in hospital he wrote a retrospective diary which he gave me in 1978. In it he describes daily events from 13th December 1916, when his battalion embarked at Southampton, through to its ending in April 1918, when he was invalided back to England. It describes the grind of daily life in the trenches, with sketches of the manoeuvres and patrols, but only occasional contacts with the enemy. The terrible slaughter that he had seen left a mark on him that influenced his social and political attitudes without dimming till the day he died in October 1981. He saw that the British working men had been sacrificed by the tens of thousands for reasons they did not understand. He became a convinced Socialist, devoting all his spare time to helping poorer people in the East End of London, where he lived until late 1939. On a more personal note, he was one of the few survivors of his sixth-form year at Rugby School, losing all the friends who mattered to him at that age. Young infantry officers died first.

On holidays or travelling anywhere in the British Isles after 1945 he would seek out the local church and, although he was a devout atheist, he would find the war memorial and sit gazing at the carved columns of names. He would point it out to me, saying, "Look. The list of names for the First World War is four times longer than for the Second War. Think about it." He did not weep but his face was frozen by sorrow and memory. In spite of the need for patriotism during the Second World War and afterwards in the confrontation with Soviet Russia, he showed me that war is never a glorious game. He would sit stony-faced, counting the names and say no more. We would leave the church in silence.

From 1940 to 1945 we left London to live with my paternal grandparents in Berkshire on the edge of an idyllic postcard village. The fields and forests round about were the scenes for frequent military exercises, practice parachute drops, and river crossings with improvised pontoon bridges where the sweet Thames flowed muddily. At the age of six I revelled in the excitement of the technology with the smell of hot machinery and the clank of tracks. After a high-altitude dogfight, too high to be seen with the naked eye, a British pilot ejected from his aircraft and floated down into a local farmyard, finally dangling ungallantly from a telegraph pole. The farmers helped him down to a hero's welcome. Shiny bunches of reflective strip metal appeared mysteriously in fields and hedgerows, which the grown-ups told me was thrown out by the British aircraft during dogfights to confuse the enemy radar. It was known as "chaff" or "window". It was impossible for a child to understand what the war was about, although my parents tried to explain how awful Hitler and Nazism were. Relatives from Germany and friends of relatives stayed with us for several years as refugees, but I did not understand that they were Jewish, nor could I conceive what was happening in Europe that had caused them to flee.

In the family jargon the First World War was "Daddy's War", and the one we were living through was "This War". I sensed that there was something very different about the nature of the two conflicts, but could not fathom why the grown-ups held these attitudes. What were the Germans doing that was so horrible? My father stayed in London, working in the Cabinet Office throughout the war, visiting us occasionally at weekends. Incendiary bombs hit his apartment twice during air raids. My mother explained that he had gone up into the roof space to extinguish the fire with buckets of sand. My admiration was boundless.

No bombs fell near us, although the city of Reading was hit several times, and I knew that enemy aircraft dropped bombs on London every night. Family friends were serving as fire wardens and I listened to terrifying stories of the flames burning close to St Paul's Cathedral. As the pace of nearby military exercises was increasing there were more American vehicles on the country roads carrying the strange sign "Caution: Left-Hand Drive". On childish walks we saw canvas tents occupied by American soldiers, who offered us chewing gum, which our nanny allowed us to accept. As military aircraft of all kinds flew overhead more often, I became self-satisfied at my ability to identify many variants, with an especial favouritism for the Lockheed Lightning with its twin fuselages. The drumming beat-note roar of four-engined bombers disturbed the nights more frequently. Only decades later did I learn of the horrific fire-bombing raids on the German industrial cities and the calamitous raids on Hamburg and Dresden.

One morning the sky was almost black with aircraft from horizon to horizon. Like migrating birds, they swarmed south in waves of undulating thunder. I lay on my back on a sun-warm brick wall in the garden gazing in astonishment at the sky and tried to count them, but that was impossible. I just stared in disbelief for what seemed like hours. Nobody told me what it meant. Perhaps nobody knew. In retrospect, I imagine that it was the air support and transport for the D-Day landings in Normandy, or perhaps the tragic parachute attack on Arnhem.

After this the military manoeuvres in the neighbourhood faded away, although one glider crashed in Farmer Nelson's field, prompting locals to hold the uncharitable opinion that the pilot had crashed on purpose rather than risk going across the Channel to land in France. It was now safe to visit London for sightseeing, and a year later the war ended with the final news of the atom bombs on Japan that produced a state of peace. Innocence was bliss.

In the first year after the war, when we were living in west London, my father discovered that there were still holding-camps nearby where German prisoners were living while they waited to be sent back to the chaos of rubble and burned-out buildings that was their homeland. Since local families were encouraged to invite the men to join Sunday lunch, we hosted pairs of German ex-soldiers on several occasions. They seemed so like us that I wondered why we had been fighting them.

I followed the news of the Cold War with distant interest over the succeeding decade, and it was impossible, even as a child growing into the early teens, not to know that there was continuous tension, suppressed violence, and occasional outbursts of conflict with the USSR or its allies. I read the daily newspapers without full understanding of what they described. I started boarding school at Rugby in 1950, where, at the age of about 15, all the boys had to join the Combined Cadet Force. The Korean War was raging, reported in detail in the British press, since many British troops were engaged in the battles.

We knew that at the age of 18 we would be called-up for two years' military National Service, and that only those who were very unfit, or employed in special professions, would avoid this. The CCF activities occupied one afternoon a week, with a further two-week camp field-training session in a wild location during the summer holidays.

As I had always adored outdoor life, climbing, adventure, swimming in the sea, and taking cheerful risks, it was a pleasure to discover that some grown-ups could encourage one to behave in this way and that I could be praised for it. There was also a touch of the spirit of John Buchan and the lifestyle of Tom Sawyer and Huckleberry Finn that appealed to me. Some of the masters at Rugby had served in the forces in the war, and one master had parachuted into Yugoslavia to fight beside the partisans, so the CCF lectures were not just theory. It was the outdoor life, the cross-country orienteering, the survival in bad weather, and the element of overcoming physical adversity that I enjoyed most.

During my last year at Rugby, in autumn 1954, at about the time of my 18th birthday, another boy in my house told me that he wanted to conduct his National Service in the Royal Marines, an elite unit in which you could climb cliffs, swing on ropes, and cross rivers on rope bridges. It seemed too good to be true. I enquired further, discovering that if I joined the Royal Marine Forces Volunteer Reserve before receiving my call-up papers, I stood a better chance of joining the Royal Corps proper for National Service. I contacted the RMFVR office in Queen Anne's Mansions, in Westminster, where they advised me to report for a volunteer training fortnight immediately after Christmas. It was winter 1954-5. New Year's Day was not a holiday in England, and most people only took one or two day's holiday over Christmas, so I was happy with these instructions.

* * *

The two-week course was held mainly at the Royal Marine Training Centre near Lympstone, just south of Exeter, in Devon, with a few days in Plymouth for rock-climbing. On the first day the store-keepers issued me with my khaki uniform, black canvas webbing belts, and denim fatigues for daily activities. The black webbing harness, belt, backpack, and ammunition pouches had to be scrubbed with black boot polish. An important item was a sewing kit, since this was before nylon socks, and you had to darn the holes in your woollen socks frequently. It was a punishable offence to get blisters on your feet. The camp was on the banks of the River Exe, near the mouth of the estuary, so that the winter wind from the sea blew over us in a nasty mixture of rain, snow, and salt with apparently no obstacle between us and the open Atlantic. The NCOs ran us through the gamut of assault courses, running in full packed kit carrying a comrade also in full kit and both carrying rifles, and long-distance marches on the moors. The rain and snow soaked through our uniforms. We learned death-slides and how to cross a river on a single taut rope doing the cat-crawl, and how to recover back onto the rope if you fell sideways and were hanging by your hands. Since I had already done many of these things with ropes rigged on trees in our family garden at home (see Chapter 3), I was in assault-course heaven.

* * *

The Royal Marines had traditionally provided amphibious troops, gunnery crew for the aft turret on large Royal Navy warships, and military police on board. We were taught the history of the Royal Corps, but not the predicament that confronted the Corps in the 1950s. During World War II the Royal Marines were, in addition to their ship-borne duties, at the forefront of innovating highly mobile and relatively lightly armed Commando Brigades, a type of warfare that was often scorned by Army strategists who could see that in a major war the land captured by commandos could not be held against a determined counter-attack. Conventional forces with artillery and armour always had to move in as quickly as possible to secure the terrain gained by speed and surprise. By the 1950s the era of great battleships was waning fast, and, despite the experience of the D-Day landings, amphibious commando warfare had still not established itself as a permanent part of the military tool-box.

The budget for the Royal Marines comes from the funding for the Royal Navy. There was continuous tension between having more fighting warships, which the admirals preferred, or paying for commando units on land that seemed to be constantly required in the sporadic regional wars. The Royal Navy constructed two amphibious support vessels in the 1950s by adapting the light aircraft carriers HMS *Bulwark* and HMS *Albion* to carry commandos, landing craft, and helicopters. When I joined the Royal Marines in 1955, I expected and hoped to qualify as a commando but had no further understanding of what the training might lead to. I did not know that there were special units trained in diving.

* * *

I attended the standard National Service medical examination and interview at the end of January 1955, after which my posting to a military unit would be decided. I was thankful to be posted to the Royal Marines, returning a few days later to Lympstone to start training as a proper recruit. The pay was four shillings a day, which made me feel rich.

We spent the next four months in basic training, mostly in the wilds of Exmoor learning infantry tactics and survival, long-distance marching, many hours of parade drill, assault courses, cliff-climbing, kit inspections, and shooting target practice. The weather was wet and stormy for the first few weeks, but soon the spring sunshine became stronger. I was the only privately educated recruit in a barrack hut of 20 men, where I quickly developed a great respect and liking for my fellow marines, many of whom were physically fit from heavy work in coal-mines, steelworks, or on farms. They had left school at age 15. The selection process for the RM had ensured that they were bright, intelligent, and self-motivated. A popular motto was "Bullshit baffles brains!". That can be roughly translated as follows: "The cunning of the underdog always allows you to survive in spite of the actions of so-called superiors who try to make your life miserable and even risk your life." The subversive spirit of the *Good Soldier Schweik* was alive and well in Hut 42, as in most military units. We were in the farthest corner of the barracks' perimeter, from which we could sneak over the boundary fence, cross the Exmouth railway line, and collect mussels from the foreshore. We cooked them over the hut heating-stove and they were delicious. In spite of their ritual assertion of independence, the recruits were determined nevertheless to be good troops and were proud of their military achievements.

The all-embracing authority of the adjutant ensured high standards of uniform neatness, polish, and discipline amongst the marines. For anomalous historic reasons the adjutant wore a dark blue RM uniform with dark blue jodhpurs and black riding boots, creating a silhouette that could easily be identified from 200 yards distance. It was reputed to be impossible for two marines to march past the adjutant, since no matter how perfect the simultaneous salutes and eyes-left he would always find something wrong with the lacing of their boots, trouser creases, buckles, badges, or the timing of the salutes. You would then be stopped, interrogated, inspected, and probably condemned to extra drill parades.

Since we could always identify the adjutant from a safe distance, we would immediately turn from the main road to escape down a side path whenever we saw him. He thus moved in a huge empty bubble devoid of marines. Did he wonder where we all were? On the daily morning parades, which involved all the training squads, the roles were reversed. We were stationary targets and the adjutant could choose whom to inspect by selecting one squad himself, taking over from its officers and NCOs. The condemned squad would still be there two hours later having opened every pack, taken their socks off, and had the insides of their berets inspected for name-tags.

After three months, our company commanding officer called for volunteers for officer training. Successful candidates would transfer to the Army Officer Training Camp at Eaton Hall in Cheshire, not far from the Welsh border. I passed the War Office Selection Board and my instruction was then in the hands of "Pongos", which was our term of contempt for the Army infantry regiments. That summer was unexpectedly idyllic. Eaton Hall, which belonged to the Duke of Westminster, was surrounded by huge landscaped gardens, obelisks, lakes, equestrian statues, and acres of rhododendrons. The grand house itself was a fabulous pile of multi-columned reception rooms, vast halls, galleries, a library, towers, and a church, all embellished with coloured spires sprinkled with gold paint in the high Victorian fashion.

The drill sergeants, from the various Guards regiments, had nicknames such as "Small Eyes" or the "Screaming Skull". The regimental sergeant major was RSM Lynch from the Irish Guards, who was reputed to be able to shout so loudly that you could understand his orders at the distance of a mile. I remember him shouting on a drill parade: "When I say "T'Shun" I want you to pick your feet up six inches and drife them down twelf!" Since marines normally do Navy drill, clicking their heels together without stamping, we regarded this as rather vulgar. But we did it as ordered.

In the gorgeous hill country of Trawsfynydd in North Wales we conducted live-firing exercises and company manoeuvres, scrambling from valley to hilltop and back again, watched by sceptical sheep who munched their grass without a qualm. The sheep-farmers had thousands of animals grazing safely on the firing ranges, presumably accepting the occasional accident. We were instructed not to shoot the sheep on purpose, and I never saw one hit by mistake. But we did hit a careless training officer with a mortar smoke grenade. It was his fault, as he was showing off that he was not frightened by nearby detonations, and he was not badly hurt. Alan Blackshaw, a famous Alpine mountaineer who was in our squad, took a group of us to one of the less diabolical rock faces in the region to coach us in real mountain climbing. In the evenings I would sometimes telephone home from the coin-operated phone box overlooking the shadow-filled valleys. The sun set while the bugler played *The Last Post*. As the tragic notes of the final drawn-out nostalgic and mournful phrase echoed back from the darkness below, I would remember my father in the trenches.

The Army still considered nuclear war to be a practical strategic combat option in which you lobbed nuclear shells 20 miles onto enemy territory and waited a little while for the smoke to clear before advancing to capture the devastated ground. Conversely, we were taught that rapidly digging a 70 cm (2 ft) deep trench and lying in it was reasonable survival technique in the area around a nuclear blast, provided you were not too close. These statements seem so incomprehensible now, when everybody knows what happened at Hiroshima and Nagasaki, that I wish I had recordings of some of the lectures we were given on this subject. I almost doubt my own memories of this carefree attitude to nuclear war, but find that it is described accurately, with an overlay of black humour, by Tony Thorne, who was at Eaton Hall the year after me, in his hilarious book on National Service entitled *Brasso, Blanco and Bull*. Only later did the concept of mutual assured destruction using intercontinental ballistic missiles come to dominate military thinking.

The training was structured by monthly squad intakes each of about 40-50 men, progressing through the various stages during four months, fighting in larger organized units, finally culminating in a ceremonial graduation pass-out parade with a military band playing Sousa marches. One of my fellow officer-cadets was Bill Hill, a Royal Marine who had already completed his undergraduate degree at Cambridge University, where he had earned a "Blue" in swimming. Bill was tall, angular, and pure bone and muscle. Perhaps because he was three years older than the rest of us, he adopted a quizzical, humorous, slightly disdainful attitude to the training, at which he nevertheless excelled. We got on well together but were not particularly buddies while at Eaton Hall. With others in the small group of Marine cadets in the squad we discussed our future chances. As the final ceremony approached, we had to face the direful possibility that we could fail to be posted back to the Royal Corps and could be commissioned into the Army. If the Royal Marine major or colonel who was observing the officer-cadet trainees judged that we had not come up to standard, they would refuse to accept us back. We would become pongos.

The two Marine officers on the training staff were recognizable by their black berets and black webbing and straps, which stood out against the sea of Army khaki. Given the possibility of relegation to the Army, Bill and I were uncertain what to expect when, a week or two before the graduation parade, a Marine officer called us into his office. During a five-minute interview, mostly taken up by formalities, the officer put a proposition before us. "Would you two be interested in diving – canoeing – parachuting – things like that?" We both answered, "Yes, Sir!" The officer handed us a slip of paper, and said: "Go to this address." There was a date on it.

* * *

Two weeks after the meeting with the Marine officer at Eaton Hall, in October 1955, Bill and I met again at Waterloo Station in London and boarded a train to Poole. On arrival at the Hamworthy Amphibious Base at dusk, we were shown to the officers' mess in the

"Big House", a down-at-heel stately home, assigned rooms in the nearby accommodation blocks, and told to report in the morning to Sergeant Richmond. As the evening darkened into night somebody was practising the bagpipes at a distance amongst the pine trees. There were deep beds of needles underfoot and an overpowering scent of pine resin filled the air. We had no idea what would happen the next day.

In the morning Sergeant Richmond took us to see the legendary Captain Peter "Pug" Davis, a charismatic officer who had led daring amphibious raids in the Adriatic during World War II, and had achieved fame as a boxer. Captain Davis addressed us sternly with a vigorous lecture. His message was forceful as he concluded.

"We don't want any comic-book heroes here! The SBS will expect everything of you! Maintain our great tradition! Not the kind of men that brag in the saloon bar! Dedicated to the Royal Corps! No showing off to your girlfriends! It's a hard life! No concessions! Do you really want to volunteer for the SBS?"

Bill and I tried to look as if we understood the reason for this harangue, but it was lost on us. One of us, and I don't remember who, finally blurted out.

"Apologies, Sir, but what is the SBS?"

Pug Davis roared at us and banged his desk.

"Sergeant Richmond, take these two young officers away, show them what the SBS is about. Don't mince your words. Bring them back tomorrow morning. And you two, make your minds up fast. I don't like wasting time. Dismissed!"

* * *

It seems almost unbelievable now that the existence of Special Forces (SF) was a secret successfully kept from the general public, and indeed from most serving military personnel, for many decades. In Britain it was the Special Air Service's (SAS) dramatic liberation of hostages held in the Iranian Embassy in London in May 1980 that alerted the public and journalists to the existence of such elite troops. After that, the other unit of the Special Forces, the Special Boat Service (SBS), hit the headlines with the capture of South Georgia during the 1982 Falklands War, and then it was the SF achievements in Iraq that attracted and maintained steady journalistic attention. This attention was unwelcome but had become unavoidable. In reality, the commandos and SF troops had been active repeatedly since 1945, but newspaper restraint in Britain was enforced by the system of censoring "D-notices", and the self-discipline of individual soldiers prevented the publication of memoirs. In practice, the members of SF have to be modest, inconspicuous, and able to blend into the background whenever possible, or they would always be detected before reaching their targets. The training has to be ruthlessly tough and exhausting to weed out those unsuited to the work, but those who survive the initiation courses tend to be quietly spoken and not flamboyant. In the SBS the deliberate practice of modesty and the absence of boasting also provides an important moral and ethical counter-balance to the privilege of elite training and the possession of antisocial skills. In 1955 I had never heard of the SAS or SBS, nor had anybody at Lympstone mentioned their existence.

Bill and I learned rapidly about the Special Boat Service. We were escorted from one sophisticated training area to another, each designed for different technical tasks. The initial course for the lowest qualification took three months, plus a parachute course. The

Figure 2.2. SBS divers wearing parachutes over their diving suits and carrying their fins as they board an Armstrong Whitworth Argosy transport aircraft, prior to a training drop into the sea. Note also reserve parachute strapped across the chest (Courtesy RM SBS Archive).

skills we would acquire included diving breathing oxygen, beach survey, and ship attack with magnetic limpet mines. There would be a one-week session where the instructors hunted us day and night across open country, and we would have to avoid capture. There was much to learn. On the following morning we again stood to attention in front of Captain Davis to confirm that we really volunteered to become “Swimmer Canoeists”.

The next formal SBS course did not start until late December, so Captain Davis instructed us to keep fit while the bureaucrats organized a parachute training course during the interim. Keeping fit at Poole consisted of long-distance morning runs on the beach, swimming in the harbour without protective suits, and canoeing around the remote stretches of Poole Bay, often sleeping overnight in the open air on the uninhabited islands. We kept fit, but it was no hardship.

We went to RAF Abingdon near Oxford to take a four-week parachute course in early November. All British military units sent men to Abingdon, so that the squad under training was very mixed, consisting of the trainees for the Parachute Regiment itself, SAS and SBS Special Forces, and a mixture of support personnel including a doctor and a magnificently robust chaplain from one of the Guards regiments. The purpose was to produce troops who could be dropped into combat with heavy kit bags and weapons, with added preparation for awkward landing environments such as woodland or desert, and the ability to parachute at night. Bill and I knew that after the Abingdon course we would receive further training for parachute descents into the sea (Fig. 2.2). I celebrated my 19th birthday shortly after arriving at Abingdon by consuming a few extra beers with my comrades.

There were three parts to parachute training: how to jump out of the aircraft, so-called exit-training; how to control the parachute and heavy kit bag during your descent; and how to land without breaking your legs or knocking yourself unconscious. The instructors told us that statistics showed that it was safer to land by parachute than to land in an aircraft. The jargon had it that the value of training was to avoid becoming a statistic.

Modern parachutes are greatly improved from the simple circular canopies that we used in 1955. Today a parachute is a mathematically shaped aerodynamic wing that can be steered so as to land on a sixpence. Although the old 'chutes could be steered slightly by pulling down the rigging lines on one side, causing the canopy to spill air out and slip sideways, the scope to do this was very limited. After practising exits and landings with a host of ingenious machines and simulators on *terra firma* we ascended in the gondola under an adapted tethered barrage balloon, from which we made two jumps over Abingdon airfield. Jumping in total silence from the stationary balloon was rather unnerving, but the first experience of floating down bird-like, swaying under the canopy, was pure exhilaration. The preparatory exercises had succeeded in Pavlovian conditioning our reflexes so that, when placed in the familiar position relative to a replica open aircraft door, signalled with a green light, and barked at with a standard instruction, we would jump without a split second of hesitation. I am not being pseudo-scientific in this description or using an analogy. It was Pavlovian conditioning, which, in the circumstances, was very beneficial to our life expectancy.

After the balloon drops we were ready for descents from aircraft onto the grass airfield dropping zone at Weston on the Green, known as the DZ, which is on the other side of Oxford. We kitted-up each morning at Abingdon airfield, with our parachutes and reserve parachute packs which were strapped across our chests. The November weather was typical, with rising mist in dawn darkness, cloudy days, and bouts of gusty rain, drizzle, and snowfall. On most mornings we would be roused by reveille before six o'clock, and then sit in the crew room for hours with our parachutes fitted, waiting for dawn and the fog to lift, while the pilots tried to catch the brief hours before the wind became too strong for a drop over the DZ. We sat for hours, usually singing. We sang slowly to fill the time. Soldiers who had a new song to add to the repertoire were popular. Since the planned jump would often be cancelled, we then spent the rest of the day in ground training on simulators.

If the weather was kind, we boarded the aircraft where we sat in rows on canvas seats, each with the static line from our parachute pack hooked onto a strong wire running the length of the cabin. The aircraft usually flew with the door open while we droned across Oxford to circle over Weston on the Green at a height of 1000 ft (305 m). The DZ was a grassy airfield churned into mud by the repeated landings of booted soldiers, with ice on the frozen puddles and a sprinkled top-dressing of snow. When dropping onto land you stayed in your parachute harness, rolled sideways, and collapsed the canopy by pulling the rigging lines on one side. After a jump, having recovered our parachutes, we climbed into open canvas-topped army lorries which drove back through Oxford to Abingdon. Sometimes we made two drops in a day if the weather was good. We had to achieve eight drops from aircraft, including one night drop, to qualify for our wings.

We jumped from Handley Page Hastings aircraft in sticks of 15 men. There were incidents. One man during a drop with kit bags allowed his heavy bag hanging below his harness to swing into the rigging lines of the chute that had gone before him. He was dragged sideways until his canopy collapsed so that he hung downwards below the other parachutist and landed head first. But they did not tell us the details. On a lighter note, the rugger-playing Guards chaplain, who weighed at least 15 stone, sprained an ankle on his fourth jump. He missed several days' parachuting and approached the last day of the course still needing four jumps to gain his wings. While the rest of us did only one jump that day, the top brass organized special transport for the padre so that he made his first jump landing in the miserable face-freezing snowy slush at the DZ, where three men picked him up, removed his 'chute, rushed him to a Land Rover and drove fast back to Abingdon. There he boarded a second aircraft, was dropped into the ice-cold slush at the DZ with the next squad, his handling team picked him out of the snow, rushed him to the Land Rover, and so on. He completed the four drops on that day without injury and we all welcomed him as a hero. Somebody really needed that padre.

Bill and I returned to Poole with our "wings" proudly sewn onto our uniforms. The swimmer-canoeist course started in the last days of December 1955 and continued through the following three months, including diving with compressed oxygen breathing gas in closed circuit rebreathers, open sea canoeing for many miles in kayaks with double paddles, unarmed combat, beach survey and mapping at night, demolitions with explosives, use of magnetic "limpet mines" to sabotage enemy ships, removal of limpet mines from ships that had been attacked, beach landings in heavy surf, and river crossings, first aid, navigation, chart reading, cooking, operational planning, and living in the wild for long durations if necessary. The squad under training consisted of marines, NCOs, and officers, including two Australians. The marines, who were all older than me, bestowed the nickname of "Young Sir" upon me when speaking in my absence. I noticed that my diving gear was always chalked with this label when prepared for use

Figure 2.3. Royal Marine SBS members in the 1950s offloading kayak canoes from a landing ship for a seagoing exercise. Note the inflated buoyancy tubes down the side of the canoes for increased stability in rough seas, and when divers are getting out of and into canoes wearing heavy diving gear (Courtesy RM SBS Archive).

Figure 2.4. Diver with oxygen breathing set surfacing and climbing into the front cockpit of a canoe (Courtesy RM SBS Archive).

and hung in the store (Figs 2.3, 2.4). The training was deliberately egalitarian since there could be no privileges, the officers having to perform as least as well as the marines at the same tasks. My nickname was meant in a kindly fashion, and the marines were never disrespectful. Perhaps the nickname even implied an extra respect on account of my youthfulness.

Winter 1955-56 was one of the coldest on record. The sea froze in Poole Harbour, and the diving training lake attached to HMS *Phoenix* in Portsmouth froze half a centimetre thick with ice. (As with all Royal Navy bases, the training base in Portsmouth is named like a ship). The lake is saltwater, which means it was below zero Celsius. When we went for surf training at Westward Ho!, in north Devon, the surf ran up the beach, and, with each wave, an additional film froze onto the thick crazy-paving ledge of ice accumulating at the top of the beach before the saltwater had time to turn round and flow back into the sea. We practised paddling the canoes shorewards into the surf from the open sea, trying to land without capsizing. The technique required the front canoeist to climb aft and sit astride the stern of the canoe with his legs in the sea so that the bow was cocked up into the air. In theory this prevented the canoe from burying itself in the breaking waves. It was not easy. Sometimes we got to the shore intact; sometimes we capsized. Either way we were soaked through by seawater. Statistics of survival of the human body in icy water at 0° C (32° F) purport to show that exhaustion or unconsciousness sets in after less than 15 minutes. As we had been shown these figures, we monitored our own condition and that of our comrades watchfully, but I can only record that we survived. Maybe the statistics are wrong, because we were often saturated with seawater at a temperature below zero for many hours at a time, in a brisk wind, and we enjoyed it. It was our job, which we were proud of. When we were diving on oxygen we had dry diving suits, with woollen clothes underneath, which were excellent protection from cold, but for the other exercises at sea we wore clothing that, although the best available, would now be regarded as inadequate.

I made my first dive breathing pure oxygen on Friday 13th of January 1956. The date was not auspicious, and I can honestly report that I did not enjoy the dive. There was snow on the ground. The ice had not yet formed that day on Horsea Lake, the mile-long torpedo-testing lake, but the water was close to freezing. The underwater visibility was a few feet of greeny-yellow haze, while my drysuit rapidly clamped round my body and seemed to be a rigid prison. My ears hurt. I descended to the bottom. With freezing fingers, I undid a screw-thread shackle attached to a sinker and brought it to the surface to complete my test. I was not enthused by the experience but determined to persevere. Later dives rapidly became enjoyable as I learned to bleed gas from my mask and hood into the suit itself so that it stayed flexible and did not compress onto my body. Practice for military attacks or beach reconnaissance required us to dive at night and navigate through the blackness without surfacing. Night dives under the thin sheet of surface ice produced a beautiful ghostly view as you looked upwards at the circle of the starry sky.

The officers were accommodated at RN base HMS *Phoenix* which was also the main centre for fire-fighting training. (The Royal Navy is famous for its sense of humour.) Long before breakfast each morning Bill and I met the rest of the squad in the dark street nearby which was thick with snow and only dimly illuminated by aged streetlights. The instructors then took us on a long run, jogging through the deserted streets of Portsmouth, where we broke off at times for press-ups and other exercises on the ground. The run sometimes ended on the bank at Horsea Lake where we stripped naked and swam to the far side of the lake and back, a total distance of perhaps 45 m (150 ft). Then we dried hurriedly, got dressed, and ran to the diving centre for a huge breakfast. It was still dark. Then the day started.

After completing the oxygen diving course, which included swimming the length of Horsea Lake while staying submerged, we returned to Poole for extensive canoe training at sea, long-distance navigation at night over 8-16 km (5-10 mi), and mock attacks on coastal military bases. Although the weather got even colder, we sometimes spent the night sleeping out in the snow in thick sleeping bags which, surprisingly, really did keep us warm. The give-away from the point of view of security and concealment was that cows would come and watch us when we cooked our breakfast while we kept warm in the sleeping bags and just poked our arms out. We successfully sabotaged an entire RAF radar station by making a sea voyage by night with landing craft (Fig. 2.3), then attacking the shore by canoe, climbing the cliffs, and placing dummy explosives on the radar aerials. The defenders did not hear us coming or going. Our pride at this achievement was dimmed when we tried to keep warm afterwards on the icy beach by lighting a fire of driftwood inside an abandoned World War II block fort. The fire burned nicely, but the smoke drove us out into the cold, and we sat frustrated in the snow watching the smoke, glowing with light reflected from the fire inside, curling up from the gun-slits into the freezing night.

The film *The Cockleshell Heroes* had been shot in the previous year and was released that winter while Bill and I were doing our training at Hamworthy. The film company hosted a party celebrating the film in the officers' mess, and several of the minor stars of the film attended, but not the leading stars Jose Ferrer and Trevor Howard. The film is a popularized version of the raid that really took place in 1942, known officially as "Operation Frankton", to sabotage German ships in Bordeaux harbour. The Royal Marine canoeists launched their five canoes from the Royal Navy submarine HMS *Tuna* at night, paddled up the Gironde estuary, and successfully placed 19 mines on five ships, which were badly damaged. Of the ten participants, only two survived and escaped overland by the Marie-Claire Escape Line. Six men were captured by the Germans and shot as spies. Two drowned when their canoe capsized in the rapids at the mouth of the estuary. This heroic but desperately dangerous act of sabotage is commemorated every year by the Royal Marines and their French hosts at Bordeaux. There is a monument at Le Verdon-sur-Mer, near Bordeaux. At a stage in the war when the Allies needed to show that the Nazi military system was not impregnable, this relatively minor raid was a great boost to morale in Britain and an infuriating irritant to Adolf Hitler.

The cold weather progressed and we found ourselves tested by unexpected events. The instructors set us a night exercise that started by launching canoes on the Wimborne River several miles inland, then canoeing downstream to rendezvous at a bridge where we would be met by a lorry to take us back to base. The trainees launched their canoes at intervals of several minutes so that everybody had to make their own way and solve their own problems. By this time Bill and I were an established kayak pair with many hours of sea canoeing behind us in rough breezy conditions. We started off cautiously, determined not to take any avoidable risks. After a while I felt as if my kayak double paddle was extraordinarily heavy, which was odd because I was used to propelling the canoe for hours in the open sea. I tried to adjust my grip to relieve the strain in my arms, but could not. Then I discovered that the paddle was inches thick in ice, and my hands, protected by waterproof gloves, were locked into the frozen mass. Because we were in freshwater, it froze much more quickly.

We paused while we both chipped at our paddles to break the encrusted ice away. In the relative silence we heard the distant rumble of a waterfall. It was not a comforting sound. Nobody had told us about a waterfall.

We started paddling again, steering towards the left bank of the river, so that we could grab onto the bank if we got too close to the waterfall. But the bank seemed to have gone missing. Our excursion extended so far to the left that I prodded down with my paddle to feel the water depth and found that it was only 30 cm (1 ft) beneath the keel. We were in a flooded field, not on the main river channel. We could hear the crashing waterfall much louder now away to our right, so we both got out of the canoe to walk through the fields in an arc, dragging the canoe, until we re-joined the river below the fall. We reached the bridge in record time, and gradually the whole squad assembled. Several marines had been less lucky than us, having either capsized in the waterfall or snagged on fallen trees and submerged branches.

Having completed our training as Swimmer Canoeists 3rd Class, some of us continued onto a SC2 course. This included parachuting into the sea (Fig. 2.2), exercises of launching canoes from an RN submarine in mid-Channel (Fig. 2.5), cutting chains and steel bars underwater with a hacksaw, searching a ship's hull for possible limpet mines planted by an enemy attacker, and learning to make underwater exits from an X-Craft, a World War II mini-submarine. During the launch of canoes from the torpedo loading hatch of a full-sized submarine in mid-Channel in February, the boat wallowed in a choppy sea so that the waves blowing in from the Atlantic washed over the curved deck. The marines skidded and slipped as they fought to hold their balance. From the conning tower, Pug Davis and I watched the canoes first launched and then recovered while Pug shouted instructions. We did much of this training at HMS *Vernon*, the RN Diving School in Portsmouth, which also gave training for Fleet Clearance Divers and minehunters.

The diving training ship at *Vernon* was HMS *Deepwater*. One day in April 1956, while we were doing work exercises like cutting chains with a cold chisel and sledgehammer on the muddy seabed alongside *Deepwater*, the Russian cruisers *Ordzhonikidze* and *Sverdlov*, on a diplomatic visit, entered the harbour to berth at the South Railway jetty a few hundred yards away. We trainee divers on deck at the time could see bulky figures

Figure 2.5. Swimmer-canoeist divers of the SBS on the deck of an Oberon Class submarine loading canoes through the torpedo hatches in preparation for going to sea and submergence. The sea is flat calm, which never seemed to happen in the real world that I experienced (Courtesy RM SBS Archive).

in overcoats standing on the foredeck and rather hoped that they were Bulganin and Khrushchev. As soon as the Russian cruisers were moored safely they put helmet divers over the side, and we could see them descending on ladders into the black harbour water. The humblest diver amongst us knew that the Russians would protect their ships constantly by inspecting the hull under water, probably rotating the screws at unpredictable intervals, and would do everything possible to deter underwater visitors.

The distance to the Russian cruisers was several hundred yards, but our training exercises were all either on the seabed on a shot weight lowered from the stern deck of *Deepwater*, or under the hull of *Deepwater* itself. There was no reason to swim even 10 yards in the direction of the Russians, and we did not. Our RN Petty Officer instructors emphasized the obvious and ordered us firmly not to swim towards the Russian ships under any circumstances. It was a strange juxtaposition of military facilities at a time of tension in the Cold War, but the Russians certainly knew that they were berthed close to the RN Diving Training School.

Many days later we were astonished to read in the newspapers about the mysterious disappearance of the diver Lionel "Buster" Crabb, followed by the gradual revelation of scandalous goings-on by the security services. By 29th April, the first of many varied official explanations was published trying to minimize the diplomatic damage. Apparently Crabb tried to swim under the Russian ships. In spite of being so close to these historic events, I have no inside knowledge of what happened, and will not waste space here speculating about the truth. There are dozens of books and hundreds of articles published about this bizarre event.

We did night dives searching the hull of *Deepwater* for dummy magnetic limpet mines which the instructors had placed there as a test. It was disconcerting to swim under the hull in absolute blackness feeling the steel of the ship on your back while the front of your diving counterlung, the flexible bag into which you exhaled oxygen, was brushing against the mud below. One night one of the trainees panicked in this situation and resigned from the course. During the three months of training several trainees dropped out quietly, but we never discussed these events, either amongst the remaining course members or with the instructors.

Shortly after this we participated in the three-day Joint-Service operational amphibious demonstration for the Imperial Defence College, dubbed "Operation Runaground", on Southsea beach, near Portsmouth. This was a huge demonstration exercise comprising amphibious tanks, parachutists landing in the sea with inflatable boats, landing-ships debouching tanks and other tracked vehicles, and mine-detonating flail-tanks onto the beach, with commandos swarming ashore, and every aspect of amphibious warfare packed into a two-hour spectacular. Fighter jets roared low overhead giving air cover to the landing. The assembled generals and field marshals from 20 or more countries watched from a large temporary viewing grandstand set up at the back of the shore.

Bill and I had minor parts in this display. There was a swimming-pool demonstration of oxygen diving equipment. The opening gambit showed the efficiency of fins and consisted of me swimming two lengths of the pool submerged wearing fins, racing against Bill swimming without fins. The fact that we did these long swims holding our breaths gave the hidden message that we were both very fit young men. I usually won by half a length, which proved it nicely. But, since Bill was a Cambridge "Blue" in swimming, there was one fatal day when, after we had had a heavy night's drinking at the Nuffield Club, Bill almost won, and the commentator had a hard time making his point with conviction. Having won by a whisker, I was helped out of the pool, very much the worse for wear, while my comrades concealed my groggy state of inebriated exhaustion.

In late May we graduated as Swimmer Canoeists, 2nd Class, after which the instructors explained to us the next stage of training. I rebelled at this point, requesting a post where I could do something useful. Out of my two years' compulsory service, I had now spent 17 months in intensive training, and I wanted to put some of it into practice. As Bill shared my opinion, we formally requested fresh postings. Two days later Bill was posted to Malta and I was posted to the Royal Naval Rhine Squadron, based in the German industrial city of Krefeld Uerdingen. The Rhine Squadron included a Royal Marine SBS unit.

* * *

In Germany I reported to Lieutenant Patrick Farncombe, RM, commanding the SBS unit, and, since the SBS did not conduct operations all the time, I was also appointed to command a Motor Launch Patrol Boat. These 20 m diesel-powered launches carried a single Oerlikon machine-gun for'ard, which was their only armament. The crew included an experienced Marine Coxswain Sergeant, two RN engine-room ratings, and a couple of Royal Marine deckhands. The Rhine Squadron patrolled the River Rhine from the Dutch border to Strasbourg, exchanging formalities *en route* with our French and American counterparts. If the Russians invaded Europe from the east, our job was to demolish or sabotage bridges on the Rhine. We also conducted exercises with landing craft transporting an American atomic cannon across the Rhine. Further training with the SBS included thorough courses in explosives and demolitions. The instructors' motto was: "Explosives are perfectly safe until you forget that they are dangerous." The art of good demolitions is to place the minimum of charges necessary in exactly the places where the cumulative damage will be maximum. It is much more difficult than it looks in the movies.

Shortly after my arrival the SBS squadron of about 20 men entered into the International Nijmegen Four-Day Marching Festival in Holland. This was not supposed to be competitive but just for the joy of marching in the company of several thousand like-minded folk. Many civilian walking clubs and military units entered the Nijmegen Marches but, as a result of previous years' experience, a rivalry had developed between the Royal Marines SBS and the British RAF Regiment. For a club or military unit to be awarded the collective achievement medal (it did not count as a military medal in the formal sense), only one person in ten was allowed to drop out or fail to complete the 30 km (19 mi) march each day for four days.

Pat Farncombe was very fit, being one of the few people I have met who could jump into the air from a standing start and turn a somersault without touching the ground and land on his feet again. We made a number of long-distance practice marches to get in training while support staff rehearsed an efficient routine with a small lorry for supplying food and water at regular intervals around the course. Pat made sure that the planning missed no details. The march would consist of a different route each day, looping out through the gently rolling countryside around Nijmegen, then back to the city. The third day would take us through what the Dutch call hills, but there were no difficult gradients.

The marches started well as we proudly carried a huge RN White Ensign and completed the circuits without incident for the first three days. After the march of the third day, Pat and I were chatting by the dining hall of the Dutch Army barracks where the marines had their accommodation. We were planning the details of the fourth day when we would change into our No. 1 Dress "Blues" uniforms for the final miles into the centre of the city. A British Army colonel came striding towards us.

"Ah, Farncombe, there you are. I've been looking for you. The Chief of the Israeli Army has been in touch with CinC British Army of the Rhine (BAOR) and the Israeli Army has challenged the British Army to a volleyball match this evening. It will be on Dutch TV. Very important. Difficult to find a team quickly as there are so many different regiments here, you know. Your men are supposed to be fit. So I am appointing the Royal Marines to represent BAOR against the Israelis this evening. 1800 hours in the barracks sports centre. I'll be there."

"Yes sir," said Pat, saluting rapidly.

The colonel strode off.

Pat turned to me.

"Nic, you heard that. Go to the barrack room and get volunteers. We need six men at least."

"Yes, Pat."

I set off towards the barrack building where the marines had several rooms on the ground floor with double-tier bunk beds. I thought that the men deserved a real rest after three days with worn feet, sunburn, near-blisters, and sheer muscular fatigue. We had set a cracking pace and our sworn rivals, the RAF Regiment, with a larger squad than ours, had lost two men dropping out from heat exhaustion and dehydration trying to match our times. The marines had good reason not to volunteer. I had an uneasy vision of the newspaper headlines: "Mutiny in the Royal Marines: Men refuse to volunteer when requested by Officer." Then: "Young Officer on Court Martial for failing to assert his authority". I faked an air of confidence.

The barrack rooms were just as I had expected. Marine domestic bliss. A radio playing pop music. One man sitting with his feet in a bowl of warm water. The sound of a shower running in the background and a booming voice singing *Roll out the Barrel*. Others lying on their bunks with their feet propped up on piles of cushions and pillows while they read magazines or wrote letters home. Drifting clouds of cigarette smoke. Corporal Gallard was writing notes in preparation for the final parade.

"Something up, Sir?" he enquired as I appeared in the doorway.

"Stand easy," I instructed. "There's no need to move. Keep your feet rested. Everyone, don't move. Corporal, we will need some volunteers at 1800. The Israeli Army has challenged the British Army to a volleyball match, and it will be broadcast on Dutch TV. BAOR has instructed the SBS to represent the Army. I need volunteers to report outside the dining hall at 1745. And the game starts at 1800."

Every marine jumped up and started shouting and pleading.

"Take me, Sir!"

"I always wanted to be on TV!"

"We'll show them, Sir!"

"We'll show the pongos what we can do!"

"And me, Sir!"

Their morale was incredible. We agreed that six men would parade for the game, where Pat and I both joined in. Volleyball was almost the national sport for Israel, so we did not win. But the game was exciting, everybody enjoyed it, and honour was satisfied. The marines were cock-a-hoop at being filmed for television.

With the volleyball match behind us and the marines in high spirits, we embarked on the final day in bright sunshine, marching fast, and with supplies of freshwater and food delivered from a truck at several stages of the route. The truck failed to arrive at one of the rendezvous and we became desperately thirsty and frustrated until we found it a few kilometres further on. For a while I thought that the marines would become depressed and the strain of the volleyball match the night before would slow them down, but we arrived in the outskirts of Nijmegen hot, exhausted, and secure in knowing that the RAF Regiment were far behind us and suffering from the heat. During the four days they had lost two men from heat exhaustion, but with a large squad, they still qualified for their medal. We quickly found the lorry with our best uniforms and washed and dressed smartly, surrounded by other military units making the same transformation. The regiments from Holland, Belgium, and a few others had brought their marching bands for these last few kilometres so that the air of celebration and triumph was overwhelming. With the White Ensign held high we crossed the finishing line and finally broke ranks to enjoy an evening on the town.

In the autumn the Rhine Squadron sailed to Strasbourg where, amongst other unlikely events, we attended a sauerkraut (*choucroute*) festival during which the captain presented the mayor of Colmar with a White Ensign. I doubt if any other RN fighting unit has sailed higher than 136 m (446 ft) above sea level flying the flag. After steaming back to Krefeld, Pat and I then took the SBS to the Möhne Tal See where we conducted a fortnight of intensive diving practice, including night navigation exercises underwater and long-distance swims submerged, breathing oxygen.

One day during our lunch break I was sitting in the sunshine on the grassy bank of the lake chatting with Barry Blair, another National Service marine. The lake was retained by the repaired dam that had been destroyed by the dam-busters in 1943, and our camp was close to the dam. Before signing up for National Service training, Barry had joined a sport-diving club in England. He told me that it was known as the British Sub-Aqua Club, or BSAC, and had been founded in 1953. The suggestion of diving purely for sport struck me as strange, almost absurd. Barry explained that you could take photographs underwater, watch wildlife, or dive on shipwrecks. People did it for fun and adventure. They breathed compressed air instead of oxygen, but otherwise, it was quite normal. Some people hunted fish with spear-guns. My doubts persisted, but I was intrigued by these tales and resolved to find out more.

I spent the winter months in ski warfare training using cross-country skis, instructed by Swedish and Norwegian Army officers at Winterberg in the Harz Mountains. Finally, in early February 1957, I returned to the UK, to hand in my kit at the RM Southsea barracks. Home at last. I decided to find the BSAC.

* * *

My military diving training showed me that diving was primarily a means of transport to get the skilled worker to the site of work. Once there, oxygen sustained life and enabled the mind and hands to do useful things. You had to learn the art and skill of diving so thoroughly and instinctively that it did not dominate your thoughts, thus freeing your mind to work. The reflexes and alarms that are needed to maintain safe diving must be kept ticking away in the background, ever ready to respond to emergencies, but the first thought must be work, the target ahead.

I had always had a fascination with flying. When I arrived in Cambridge in October 1957, I applied to join the University Air Squadron, which was a unit of the Royal Air Force Volunteer Reserve (RAFVR). How could I refuse the temptation of being paid to learn to fly? I passed the theory exams and the psychological and co-ordination tests but was told that the RAF had to prioritize training for those undergraduates who had not yet done National Service. I could see the reason for this, and gave up my hope of learning to fly, focusing instead on the promotion of the scuba diving club and the first expeditions to Apollonia.

Three years later the university accepted my application to study for a doctorate aimed at explaining how earthquakes and sea level changes had influenced the coast of the Western Mediterranean during the past million years. (See Chapters 4 and 6.) Since National Service had been abolished in 1960, I applied again to join the Air Squadron, and this time they accepted me. I was my own boss since I did not have to attend academic tuition at set hours and so found it easy to bicycle out to the

RAFVR crew room at Marshall Airfield, 5 km (3 mi) outside Cambridge. Then I could fly whenever the conditions were calm enough. I often read research papers or drafted bits of my thesis while sitting at the airfield drinking tea, waiting in the crew room for a break in the weather.

My three years flying single-engined Chipmunks further influenced my attitude to training for hazardous tasks, the role of delegation and control, and the extent to which responsibility should be divided between the man/woman "at the coal-face" and the bureaucrats higher up the hierarchy, often at a desk hundreds of miles away. I enormously enjoyed flying and found aerobatic manoeuvres exhilarating, especially a repeated vertical figure-of-eight achieved by a half-loop and roll-off-the top which could be repeated as often as one wanted before getting bored. Aerobatics in bright sunshine, while Cambridge was cloaked in mist and cloud below, was a delicious winter luxury. I returned to the ground refreshed and happy. Flying requires the same combination of motivation, ability to manage risk, determination to do a good job, and the ability to survive in a strange environment, that is required for working diving. It is the parallels with diving management and the influence that it had on my management of diving safety in the UK marine scientific community that became so important later in my career.

I noticed that the RAF was content to allow an individual, after 8-10 hours of supervised flying with an experienced co-pilot instructor, to go solo and fly away with thousands of pounds worth of government property. In a few seconds of mistakes, the novice pilot, or any pilot for that matter, can write off a vastly expensive machine while destroying several human lives. But experience has shown that you can train and then trust individuals to fly safely, provided that there is a background infrastructure of weather forecasting, radio communication, control towers, maintenance of the aircraft machinery, and so on.

As we will see later (Chapter 8), both oil-field commercial diving and Royal Navy ship's maintenance diving tend to treat the diver as a subordinate of the officer on deck, so that the diver must obey orders and not have too many ideas or try to make original decisions on his or her own. I observed how the RAFVR achieved a balance by trusting the pilot who, while flying, scanned half a dozen instrument displays continuously, as well as the outside view from the cockpit windows, using that information, with good previous training, to control the aircraft throughout the flight. I am eternally grateful to them for the experience. Of course, I had already been exposed to the SBS spirit of extreme independent delegation, but that was in the context of anticipated clandestine warfare where higher risks were accepted as routine.

* * *

My next collaboration with the military arose from experiments to improve divers' intelligence when working in the depth range of 40-60 m (131-200 ft). Diving deeper than 10-20 m (33-66 ft) breathing compressed air, which is a natural mixture of oxygen and nitrogen, causes inevitable deterioration of mental ability due to nitrogen gas narcosis. This condition is quickly reversible, so that mental alertness decreases gradually with depth, and increases again as you ascend. Popular understanding of this problem suggests that narcosis only occurs at great depth, but in reality, it creeps up on you gradually and is unavoidable. At depths greater than 40-50 m (131-164 ft) the effect is so strong that divers breathing compressed air only work slowly and inefficiently, while deeper than 60 m (200 ft) it gets difficult to do anything useful at all. This effect determines how much authority you can delegate to the diver at different depths, and how you design tasks when the divers are breathing air.

In 1964, I conducted two sessions of diving work at a depth of 50 m (160 ft) off the coast of Cornwall between Eddystone Lighthouse and Polperro. The divers were undergraduate members of CUUEG who were measuring the physical conditions on the seabed that determined the junctions between moving bands of gravel and sand. Arthur Stride at the National Institute of Oceanography (NIO) had detected these widespread features using sidescan sonar and wanted close-up measurements of what was going on. I based the diving team at Polperro, in Cornwall, and we dived almost daily from hired fishing boats. The rocky harbour is minute, but provides good shelter in most wind conditions, and is close to the seabed gravel strips discovered by Arthur. The tidal currents were strong, and the weather choppy with occasional storms. I designed and built a small electric diver-towing tug so that we saved effort from swimming, and saved breathing gas while working at depth, but we were still inefficient at 50 m (164 ft) from narcosis. We obtained some encouraging measurements and photographs by diving, but the combination of nitrogen narcosis, cold, and poor visibility in the murky green-brown waters had reduced the divers' intellectual ability and manual dexterity so much that work was very slow, and we abandoned some experiments.

Navy scientists and diving doctors had already discovered that by diving with a mixture of helium and oxygen instead of air (which is made of nitrogen and oxygen) you eliminated narcosis because there is then no nitrogen in the breathing gas. But helium is very expensive, and heliox breathing systems had always been used with expensive and heavy commercial or Navy helmet-diving systems with a powerful ship for gas supply. I decided to see if it was possible to use heliox in a lightweight diving set like sports scuba, or an adapted rebreather. In the latter type of diving gear the gas is breathed out into a rubber bag and then inhaled back in again after removal of the exhaled carbon dioxide, so that the diver gets to use much more of the available oxygen, instead of venting it straight out to the sea after every breath. The oxygen diving sets that I had used throughout my SBS training were rebreather sets, but they used pure oxygen without either nitrogen or helium as a diluting gas. You cannot use this safely below a depth of 10 m (33 ft), because pure oxygen becomes poisonous at high pressure.

I sought military contacts in the UK to see if existing equipment could be adapted for my purpose. At the Royal Naval Physiological Laboratory (RNPL) in Portsmouth, I explained my requirements to Dr Val Hempleman, a research physiologist responsible for the RN decompression tables, and to his boss Dr H. J. Taylor. Together with some of their technical staff, we calculated that the existing RN scuba breathing set known as the Swimmer's Air Breathing Apparatus (SABA), which had very large cylinders, could be filled with heliox as the main breathing gas, which you could then supply into a rebreather bag by a pressure reducing valve. This would provide 20 minutes or more of diving at 60 m (200 ft), with plenty of reserve for the long slow ascent. During the prolonged decompressing that would be needed we planned to have a separate supply of pure oxygen which the diver could feed into the same bag when reaching the depth for stops. The divers would need to turn off the heliox supply before turning on the pure

Figure 2.6. Assembling the helium-oxygen (heliox) self-contained breathing set. Note the full-face mask connected by a single hose to the cylindrical canister that is the soda-lime absorbent to remove CO_2. The oxygen cylinders are encased in rubber sockets just beyond the soda-lime canister. The large grey twin cylinders are from the Navy air-breathing set known as SABA but have been filled with heliox, not air.

Figure 2.7. The author wearing the heliox set designed by combining the front of an oxygen breathing RN diving set, and the back of an RN SABA set, with two constant mass reducers to provide the correct gas flow. The pure oxygen was only used for decompression stages. Note the nose clip to make it easier to clear the ears when descending. Once at depth you could knock the nose clip off as it was uncomfortable.

oxygen at the right depth. All the components were standard RN supplies, but we would join them together in an unconventional way (Figs 2.6, 2.7). I also did the calculations that showed that the ideal gas mixture would be 25% oxygen mixed with 75% helium.

Val Hempleman calculated a new set of decompression tables for this gas mixture but required the whole team to be tested in a recompression chamber at Gosport breathing the gas at different depths for different durations to prove the tables. Nowadays much of this kind of calculation would be done by well-proven computer models. In the 1960s all new decompression research was first tried out on goats, and then cautiously on volunteer divers. Goats are sufficiently close to humans in body mass and general anatomy to serve usefully as test animals. The new tables for our heliox dives were only a variant on an existing table, so Val did not test them first on the goats. I developed a bend on one of the dives and had to be recompressed, so Val increased the stops by a small amount.

I recruited from CUUEG a team of six volunteer undergraduate divers to train on the complex breathing set in the murky lakes around Cambridge, then to undergo the experimental decompression table tests, before demonstrating the equipment at sea. Two further divers were myself and Dr Alan Baddeley, a research psychologist at the Applied Psychology Research Unit in Cambridge, who planned to measure the performance of the divers using the different gas mixtures at different depths so as to define exactly the benefits of using heliox for skilled and intellectual work underwater. Baddeley later became a Fellow of the Royal Society and professor at Bristol and York Universities. He had the rare ability of being able to study people's brains in a busy hazardous situation without making them feel nervous, distracted, or apprehensive. He never disrupted the smooth running of the diving team, which was essential when we were diving in the open sea at 60 m (200 ft). It was a great help that Baddeley was a diver himself, taking part in the most arduous dives.

The project was a success, and in summer 1965 we tested the innovative diving sets for two weeks off the coast of Malta to conduct the psychological measurements, then used them for a further two weeks off the coast of Cornwall to make the geological observations needed by Arthur Stride at the NIO.

The dives to 60 m (200 ft) in the clear indigo-blue luminescent sea outside Valletta Harbour, Malta, were truly enjoyable. It was just plain fun to sit on the seabed working out the complex puzzles that Baddeley had set us, transferring nuts and bolts from one metal plate to another, or doing mental arithmetic and scribbling results on a plastic pad, working against a time deadline. The support ship that we used in Malta was a 100-ton Motor Fishing Vessel, or MFV, provided by the Royal Navy, and they provided the same type of vessel when we dived from the Navy base at Plymouth. The hardest part of the work was using a hand pump to boost the pressure of the big heliox cylinders, at which we toiled for many hours each day to get the large diving sets filled with gas.

In spite of the complications of the heliox diving set, with the need for the divers to change manually from one gas mixture to the other during the stops (Figs 2.8, 2.9, 2.10), there were no mistakes or emergencies with the set during the training phase of the project, nor during the open sea trials in Malta and off Plymouth. To any diver who has been below 40 m (131 ft) on air it is a revelation to dive on heliox (or the more modern helium+nitrogen+oxygen trimix) since it feels as if you are seeing the seabed clearly for the first time. The sense of relaxed control and alertness is almost hyper-realistic, as if

Figure 2.8 (top, left). Diver kitted-up and ready to dive to 60 m (200 ft). The diver breathed the heliox mixture for two minutes before getting into the water so that the gas in the counterlung would reach equilibrium with the loss of some oxygen that was being converted to CO_2 and O absorbed in the soda-lime canister.

Figure 2.9 (top, right). Diver entering the water wearing the heliox set. It was quite rugged and trouble-free and, although it required careful maintenance and adjustment of the gas flows, could be used in a full working environment, not just an experiment.

Figure 2.10 (left). Two divers ascending from 60 m (200 ft) depth on different decompression schedules, one breathing heliox, the other air.

one is on some sort of drug, the polar opposite of narcosis. Only then do you realize that previous deep dives have required you to fight back the disorientation and memory loss of nitrogen narcosis. We demonstrated that intelligent divers in a scientific environment can be trained and trusted to make many complex decisions for themselves. We also showed that scholars with specialist training can work efficiently breathing heliox, using their training and memory almost as if they were on dry land. Delegation to the well-trained diver works.

* * *

The military connection went cold for a few years, then in 1969, when I was working at the NIO, I received a telephone call from a colleague, Lionel, in the Foreign and Commonwealth Office, asking if I knew where Rockall was. I confirmed that I knew the location of the rock pinnacle in the open Atlantic that appears regularly in the marine weather forecasts as the focus of storms. It is a volcanic pinnacle 390 km (240 mi) west of the Hebrides. Lionel explained that the UK needed to build something useful on Rockall.

The United Nations Law of the Sea was still under debate, but it would only recognize a claim for territorial waters, mineral rights, and fishing zones around islands that someone lived on, or where there was a building or structure of some kind. An uninhabited rock far out in the ocean could not count as part of a country for economic rights and benefits. Rockall was claimed for Britain in 1955, and the Royal Navy first landed on it back in 1810, so there was no dispute about that. But we did not designate it as part of a British county, and nobody lived on it. We needed to show that it was being used in some way, that it was linked directly to the UK mainland, and that it had appeared in acts of administration.

I wrote a report for the FCO showing the feasibility and usefulness of building a lighthouse on Rockall, or installing a radar beacon, scientific instrumentation, a weather station, a wave recording oceanographic station, geophysical earthquake detection, equipment for measuring variations in the magnetic field, and several other less practical options. A year later I received another telephone call from the FCO. They had decided to build a lighthouse on the rock.

In 1971 the Army Royal Engineers used Wessex helicopters to drop engineers on the rock where they drilled holes filled with explosives to blow the pointed top off. The Institute of Geological Sciences (IGS), now the British Geological Survey, also obtained a sample of the aegirine granite, known as rockallite, for laboratory analysis. The new flat triangle on top of the rock was drilled to take the fixings for the lighthouse. Installing the light would help to establish that the rock is part of the United Kingdom and would permit the island to be included in the District of Harris in the County of Inverness. This was done legally in Parliament by the Island of Rockall Act 1972, reinforcing the UK government's position for seabed rights in the surrounding seas.

A few months later the FCO invited me to a meeting at the Ministry of Defence. There was an inter-departmental consultation to plan the building of the lighthouse on Rockall, but they also wanted to use the ship-time for as many activities as possible. They invited me to develop a scientific programme using divers round the rock in a practical way, working from a Royal Fleet Auxiliary helicopter support vessel.

The NIO where I worked had a small diving team (see Chapter 8), but a project 435km (270 mi) offshore working from a large naval ship in stormy oceanic conditions near multiple rocks and pinnacles that were poorly charted was a challenge beyond any plans that we had in mind. The waves were often 10 m (33 ft) high or more. Working so far offshore meant that to dive safely we needed to have a recompression chamber on the ship to treat possible cases of bends, while navigation and positioning of the divers and plotting objects on the seabed would be difficult, as there was no frame of reference. The top of Rockall provided one survey point, but there was no easy way to measure both range and bearing accurately, and no second point to make a survey triangle. We decided to use a Decca trisponder radar range-finder from the dive boat to the support ship, and the ship's radar to fix its position relative to Rockall. Again, modern familiarity with GPS makes the difficulties hard to imagine. Our errors in position for a dive or echo-sounding measurement could be tens of metres.

Working together with the IGS in Edinburgh, who maintained a small research diving team led by Dr Paul Binns, we had enough skill and experience to conduct an effective programme of exploration and rock sampling round the base of Rockall itself, and on Helen's Reef and Hasselwood Rock nearby. The whole complex of pinnacles rising to the surface of the sea from a water depth of 100 m (330 ft) or more is the remains of a series of volcanic vents and injections of lava from deep beneath the Rockall Plateau, which itself is the largest microcontinent in the North Atlantic Ocean. The violence of these geological events was part of the rifting of the Atlantic Ocean tens of millions of years ago, tearing Europe away from North America, during which Rockall Plateau separated from Greenland. Although the rock composition of the top of Rockall was now known, no one had sampled or dated the surrounding volcanic reefs.

Helen's Reef gets its name from the brigantine *Helen*, which ran aground and sank there while *en route* from Dundee to Quebec in 1824, drowning most of the passengers. In 1904 the steamer SS *Norge* also crashed into the reef with the loss of 635 lives. Although modern navigation aids and charts were much more accurate by the 1970s, the proposed lighthouse was perhaps not totally pointless.

I was in charge of the diving project, and my chief diver was Stewart Willis, supported by divers Roger Clements, Ray Peters, and Garry Mardell, all from the NIO. The IGS provided two divers, Dr Paul Binns and John Butler. My visit to the MoD revealed that we would be working from the 9000-ton Royal Fleet Auxiliary (RFA) vessel *Engadine*, which carried two Sea King helicopters, and had stabilizers to help make deck landings safer in rough seas (Fig. 2.11). The project planning was code-named Operation Top Hat. A surprise participant was Michael Bizony who had served in the RNVR during and after the war and had swum from the Fleetwood trawler *Bulby* to Rockall in 1948. The NIO had recently purchased a two-compartment recompression chamber, which we took by lorry to Portland and loaded onto RFA *Engadine*. The two teams of divers then joined the ship a week later at Liverpool as she steamed north towards the North Channel between Scotland and Northern Ireland. The MoD had planned the project for the two weeks at the end of June when the likelihood of storms and strong winds was at its lowest. We carried a very senior RN meteorological officer on board, who, as it turned out, had not been to sea for many years. Climate statistics are no respecters of seniority.

As we passed through the North Channel, we met the dramatic force of the Atlantic. The wind strengthened rapidly and the ship began to pitch and roll despite the stabilizers.

Figure 2.11. Sea King helicopter folded up for storage in its hangar on board RFA Engadine during the work on Rockall in 1972. The red cylindrical drum of the stumpy lighthouse to be installed on Rockall is just visible on the deck above the hangar.

The met officer, who was the only person to become sick, retired to his cabin. The storm got worse as we steamed into the open ocean, and next day the ship's doctor came to consult me.

"Nic, I'm worried about the met officer. He is really very ill. I have tried giving him some strong pills, but he vomits them up again. I have given him an injection to help him sleep, but I think he should be flown back to land if possible. Do you need him essentially on the project, or can we send him ashore?"

"His health comes first," I said. "I am sure that the captain and his senior officers can manage with the usual meteorological shipping forecasts and data from MoD. Please feel free to arrange the best for your patient."

We had two weeks to do our work on Rockall, during which time the wind blew Force 10 gusting Force 12 for ten days out of the fourteen; that is winds of 100 km/h (60 mph) most days, with gusts to 116 km/h (70 mph). We did have four days of relative calm when at Rockall itself, but not consecutive days, and when winds of this speed die down the waves take many days to dissipate, so they continue to roll by as round-topped swell 5-10 m (16-33 ft) high from trough to crest. After my discussion with the doctor, he consulted the helicopter pilots, but the weather was too bad to fly, so the met officer had to remain on board. He bore his suffering nobly, and on the occasions when the ship's motion became more tranquil his cabin door would open, and he emerged wanly for a cup of tea and some biscuits. As the storm raged again, he returned to bed. Rather unkindly, someone compared him to one of those Swiss doll chalets where figures emerge to tell when it will rain or not and then go back again when the weather changes.

The captain ordered emergency lifeboat drills on the second day at sea. We huddled on the steeply angled foredeck in our life jackets, ducking our heads against the wind and spray as water gushed into the scuppers. The ship almost buried her bows in the colossal waves and tons of saltwater flew into the air so that spray was soaring over us high up onto the bridge windows. The wind howled and seemed to tear sheets of white water from the surface of the waves as if stripping off a bandage from a black wound. The crests of the waves roared by level with the deck rails, and smaller waves crawled over them as if the wind could not move the water fast enough and was breaking the very waves themselves into pieces. It was difficult to know whether to be exhilarated or terrified.

We returned to the comfort of the wardroom where I played chess with Michael Bizony, who had taken on the duty of expedition bird-watcher. The Fischer-Spassky match was just starting in Reykjavik, from which we received the results by radio with great anticipation. The next day, in the late afternoon, we arrived in sight of Rockall where the wind had dropped to nothing. Complete stillness. Not peace, but an unreliable armistice with the storms. A temporary cease-fire. The waves rolled by as ever, while the ship pitched and climbed the slopes, but the air was clear and the tiny pinnacle of Rockall gleamed in the sunlight. White foam boiled around the base of the rock, surging and falling with the swell.

The engineers responsible for installing the lighthouse came from the Trinity House General Lighthouse Authority, and they were escorted on their excursions to the rock by two Royal Marine cliff-climbers. The helicopters were scrambled and the navigation expert who was due to install a radar reflector on the rock took off immediately, followed by lighthouse engineers, and the two marines (Fig. 2.12). The radar engineer was roped down on to the tiny flat shelf known as Hall's Ledge but

Figure 2.12. Helicopter hovering over Rockall on a day when there was a rare calm sea. The round barrel-shaped structure that will contain the automatic light has already been lowered onto the Rock. Bad weather delayed the next stages.

was so alarmed to see the crests of the waves roaring by below him that he had to be winched back again. The Trinity House engineers bolted geological instruments into the rock with huge bolts, and then they fixed the mountings and cylindrical drum for the light onto the bedrock. As darkness fell at the end of a long summer day the helicopters brought the tired teams back to *Engadine*.

Then the next storm hit us. Captain Rutterford called me up to the bridge.

"We've got the weather forecast, and there's one low-pressure system after another coming in from the west. I've been speaking to a trawler skipper who is a few miles further west from here, and, in his words, 'The lows are coming through like a string of sausages.' For everybody's sake, I think we should turn back to shelter in the lee of the Hebrides."

"How long would it take to get into the lee, and how long to return to Rockall if the weather improved?" I asked.

"Eighteen hours at good cruising speed, but of course the run back before the wind with a following rough sea would not be at full speed. We might have to gamble on the forecast, steaming in moderately bad weather both ways."

"So if we were sheltered in the Minch Channel, between the Hebrides and the mainland, and if the weather improved, or the forecast looked good for the next day, it would take at least 12 or 18 hours to return close to Rockall?"

"Yes."

"The lull between storms is likely to be half a day, or a day – say 24 hours – so we would have almost no chance of ever being close to the rock when the sea state enabled us to work. But my team is not the most important here. The engineers installing the light must take precedence. What do they think?"

"I'll talk to them next. But you seem to have worked closely with the Foreign Office. You think we should stay on station close to the rock?"

"Yes, it is essential, unless it is a real risk to the safety of the ship."

"We are perfectly safe. I can steam on track into the wind for three or four hours, and then go about and steam back downwind. The ship will roll like hell while we go about, but otherwise, life will be comfortable. Under present sea conditions, my job is to satisfy you people running the projects. There is no risk."

After further consultations, this was the plan we adopted. Bizony and I played chess. The others read books from the ship's library. The met officer stayed in his cabin. Everybody hunkered down for the duration. From the bridge the scene verged on apocalyptic. The waves tore by, as the ship's stabilizing system fought to keep her on a steady keel, but occasional violent lurches and juddering of the hull showed when the machinery could not keep up with the stresses of the storm. Incredibly, we could sometimes see trawlers through the rain and spray. Their hulls disappeared completely between the crests of the waves. I watched them with a sense of anxious horror when they vanished sometimes for the passing of two or three wave crests, and then rose again as tiny waterlogged black silhouettes fighting to stay afloat.

As Captain Rutterford had predicted, going about was a messy business. To warn everybody what was going to happen the officer on the bridge would order a broadcast message on the ship's tannoy or "pipes". "Ship going about in five minutes. Going about in five minutes." Anybody doing anything delicate, the cooks in the galleys, workshop engineers repairing a helicopter, or just a crew member carrying a cup of tea, would hurriedly find a way of fastening everything down, getting into a position where he could hold onto a rail or bulkhead fitting. "Going about now!" As *Engadine* swung slowly round broadside onto the wind and waves she would start to roll dramatically, and the stabilizers were turned off simply to protect them from the stresses. For several minutes it was impossible to stand upright without holding onto some fixture, until the ship gradually turned downwind and adopted that strange surging motion where the waves pick up the stern and hurl it forward until the crest of the wave slides past under the hull and you suddenly find the bow struggling uphill and the motion slows down abruptly as the hull squats deep into the trough. And then again. We would settle down to the motion of running with the waves, until, a few hours later, there would be the next warning.

During two weeks we had another three separate days of calm with no wind. But the swell, of course, continued with waves of 5-10 m (16-33 ft) or more in height from trough to crest, and some of the waves sent spray breaking right over Rockall even when there was no wind. In a few calm hours, the Trinity House engineers suspended the light mechanism and batteries of the lighthouse under a helicopter and lowered it into its rugged cylinder. That was the priority. When the geologists visited the rock, they found that all their equipment attached on the first evening had been ripped away, although it had been bolted on at least 15 m (50 ft) above still water level.

On the calm days, while the engineers installed and wired up the electronics, batteries, and timing mechanism inside the lighthouse, we started echo sounding around Helen's Reef and planning where to dive for rock samples (Fig. 2.13). We had estimates of tidal current direction and speed, but the wave height was the most dangerous threat.

Figure 2.13. Divers discussing plans on board RFA Engadine in 1972. We are working from the flat deck on the roof of the helicopter hangars, quite high above the sea. The divers, the dive boat, and their equipment had to be lowered from this height. The diver wearing the orange Adjustable Buoyancy Life Jacket (ABLJ) is Paul Binns; the diver on the extreme right is Ray Peters.

Although there were no white crests to the swell waves, the water surface still rose and fell with a force that could throw a small boat onto the rocks, or drag the divers' shot-line into deeper water. *Engadine* was powerful, and the helmsmen were used to navigating accurately to support the helicopters, so Captain Rutterford had no anxiety about keeping station within half a mile of the rocks. I stayed on the bridge with a pair of binoculars and two radio sets, watching the ship's workboat and a large inflatable Gemini motoring over towards Helen's Reef. Since the swell was still running there was a turbulent surf over the crest of the reef, but the survey team in the small boats were so low down that they could not see it. A helicopter flew over the reef to drop an orange smoke flare onto the surf patch each time that we sent the dive boats out so that they could steer towards it.

We made numerous echo-sounding traverses across Helen's Reef to survey its shape, navigating the tracks and plotting positions by radar range and bearings from the ship to Rockall, and trisponder radar range from the workboat to the ship. Given the delays in writing down numbers that should have been simultaneous, and the intermittent visibility because of the wave crests, the plots were rather rough, but a picture of the steep and rugged stepped topography of Helen's Reef gradually emerged. The crest of the reef extended north-south, being shallower than 20 m (66 ft) for much of its length, and rising to within a metre (3 ft) of the sea surface at several points, although this was difficult to estimate exactly since we never had a calm, wave-free day. White breakers boiled over the shallowest points and the helicopter pilots could see exposed rock emerging in the wave troughs. The eastern face dropped steeply to 40 m (131 ft) depth, with a more gradual slope to the west. Spurs and steep-sided bluffs projected from the main reef.

Finally, on 27th June, the weather was good enough to dive. From the dive boat we placed a float and sinker shot weight on the reef where it would not be swept away by the waves. The helicopter pilots reported sighting sharks within a few hundred metres of the reef, but during several dives the divers never saw them. I settled on the wing of the bridge in the sunshine, clutching two radio sets and receiving reports of the action (Figs 2.14, 2.15).

The divers worked in pairs and plunged into the oceanic water albeit slightly clouded with bubbles while the helicopter hovered nearby, providing commentary from above. The visibility was 15-20 m (50-65 ft) once the divers got away from the foam and bubbles. Three dives were close to the shallowest pinnacle in 10 m (33 ft) depth, and two further dives 100 m (330 ft) to the south in 29 m (95 ft) depth. The divers worked with heavy club hammers and cold chisels to break off flakes or chips of rock, which they put into numbered plastic bags, and the largest sample obtained was about 10 x 8cm (3 x 4 in). This was big enough for the geochemists to do chemical analysis back on shore and make potassium-argon radioactivity measurements to calculate the age of the rock. Altogether the divers managed to smash off 12 fragments of rock from the five dive sites, which was sufficient to justify the whole diving project.

Each time the helicopters took off or landed a special "pipe" was announced throughout the ship to ensure that no rubbish was loose or was thrown overboard that might be sucked into the jet engine air intakes. The phrase was so stark and insistent that it has stuck in my memory like a magic spell, a mystic refrain:

"Helios landing. Ditch no gash. Ditch no gash."

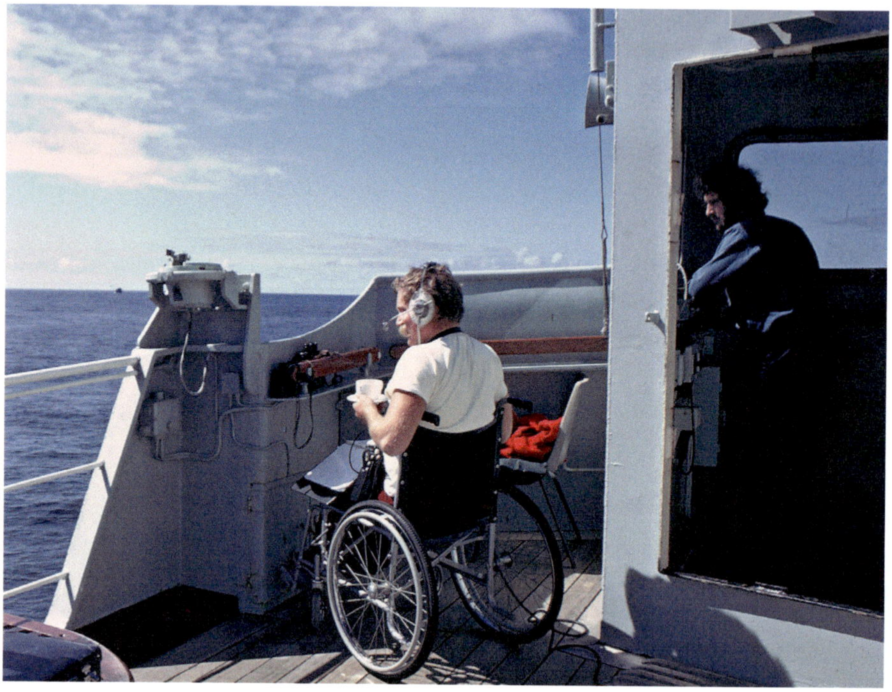

Figure 2.14. The author on the bridge deck of RFA Engadine monitoring the progress of diving on Helen's Reef, near Rockall. The black dot on the horizon to the left of the bridge prismatic compass is Rockall. The author has two radio sets for talking to the dive boat and the helicopter pilot.

The analysis of the rock samples after the cruise showed that the age of the Helen's Reef was 81 million years; but Rockall itself is only 52 million years old, so the geologists discovered that there had been multiple phases of volcanicity as Europe and North America were torn apart, breaking Rockall Plateau away from the surrounding continental crust.

We returned to Liverpool with a following storm and staggered ashore still swaying as if counteracting the ship's motion. This unsteady feeling lasted for several days. The major British newspapers reported the installation of the light. We even achieved fame by being lampooned in *Telegraph* humourist Peter Simple's "Way of the World" column with his usual melancholic twist. He mourned the loss of one more symbol of wildness, loneliness, and romantic horror. To his distress, we had domesticated Rockall.

* * *

The lighthouse that had been so ostentatiously installed on Rockall in 1972 only performed sporadically, and was soon abandoned as not useful. Even during the project to install the light, I had been astonished by its apparently flimsy construction. The protruding rim of the top cover, combined with the lack of streamlining, seemed to be an invitation for the upward surges of colossal waves to force jets of water into the drum-shaped container. I do not know how it did fail, but it was, in the event, not strong enough to protect the electric components inside. In 1975 IGS decided to put a seismic

Figure 2.15. Rockall with the lighthouse installed on top. The rock is 18 m high. The flat base for the lighthouse was made in 1971 by using explosives to blow off the sharp pointed top of the rock. Winter storms drove waves over Rockall and damaged the light, so we went out again in 1975 to work on the same site. Inset: The top of Rockall showing the light in position on the platform flattened by the Royal Engineers. Note the grooves left by the rock drills used to insert the explosives. The patch of white water beyond Rockall is caused by another volcanic peak known as Hasselwood Rock.

recording package in the lighthouse cylinder, so DoT Lighthouse Division contracted with the RFA for two working days to remove their light while IGS installed the recording seismometer and reinforced the structure. RFA *Olna*, which carried one helicopter, was available for two days for this work, during which diving to obtain more samples from Helen's Reef had to be done in the same limited time. *Olna* was a 36,000-ton Fleet Refuelling Tanker of such vast proportions that launching the dive boats was going to be like abseiling down a mountainside (Fig. 2.16). The work on the seismometer and the diving project were added to the existing deployment for fuelling other warships during a training exercise. There was no funding for an extra day of diving, so we would have to work within the schedule for the seismometer installation. The instructions to

Figure 2.16. RFA Olna, the 36,000-ton tanker that provided the support ship for divers working around Rockall in 1975. The dive boats had to be launched from the same height as the helicopter deck, aft of the funnel. The numerous cranes and heavy pipes amidships are for transferring fuel to warships while at sea.

conduct this project gave us only two weeks to prepare, and the only available divers at the NIO were Roger Clements and Martin Saunders. I arranged with the diving officer of HMS *Tiger*, also participating in the exercise, that we would give their ship's divers some extra practice by diving them on Helen's Reef. This was agreed. On board *Olna* we installed the recompression chamber, in which I gave the RN divers practice exposure to 30 ft (10 m), as most of their dives were usually very shallow.

As soon as we had boarded *Olna* at Rothesay in early July Captain Taylor asked to have a briefing discussion. The first engineer was also present. I could see from their tense faces that they wanted a very detailed review of how we could work so close to Rockall or Helen's Reef.

"Welcome aboard," said Captain Taylor. "I want the first engineer to join us in this discussion because he has some serious problems to contend with. We know in outline what you want to do near Rockall, but the safety of the ship must come first. Your operations come a long way down the priority list after we have fulfilled our duties in providing fuel to RN ships. Although the positions of Rockall and Helen's Reef are officially shown on Admiralty charts, the water depth around the rocks has not yet been surveyed to normal chart standards, and I cannot possibly take the ship close in. How close do you want to go to the rock, and how do you plan to launch the divers?"

"I appreciate that a ship of this tonnage and windage cannot hold station close to Rockall, and I would not expect you to do so," I answered. "But we do need to get within a mile or two of the rock, or the same distance from Helen's Reef, to lower dive boats when the ship is not moving. Either the ship's workboat, or inflatables or RIBs, can be lowered with the divers on board, and then picked up later, maintaining radio contact with the coxswains. The screw must be stationary when the small boats are close to the ship. Apart from that, you would be able to steam away from the diving area for a further mile or more, and only return when the divers need to be picked up."

I thought that this was a reasonable negotiating position, but the first engineer was straight-lipped and frowning. He explained how the large turbines that powered *Olna* could not be stopped and started like the diesel engine of RFA *Engadine*, and that when the turbines were stopped the massive weight of the turbine shaft caused it to sag, risking contact between the rotating blades and the static blades of the turbine. The only way to prevent this was for the engine-room staff to insert chocks under the shaft to prevent sagging. Once that had been done, and the shaft had cooled, it would take half an hour or more to warm up the turbine slowly to turn the propeller again. After discussion, we agreed that the ship's screw could be stopped for as much as 20 minutes before the turbine either had to be turned for a few minutes to keep it hot, or the chocks were inserted. I agreed that we would launch the divers directly loaded in the boats, and that we would always complete this within the 20 minutes, or find a way to keep the divers safe while the engineers either inserted the chocks or turned the turbines.

We began to relax. A solution was in sight. A steward served coffee as the discussion became more informal and discursive.

The second season at Rockall was delightful summer weather while RFA *Olna* conducted several refuelling transfers to other naval vessels on exercise in the Rockall area. We were impressed beyond measure by the skill of refuelling a warship at sea, so-called RAS (Figs 2.17, 2.18). Because we were using the tanker during a military training exercise, we had been given initial instructions that there was only one day free to dive on the reefs. The working relationship with the officers and crew of *Olna* became so efficient, with mutual respect, that they arranged for three days to be allocated for diving, while still managing to meet their formal commitments of refuelling the warships.

Rockall showed up on the radar just after midday on 7th July. The weather was perfect so that the five RN divers from HMS *Tiger* flew over by helicopter and Roger and I ran the team through the recompression chamber drills, after which briefing the team and practice drills continued until midnight. The next day, when the boats were launched quickly and smoothly, the first engineer became much more relaxed. *Olna* even kept station less than 2 km (1 mi) from Helen's Reef, but that was partly because the weather was so calm. The helicopter picked up the reef quickly and dropped smoke flares to guide the dive boats where to drop the shot weight. We repeated the drill on three different days, with divers descending 15 m (50 ft) deep into heavy kelp beds and successfully breaking off rock chunks on every dive. The summer days were long and we exploited every hour that we could. The helicopters sometimes reported seeing sharks, but they did not come near the divers.

On 17th July the weather forecast was Beaufort Force 2-3 freshening to 4-5. There was a 2 m (6 ft) swell running with white horses beginning to break at the crests. From the bridge I watched the helicopter pause to hover over the reef. The 11 m (36 ft) workboat and Gemini inflatable with divers on board were heading for the reef to try to get at least one dive in before the sea was too rough. The helicopter pilot dropped the smoke flare and the dive boats motored towards the smoke.

"Lot of sharks today," came the voice of the pilot on the radio.

"Thank you," I replied. "Let me know if they are getting close to the divers."

"There's a problem, Sir. One of the sharks is eating the smoke flare. He doesn't look happy!"

"We'll cancel this dive!" I turned to the bridge officer. "Call the workboat and cancel the dive."

"Dive cancelled!"

Figure 2.17. Helicopter on the aft deck of the cruiser HMS Tiger. In 1975 the divers worked from the Fleet Refuelling Tanker, RFA Olna, and divers from HMS Tiger joined the team. They were ferried between the ships by helicopter each day.

Figure 2.18. Refuelling at sea (RAS) between RFA Olna and HMS Tiger, during work in the Rockall area in 1975. Note the massive oil pipeline suspended between the two ships. Divers from HMS Tiger joined the diving team to take geological samples off the submerged rocks of Helen's Reef.

The unhappy shark, belching orange smoke, thrashed around while the weather deteriorated. Apart from this curious but minor setback, the 1975 Rockall trip was a smooth success, with numerous rock samples transferred to IGS for dating and analysis. All thanks to amazing support from the master and crew of RFA *Olna*, and full co-operation from RN ship's divers.

My last example of military collaboration was with the Royal Australian Navy when, in 1982, we worked for six weeks based on the Larrakeyah Barracks near Darwin, diving for a month on the Cootamundra Shoal 320 km (200 mi) offshore in the Arafura Sea. We worked from a specialist dive support vessel called *Reef Explorer*, and I describe that pioneering project in Chapter 10.

* * *

I have spent much of my life responsible for the training, leadership and safety of divers, both as sports divers and as employed scientific divers. My experience in the SBS, the RAFVR and other projects jointly with the military helped me to develop a style for managing safety and achieving results.

For most of the time after basic training at Lympstone, including that at Eaton Hall, we learned skills that required theoretical knowledge, and the ability to remember details and then calculate the next action. We often carried out these cerebral activities against a background of discomfort, stress, and distraction. In the SBS the conditions for action were potentially hazardous. We knew that the instructors could do everything they told us to do. There is no point in shouting at a trainee who is converting minutes of arc on a chart into travel time in a kayak while allowing for wind direction and tidal currents, or screaming at a trainee loading carbon dioxide absorbent into the canister of an oxygen diving set. In the SBS squad, many of the older marines who had seen active service in war zones would have resented being treated like raw recruits. The instructors reacted very quickly if a mistake looked as if it was going to endanger someone, but otherwise mistakes were corrected in a learning mode, more in the fashion of an experienced school teacher being stern with a slow pupil. Any one of us knew when we had made a stupid mistake, and that was humiliating enough in front of our comrades without being shouted at.

Special Forces operate in small units far from senior command. They are given instructions and told to go and carry them out. Individuals must then take responsibility for their actions and decisions. Delegation and trust by senior authority is total. In the event of careless failure or negligence, or some form of deliberate action contrary to instructions, reprimand and disciplining would be severe. Everything depends upon the individuals and their training. The individual must be trained and then trusted. Modern encrypted communications and GPS have changed the military details since the 1950s, but the principle of training and delegation is strong.

3

Childhoods and Family Life

The grown-up held my hand as we walked down the cliff path. It was winter. I was muffled up in layers of wool. To my right, the edge of a cliff tufted with dead grass fell away to the sea. I could hear waves breaking and clattering on the cobbles far below. The sky and the sea were grey and black and flecked with silver. The grown-up arm extended upwards above my head towards a shoulder near the sky. I don't know who the grown-up was.

A broad building appeared below us where a road came down the valley to the brink of the cliff. We went into the building where there were many people larger than me and some children. I was alone for a time. No memories. Some bright lights. The grown-up came back for me. "What was on that cake?" I demanded. "It was covered in coloured dots." "They are called 'Hundreds and thousands'" said the grown-up. "That's not a name. It's a number." I was frustrated. The dots ought to have a name.

We started back up the gravel path towards my grandfather's house. Looking up the hill into the evening sky I saw that there was a man poised on the roof taking down the metal weather vane. It was Mister Skur, my grandfather's gardener. "Why is Skur taking the weather vane down? It is beautiful." "The war has started, Darling," said the grown-up. "The king needs all the iron he can get to make guns and tanks. All the railings from gardens, and the iron from signposts and weather-vanes, will be melted down for the war." I felt cheated again. It was wrong.

My maternal grandfather, Stanton Coit, was a celebrated humanist philosopher who, on his retirement from active writing and public speaking in 1935, bought a bungalow at Birling Gap, on the south coast (Fig. 3.1). It was on the brink of the cliff. He then read an article in a national newspaper bemoaning the fact that the south coast was being ruined by infestations of bungaloid growths. To prove this wrong he built a three-storey tower onto the bungalow. The tower was crowned with a weather vane with a figure of Father Time on one side, and a curlicued arrow on the other showing the direction of the wind. I liked it. The king had no right to take it away and melt it into guns.

The declaration of war with Germany was on 3rd September, 1939. Gilbert Flemming, my father, a civil servant who had been seconded to the Cabinet Office, knew how precarious the situation had become for Britain. While he stayed in London the rest of the family – my mother, my sister Sara, and I – moved to Birling Gap to escape the probable bombing of London. Groups of grown-ups often crowded round the wireless listening to the news. I had no understanding of why they did this. Soon, however, we

In: Flemming, N.C., 2021, *Apollonia on my Mind. The memoir of a paraplegic ocean scientist*, 101-136. Leiden: Sidestone Press.

Figure 3.1. Myself aged close to my third birthday with my maternal grandfather Stanton Coit and his dog Winkie at the house in Birling Gap, very close to the sea.

realized that the south coast was where the Germans would land if they tried to invade, so we moved from there to my paternal grandfather's house in Berkshire early in 1940.

My mother later assured me that having been born in November 1936, I heard King Edward VIII's abdication speech on 11th December. That is probably true. While I have the faintest memories of the house in Stratford, East London – the smell of linoleum flooring, a pale-blue musical box with angels on top that slowly rotated, and an illustrated book describing two dogs called Bunty and Bianco – my first detailed memory of a linked sequence of events, a narrative, is that walk to and from the children's Christmas party at the hotel in Birling Gap, shortly after my third birthday.

At Birling Gap my grandfather's butler, Lanchbury, usually served meals. On one occasion, when we had fishcakes for lunch, I unexpectedly found a fresh green pea in my fishcake and was delighted by the discovery. After that, whenever fishcakes were served, just one of them would have a pea hidden in it and I would be agog with anticipation until we found who had got it. Lanchbury always expressed surprise too, pretending that it was a chance mistake.

* * *

My paternal grandfather, Percy Flemming (Fig. 3.2), was a retired eye surgeon, whose wife, Elizabeth Wood, had been born in the Shetland Islands, becoming one of the earliest qualified women doctors in London in the 19th century. They are grey ghostly figures in my memory, shapes in winter, silhouettes against the windows in the room

Figure 3.2. My paternal grandfather Percy Flemming was an eye surgeon working in London. He retired to a house in Basildon, Berkshire, where my mother, I, and my siblings Sara, Mandy, and John, were evacuated to avoid the bombing in London during the war.

where they took breakfast, drinking coffee brewed from a magical pewter percolator with a translucent domed lid. It bubbled hypnotically. Grandpa Flemming died of cancer within the year, and I saw his limp collapsed body being carried out of the house. Granny Flemming was a keen gardener who always had hanks of coloured raffia in the pocket of her gardening apron. I loved to sit with her, sorting out the good and bad strawberries or top-and-tailing gooseberries. When she died a few years later, I asked my mother whether the angels had come to take her away. She said they had.

The house, which was called "The Firs", was on the edge of the village of Upper Basildon, near Pangbourne. I observed with displeasure that the frontage onto the road exhibited hundreds of little metal stumps like rotting teeth where the iron railings had been cut away to make tanks and guns. The garden was extensive, with ill-kept lawns, vegetable patches, ragged rose beds, an old stable block, a decrepit cast-iron water pump that never quite produced water, and fruit trees, and it backed onto farm fields on three sides. For playmates, I had the company of the cook's son, Michael, my younger sister, Sara, two evacuees from the London bombing, Peter and Colin, and after their birth in February 1941, my new siblings, the twins, Mandy and Johnny. Visiting grown-ups and children arrived and departed unpredictably. I did not understand the causes of this flux of refugees but enjoyed the arrival of the Samson family who greatly enlarged and enriched our lives. They called to each other by whistling the six-note opening phrase of the first movement of Beethoven's Eighth Symphony. The habit fascinated me, although I had no inkling then as to the origin of the music. It seemed mysteriously different from the popular tunes of the moment, *Pistol Packin' Mama* and *Clang, Clang, Clang Went the Trolley*. These were the tunes we boys whistled when the grown-ups could not hear us.

When we arrived at the Firs in early 1940 most farm machinery was horse-drawn. By the time we left in late summer 1945, most farms were using tractors. During these years I read a favourite book called *To be a Farmer's Boy* several times, but, by the end of the war, it described a way of life that had almost disappeared. In the early winters, I would tramp beside the ploughman for hours, watching the ploughshare cut and curl the endless shimmering helix of clayey soil as the smooth metal blade lifted one side, turned it over, and deposited it on the other in immaculate rows. The huge muscles of the horse's legs and the shaggy hooves moved with a solid tireless rhythm. In the summers we picked wild strawberries, and later blackberries. Apple and pear trees fruited in the garden, greengages, and plums, followed by hazelnuts in the autumn. We explored the nearby woodlands, climbed trees, built houses out of branches, and fell into ditches. In spring we collected frogspawn in jam-jars and watched the tadpoles grow and sprout legs, but they never grew into frogs. We jumped into puddles and learned how to stretch a blade of grass and blow on it to make a reedy screech. We helped to feed the chickens, which were kept to supplement our diet and collected the eggs. We ate the tough old birds when they could no longer lay. They had to be boiled for hours.

The River Pang that trickled into the Thames at Pangbourne was a lovely place to catch tiddlers and sticklebacks. At the annual fête in Yattendon we threw ping-pong balls into jam-jars to win goldfish in large bowls of water. We always tried to win a pig by bowling heavy balls at ninepins. The pig was there for all to see, and our mother was terrified that we would one day win it. We never did. In the late autumn the wheat that had been harvested and stacked in summer was threshed by a huge clanking machine with numerous spinning pulley wheels driven by a belt connected to a steam traction engine belching smoke and steam into the air. The smell of smoke, wheat dust, and straw mingled as the grain poured into the sacks.

We learned to recognize and name many birds and butterflies. I loved the names yellowhammer and greenfinch. Local boys collected birds' eggs but the grown-ups told us that this was wrong. Some trees defied climbing, which annoyed me. I could not see why God had bothered to make trees that you could not climb. There was a monkey-puzzle tree in a garden on the way into the village, and that was so odd that not climbing it was reasonable. I forgave that tree. But a tall dense cypress in our garden was an insult to my sensitive nature: it looked quite normal, but the branches were tightly packed, curving into each other with dozens of interlocking tangled twigs so that it was impossible to obtain a grip or find a foothold. As a tree, it was a complete failure. A design error. I felt pain and resentment whenever I walked past it. I avoided it as much as possible.

We children ganged together to dig a hole to Australia, which the grown-ups watched with condescension for a week or so as the hole got deeper. When it was so deep that a child was invisible at the bottom digging away with a garden trowel we were ordered out and the pit was filled in again. Grown-ups had no sense of adventure. At night we tried to see if you could stand under one star on one side of the garden, and then run to the other side and stand under a different star. But they always looked the same, which puzzled us.

While exploring nearby fields with assorted children I slipped while vaulting a barbed-wire fence and gashed my left knee. Blood flowed down my leg into my small gumboots. We ran home quickly, and by the time we had found my mother there was a strange squelching sound from my boot. I realized with pride that the boot was part full of blood. My mother did not appreciate this achievement, which I felt was unfair. Not everybody could produce a boot-full of blood. The local doctor was called quickly, and I watched with fascination as he stitched up the cut with black thread and tied the knots.

* * *

Our mother, Virginia, was one of the three daughters of Stanton Coit and Adela Ganz, a German Jewish widow who already had three children by her first marriage. The two groups of children, although spanning a wide age range, had grown up together, spending most of their time in London in the early 20th century, but travelling regularly to visit family and cousins in Germany in the inter-war period. My grandmother had been an active suffragette in London, corresponding with Bernard Shaw and other members of the Fabian Society circle, and Virginia told us of how she carried banners in marches demanding votes for women. There were books on great English women in the house, while our mother made serious and overt efforts to minimize the assumptions about what would now be called gender stereotyping. She cheerfully admitted that there must be some genetic factors that could not be overcome since the boys still tended to play with machines and the girls with dolls. In fact, while at the Firs, we played very much as a gang, with old-fashioned games like grandmother's footsteps, blind man's bluff, hide-and-seek, and tag, or exploring the local woods and meadows. While the twins were still very small, my mother managed the household, ensuring that Nanny also looked after Sara and me, that meals could be assembled by Tabby, the cook, for the various visiting groups, and that we complied strictly with wartime regulations on food rationing and black-out at night.

My parents, who were atheist Socialists, held different beliefs from most other adults and children that I met. I did not know the word atheist, nor indeed Socialist, but when I was about six or seven years old they had told me that they did not believe there was a god in a place called heaven. They said that Jesus was a good man. They explained that doing good here on earth, being kind and thoughtful to other people, helping other people when one can, taking responsibility for one's own actions, and being generous with one's possessions and advantages, encouraged people to work together and live together more happily. It helps to avoid quarrels, confrontation, fighting, and warfare. Everybody is better off as a result. This seemed to me a reasonable point of view and still does. Scholars, sociologists, and commentators have written thousands of books on philosophy, ethics, morals, the Enlightenment, sociology, justice, religion, humanism, rationalism, existentialism, and epistemology linked to this simple concept, and I have read a few tens of them, including Plato, Seneca, Voltaire, Lecky, Bertrand Russell, Albert Camus, and recent commentators such as Amartya Sen, Anthony Pagden, Charles Taylor, and many others. With all the complexities and caveats, I still find it a reasonable principle by which to live.

We children were somewhat emotionally detached from our parents during the years 1940-45. This was amplified by my father being in London most of the time, and by my mother's occupation with running a complicated household. That does not mean I was lonely: life was almost communal. With three younger siblings, Sara, Mandy, and Johnny, our Nanny, Tabby the cook, Michael the cook's son, evacuees from London, Jewish refugees from Germany in transit, the Samson family, my father's elderly aunts staying in remote upstairs rooms where they made tatting and embroidery, an occasional gardener, the neighbouring farmers, and soldiers camping in tents in the woods, there was constant social activity. Nevertheless, my father was seldom able to visit home even at weekends. My mother spent much of the time in her own tidy "grown-up" rooms, organizing the household so that we seldom spoke with her. We came to know and spend much more time with our parents after 1945, when we moved to London.

* * *

During the war the winters were very cold and snowy, and the summers hot. This is not just childish idealization, and meteorological records bear it out. At harvest time the horse-drawn reaper-binder circled the wheat-fields in ever-decreasing loops until only a small area of grain remained standing. The rotating arms of the reaper swept the standing stems against the oscillating cutter so that they fell onto a moving canvas belt, which carried them to an accumulating sheaf. This was tied automatically by a huge swinging iron needle and string, before pitching out onto the stubbly ground. We children followed the binder collecting the sheaves into stooks of six, propped leaning against each other to dry in the sun.

As the binder bit into the last island of standing wheat we all knew what would happen. Rabbits ran out in every direction, scattering desperately to avoid dogs and the stones we hurled. I never saw anyone catch a rabbit, but we always tried.

Miss Hadfield's kindergarten school on the edge of Pangbourne attempted to educate me. I remember the process with pleasure. Somehow, as I had already achieved a primitive facility with reading, I quickly pranced through brightly illustrated books with titles like *The Little Red Engine*, and simplified bible stories like "The Good Shepherd". The school book I remember most vividly was *Tales of Troy and Greece* by Andrew Lang, although this was probably in my last year there, when I was seven or eight years old. The book retold the Homeric poems and Greek myths in simple language. The legends and myths captured me completely as if they were true stories about the next village. I knew that this was not the case, and I knew that the stories of magic, spells, curses, haunting, heroism, gods, tragic deaths, cunning, one-eyed giants, and sailors turned into pigs were no more rational descriptions of fact than were Grimm's fairy stories, which I also enjoyed. Nor did I bother to enquire where Greece was, nor how long ago the events in the stories were supposed to have occurred. I accepted the stories to be understood and enjoyed on their own terms, describing their own world. Countless images, the Minotaur in the maze, Medusa's head with writhing snakes, the Trojan Horse, and Ulysses wandering the earth with an oar over his shoulder sank deep into my brain, where they settled like stones on the bottom of the sea. They are still there.

One of the teachers pinned up a postcard-sized colour print of Turner's *Fighting Temeraire*. I was stunned by it, scarcely believing that anything could be so beautiful. I would walk away guiltily from the picture as if afraid to stare at it for too long and then sneak back later to see if it was still there, unchanged. I was relieved to find that it did not change, and nobody took it away. I did not dare to tell anyone the effect the picture had on me in case I was in the wrong. Was such pleasure forbidden? I still love that painting, but how the little postcard could have had such an impact on me is almost incomprehensible.

There were many trees in our garden that were good to climb, which I relished on windy days when you could hang onto high branches swaying like the mainmast of Columbus's *Santa Maria*, pretending to be the lookout in the crow's nest longing to be the first to see land. I did not like Columbus himself, but I wanted to be that man in the crow's-nest seeing farther and first. An ancient apple tree by the cart shed was the perfect place for a loop of rope that provided a fake crow's nest so that I could let go of the tree with both hands and pretend to look through a telescope.

The Columbus motif recurred. We children enjoyed cutting out complicated paper models of houses, towers, castles, and ships, and glueing the pieces together with a foul stinking glue called Croid. It smelled like a tanner's yard on a bad day. The cut-out I liked most was a paper model of the *Santa Maria* and I read a children's version of Columbus's diary with interest. I was impressed that he concealed from his sailors the true distance that they had already sailed to avoid frightening them that they would fall off the edge of the Earth, but I felt that it was dishonest. I read many children's books, amongst which there were often biographies of famous people such as Helen Keller, Sebastian Bach, or Lawrence of Arabia. I acquired several small books for identifying butterflies, rocks, and mushrooms. My parents gave me books with titles like *How to draw trees*, *How to draw ships*, or *How to draw steam trains*, which I referred to often in my frequent drawings. Where all the paper came from in wartime is a mystery, but I was never without a scrap of clean paper to draw on. My sister Sara and I were both at an age when drawing and painting with poster colours was most enjoyable. The twins were happy to scribble, so that all four of us children drew, scribbled, and painted whenever we could.

As the war drew to an end, we made short family visits to London. My father and I visited his other elderly aunts in Farnham, one of whom, Great Aunt Ursula, had been a well-respected painter in the early decades of the century. She lived on Castle Street, accompanied by a talkative parrot, and overshadowed by the looming masses of the Bishop's Palace and the Norman keep. I adored her studio which was still stacked with half-finished or rejected canvases reeking of oil and turpentine spirit. They radiated the lost culture of Venice in the early 1900s or still-lifes with cut glass vases piled with roses. When my father was visiting the Firs towards the end of the war, he took me to hear a string quartet performing at the village hall. I was mystified but entranced. Why did four people play that music? How did they know what to play? We walked back to the house on a moonless night with the Milky Way shining like a river meandering across the sky. My father explained that the glow was caused by stars very close together. The icy-cold air stung in my nostrils as I stared upwards, but I was not convinced that there were stars in the continuous pale glowing light.

During my last term at Hadfield's school, I used to stop after lessons at the blacksmith's in Pangbourne to watch the smith heating the iron horse-shoes to a rich red glow, before beating them with a massive hammer to perfectly match the shape for each hoof. I loved the smell of the furnace, the singeing hoof, the noise, the sense of pure skill, and on some days the smith would let me work the bellows.

* * *

In spring 1945, just before victory in Europe, my father bought a house in London on Chiswick Mall overlooking the Thames. During the summer the family moved into the large, cold Victorian house with half the windows blown out by bombs and boarded up with strange powdery synthetic panels. We lived for the next year with cold air leaking through the gaps.

Living on the bank of the Thames was a frequent source of excitement and minor dramas. The river was much grander than at Pangbourne, with a huge tidal range flooding the street at high tide and vanishing to the other side of a muddy willow-covered island at low tide. We would watch while tourists or sightseers wandered onto the island at low tide only to find themselves cut off or almost marooned when the water came back. We would wait until the last minute as the strangers hesitated before we rushed across the mud to help them paddle back through the water, using boulders as stepping stones. We always got there just before it was really too late. That way one could feel naughty and heroic at the same time. At the major spring tides, the water flooded across the road, even sometimes into our cellar, leaving behind on the road a stinking sludge of mud mixed with many tons of driftwood as it ebbed away.

We collected the driftwood to burn on the fire in the drawing-room, and we were always proud when we found huge baulks of timber. Size was everything. I found that a half-brick tied onto a length of washing line made a perfect lasso with which to catch the biggest chunks of wood as they drifted past in the river's current. Some of these had to be sawed up with a two-handed crosscut saw to burn on the fire. As my father and I worked together cutting these thick timbers, he pointed out that if I caught more smaller logs they would not need so much sawing, which would save time. I found this logic somewhat lacking in frontier spirit and even a bit ungrateful.

There were towering chestnut trees at the back of the garden, beyond which were several derelict sheds and a deserted coach-house with a pigeon loft. In the sheds we found piles of thick rope, heavy planks, and old copies of *The Times* from the 1930s. Everything smelled musty and old. I read the papers with a curiosity that merged into guilt. What could they mean? Why were they there? Gradually we found more and more ingenious ways of using ropes to build bridges between the trees, to make large swings, and single ropes that you could crawl across lying on your tummy. With timbers and planks from the river we built platforms high in the chestnut trees, connected from one tree to the other by a bridge of two ropes. You walked on one rope and held onto the other which was arranged at about chest height. There was also a platform in the biggest apple tree, and a rope from there to the mulberry tree. From the platforms in the chestnut trees, you could watch the steam lorries being loaded with beer barrels in the yard of Fuller's Griffin Brewery on the other side of the road. We built two substantial timber houses at ground level, of which one nominally was for the boys, and one for the girls, but we mixed back and forth pretty freely.

The Thames was muddy, but initially the concept of harmful pollution did not trouble people, so that children from several houses along the Mall would swim happily after school. Soon, though, the fear of chemical pollution and poliomyelitis resulted in a ban on swimming, but not before I had managed to swim from the road to the island and back at high tide. There were two large open-air swimming pools nearby with high diving boards which became a staple for holiday games of swimming and jumping when the river became a health risk. Also, we found old oil drums in the junk-yard at the LEP, a fabulous zone of post-war scrap and derelict vehicles. We tied barrels together with planks to make a useful raft so the river was still fun. The raft could carry one person

Figure 3.3. Sara, Nic, John, and Mandy on a branch of the mulberry tree in the garden at the Hollies, in Chiswick, 1948. The mulberry fruit were very good to eat, but the juice made an indelible stain on skin and clothes.

with an improvised paddle quite safely, and it enabled me to capture even larger chunks of driftwood, to my father's dismay.

We bicycled around the local streets, especially down steep hills just for fun. By modern standards the four children were allowed to roam freely and widely to the point of neglect. But it felt entirely right at the time. We would slide down a muddy bank on the island, wait for the tide to come in, launch the raft, catch some wood, then get washed and clean for tea, and sit politely like good little boys and girls. The grown-ups did not object to our escapades, nor did we feel either restricted or neglected. Falling out of trees was normal. There were inevitably some crashes, and Mandy broke her leg jumping off the front wall of the garden down into the street, while on another occasion Johnny broke both wrists jumping to catch a branch on the mulberry tree, which he missed (Fig. 3.3). Somehow we all survived.

There was a large brick-built conservatory facing the river, entirely walled in glass on three sides. It contained huge heating pipes, now stone cold, that looked as if they came out of the *Titanic*. Since the old boiler had collapsed to a rusty heap during the war, the pipes had been covered over in shelving, and this beautiful space, with endless atmospheric ever-changing views of the Thames, became my workshop, complete with a professional carpenter's bench. Decades later I discovered a wistful misty oil painting done by Asphodel Fleischman in 1944, showing the view upstream at low tide, with a horse and cart standing in the shallow water. It could have been painted from our front garden. It was a scene that I had often attempted to capture myself. The majestic bend of the river, with the swirling light on the flowing water, was an attraction to painters who occasionally set up their easels on the shingle foreshore in front of our house.

The internationally known artist Julian Trevelyan, a near neighbour, often painted the island, Chiswick Eyot, in his stark angular style.

In September 1945 I began my first term at Westminster Under School. There is an archway just to the right of Westminster Abbey that leads into Dean's Yard, from which, on the left, a smaller arch leads into Little Dean's Yard. This is the home of Westminster School. In 1945 the preparatory school, Westminster Under School, was also in Little Dean's Yard. Since you could hear Big Ben strike every quarter hour there was no excuse for ever being late. At dusk in winter a bent old man walked round the squares with a long pole with a flame at the tip. He turned on the gaslights that glimmered wanly in the thick mist and fog in a manner worthy of Sherlock Holmes or Oliver Twist.

With 50 or 60 other small boys I accepted that morning chapel was held in the vast space of Westminster Abbey every morning, that masters in black gowns and mortar boards scurried between the buildings of the main school, and that we ate lunch with the main school in a medieval beamed hall, dark with smoke and varnish, preceded by three gowned scholars reciting a long Grace in Latin. The temptation now is to make comparisons with Hogwarts, which is reasonable. It was another world. I say that I accepted the situation, as of course I did, since the alternative would have been to run away and take the Underground back home. But I was awestruck by the scale of the buildings, the towering trajectories of flying buttresses propping up the Abbey, the rigmarole of the Latin Grace, the sense of ritual and tradition that permeated everything, the booming notes of Big Ben, the ancient pews in the Abbey itself, and the long gold-painted choir stalls where the main school sat for the morning service. The Under School sat on wooden chairs in the north and south transepts, directly under the central tower. As I became accustomed to this dramatic location, I noticed that sparrows were flying about high under the ribbed and vaulted roof. I also became interested in the way that the candelabra swayed with an incredibly slow swing on the long cables that suspended them from the vaults above.

Soon the unique historic situation of Westminster School became the routine of daily learning and break time, bells, lessons, and then more bells and more lessons. We played cricket and football on the grass of Vincent Square, but there was only a narrow concrete yard for the morning break. In the classrooms each gnarled wooden desk had an inkwell recessed into the corner, into which we dipped our pens using a wide relief nib. Like the classrooms themselves, even the corridors smelled of ink, linoleum, and paint.

There were bombed and burned-out buildings everywhere in London, some shored up with wooden or metal props, others still with teetering overhangs. Wrecking machines were in action destroying unstable and irreparable buildings. New houses and offices sprang up. From the windows of the Underground train *en route* to Westminster I watched the destruction and resurrection year by year. Several bombs had hit Westminster School and the Abbey itself. Scaffolding soon enveloped the gutted sections of the school, where workmen toiled with bricks and cement to repair the damage. The multi-storey wall next to our concrete playground fronted the shell of an empty structure which was soon encased in steel scaffolding rising from the playground as men worked overhead. One day I was gossiping with my chums Perret, Goodyear, and Meadows (we never used Christian names in those days) when I heard a sharp thumping crack just behind me. I looked down and saw a half-brick splintered on the concrete, with black and yellow dust scattered around it, just behind my right heel. I looked up. It had fallen from high up, past two or three storeys of gaping empty windows, and I realized that if I had been standing 6 inches (15 cm) further back I would be dead. I kicked the crunchy fragments back under the scaffolding, but nobody else seemed to notice. Gossip continued.

* * *

The routine of travelling to Westminster School by Underground during term times alternated with the big family summer holiday, creating an annual cycle. At Westminster the boys were used as a rent-a-crowd for state occasions such as the Opening of Parliament, the Victory Parade, and even the royal wedding in 1947. We marched out clutching Union Jacks for these events, which we enjoyed as a break from classes. At home my father devised a series of summer holidays in remote locations that probably recreated something that he remembered with pleasure from his youth. There are photographs of him as a teenager with a bicycle, and another of him jumping over a stream, which could have been me from my childhood (Figs 3.4, 3.5, 3.6). The summers of 1946 and '47 we spent in Scotland on a farm at Lagganulva on the western side of the Island of Mull, a few hundred yards from the shore of Loch Tuath. The journey alone was an adventure, including the night sleeper train to Glasgow through the flaring industrial towns of northern England belching smoke and flames into the night. Then the single-track train through the mountains to Oban, a ferry steamer to Salen, and finally a taxi ride on gravel roads to the Atlantic side of the island.

Since the twins, Mandy and Johnny, were now five years old, we played together, damming the nearby streams with boulders to make lakes, or collecting seashells and

Figure 3.4. Gilbert Flemming, my father, in 1907. His father, Percy Flemming, was an enthusiastic cyclist and photographer of churches and ruined monasteries.

Figure 3.5. Gilbert Flemming, my father, in 1911. From various conversations with my father, I know that, as a child, he had enjoyable family holidays in open country, but I do not know where this picture was taken.

Figure 3.6. John, Mandy, Sara, and Nic during a holiday in about 1950 playing in the mountain stream near Watendlath in the Lake District. A favourite game was to use large boulders to make a dam so as to cause a waterfall.

coloured pebbles on the beach. There was a cave in the foot of the hill at one end of the beach which played into all my fantasies, from Oscar Wilde's fisherman with his lost soul to Enid Blyton's Adventure Stories with secret tunnels to islands. I had a small pocket book, *The Observer's Book of Geology*, which I used to identify numerous types of granite, basalt, and quartz on the beach, carrying many of my samples back to the farmhouse where I examined them through a magnifying lens.

I was nearly 10 years old for our first trip to Mull, where my parents allowed me to wander and climb on my own in the windswept hills behind the farm. I climbed for miles over tussocky grass, patchy bogs and mosses, finding abandoned dry-stone cottages with their roofs missing, overgrown with grass and bracken, and nearby the skeletons and skulls of dead sheep and deer. Desolation reigned. I accepted but did not understand. Families, farmers, children had lived high up in these mountains long ago, and had vanished. You could feel the ruins being sucked back into the earth. Most of the animal bones had been picked clean by scavengers, but I proudly carried a complete sheep's back-bone of interlocked vertebrae back to my parents. I displayed the bones of various animals and the crystals, rocks, and seashells from the beach in a bookcase when we returned to London, where I arranged them in my "museum".

On Mull I explored mountain streams and wild gullies where small waterfalls plunged over the rocks in repeated cascades as they twisted down to the sea. At the last cliff a waterfall crashed from rock to rock, swirling in the wind down to the cobbled beach of Loch Tuath. The spray blossomed seawards, lit by the dappled sunlight. I looked downwards on a rainbow. The anomaly delighted me. I gloried in the feeling that this is what I expected life to be like: this is what it is like, this is right. It was exciting, reassuring, and then calming. I was at peace with myself, and with those around me when I returned to the farmhouse. My father always checked that I had strong footwear, not gym shoes, before embarking on my treks, that I had a chocolate bar in reserve (sweets were rationed) and, so far as I can reconstruct, that I had a pocket watch. A wrist watch was a very expensive item that boys were only allowed to wear in their early teens: up to that age you carried a bulky but very cheap pocket watch on a strap attached to your trouser belt. From the strict look on his face, I sensed that my father knew the risks in my treks, even if I also felt that his anxiety was unnecessary, and that he trusted me to survive. I accepted that trust without much worry, and indeed felt carefree, but I realized that it was there. My parents considered it safe for me to wander off in the hills, and I was happy to wander.

There was a sweeping view from heather moorland high above Lagganulva. Below me lay the tiny white-painted farmhouse half hidden by a glade of trees, surrounded by small fields of hay and oats. Close by, to the southwest, lay the green islands of Ulva and Gometra with their stepped profiles, which my father said were caused by volcanic lava flows long ago. The sun beat down on the silky calm waters of Loch Tuath stretching far off towards the open sea. On the oceanic horizon were the twin islands of Lunga and Bac Mor, and just visible as grey streaks in the distance, the islands of Coll and Tiree. To the north the peak of Beinn na Drise rose above successive heather-clad ridges, and in the distant south, across Loch na Keal, the swelling curves of the tallest mountain on the island, Ben Mor. This mountain disappointed me because the tallest peak should have had a Matterhornish swagger, while in reality, it was a gently swelling dark green curve that looked like the South Downs over-inflated with a bicycle pump.

In the afternoons we usually made a family excursion, which had to be more modest to allow for the young twins. One day we did climb Beinn na Drise *en famille*, which everybody felt was a great achievement. My father doled out portions of chocolate, one square each, when we reached the summit.

The farm produced hay fodder for milk cattle, male calves were sold for slaughter, oats were sold to market, and the cows not in milk were ranched with common grazing in the hills. Sheep also ranged free in the hills and mountains, being rounded up with dogs when they needed to be dipped in protective chemicals or sheared close to the farmhouse. There was a bull by the name of Monty who roamed free with the cows in the hills, presumably to their mutual satisfaction, but for reasons that nobody could fathom, Monty would occasionally arrive down at the farmyard where his presence was seriously dangerous. We participated in every activity of farm life wherever we could, roaming with the shepherd to collect the sheep from the hills, raking up hay to pile on tripods of sticks to dry, and searching for the eggs laid by chickens and ducks. Sara and I were happy to swim in the sea, while the twins paddled and splashed.

Successive family holidays explored other remote tracts of the British Isles, but they never quite attained the anarchic wildness and freedom of Mull. We stayed in farmhouses near the coast of Somerset, in Arkengarthdale, Yorkshire, where I explored abandoned lead mines, at Watendlath in the Lake District, flanking a dark peaty lake, and in adapted Coastguard houses on Chesil Beach, Dorset. As I and my siblings grew older, we were able to swim together, go on long bicycle rides, or engage in long hill walks organized by my father with military precision, each of us equipped with oilskin waterproof coats and large supplies of chocolate bars. Sometimes these holidays included friends of our parents with their children.

In Chiswick, the annual calendar revolved around the panoply of the Oxford and Cambridge Boat Race on the Thames. Our house, just opposite the upstream end of the island called Chiswick Eyot, was about halfway along the course, so that, from the roof of the conservatory, we could watch the crews heaving on the oars and struggling with the cross-winds and tides, as they sweated and struggled. They rowed past us in full view from the Bemax factory near Hammersmith Bridge almost to Mortlake, just short of the finish. The course is over 4 miles (6.5 km) long, and typical times are around 17 to 19 minutes. We always held a family party with relatives, the Samsons, friends from school, and many others clutching mugs of orange juice, and, in later years, beer, we screamed ourselves hoarse to encourage the rowers.

In the absence of widespread ownership of television sets, the crowds of spectators on the banks on both sides of the Thames were biblical in scale. Hawkers and hucksters strode hither and thither shouting their wares and selling dark blue and light blue rosettes, programmes listing the names of the crews, soft drinks, flags, and souvenirs. Since the great majority of Londoners can have had no special allegiance to one university or the other, and scarcely knew what a university was, their vocal enthusiasm and apparent preference for either crew was amazing. In our house we had the radio switched on to listen to the start of the race from Putney. As the crews came into sight the bellowing from the crowds below rose to wild screams and shouts. On the far side of the river the crowds were packed densely, with many attempting to run or ride bicycles along the bank. We always hoped that the race would be close so that there was some reason to motivate the trailing crew to try harder and to close the gap, but often there were two or three lengths between the boats by the time they reached us, so that the winner was already decided. One year, 1951, when the weather was peculiarly windy and gusty, the Oxford crew sank, and the race was abandoned before coming into our view. No matter, we adjourned as usual to the garden where a delicious picnic was spread out on rugs on the grass, while children climbed the trees and swung on ropes.

* * *

During my last year at Westminster, that is 1949-1950, I sensed that the school had come adrift and was not performing properly. Maybe it is typical as one grows out of an establishment. I learned very little during that year, with the result that I did badly in the Common Entrance exam for Rugby. I passed into the bottom class, LM3, and I can recall sharply the expression of dismay on my father's face when he saw the letter confirming my miserable exam results. But I soon made up for my poor performance.

During one's first year at Rugby, the masters applied a competitive principle to help find the best level of teaching, both for the class which you belonged to, as well as the set-streaming in mathematics and languages. At the end of the first week of term the top boy moved up one class or set, while the bottom boy moved down one. At half term of the first term the process was repeated, and again at the end of the first term. At the end of one week I moved up one class, moving up again at half term. Thus, at the start of my second term I was being taught in the highest class into which the brightest boys usually entered the school.

This procedure sounds brutal, almost destructive, by modern principles of education, but it helped me to settle in amongst other boys with whom I felt on a par, and with teachers whose lessons I respected and enjoyed. I am not defending the system, and have nothing to compare it with, but we all accepted it, and did not feel offended or demeaned by it. In spite of its apparent short-term abruptness, the system was self-correcting because anybody who was obviously out of place at the top or bottom of a class or set could always be moved at any stage of their career. A boy might be held back at one point, and accelerate later. Minds, hormones, and genes do funny things at that age.

The great majority of boys attending Rugby School had been at boarding preparatory schools. One other new boy in my house, Bradley House, had, like me, been at day school in London. Julian Harris and I formed a bond or understanding that the insular, popularity-chasing nature of the boarding prep-schools was something we did not share. Somewhat arrogantly, we felt that we had spent more time amongst adults, and were the better for it. Later, when I shared a study with Julian, his interest in pre-1940s jazz introduced me to an unexpected musical and social world, and to the rudiments of racial history in the USA. I settled into the grind of lessons, three afternoons of compulsory sport each week, piano lessons (at which I was not gifted in any way, but half enjoyed), and extra art classes, which I greatly enjoyed. The boys of each house wore different coloured caps, blue and white stripes for Bradley, pure scarlet for School House, and others of green and white, brown and red, and so on, so that the boys streaming in to chapel looked like a field of wildflowers blown by the wind.

British public school education is pilloried mercilessly by writers and journalists for its reputed brutality, philistinism and potential for homosexual abuse. Since most career writers are necessarily thoughtful, observant, perhaps introverted, reflective, or sensitive, and often bad at sports, it follows statistically that most people writing about boarding schools write with recollected distaste, even horror. Several successful writers went to Rugby and survived, including Charles Dodgson (a.k.a. Lewis Carroll), Arthur Ransome, and Salman Rushdie. I cannot defend the system in its entirety, but can only record what I experienced. Several boyish codes of principle applied strongly in Bradley House in a way that may seem absurd today. For a boy to hit or bully someone smaller than himself attracted scorn and contempt for the bully; for two or more boys to gang up on a third was regarded as contemptible. The masters and prefects similarly let it be known that bullying was unforgivable. Punishment was swift. The housemaster had authority to beat a boy with a cane for extreme offences, but that was unusual.

* * *

In the summer holiday of 1952, my parents arranged for me to spend a month as an exchange guest with the French family Castier who had a large apartment in Paris, and a summer retreat on the island of Belle-île-en-Mer, off the coast of Brittany, near Quiberon. It was not a direct exchange, since two of the Castier children had stayed a few weeks previously with our friends in London, and I had briefly met Milène, who was several years younger than me. Monsieur Jules Castier was the famous French translator of English novelists such as Aldous Huxley and Somerset Maugham.

I travelled by trains and ferry to Paris, where, after a few days at the Castier apartment, I departed for Belle-île, escorted by Mme Castier and Milène. We took a train to Quiberon, where we boarded a wooden converted fishing boat which served as the ferry to Belle-île. My gaze swept everywhere as I tried to absorb the stream of colour, noise, and novelty. The ferry chugged towards the distant island which gradually came into focus as a towering cliff, half cloaked with trees, which then resolved into etched details of crags, turrets, fortresses, gun-slits, battlements, and crenellations piled up towards the sky. Overhangs of rock and vegetation made deep shadows, which increased the aura of menace. At the foot of this grand and overwhelming fortification lay the small harbour of Le Palais, crowded with fishing boats, dinghies, and a few fancy yachts. The town itself lay in the shadow of the fortress, opening onto the quays of the harbour. This first image of Belle-île haunted me so much that I later painted it from memory using ink and watercolour at the extra art classes at Rugby.

We loaded our suitcases onto an aged taxi and were driven 1.5 km (1 mi) north onto the exposed granite headland of Pointe de Taillefer. A solitary house stood four-square on the ridge of the headland so that the ground sloped down steeply on two sides towards the open sea, and on the third side into a sheltered cove. The house was mainly stone-built, white-painted, with pale-blue window shutters further bleached by wind and salt spray, but with pale timber sections around the upper third floor under the eaves. There was no garden or cultivation, and the dusty scrub which covered the slopes stopped only where the repeated footsteps of the occupants walking round the house had trampled the dirt hard and flat (Fig. 3.7).

Madame Castier hustled Milène and me into the house, where I was introduced to an older brother, Pierre, in his twenties, and another brother, Jean-Claude, who was about 18. In the following weeks a younger friend of Jean-Claude's called Jean-Michel arrived, and then Fifi Castier who was just a year older than me came back from London. At

Figure 3.7. The house at Pointe de Taillefer, on Belle-île-en-Mer, where I spent most of a month in 1952 as a guest of the Castier family. The house was isolated and remote, but I used to bicycle from there all over the island. The cave under the house goes through the headland and opens onto the sea at the other side. The house is now known as the Villa Jade.

the start of the holiday there was nobody of exactly my age, and I felt rather isolated, but gradually the age gaps filled in and I was accepted into the gang. Jean-Claude was a superb flautist and sometimes I would return from treks across the island to hear his flute music blowing in the wind with the rustle of the sea waves as a background accompaniment. I would lie down in the grass just to listen to his music. Everybody enjoyed painting with watercolours, and we would sometimes work together, discussing in French our styles and methods.

That first afternoon I unpacked my clothes in a bedroom on the second floor, with a small balcony overlooking the channel towards the mainland and Île de Groix, Île de Houat, and Hoëdic. I found on a shelf a copy of Aldous Huxley's *Crome Yellow*, translated by M. Jules Castier as *Jaune de Crome*, which I read slowly but completely over the course of the first two weeks. A sparse scatter of fishing boats dotted the sea towards the northern horizon. This initial view, though beautiful enough, did not prepare me for the first night when I came up to bed after dinner and saw the lighthouse beams from every island and rock cartwheeling across the sky, flashing and pausing with a hundred different signals and messages of danger for ships that might navigate those waters in the dark. The susurrating murmur of distant small waves rose from the darkness below the cliff. I noticed after a few days that the most bountiful fishing for sardines was done with large nets at night, so the fishermen were always at risk.

Breakfast was large bowls of milky coffee, into which we dipped crusty chunks of baguette. These were followed by plates piled with fresh sardines straight from the dock and lightly fried in olive oil. That first morning I started to explore outside the house, where I found an abandoned cellar on the side facing towards the tip of the headland. There were broken bicycles, piles of timber, some tools, and a large cracked wall mirror in a dark wooden frame. I looked at myself in the mirror wondering who I was.

A path twisted between outcrops of rock down to the cove, where a shingle beach promised good swimming. A dark-clothed man bent over the gunwales of a fishing boat, working hard with tools, paint pots, brushes and scrapers. A boy of about my age stood by, occasionally helping the man, who I guessed was his father. The boy turned to me with surprise and came across the beach. I stammered my name and said "Bon jour" with my best accent. His name was Michel, and, with very few words, we began to explore around the cliffs. Climbing over boulders on the flanks of the headland, Michel showed me a cave which narrowed back into a tunnel going right under the house and opening out on the far beach facing across the wide gulf to the mainland, eight miles distant.

Michel and I explored every corner of the cave with its side tunnels, and he dragged out from under a rock a battered World War II shell case. He gestured for me to wait while he ran back to his father's boat, returning with a pair of pliers and a cigarette lighter. With some effort he extracted stiff slabs of explosive propellant from the shell case, and grinning gleefully, applied the flame of the lighter. The yellowish slabs burst into brilliant flame that burned with a hissing glare like gigantic birthday sparklers. Michel gestured around and I saw that there were many other shell cases half-buried amongst the boulders. I wondered if the Castier family knew that their house was located above an ammunition store, although I noted with relief that the propellant did not actually explode. I thanked Michel for the entertainment and we parted.

With the tools that I found in the cellar I assembled a working bicycle. It was simple and useful but had no changeable gears. The island was planed-off and fairly flat on top, but dissected by many steep little valleys. Belle-île is only 18 km (11 mi) long and 10 km (6 mi) wide. I started to explore, ranging farther away each day. Nobody asked where I was, or who I was. As I never wore shoes I sometimes walked the mile into Le Palais barefoot on the dirt road. I had no map of the island, but gradually accumulated a visual and verbal vocabulary of the barbaric place-names that had no connection with the French sounds I had learned at school: Locmaria, Loctudy, Quinénec, Kerguec'h, Borlagadec, and Samzun. On every shore the cliffs and outlying stacks of sheer rock rose above the Atlantic waves in defiance. There were caves everywhere. Sometimes I would swim in a deserted bay before drying out in the sun. In the fields there were standing stones and dolmens that reminded me of Stonehenge. I cultivated rootlessness: the pure observer. Absorbing the terrain and the landscape at every point, as if I or it was three-dimensional blotting paper. Every evening back at the house just after dusk I walked half a mile with one of the family members to a farm to collect a pail of fresh milk.

My solitariness during the day was something I savoured, as I wandered across the island and back again like a bird driven by the wind, conscious also of my sympathy with Ulysses carrying his oar over his shoulder, or Huck with Jim on their raft drifting on the stream. Sometimes after walking barefoot into the port I would sit on the dock wall at Le Palais watching the world go by, especially the young women, feeling like an invisible observer, although with conventional teenage angst, I wondered what the girls saw in me as they passed.

A summer thunderstorm blasted the Taillefer headland with shuddering gusts of rain and salt spray that howled and roared for a whole day. The house creaked and shook as the water streamed down the windows and the waves reverberated in the cave below. The family and guests were cooped up all day with a friendly social intensity. It was so gloomy that we had to switch the lights on. I was reading Jules Verne's *Les Cinq cents millions de la Bégum* in a richly illustrated volume with full-page engravings of much fancy and imagination. I was captivated by the pictures. The story, not well known today, is prescient in many ways, even politically. We did puzzles, played card games, and read the newspapers and *bandes dessinées*. The papers were full of the scandalous Drummond murders, in which a British family on holiday had been slaughtered in the south of France. Maurice Thorez, a Communist trade unionist was often in the headlines. In the midst of the cramped sociability and chatter of numerous parallel conversations, the cat produced a litter of six kittens in the corner of the living room next to the fireplace. Happiness abounded. On the long dining table there were rectangular plastic blocks by each place-setting on which we rested our cutlery between courses. Over lunch during the storm, I constructed an improbable tower of these blocks balanced with irregular extensions and large overhangs. Fifi laughed with delight and rushed round the table and kissed my cheek. Madame scowled.

After ten days of my foot-loose, image-soaked, and delightful holiday, Mme Castier approached me, speaking in French.

"Nicholas, you are supposed to be learning French during your visit here, but you very seldom talk. I can see that you are reading a lot, newspapers, books, comics, but you must talk more."

I could not construct an adequate excuse, so just looked embarrassed.

"I have decided," continued Madame sternly, "that we will read Molière plays together, and we will discuss the plots and characters."

I agreed to this arrangement, and henceforth spent one or two hours on most mornings improving my pronunciation by reading Molière, then struggling to express my thoughts, but I could not understand how his plots of personal jealousies and comic failings could be compared with the scope, power, and richness of Shakespeare. I kept this opinion to myself.

Towards the end of this magic holiday, we made a family excursion to the west side of the island to see the famous Grotte de l'Apothicairerie, a cave and rock arch of gigantic scale. I scrambled down the cliff, working my way seawards around the base of the outer end of the arch and then back to the mainland again. I fondly imagined that the girls were impressed. Perhaps they were. After a hilarious picnic with much teasing and badinage, we bundled into cars and onto bicycles and I rode back across the island with Milène sitting on the bicycle cross-bar singing all the way. The final return to England was a solitary but fascinating journey by train from Quiberon to Saint-Malo, then the night ferry to Portsmouth, and finally train to London. I felt grown-up again.

* * *

Belle-île was and is a pure defined memory of sensation that had firm boundaries of time and space like a crystal that dropped from the sky, and then evaporated and disappeared after four weeks. Never question or explain a magic gift. Many decades later I took our son Peter for a short holiday in Paris, during which we visited the

Musée d'Orsay where I was unexpectedly confronted by Monet's painting of the Grotte de l'Apothicairerie. I wept with happiness.

* * *

Since my parents continued to try to inculcate an international, cultured spirit in their children, our mother took Sara and me to Florence in the summer of 1953, followed by a tour of the local towns of Siena, San Gimignano, and Pienza. The families of various friends of our parents had similar plans, so that we met several half-known people, bumping into them in the streets. After visiting the Medici tombs, the Duomo, and the Uffizi Gallery in compulsory mode, I began to wander on my own and pick and choose the things I wanted to see. The torrent of images, dates and names of the Renaissance poured over me, but I found a few items that obsessed me and became fixed points. I recall a small gold statue by Donatello in the Uffizi that I returned to again and again. It seemed to contain more than the overwhelming beauty of Michaelangelo's *David*, although it was less than 30 cm (1 ft) tall.

* * *

Several of the boys in my house at Rugby, Bradley House, were enthusiastic climbers, and we climbed trees, drainpipes, walls, rooftops, and into and out of the school at night. There was one myth-making and obsessive summit that nobody had climbed. This was the Hillmorton radio masts, a few miles southwest of town, visible from everywhere, dominating the far skyline. We discussed the possibility of climbing the masts, which were a constant reproach to us, a provocation, looming dim, grim, and grey in the distance, so that we could never imagine that anyone would, in reality, climb them.

During the winter 1953-54, I decided that this was unacceptable: we must climb the masts. Honour was at stake. Claude Marshall, a fellow Bradley boy, and I sketched out a secret plan of how we could achieve the ascent. We were already adept at climbing out of the locked house at night, breaking into the bicycle shed, and then getting back into the house after locking the bike shed. The question was the timing: how long would it take to get to Hillmorton and back; how fast could we climb the most accessible mast? There were 12 masts, each 250 m (820 ft) high, almost as tall as the Eiffel Tower, and we did not know what sort of ladder we would find. We assumed that the engineers must use a fixed ladder to service the aerials at the top. The masts were the tallest structures in Britain at that date, presenting a challenge that could not be refused.

The Hillmorton radio station comprised a circular cluster of 12 colossal metal towers supporting a circular transmitting aerial that had been built for long-range very low frequency radio communications in 1926, and later, in the 1950s, adapted to convey orders to nuclear submarines (Figs 3.8, 3.9). Each mast, triangular in cross-section, was made of a lattice of steel girders, which tapered to a point at the bottom, where it rested on a huge hemispherical bearing. This in turn was mounted on a large rectangular inspection frame, with a solid support in the ground. Widely spaced metal guy wires held each mast upright, and from the top the ceramic insulators took the weight of the aerial itself suspended within the circle of towers. The transmitter signal-generator

Figure 3.8. The Hillmorton Rugby radio masts for very low frequency longwave communications. I climbed one of these masts at night with a friend from school. The masts were decommissioned and destroyed in 2007.

Figure 3.9. This picture shows the lower part of the mast support, before we reached the base of the triangular mast itself. We climbed the mast at night in 1953. The masts were 250 m (820 ft) high, the tallest structures in the country.

building was nearer one edge of the circle, so we had to find an accessible tower as far as possible on the other side.

The distance from Bradley House where we lived to the Hillmorton masts was about 5.5 km (3.5 mi). We timed the ride on bicycles in daylight, taking a turning where a small farm road led past one of the masts at a distance of less than 90 m (300 ft). At the next new moon we timed the journey more carefully at night, after camouflaging our faces with charcoal. For this reconnaissance we hid the bikes in a farm ditch and climbed through the hedge, so as to approach the tower across a grassy field. The square frame at the base consisted of thick steel beams that would be very difficult to grip and, even if one could climb it, the next stage presented the climber with an overhang (Fig. 3.9).

Then I noticed a projecting bar or beam, circular in cross-section like a household heating pipe, that hung down from the inspection platform at the base of the tower itself, just below the ball bearing. I jumped up and grabbed it with both hands, then pulled myself up as if climbing a rope. The beam was vertical for about 1.8 m (6 ft), before curving in towards the tower. I scrambled along it, and was able to climb a low railing onto the inspection platform. I called down to Claude, who followed up the same route.

A ladder led up past the curved bearing, onto the base of the latticed tower, where there was a second railed platform. The main tower had a triangular cross-section, and sure enough there was a ladder bolted onto one of the corner beams going straight up into the night sky. We timed ourselves to climb 100 rungs, and then fled the scene as quickly as possible, cycling back to Bradley and an innocent sleep in the dormitory. The next day we did the sums: we added the times for all stages and it was feasible. The ascent of the tower from the second inspection platform up to the top would take at least 20-30 minutes. It was essential to set a steady pace and not get tired or, worst of all, risk cramp, which would be very dangerous. If we attached any flag or visible display to the top of the tower, gratifying though that would be, it might be seen by staff at the transmission station before we got back to Bradley House. Instead we would take a pot of paint and leave a message on the railing at the top of the tower.

We waited for another new moon, and while the dormitory was deep in sleep, crept out at one o'clock in the morning to camouflage our faces and collect gloves, the paint and a paint brush. Everything worked to time, so that we arrived at the base of the ladder inside the tower within the time limit. It was a cold, brisk night, slightly overcast, but with many stars visible, and with that strange illumination which seems to pervade even the blackest night once one's eyes are dark-adapted. Now for the big ladder and the long climb.

The ladder was bolted onto the corner pillar of the tower, with a vertical strip of metal supported at wide intervals by a circular brace or loop of metal making a crude back support if one needed to stop for a rest. Apart from this single back-strip, perhaps 15 cm (6 in) wide, the ladder was completely exposed to the wind and rain, with no restraint or enclosure if you slipped or fell. We now had to go straight up the equivalent of a modern 80-storey building. The trees and bushes below us faded into a uniform darkness as we rose from the earth. Iron rung by cold iron rung we climbed towards the stars.

We took less than half an hour to reach the round platform at the summit of the mast, where an array of huge ceramic insulators slanted out into the night, connecting to the aerial some 3 m (10 ft) away from the railing. The aerial was made of multiple wires spread into a cylindrical array by spacers, and frayed bristles or spurs on the wires gave off an eerie flickering light, almost small sparks. These might have been some sort of Saint Elmo's Fire, or it may have been generated by the flickering voltage of the transmitted signals. We paused in wonder to contemplate the world, and to get our breath back. For about a quarter of an hour we gazed around and enjoyed the achievement, the exhilaration of looking down on the rest of the kingdom. We could see the glow from clusters of street lights in the towns far away stretching to the horizon, and the clouds above us seemed almost closer than the ground below. The air was very cold and fresh, but we did not feel cold.

After a brief discussion about the most incisive and condemning message to write on the railings, I painted the following words: "You need better guards on this place".

Far earlier than we had anticipated, a gleam appeared in the eastern sky, which was soon bright with the coming dawn. The sun must be just below the eastern horizon. How could I have miscalculated the time of sunrise so badly? Leaving the top platform reluctantly, we scuttled down the ladder as fast as possible, skipping some of the rungs in our haste, and were soon wrapped in total darkness again. Only at the bottom did I realize that we had ascended so high that we had been able to look over the rim of the world to the east and into the next morning. On the ground it was still pitch dark. Problem solved. Relax. But don't waste time. We sped back to Bradley.

The story gradually leaked out over the next year, but we were never caught or punished.

Some years later an engineer who had been associated with operating the Hillmorton radio station told me that it would be lethal to try and climb the masts. I feigned innocence and said nothing about our exploit. Perhaps he had said this to deter enthusiastic climbers, or perhaps it was true. If it was true, had Claude and I just been absurdly lucky? Was the transmitter not switched on that night? Was the initial jump up from the ground to the bent pipe the factor that had saved us? After this I did wonder what radiation we had been exposed to, but perhaps the very low frequency of 19 kilohertz meant that the interaction with human flesh and tissue was minimal. Modern websites on the Hillmorton array include reports and technical explanations from the 1930s and '50 s, showing that engineers did sometimes work on the top platforms while the aerial was transmitting. I have not developed cancer (yet) and have fathered two healthy children, so the radiation damage must have been small. Engineers dismantled the towers and destroyed them in 2007.

* * *

My father bought me a kayak with double paddles when I was in my mid-teens. I explored the River Thames several miles upstream and downstream from Chiswick, covering much of the course followed by the Oxford and Cambridge Boat Race. A few years later he bought John a small sailing boat, so that we both developed a familiarity with boats and water on a modest scale.

Throughout my teens I read avidly, though not with any particular intellectual focus, and the fact that I can remember many of the titles suggests that there is a residue of the meaning stuck in remote corners of my brain. The change of taste and reading from 13 to 18 years old is enormous, but rather strangely most of books that I found to hand continued to be traditional, even old-fashioned, though new books for young people must have been published since the war.

In no particular order, although one could attempt an age gradient, I read the whole range of John Buchan, including *Memory Hold-the-Door*; *Tom Sawyer* and *Huckleberry Finn*, *Peter the Whaler*, *Mr Midshipman Easy*, *A Tale of Two Cities*, *Oliver Twist*, Harrison Ainsworth's *Old St Paul's* and *Tower of London*, both the *Alice* books, most of the Dorothy Sayers' Peter Wimsey stories, one of the Hornblower adventures, Dylan Thomas, Fitzroy Maclean's *Eastern Approaches*, William Grey Walter's *The Living Brain*, Richard Hillary's *The Last Enemy*, *The Cruel Sea*, *Casino Royale*, *Night Flight* (*Vol de Nuit*), *The Little Prince*, *Treasure Island*, *Dr Jekyll and Mr Hyde*, *Kidnapped*, *Mathematical Recreations and Essays*, some of the Sherlock Holmes stories, *The Pit and the Pendulum* and all the *Tales of Mystery and Imagination* by Edgar Allan Poe, all the Arthur Ransome books, a biography of Scott of the Antarctic, Edward Wilson of the Antarctic, one or two of the Biggles books, *The Last of the Mohicans*, *The Little World of Don Camillo*, *Around the World in Eighty Days*, *The Brothers Karamazov*, several Graham Green novels, including *Brighton Rock*, much of Aldous Huxley, the pre-war Evelyn Waugh books, *Four Quartets* (which I usually carried with me when travelling), some William Blake, *Germinal*, *Kim*, George Orwell's *1984* and *Animal Farm*, one or two by Raymond Chandler, *I Claudius*, *The Time Machine* and *The War of the Worlds*, *Mathematics for the Million*, a life of Leonardo entitled *Man with Wings*, some Spender and some Auden, *Darkness at Noon*, *Heart of Darkness*, *Nostromo*, *Lord Jim*, and one or two other Conrad books, and more.

Reviewing this list which I have written from memory I conclude that books for so-called "young adults" were still few and far between. I am sure I have forgotten many books that I read, and perhaps the selection that I do recall is a biased sample that gives everything away.

I intensely disliked *The Scarlet Pimpernel*, Agatha Christie, and *Dr Dolittle*. I had a casual dislike for Walter Scott, Somerset Maugham, and the Hornblower books.

When I reached the Science 6th Form at Rugby, I automatically became a prefect, as did all members of the 6th. The idea was to give all the senior boys the experience of authority and responsibility, not just to reward the usual noble suspects. The authority granted to prefects to maintain discipline in the house would probably be regarded now as low-key institutional psychological bullying, and this may be the source of many unpleasant memories for some writers. My housemaster and many of my friends commented that it was a strange system that rewarded a well-known breaker of rules with the task of maintaining rules. But I accepted the role with aplomb, doing my best to balance the need for some discipline in the house with a strong understanding of the impulses of young rule-breakers and a sympathy for them. Since Rugby took the sensible view that scientists should not be illiterate, we were taught English literature by Timothy Tosswill, who was known as a published poet. When we "did" *Othello*, he told us on the first day to go away to read it right through as a good story, and then we would start to discuss it at the next class. The subjects that he set us for essays were always stimulating, and I adored his lessons.

* * *

During my last year I had a nasty accident. I was running downstairs from my study to get the evening mug of cocoa before prayers in Bradley House. I was in the habit of jumping a whole flight at a time, throwing myself forward to grab the railings halfway down, and then swinging through with my legs to alight on the next landing without touching the steps at any point. On the last flight, as I swung through to land on the flagged stone floor, I caught my elbow against the railing and felt a spasm of numbness. Thinking that the feeling would pass, I began to walk towards the trolley with the cocoa. The numbness spread up my arm to my head until I fainted, collapsed backward, and smashed my head into the stone floor.

When I recovered consciousness, I was lying on a bed in the matron's room, surrounded by my friends and the housemaster talking fast into the telephone. Soon I was in the school's sanatorium under the strict eye of the school doctor. I had probably been unconscious for 10-15 minutes. The next day an X-ray showed that there was an 8 cm (3 in) fracture down the back of my skull. However, I felt strong and relaxed, and after a few days, I was sent back to classes as normal. My parents visited for a day while I was still in the sanatorium, looking very concerned, but also concluded that there was nothing to fuss about, and went home again. There seemed to be no after effects.

The school doctor, who was worried that the concussion might have caused deeper brain damage, sent me to London to a hospital where they could apply the new technique of electroencephalography, or EEG. I took the train to London on my own, presenting myself at the hospital laboratory where the doctors conducted the tests. Two scholarly-looking doctors sat before banks of dials and flickering needles on a wall of electronics that was probably a very primitive computer. A multi-pen recorder was standing on a large table, surrounded by switches.

"Come in," said the first doctor. "I am Dr Dalgliesh, and this is Dr Van Steenberg. Nurse Brampton will assist with the electrodes. She is visiting us from Canada to study the methods. Please give me a brief summary of your accident."

I rattled off a description of my fall, but was mentally trying not to look at the nurse who was every 17-year-old boy's extreme fantasy of a desirable and voluptuous young woman. Single-sex boarding schools may have some advantages, but a relaxed attitude to the opposite sex is not one of them. Dr Dalgliesh made notes.

"Please lie on that couch, relax completely, and Nurse Brampton will attach the electrodes to your scalp. Don't be worried. The voltage is very low and is not dangerous."

I lay down and Nurse Brampton bent over me to start fixing a dozen small metal patches with adhesive paste to my head. Our eyes never met, but she knew perfectly well what she was doing to me as she reached over my body, and I was looking down her loose nurse's smock and shirt to her brassiere. I thought of icebergs and declined Latin nouns and tried to remember Maclaurin's Theorem and de Moivre's Theorem and recited as many digits of pi as I could recall. My physical reactions were under control, but only just. As the electrodes were stuck on my head the sound of the pen recorder scratching away got louder, and needles jerked and swung on the great computer. The doctors, who were behind me, seemed to be adjusting switches and changing connections as if the EEG machine was not performing properly.

"Nurse, please re-locate electrode 7, it is not picking up the alpha waves smoothly. A bit to the right should do it."

Nurse Brampton leant over me again, and more needles jumped and pens scratched and scribbled.

"That's about it, I think", said Dr Dalgliesh. "We can start the tests now."

The doctors gave me some arithmetic to do in my head, asked me to repeat phrases and solve simple logical problems. I had to remember lists of things, or repeat them in reverse order. Some tests were with my eyes shut, and some with my eyes open. The whole procedure took about 20 minutes. Long strips of printed paper had spewed out of the recorder which the doctors gathered up into boxes while the nurse approached me to remove the electrodes. At least for this torture session my sexual fantasies were not recorded on every dial and pen in the laboratory, but the effort to control myself was no less strenuous.

"Thank you, Mr Flemming," said the doctor. "We have some excellent data. It will take a little time to analyse the details. We will inform the Rugby School Sanatorium when we have the full report ready."

"Does the brain pattern look all right?" I asked. "Is there any obvious sign of the skull fracture or concussion?"

"As far as we can see it is perfectly normal, but we do need to work on the statistics. There are many different types of brainwaves that must be analysed."

I took the next train back to Rugby, hoping that the statistical details would not produce a printout "The patient is lusting after the nurse." As soon as possible I bought a copy of William Grey Walters' book *The Living Brain*. Amongst many useful bits of information on brainwaves and EEG it also contained the mechanical drawings and electronic circuit diagrams for a robotic tortoise that David Southward and I then built with a great deal of technical glee. Our robot Tortoise would steer itself round obstacles, but in spite of fitting it with a sophisticated photo-receiving cell we could not persuade it to home onto a bright light, which it was supposed to do.

By the time I left Rugby I had devised two maxims or principles that I felt provided a bit of guidance or warning as I progressed through the chaos of life. Firstly, really dangerous accidents strike from out of the blue when you do not even think that the situation is dangerous. In contrast, when you plan to do something that you know is potentially dangerous, you can rehearse and anticipate all the most likely hazards, and plan to avoid them, and even if something unexpected does occur, the state of enhanced preparedness and expectancy should enable you to cope. Secondly, I never wanted to feel that the great days, the happy days, the sense of achievement or success were something in the past. If I ever felt that "Those were the days," I determined to go out and do something to prove to myself that "Today is the day."

* * *

My childhood was enjoyable and helped to launch me on an exciting life. I left Rugby School in December 1953 and spent the years 1954 to early 1957 in the Royal Marines (Chapter 2), followed by Cambridge University from 1957 to 1960 as an undergraduate. Natalka Czartoryska and I were married in 1961, but we separated and divorced in 1977, having had no children. Jay Kleinberg and I married in 1979 and decided that we wanted a family. We have two children who are now grown up, and I am delighted to be a grandfather. Happy memories motivated me to try and pass on such happiness to my children, and now to our grandson. At the age of 84 I enjoy the company of children.

I married Jay in September 1979 (Figs 3.10, 3.11). We held the wedding reception in the Crown Inn, Chiddingfold, which claimed to have been a stopping point for Chaucer's

Figure 3.10. Jay and I before our wedding reception in September 1979.

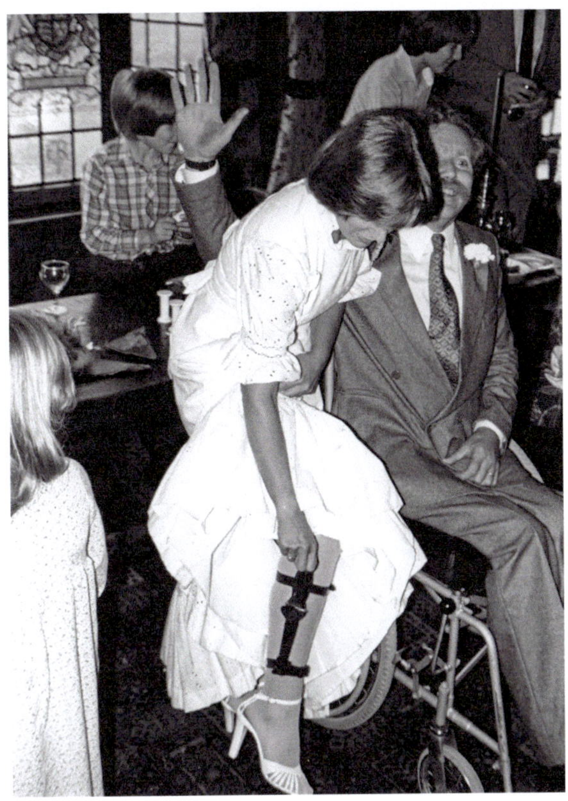

Figure 3.11. Jay prepares to cut the wedding cake with a diving knife that she had been carrying strapped to her leg, concealed by her wedding dress.

Figure 3.12. Jay and I diving at Sharm el Sheik 1979 on our honeymoon (Photo Howard Rosenstein).

pilgrims to refresh themselves on their way to Canterbury, and boasted a new wing built in the 17th century. The party included friends from the worlds of science and diving, some of whom had travelled a great many miles to join us. Jay created a sensation by cutting the cake with a diving knife which she had strapped onto her leg under her wedding dress.

Jay, who is an American citizen (now also with British citizenship), is a professional research historian and university professor, publishing at an international level, and having a busy teaching schedule. She grew up at Great Neck, just outside New York, and had greatly enjoyed the US tradition of extended children's summer camps from the age of four onwards. As a university student she had played soccer, and had been a guide or coach at children's camps where she passed on her love of the outdoor life. She enjoyed travel and camping on her own in the wild national parks. She and I drove across Europe on our honeymoon, visited remote Greek archaeological sites, swam underwater, and spent a week diving on the Israeli Red Sea coast (Fig. 3.12). We have a lot in common in our attitudes to risk, exploration, adventure, and the intellectual life of research. Jay had published a very successful book on 19th century Pittsburgh steelworkers' society and family life, and was developing a strong programme in women's history in the USA which was later taken up by Brunel University in England. However, when we got married and Jay first moved to England in 1980, she had no job offer in advance, so that we subsisted on my income of about £11,000 per year.

We bought a picturesque small house on the banks of a mill pond, just 10 minutes' drive from the NIO where I worked. Kirsten was born in late December 1981, and Peter was born in January 1985. Adding the needs of children for support and love into our complicated academic world of travel, research, and risks at sea, was a challenging new departure. It was obvious that two academics with 14-hour-a-day workaholic routines had to make some adjustments when bringing up young children.

* * *

Jay and I were determined to provide Kirsten and Peter with a sense of unlimited choice and possibilities. Life should be fun, but also intellectually and physically challenging. My parents, who were very conventional in most of their views, had tolerated and encouraged my choice of a totally unpredicted and undefined career, and I wanted to ensure that our children felt similarly trusted to make their own decisions. Jay and I needed domestic help, a nanny or au pair, to share the work of looking after a baby, if we ourselves were to continue our careers. The nannies and au pairs looked after the children until they reached their teens, and usually worked with us for a contracted time of about a year on each occasion. I didn't resent being attended by a sequence of nannies when I was young, and I don't think our children did either. We bought books on babies and child care, but since they seemed to contradict each other on so many issues, agreeing only on things that were pure common sense, we did not rely on them for very much.

Kirsten was born during a snowstorm and spent her early days at home in a bouncy-cot under the Christmas tree. Jay and I took Kirsten with us for a diving research expedition in North Australia (Chapter 10) when she was five months old. For a month she lived on a research diving boat surrounded by divers and the ship's crew. She was the official ship's baby and even had her nappy changed by a brave officer of the Royal Australian Navy while Jay was on a 60 m (197 ft) deep dive. In the first three years of her life, before Peter was born, Kirsten had visited Australia, Greece, the USA, Israel, and Italy. She loved Italian food, seeming to feel at home everywhere. On a swimming holiday in Corfu in 1984, Jay was snorkelling down from a small fishing boat to photograph the cliff, before surfacing with the underwater camera. As Jay broke the surface Kirsten, aged three, leant forward and said "Hand me the camera Mummy. I'll look after it so you can get back into the boat easily". Jay declined the offer, with many thanks, as the camera was both heavy and expensive. On a later trip, also in Greece, at the age of six, Kirsten became best friends with a little Greek girl of the same age, with whom she could not exchange a single word, but they went everywhere together, accompanied by a small donkey.

Jay is a dedicated feminist and, although we employed help with the children, we shared a lot of the tasks of infant care that traditionally would have been almost exclusively done by the mother. As required, I bathed and washed both children, changed nappies, gave them bottled milk when needed, often read stories to them in their cots, and sang them to sleep. I always did the family's weekly shopping. Jay and I both tended to work late after the children were in bed. Jay is an excellent cook, and partly from shortage of money in the early days of our marriage, and partly because I am paralysed in a wheelchair, she did most of the housework and kitchen work. I was good at washing up and cooked simple meals when necessary. Jay did most of the gardening, choosing what flowers and vegetables to grow, though I was good at clearing scrub, uprooting surplus bamboo plants, or chopping logs for the fire. Jay returned to work a few weeks after giving birth to Kirsten, and similarly after Peter's birth. On occasion, I accompanied Jay with Peter to her office at the London School of Economics to give Peter his bottle of milk or change his nappy while Jay was lecturing or attending meetings.

Peter was born in January 1985 in another snowstorm. Both children started to speak around age 17 months, and I plotted the words they used each week so that I could understand their increasing vocabulary with its grammatic limitations. The rate of growth was exponential, with a doubling time of about 5-8 weeks. When the vocabulary reached 150 words the number of new words added each day became impossible to identify or record, and I gave up. In both cases, the first 100 words were almost all nouns, but the children did not choose the same nouns. There were many differences once they got past the essentials of family and food. Sentences with verbs and adjectives came later. All children ask thousands of questions, which Jay and I did our best to answer. We grew chestnuts in jam-jars so that you could see the roots; changed the colour of stewed plums and red cabbage with lemon juice and alkali soda; Kirsten and I plotted the flowers in bloom against the date and the temperature each day, or painted landscapes together, while Peter and Jay collected mussel shells on the beach in Dorset and cooked them for supper, or found mushrooms that we could eat. Having children is an education for the parents. Peter wanted to know if you could swallow upside down. He hung by his feet from a climbing frame and ate an apple. You *can* swallow upside down. Peter and I built replica machines of Hero's steam engine and a Wimshurst machine that made sparks an inch long. Both children learned cooking from Jay, which they enjoyed. They won't starve.

We made wooden blocks painted in bright colours, on which I painted big letters. Both children learned to read by a process of osmosis, and old-fashioned "cat sat on the mat" phonics. They grew up just before universal access to children's computers so that books were the main medium of study and enjoyment, but by 1987 we did have an Amstrad (a primitive desktop computer) in the house. Jay managed the social life with neighbouring children of the same age, and the contacts with their parents, especially the organization of birthday parties.

There were many good kindergarten and primary schools in the Surrey area where we lived, and a range of excellent secondary schools, so there was no need to consider boarding school as even a distant possibility. I felt no loyalty to Rugby, and cannot recall a single event in my long career where I have, or could have, gained an advantage from "the old school tie". My social background is obvious from my accent in an English context, but I would not normally mention my school or university unless asked to do so.

I am not sure if there is a gene for wanting to climb things and having the skill to do so. But from the very earliest stages of Peter's ability to walk, he showed a compulsion to climb things: trees, climbing frames, drain pipes, lamp posts, ropes, walls, cliffs, anything. Many a time while out shopping or waiting for a car I found myself going through the same dialogue with some busy-body grown-up who was just passing by. On

Figure 3.13. Peter, Kirsten, and Morgan the dog feeding wild ducks. The lake of Enton Mill is just visible in the background (Photo Jay Kleinberg).

seeing Peter perched high above the ground, the busy-body would first allege that the wall or tree had not been put there for Peter to climb, then that he would damage the wall or tree or lamppost, and then that he might hurt himself. I agreed with all these comments, but added that he would then learn to climb more carefully. This inherited climbing skill is all the more curious since Peter has only ever seen me in a wheelchair, and I had never told him of my climbing exploits when I was younger.

Life in our house by the pond provided frequent diversions (Fig. 3.13). The railway ran so close by that after a mail robbery in a nearby town the villains threw the empty sacks out of the train window and they fell into the pond. The children were fascinated when the police came to search for the incriminating evidence. One winter several trees fell across the pond in a storm, and we took our little rowing boat out and I cut up the trees with a chain saw. The wildlife was rich with variety: herons, many species of ducks, crested grebes, coots, and on one occasion a swan. When the pond froze over in winter the children, accompanied by the au pair, would slide and pirouette and fall over on the ice again and again with endless energy and laughter.

* * *

At the age of seven years, Kirsten accompanied me to a BSAC diving exhibition where there were instructors who would give people an experimental dive in the swimming pool to see if they liked breathing from an aqualung and were happy underwater. Kirsten, after being helped into the equipment by instructors, made a very relaxed dive (Fig. 3.14). Somebody who dived in these circumstances merely to please their parent would have shown some nervousness or strain. But Kirsten took to the underwater world naturally, and she was happy. Kirsten greatly enjoyed riding, owning a series of

Figure 3.14. Kirsten, aged seven, trying on scuba diving gear at a British Sub Aqua Club exhibition. Children this age are not encouraged to dive with scuba in open water, nor to dive deep, but there is no harm in discovering at an early age that scuba can be enjoyable.

ponies into her late teens, until studying for exams had to take priority over mucking out the stables. She is a good diver, qualified to HSE Part-IV (Chapter 8), and still takes diving holidays to explore the sea, as well as providing architectural survey and drawing skills to assist the Pavlopetri underwater city survey (Chapter 6).

The children asked different kinds of questions and needed different kinds of support. Kirsten was obviously artistic, but she did not want to paint portraits or landscapes. She loved machines as objects of pure form and colour. We found books of paintings by Charles Sheeler, who, in the 1920s, painted majestic images of the Ford motor factory, but also hydropower station turbines and other vast machines. Other paintings from the inter-war period, especially Russian ones, showed the grandeur of industrialization, with heroic workers pointing to a fabulous future. None of these paintings captured the mood that Kirsten wanted. She was fascinated purely by the strange shapes, colours, and textures of machines, which manifestly had a purpose, and were not natural, but the purpose itself did not need stating or explaining. She loved to paint huge abstract compositions of bits of car engines or ships. We collected car parts from wrecking yards so that she could paint them in arrangements that were aesthetically interesting, but with no engineering reality.

Throughout the 1980s and '90s I was travelling abroad on work several times each month on trips that were necessarily solo (Chapter 9). On one occasion I was in the kitchen discussing with Jay my next journey when Peter, who was six or seven at the time, suddenly threw himself on the floor, clutched my ankles and pleaded with me like a bad actor from a 1920s silent movie with a printed curlicue text on the screen:

"Daddy, Daddy, surely you know that if you keep going away like this it will harm Kirsten and me and damage our psychology!"

I looked down at him in astonishment, and he slowly turned his head upwards, grinning wickedly at me, then let go of my feet, laughing with triumph at having fooled me. Both children accepted my repeated travels, which were usually only three or four days at a time, though sometimes longer. A mixture of presents from strange places, and a sense of adventure, supported by stories and souvenirs, with occasional telephone calls from China or Chile, seemed to keep them amused or help them to accept that I needed to travel.

When Kirsten and Peter were aged between 5 and 12 years we realized that a beach in Britain was the same as a beach in Morocco or Israel from their point of view, so that there was no reason to spend many thousands of pounds just because it was the done thing for a status holiday. We had a small family cottage in Dorset which I shared with my siblings, and in the 1990s we spent many happy holidays there. Jay took Kirsten and Peter on long bicycle rides, while I met them with the car at the other end of the route where we had a big picnic. I would drive the car as fast as possible through fords in the streams on the minor roads because the children liked the big splash. You could wade along the streams from one village to the next. Jay walked with the children the 11 km (7 mi) along Chesil Beach from Abbotsbury to West Bay, where they then swam in the sea. I love seedy Victorian holiday resorts, piers, candyfloss, Orwellian vulgar postcards, and unpretentious bad food, so these trips pandered to my own sense of self-indulgence.

* * *

Jay travelled with Kirsten to New York for a short holiday, and no doubt the impact of the Manhattan skyline had its effect. Kirsten was developing a fascination with architectural form on the grand scale. For her 17th birthday, Kirsten and I went to the new Guggenheim museum in Bilbao where we admired the strange metallic curved surfaces designed by Frank Gehry. Another architectural trip was a day's escorted excursion round the new elliptical terminal buildings of Charles de Gaulle airport in Paris. The terminal manager explained how the spaces were designed to be coldly hostile in the check-in area so as to discourage accompanying family members from lingering, and then became more and more comfortable and welcoming as the real passengers moved towards the departure lounge. Kirsten learned about the psychology of a public building.

Peter was extremely logical and literal, and was, at various ages, obsessed with pure mathematics, irrational numbers, transcendental numbers, logic, philosophy, and languages. Before he was six, we passed the time on long car journeys by solving mathematical puzzles and quadratic equations in our heads. At the age of eight he objected to writing with gaps between the words because he said that there were no gaps when he was speaking. He insisted that I justify the use of gaps. I had to write to Professor Alan Baddeley, an expert on applied psychology, to find the answer to that question. Peter was very impulsive, tending to focus on what interested him most, and ignoring routine work that he was supposed to do. His teachers were frustrated by his habit of doing well only when he wanted to. Jay and I did not intend to send either of the children to boarding school, but, in an echo of my own childhood, the preparatory school where Peter was being taught as a day-boy recommended that he should try for a scholarship at Winchester College, which he did obtain. Thus, like me, he spent his teenage years in an old-fashioned British boarding school with traditions going back centuries. By the time he went to university he had decided to study theology.

* * *

In 1999 there was a total eclipse of the sun that passed over southwest England from Cornwall to Dorset and then cast its shadows over the English Channel before moving into northern France. A total eclipse enables you to look straight at the blacked-out sun when the brilliant silver fronds of the white-hot solar corona are visible, twisting and turning against the dark sky. On the web I found a company selling near-black welding glass that was guaranteed to be safe for watching the increasing eclipse before it was total, so I ordered four pieces, then booked a trip on a cross-Channel ferry that would be directly in the path of the eclipse. We drove to Portsmouth and left the car at the dockside while we made the journey to Le Havre and back. The ferry company had anticipated the public interest, and created a special eclipse-themed lunch menu and eclipse cocktails. Our family had a wonderful party on the ship, during which we watched through dark glass as the moon nibbled away at the diminishing sun so that we could finally stare directly at the black circle surrounded by quivering silver snakes. Medusa-like and cool.

As Kirsten and Peter reached their mid-teens they began to take summer holidays with their friends, so that the routine of a tidy family fortnight in the country cottage came to an end. Jay and I did not wish to lose contact with them at this interesting age. At one level it was essential to let go of them, allowing them to develop independence; at another we wanted to share new and more grown-up travel experiences. We proposed adventurous family holidays that would probably have been beyond their own ability to organize or afford. For the next six summers they arranged their teenage escape holidays for themselves first with their friends, after which we enjoyed a family adventure expedition.

These family trips began with a snorkelling holiday by the beach at Frangokastello in southern Crete in 2000, complete with diving and digging up a bath sponge that we still use. The following year we went to St Kitts in the West Indies, to stay at a hotel which provided a scuba training school. Kirsten and Peter both completed the basic course quickly, while the dive supervisor asked me to conduct some tests in the pool to demonstrate my competence. I complied willingly. He was right to ask me. We dived *en famille* on several wrecks, both ancient and modern. Jay and the children went on adventurous excursions in the hills, where they climbed a large jungle-covered volcano, and at the summit they could look down into the caldera. In 2002 we spent two weeks on a converted Russian research ship exploring the glaciers and walruses of Svalbard,

Figure 3.15. Jay with Kirsten and Peter during a diving expedition off the island of Corfu in 1987.

circumnavigating most of the archipelago, and seeing vast cliffs completely covered in screaming seabirds. Next, a drive along the whole Mediterranean coast of Libya visiting underwater ruins (Chapter 11); then a tour of China, including four days on a ship on the Yangtze River and a visit to the terracotta warriors at Xian; and finally two weeks on the volcanic island of Rodrigues in the Indian Ocean in 2005. To add to the sense of anything-is-possible adventure, I took the children once to Lyneham airport, home of the British Gliding Association, where they were both able to fly in gliders as a try-it-and-see experience for beginners.

In spite of the occasional conventional crises, our children made things easy for us. They accepted our way of life, and tolerated our eccentricities and commitment to work, while effortlessly adapting to travel, adventurous holidays, rock-climbing, strange food, boats, diving (Fig. 3.15), climbing volcanoes, and skiing. They might have demanded more daily personal attention, but they did not. In their teens and later their lives diverged from ours in many ways in terms of attitudes, views, and priorities, but they always entered into our lives with full acceptance of the way we wanted to live. There was no sense of rebellion, or that they constructed their lives in a deliberate rejection or defiance of ours. They had and have their own lives, their own styles and range of friends and interests, but there was much that they were happy to share with Jay and me.

I don't wish to over-simplify or whitewash the complications of bringing up two children with strong characters and very different natures and temperaments. We suffered the usual childish setbacks and afflictions, tantrums, disobedience, repeatedly forgetting simple chores, childhood diseases, myalgic encephalomyelitis, hyperventilation, under-age sneaking into bars and night-clubs, conflicts at school, some unfortunate friends, and so on. But our children survived, and so did Jay and I. These things are normal: it is the development of unique character that is interesting.

Jay has shown untiring support for me and helped with the inevitable extra chores caused by my disability (Chapter 7). We have done our best to prepare our children for the modern world, but an extra burden certainly fell on Jay. We have both felt that our relations with our children have been distinctly different from that of our own childhoods. While not falling into the trap of thinking that one's young child is one's best friend, there has been an openness and friendliness in discussion and sharing problems and difficulties that we both feel is quite different from that of our relations with our parents. This has not been due to any theory of upbringing but was probably the cumulative effect of cultural changes in the 1960s, and the increased informality of the later decades. The children's friends also showed much more open displays of friendship and affection between each other than when I was young, and this attitude had its influence on Jay and me. As I said, having children is an education for the parents. Kirsten and Peter are in their thirties now, and their careers are totally in their own hands. It is a joy to see them prosper.

4

Caves Beneath the Sea: Palaces of the Sea King

I sat on the seabed at a depth of 10 m (33 ft) off the west coast of Corfu, in the bay just south of Paleokastritsa, where the Club Mediterranée instructors conducted their open-water tests for beginner divers. It was July 1957. Paleokastritsa was famous as the first exclusive diving holiday centre, where Chief Instructor Marcel Contal supervised training and safety. I had already experienced more than a year of diving on oxygen in the Royal Marines before converting that training to civilian compressed-air scuba techniques through my membership of the British Sub Aqua Club in London. Compared with most of the other trainees at Club Med, I was an experienced, almost professional, diver.

It was before sunscreen skin creams, before wetsuits, before single-hose regulators, before divers' cylinder-contents pressure gauges, and before divers' life jackets. In minimal swim trunks and bikinis, the class sat on a rock terrace at a depth of 10 m (33 ft) lit by the translucent filtered sunlight that glimmered off the pale rocks and glinted on the ultramarine blue shoals of tiny darting castagnette fish. The instructors carried out the advanced manoeuvres of removing masks and replacing them; removing mouthpieces of the twin-hose Mistral regulators and replacing them; and finally taking off the tanks and replacing them. Some of the dozen or so pupils struggled or had to repeat the exercises. I did the tests and had time to sit back, outside the main group, to enjoy the view, thinking about what I saw.

My eyes registered the flat terrace that met the cliff behind the trainees at a sharp junction like the floor and wall of a room. There was a slight curved scalloping where they met. The terrace continued along the cliff and disappeared into the blue mist to left and right. It was not natural for the join to be so sharp, straight, and horizontal. I did not form this as a question but as an unexplained anomaly. I watched bubbles floating up from the other divers. The domed bubbles expanded, twisting, oscillating, shining, breaking in two, with multiple internal reflections and bursts of foam at the surface. My mind wandered, but the anomaly of the cliff and terrace lurked somewhere in my brain.

I surfaced and boarded the diving caique with the rest of the class, removing my gear while the Greek boatmen heated the engine with a blow torch, thus persuading it to ignite the viscous fuel. A hefty hand crank of the flywheel, and with a violent "put...put...put" the motor started, making the boat vibrate. It leapt forward with small accelerations at each detonation of the fuel thumping the propeller. We cruised back to the club

In: Flemming, N.C., 2021, *Apollonia on my Mind. The memoir of a paraplegic ocean scientist,* 137-180. Leiden: Sidestone Press.

village. There were about 50 *gentils membres* or guests. We lived in circular straw huts in a glade by the sea, shaded by olive trees and cypresses. Loudspeakers hidden in the trees played classical music in concerts carefully planned to satisfy discerning tastes each afternoon. The cicadas in the trees maintained a merciless clicking throughout the Mozart wind octets or Bach's Preludes and Fugues.

The restaurant was above the shore on a terrace in the hillside, and the food was perfect. Local fish was often on the menu. The wine was excellent. Life was good. Spearfishing was still socially acceptable, provided that the hunter was not wearing tanks. That was regarded as unfair on the fish. I believe that at this date the British expatriates in Sri Lanka, who were reputed to be very successful and sporting spearfishermen, maintained that a true gentleman never shoots a fish smaller than himself. This rule did not apply at Paleokastritsa, but fish of 5-10 kg (11-22 lb) were quite common. I felt comfortably at home in the village, where almost everybody else was French, since my spoken French had improved greatly since my days on Belle-île.

As our underwater skills improved, the groups of more adventurous divers voyaged in the caiques to huge caverns that we could explore by diving. There were soaring arches of rock beneath the sea and whole islands permeated with submerged caverns so that you could see a blue glow of light shining through from the other side. In some places there were rows of caves with their entrances all at the same depth, like the succession of ogive arches and pillars of a buttressed cathedral. The divers loved the caves for their richly coloured encrustations of sponges, the coral, and the shoals of small blue iridescent fish. In the deep shadows you could shine a torch on the cave wall which would spring to life, dripping with neon-bright colours of scarlet, orange, and yellow. We refrained from picking the ruby coral so that we left the beauty for the next visitors. I tended to wander off to explore a bit but always stayed within the visibility range of the instructors. Usually we went only a few yards into a cave, hovered in the darkness, and then turned to swim out again. I began to notice that several of the caves had side tunnels, which turned upwards at the back like a chimney. Sometimes these chimneys curved over and opened out on to the cliff face higher up. Looking up at the back of the cave roof you could see the blue light filtering down the chimney, glowing faintly.

When I found a large chimney or rising tunnel, I swam slowly up it towards the daylight, feeling my way carefully with fluttering kicks of my fins. The rock walls narrowed, closing around my shoulders so that I brushed against the encrusting shells and sea-urchin spines. I managed to squeeze out through the window in the cliff. My tanks almost jammed between the narrow walls. Then I was in open water. I was still 10 m (33 ft) below the sea surface but had climbed up inside the cliff within the chimney.

What on earth could punch a big hole at the base of a cliff 20 m (66 ft) under the sea, then carve a chimney inside the cliff, and break out through the cliff face in a "window", all still under the sea? How? And why? I was not thinking of solving a scientific problem, but just conscious that there was an anomaly. A mistake in nature. There was something going on that was inexplicable. The instructors were very tolerant of my excursions, and I tried always to keep as close as possible to the group, to converge with them as soon as I had finished one of these "chimney" ascents. I always obeyed the instructors' orders, and they gave me a bit of extra licence.

The crews of the caiques were fishermen who relaxed in the lunch breaks by congregating on one boat where they sang songs of eerie melancholic dissonance. The songs were mostly slow tempo, and, if the Greek words had any relation to the melodies, they told of loss, sorrow, passion, slaughter, treachery, unrequited love, and the honour of revenge. Sometimes two men would stand holding a kerchief between them, dancing very slowly with pirouettes turning cautiously and gracefully in the tiny deck space. The fishermen knew the coast in all its deceptive bone-white beauty. Often, near the steepest cliffs, when we seemed to be safe in deep indigo-blue water, you could see a submerged pinnacle rising almost to the surface, glimmering palely, threatening to wreck the unsuspecting navigator. These pinnacles were described by the French instructors as *secs* or *secca*. Why were they there? Since the fishermen knew every *sec* for many miles around, we were never in danger.

Back at the bar, or in the restaurant over a meal, I chatted with the others about our exciting dives. To my surprise, at the start of the second week, a novice *gentil membre* arrived whom I knew. He was Ewen Fergusson, a young man who had been at Rugby at the same time as me, a few years ahead, and both a notable scholar and head of school as well as a brilliant athlete. His distinguished career later led to his posting as Ambassador to France. He was travelling with a colleague from the Foreign Office, and we soon formed a busy trio, eager for more diving adventures. By this time I had done a lot of snorkelling and thought that I might be able to dive down to 30 m (98 ft) on a single breath. With Ewen and his friend we measured out 30 m of rope, then swam out into the bay, where we lowered a stone to the bottom at exactly 30 m. I hyperventilated on the surface before diving, and then swam as fast as possible straight down the rope, kicking my fins rapidly. I was surprised when I felt my rib cage compressing to its minimum size, then locking rigid, while my guts started to compress, pushing up under the ribs. This I had not anticipated. Also, I realized that because the volume of my lungs had compressed so much, my whole body now had negative buoyancy.

I hit the seabed going fast. My knuckles crumpled into the mud and stones. I grabbed a fistful of gravel and started to kick my fins upwards. The water was dark dusk-blue, and the silver surface of the open air was a long, long way up. Because I was so heavy, I had to kick all the time, generating carbon dioxide in my lungs, and I could not rest for a second because I would start to sink back. As I ascended, my lungs gradually expanded again so that I attained neutral buoyancy. The last few metres were traversed in an explosive burst of frenzied kicks and near unconsciousness. My lungs were screaming for fresh oxygen. As I shot out of the water like a Polaris missile, I opened my mouth and gasped for air. It took me a few more breaths to recover enough to show Ewen the handful of gravel. We swam back to shore to celebrate in the bar.

Breath-hold diving has been done for millennia by sponge-divers and purple murex-divers, and my dive to 30 m (98 ft) was no record even in 1957. It was a personal best, attempted out of personal curiosity. In the following decade, competitive spearfishermen dived deeper and deeper, causing an unfortunate spate of fatal accidents. Doctors brought this under control by careful medical supervision and increased understanding of the physiological effects as the lung volume contracts up inside the rigid rib cage. By the early 1970s, divers like Jacques Mayol were breath-hold diving to over 60 m (200 ft), and the present world record is of the order of 120 m (400 ft). That breath-hold diver was submerged for more than four minutes.

A few days later I tried another inadvisable manoeuvre. I had got into my head the desire to share a single aqualung with a buddy, with both of us breathing from his regulator, and to swim downward together to 40 m (130 ft) depth, and then for me to swim away from my buddy so that I could enjoy the feeling of being like a fish with no heavy equipment. I would have no heavy tanks. Just me, calm, and the deep sea. This would mean passing the regulator mouthpiece backward and forward several times as we descended, until I would swim away after taking a last good big lungful of air from the tank and make a very slow ascent. The instructors did not object, so the next day I made the experimental descent with a buddy diver.

When we had reached 40 m (130 ft) we stopped to relax. There were several other divers in the water breathing from their tanks, and an instructor hovering somewhere in the background. The deep blue descended infinitely below us. I took one final big lungful of air from the Mistral regulator before swimming away. The freedom was amazing. No straps, no tanks, no harness, no weightbelt, no regulator mouthpiece (and no diving life-jacket, since they had not been invented yet). My body felt light and sinuous, truly weightless, with every muscle able to move without stress or contact with anything other than water. I seemed to slip through the sea like a fish and feel the cool contact over my whole body like a communication, a touch. Other divers seeing me finning gently, cruising past them at 40 m (130 ft) with no tanks looked surprised. A beautiful diver whose billowing waist-length red hair was magenta-purple in the blue water undulated past me. I had often admired her. I spun and cartwheeled slowly with pure joy.

After swimming a few dozen metres so that I was well clear of my buddy companion I turned slowly upwards with a lazy kick of my fins. As I rose the air in my lungs expanded and I gained buoyancy so that I could accelerate with no effort. I opened my mouth as if whistling, to let the air escape: all divers learn to do this during free ascents. If you do not breathe out your lungs will expand and burst.

I was now doing a sort of upside-down parachute ascent. With no effort I controlled the rate of exhalation, maintaining a slight buoyancy. Since the air was flushing out of my lungs all the time the carbon dioxide was being washed out, and I had no desire to inhale. The sense of pure control and effortless movement upwards through the ocean was restful and calm. I was completely relaxed, not needing to move a muscle other than to regulate the air flowing continuously past my lips. Fish and divers drifted past as I floated towards the light. The shades of blue shifted from cobalt to silvery sky blue and paler still. The temptation to prolong the experience for as long as possible was irresistible. But if I breathed out too much and stopped the ascent the air in my lungs would start to build up CO_2, causing me to feel the necessity to breathe, which would be disastrous. All too soon I reached the surface and breathed out the last bit of air that I had inhaled at 40 m (130 ft) depth. The sun seemed too bright.

Independently, the Royal Navy later developed and greatly improved this technique of steady free ascent with air inhaled at depth as a means of escape from submarines. Ronald Flanagan was the brave man who made the trials of the very risky procedure from extreme depths.

By the time I left Corfu the anomaly of the caves and pinnacles was firmly lodged in my mind. It was at the back of my mind, I admit, with no urgency, but it was unforgettable. I just knew that there was no force under the sea now that could create huge vaulted caves, flat terraces, vertical cliffs, rock arches, tunnels, chimneys, or the rock pillars called *secs*. But they existed. There was something wrong.

* * *

During my second year at Cambridge University, after the mapping of Apollonia and my visit to the NIO, I had learned that coastal earthquakes and global sea level change could be equally responsible for Apollonia being under the sea. I had also read about the ice ages that had occurred tens of thousands of years ago, causing the sea level to drop more than 100 m (328 ft). In the Easter vacations in 1959 and '60, the CUUEG divers went to Weymouth and Portland to make training dives in the open sea. British seas are cold and murky at that time of year, but novices need some practice at sea after training in the swimming pool during the winter. The cliffs of the Dorset coast are famous for their pinnacles and arches like Durdle Door and Old Harry's Rocks. I began to think that the cliffs, caves, terraces, arches, and chimneys that I had seen under the sea in Corfu had the same origin as those that were the scenic tourist attractions of the Dorset coast. Only waves breaking and thundering against a cliff at the sea surface had the force to erode solid rock. Was it possible that cliffs created and eroded by low global sea level during the ice ages had been inundated and preserved on the seafloor? Or would they have collapsed into rubble and debris as the storm waves at the sea surface rose over them? I began to see the same features everywhere on the cliff coasts of England. I combed through the photographic library of the IGS (now the British Geological Survey) in the gorgeous old mahogany-panelled cabinets of the Geological Museum in South Kensington seeking pictures of coastal caves in different types of rock. I searched stretches of coast from Mull in Scotland to Devon and Cornwall, always studying the shapes of caves and the junction between the cliff and the eroded terrace at its base. Again and again I saw the same features that I had seen under the waves around Corfu. The caves sometimes had "blow-holes" fractured high up in the cliff by the explosive pressure of waves breaking inside them. These were the "chimneys" that I had explored underwater in Corfu. By 1960, when submitting a chapter for a book edited by Sir George Deacon, I could write with conviction:

> There is yet another branch of submarine archaeology that we have not mentioned at all – the study of caves. Although little has been done in this line, my own hope is to find signs of human occupation in caves down to depths of 150 ft or more. A cave at a depth of a hundred feet below present sea level would have been high and dry from 70,000 BC to 18,000 BC and could easily have been inhabited for that period. The prospect of excavating submerged cave dwellings, and thereby linking the archaeological periods of the stone ages with eustatic (global) changes of sea level, is one of the most exciting possibilities of the near future. (Flemming, N., 1962, 'Sunken cities and forgotten wrecks', in G.E.R. Deacon, [ed.], *Oceans*. London: Hamlyn, pp.123-171 [quotation p.171]).

Eureka!

All the caves deep down under the sea off Corfu could have been caused by waves in ancient seas when the level was lower than at present, and some of the caves could have been inhabited by Stone Age people during the ice ages. How could I discover if this idea was true?

* * *

Writers as varied as Jules Verne and Oscar Wilde have used submarine caves as symbols of mystery, danger, magic, and the entry to hidden kingdoms. I have found Wilde's story *The Fisherman and his Soul* to be the most perfect example of the genre. A passage from that story haunts me in its fantasy of the underwater world:

> And (the mermaid) sang a marvellous song. For she sang of the Sea-folk who drive their flocks from cave to cave, and carry the little calves on their shoulders; of the Tritons who have long green beards, and hairy breasts, and blow through twisted conchs when the King passes by; of the palace of the King which is all of amber, with a roof of clear emerald, and a pavement of bright pearl; and of the gardens of the sea where the great filigrane fans of coral wave all day long, and the fish dart about like silver birds, and the anemones cling to the rocks, and the pinks bourgeon in the ribbed yellow sand.

Rather like Peter Simple regretting our desecration of Rockall by taming it and removing the gothic horror of its ancient legends, I felt shame at the thought of using underwater caves for pragmatic or scientific purposes.

I proposed that a systematic mapping of many underwater caves would reveal logical patterns, with most caves aligning up at distinct depths, and that these would indicate the long periods of stable low sea level during the ice ages. I expected also to find groundwater channels, underground rivers, with signs of freshwater often flowing into the back of caves. That should produce extra tunnels to complement the "chimneys" or blow-holes that I had already discovered. With luck the divers might also find stalactites and stalagmites that had formed when the caves were above the sea level, when the sea surface was even lower. These observations from below the sea would add a mass of detail and consistency to the very crude models that already existed for the rate of rise of sea level after the Ice Age, revealing how the continental margins were flooded. Since there had been several separate glaciations during the last million years, when the sea level dropped to 100-135 m (330-443 ft) on each occasion, there could be caves and shorelines from different times, each containing stalactites and stalagmites of different dates. I proposed this research as part of a dissertation topic for my PhD at Cambridge, and it was accepted.

The western Mediterranean, which was more accessible than the east, had many marine research laboratories dotted along its coast that I could use as bases for diving. Also, the western Mediterranean was less prone to earthquakes, so that the alignment of caves relating to a given sea level would not be distorted. I planned to start at Gibraltar, working slowly around the coast to Italy over the course of three years, diving in the Mediterranean for a few weeks or months during each year, and working on the data back at Cambridge during the winter. I would support the fieldwork by wide reading in the nascent literature on global ice age sea level change, which was beginning to develop in the geomorphological research communities in several countries. Dr Richard Hey in the Department of Geology was my supervisor, but I had an office in the Department of Geography, which was renowned for coastal research, where they also had two hydraulic wave tanks for conducting experiments with waves breaking on different materials. I wanted to see how waves actually attacked cliffs and the patterns of how they broke inside the caves using models in the wave tanks. Richard Chorely, who was a British pioneer of quantitative geomorphology, became one of my mentors. He claimed that he had never used a photograph in his lectures, only measurements.

I started with a small team, diving around the British Royal Navy base at Gibraltar in 1962, and then progressively visited the French Marine Station at Banyuls-sur-Mer near the Pyrenees, various French laboratories along the Côte d'Azur, from Marseille to Villefranche, the zoological station at Naples, Palinuro in southern Italy, and a few dives in Sicily and the Aeolian islands.

Dr John Waechter of the Institute of Archaeology of University College London was famous for his prehistoric excavations from 1948-54 in Gorham's Cave on the east coast of Gibraltar. He had been a Hurricane pilot in the RAF during the war, with a moustache to match, and took a practical view of getting things done. Gorham's Cave is located at the base of a colossal, high limestone cliff facing eastwards onto Governor's Beach, which can only be approached from the sea, or by descending the cliff on ropes (Fig. 4.1). The prehistoric deposits of bones in the cave going back at least 100,000 years made

Figure 4.1. Gorham's Cave facing onto Governor's Beach, on the east side of Gibraltar. The deposits in the back of the cave contain archaeological remains dating back more than 100,000 years. The people who occupied the cave could only have done so when the sea level was much lower.

no sense unless the sea level was lower, allowing the occupants to hunt on the land which is now the seafloor in front of the cave. The prehistoric inhabitants of Gorham's Cave, and several other caves in Gibraltar, were Neanderthals. The first discovery of a Neanderthal Skull was in Forbes' Quarry at the foot of the north face of the Rock in 1848, but this was not published until 1865, nor was it classified as Neanderthal until 1909, many years after the discovery of a similar skull in the Neander Valley in 1856.

Since the waves now lap close to the mouth of Gorham's Cave, Waechter had independently seen the potential importance of underwater archaeology, hoping that divers could study the submerged land where the occupants of Gorham's Cave had hunted for their food. At the age of nearly 50, he had just learned to dive but did not have the experience to organize or manage a diving team. He offered to recruit and pay for a diving team to work around the Rock for three weeks if I would do the planning and dive safety management. After a reconnaissance dive off Gibraltar during the summer of 1961, we did the main work over Easter in 1962.

Accurate navigation and position fixing was still a primitive art compared with modern GPS, but not as difficult as it had been at Apollonia. In Gibraltar we could charter a small powerboat with a forward cabin and make position fixes from the boat relative to the shore. I used a method known as horizontal sextant angles. This is a technique used by coastal surveyors and early chart-makers going back to the 18th century. Notwithstanding its ancient pedigree, it can give accuracies of a few metres at distances of several kilometres from three known fixed points on shore. In 1962 it was still one of the best coastal survey methods if one was out of range of the Decca navigator electronic coverage. We purchased a Ferranti Ferrograph pen-recording echo sounder to measure the water depth and seabed profile on a series of tracks radiating out from the shore into deep water in all directions, terminating in about 100 m (330 ft) of water. This scheme would reveal any terraces, cliffs, or fossil beaches, which we would also dive on to study in detail and take rock samples from those shallower than 60 m (198 ft).

* * *

Waechter recruited an eclectic team of divers, some of whom joined us directly in Gibraltar. The team included me, John Schofield, who was an electronics engineer from Imperial College Diving Club, Herb Greer, an American photographer and writer with diving experience in Turkey, and Waldemar Illing, known as Vladi, a very experienced diver. The governor of Gibraltar gave the project his blessing.

On a grey Sunday morning in March, with a biting wind blowing rain off the Essex marshes into a maze of iron warehouses and derelict cranes, Waechter, Schofield, and I met at Tilbury Docks to board the British India Steam Navigation Company ship *Uganda*. During the next four days, as the ship wallowed through a stormy Bay of Biscay, Schofield and I worked through our preparatory reading and paperwork. The gigantic mass of the Rock emerged from the morning mist and drizzle as *Uganda* dropped anchor in the bay. The levanter wind was blowing clouds over the summit ridge while blustery squalls whipped flecks of foam off the sea. This was not good diving weather. My spirits were revived by the vintage steam tug with polished brass fittings and mahogany woodwork that came alongside to ferry passengers and luggage to the shore.

Our accommodation was scattered. Waechter and Schofield stayed at the Continental Hotel while I stayed with a friend known as Tony the Cliff-climber because his ambition was to climb the north face of Gibraltar, but he could never get permission to do so. Illing and Greer stayed in other small hotels when they arrived. We made many friends over the following weeks, including Corporal Jim Brazier, who had the responsibility of looking after all the tunnels in Gibraltar; Mr Lamb, who looked after the spare gun barrels for warships and provided us with compressed air from the torpedo sheds; Johnny Stagnetto, who owned an upmarket bar and club called The Barnacle; and Mr Lucas, who provided me with very large-scale maps of Gibraltar. I was invited with Waechter to join the guests for cocktails on the visiting Royal Navy submarine *Alaric*, during which party I found that one of the officers was John Marshall, the older brother of Claude Marshall, with whom I had climbed the Hillmorton radio masts. We drank a "horse's neck" in memory of that event. The other local guests on the submarine provided further introductions around Gibraltar.

While Waechter pulled strings and used his influence to obtain boats for our work, Schofield and I organized our gear in the store sheds provided for us. Waechter's efforts produced a 4 m (12 ft) inflatable dinghy with outboard motor on loan from the Royal Army Service Corps Water Transport Company and, a few days after that, a very useful 8 m (25 ft) motor launch with a sturdy diesel engine and covered cabin and foredeck. The launch was loaned to us by the Dean of the Cathedral of Holy Trinity, Diocese of Gibraltar. It officially belonged to the Mission for Seamen and was called the *James Johnson*. I always thought of it as the *Reverend James*. Schofield and I had already made exploratory dives from the shore around Rosia Bay when Greer and Illing arrived to bring our team up to strength. Greer was an excellent photographer who had recently published a documentary book on the French war in Algeria that had been well reviewed, and he looked the spitting image of Lenin, with a shaved head and small beard. The similarity was so striking that Russian sailors who were visiting Gibraltar to spy on NATO exercises could scarcely take their eyes off him. Illing was a strapping athletic and imperturbable young man who was always ready to work on the next urgent task, filling air cylinders, fixing the outboard, or splicing a rope, as well as being a very good diver. We were an oddly assorted group but got on well together.

Our early dives from the dinghy revealed numerous small caves and crannies around Rosia Bay and the Three Sisters Rocks down to a depth of 20 m (66 ft), but they were mostly just weak fractures in friable and fragile rocks, not heavy erosion indicative of ancient coastlines. We explored the inner corners of the caves using a floodlight powered by a car battery on the boat. The water was cold after the long winter, so we needed our wetsuits (they had become available since 1960), and visibility was miserable by Mediterranean standards, just 5-10 m (16-33 ft). The weather was unreliable with quiet sunny spells broken by sudden squalls or hours of light wind and drizzle. We rapidly learned the dangers of the local currents and learned to nurse the dinghy through the choppy seas. Schofield and I made a dive off Little Europa Point to 25 m (82 ft), descending a steep rocky scree to a gravel bottom rich in undulating orange and white sea pens fluttering their translucent white polyps in the tidal stream.

At the southern end of the Rock of Gibraltar, there are several broad flat limestone terraces, descending like vast steps from the summit of O'Hara's Battery. These are ancient coastal abrasion platforms and are testament to the slow steady uplift of the

Rock due to the plate tectonic movements between Africa and Spain over many tens of millions of years. Each terrace is a previous shoreline, and the lowest one is the Europa Flats at a height of 30 m (98 ft) above present sea level. We needed to dive to the west-southwest from the cliff 200 m (660 ft) short of Great Europa Point to see if there were submerged terraces and possibly caves in that direction.

Dawn next day blossomed into a beautiful breathless morning as the admiral's flag hung motionless against the mast on the dockyard tower. We headed the inflatable towards Europa Point. Illing and I dived while Schofield tracked our bubbles from the dinghy. The sea was smooth with a calm forecast. Illing carried a ranging pole and I had a drawing board as we descended steadily over a rocky slope to 7 m (23 ft) depth where there was a low undercut cliff dropping to a slope of sand at 22 m (72 ft). A useful start. The water was rich in fine sediments giving it a pale khaki colour intermingled with fluffy particles of silt and plankton. Steering on a compass bearing, we toiled across a sandy plain, which grew more stony as we progressed. By watching our track across the seafloor I could tell that we were drifting south, while from every stone the heliotrope fronds of pastel-coloured gorgonians fanned out, vibrating tautly in the current, and the orange cylinders of the sea pens bent before the flood. Illing and I frequently exchanged the "O" hand sign for "all OK".

Suddenly a colossal shape loomed out of the darkness, which I thought at first was a shipwreck. But soon I saw vast overhanging slabs of rock ledges, stacked one above the other extending upwards towards the surface, creating a tall cliff overgrown with gorgonians and fanworms. An octopus recoiled into its lair and a silver fish, which the locals called *leche*, flashed by in the shadows. My heart pounded with excitement. This is what I had come to find, the cliffs and caves from thousands of years ago buried in a tomb of water.

I finned rapidly up the cliff face watching my depth gauge to measure the height and then descended again to find Illing delivering the *coup de grace* to a large langouste (spiny lobster). The top of the reef was at 10 m (33 ft) depth and the bottom of the cliff at 20-22 m (66-72 ft). We explored along the foot of the cliff and found one cave after another, but not in the tidy shapes that I remembered from Paleokastritsa. When Neanderthals were hunting around Gibraltar they may have sheltered in these caves. We would have to measure and map each cave separately, but we had been submerged for 35 minutes and would soon be out of air. I signalled to Illing that we must ascend to return to shore on the surface.

On the surface we found a choppy sea running with the crests of the waves beginning to break in white foam. The wind whipped the water and the dinghy was nowhere near us. The weather had broken dramatically. Schofield was trying to follow us in the dinghy, but had lost our bubble trail in the rough sea and was halfway between us and the shore looking in the wrong direction. Illing waved the ranging pole while we both shouted and whistled, but the noise of the outboard on the dinghy prevented Schofield from hearing us.

"We'll have to swim for the shore," I said, "and I'll have to stay on the surface."

"There'll be less current on the bottom," said Illing. "but we'd better stick together."

I rolled onto my back to swim most efficiently with my fins, as we headed for Gibraltar, keeping our bearings by looking at a fixed point on the Spanish coast. We got separated by quite a bit, but I could always see the tip of Illing's ranging pole bobbing

Figure 4.2. The author in one of the caves on Vladi's Reef in 1962. Gorgonians grow on the roof of the cave. Note gloves to protect from abrasion on the rocks, also twin-hose regulator, depth gauge and Rolex watch on left wrist, and compass, partly concealed, on the right wrist.

above the waves. I was worried that Schofield would raise an alarm and say that we were lost. He kept searching on the track where we should have been, but we had been carried much further south. I turned over to look at the Rock and saw that I was about halfway back to shore, but still being carried south, and might even miss Europa Point completely. Africa next stop. I angled even more into the current and swam as strongly as possible for the long haul.

Finally I clambered ashore through the surf and boulders, took off my equipment, and waded back to give Illing a hand, as he struggled with the langouste and the ranging pole. I helped him off with his kit. We shook hands warmly. It did not seem a stupid thing to do. We knew we had had a close scrape. Children and passers-by who had gathered around us joined in the shouting and waving until Schofield saw us. He came to pick us up in the dinghy. We named the newly discovered reef "Vladi's Reef", and it is known by that name to this day by divers in Gibraltar and also in academic publications (Fig. 4.2).

As soon as we had the diesel launch, we started running long echo-sounding traverses radiating out from Gibraltar in all directions, making a total of 14 bathymetric depth profiles in two weeks. When the currents were slack we dived on the ridges and reefs that we discovered, making many more dives on Vladi's Reef, where we sketched some of the caves. Two huge lead Roman anchor stocks were lying amongst the boulders at the foot of the cliff, close to two vast iron anchors of a much later date, that looked as if they might have come from a ship like Nelson's HMS *Victory*. Ships had been losing their anchors on this unexpected reef for thousands of years. Waechter persuaded the harbour master to bring a tug with a bow lifting winch out to Vladi's Reef, and after a succession of abortive dives and failed lifts when the anchor snagged on the reef in the strong current, I succeeded in attaching a heavy steel hawser to the centre of a lead anchor, which we raised triumphantly to the sunlight (Fig. 4.3). That Roman anchor is now in the Gibraltar National Museum.

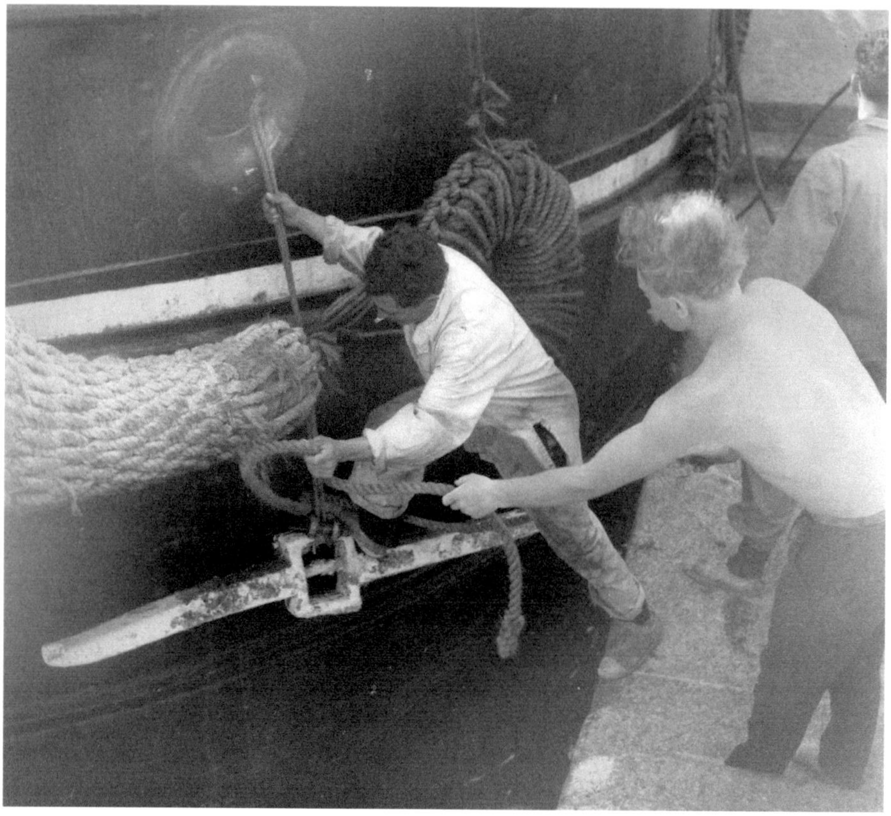

Figure 4.3. Roman anchor found on Vladi's Reef being brought ashore on the dockyard boom defence vessel. Note the heavy rope and shackle needed to lift the anchor over the cliff edge.

CUUEG divers studied Vladi's Reef again in 1969 and a multi-national group assembled by Geoff Bailey in 2005 continued the research. Although the caves gave an excellent indicator of a stable ice age sea level at 20 m (66 ft) below present sea level, and one of the caves showed signs of terrestrial sediments and accretions, we could not find signs of prehistoric occupation. In front of the caves there are fallen blocks from the cliff above, and these are cemented to the seafloor by biological concretions so that heavy rock drills and lifting tools would be needed to expose the original ground and possible terrestrial fossils. There are numerous caves above sea level in Gibraltar, including the famous St Michael's Cave, and most of them were created by erosion of the pervasive massive limestone, or its dissolving over time. Some of the caves that we found underwater were in the limestone, but others were in fallen screes and consolidated conglomerate, and yet others in a crushed and cemented rock called tectonic breccia (Figs 4.4, 4.5).

In 1962 our dives continued around the east side of the Rock, exploring numerous caves penetrating 10 m (33 ft) or more into the cliffs (Figs 4.6, 4.7). Illing in particular had the determination and skill to worm his way into the furthest interiors of the caves that we found on the east side at depths around 10 m (33 ft), but I always wondered how such a broad-shouldered man could do it. He was exhilarated to find the limits of the caves, which sometimes expanded into larger chambers, and he surfaced bursting with excitement after one dive with Schofield.

Figure 4.4. Diver working on survey and excavation of one of the caves on Vladi's Reef in 2005. A taut reference wire is extended across the front of the cave, and frames marked in 20 cm grids are ready to be placed inside the cave. The diver is breathing an oxygen-enriched mixture, nitrox, to reduce nitrogen narcosis and extend bottom time without excessive decompression. The hosepipe in the foreground is used to power an airlift dredge to remove sediments.

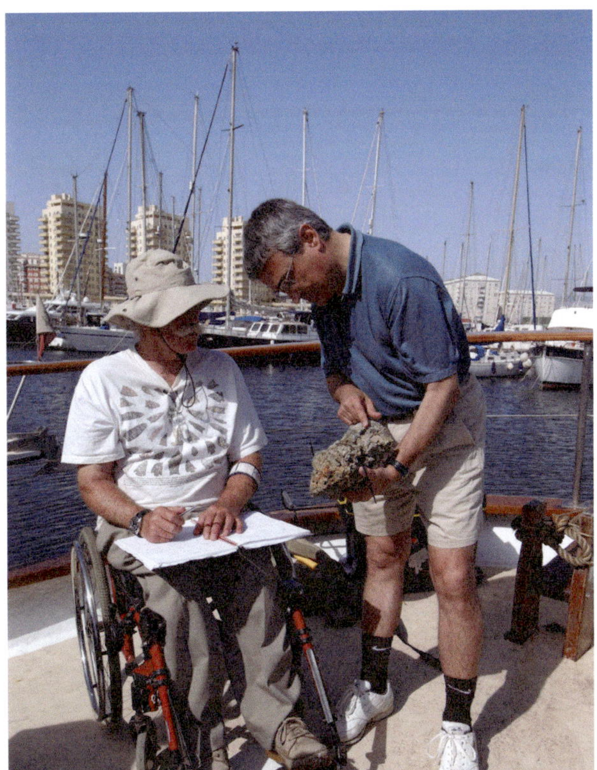

Figure 4.5. The author and Joaquín Rodríguez-Vidal on board the diving boat at Gibraltar discussing the rock samples from Vladi's Reef at the moment that they both separately concluded that the whole reef consisted of tectonic breccia, that is crushed and half-melted rock created by tectonic faulting (Photo Clive Finlayson).

Figure 4.6. Divers on the east side of Gibraltar in 1962, preparing to dive offshore from Governor's Beach. We had been asked to search for signs of freshwater seeping out of the rock in this area.

Figure 4.7. Cliff on the east side of Gibraltar just south of Governor's Beach. Note the deeply incised double notch cut into the limestone at the water level. A few metres above the water line there is an undercut in the cliff which indicates the sea level at the last warm period, the interglacial 125,000 years ago; much higher up the cliff is the road from which we abseiled down onto the beach in stormy weather.

"A real monster cave," he laughed. "Full of big fish, full of fish! Leche, a big conger eel. John, did you see that conger? Oh, he was so big! And the gorgonians, oh my!"

He handed up to the boat a bunch of gorgonians as bright as dahlias, red-and-yellow, purple, magenta, and white. For a man who had dived in exotic climes and worked with George Bass in Turkey, Illing had the wonderful ability to extract joy from dives that some people would have found dark, miserable, or even claustrophobic. The dives on the west side of Gibraltar were often limited by the strong winds off the Atlantic, so sometimes we took our gear round by taxi to avoid the wind and currents off Europa Point to dive on the east side. After one such dive, when the taxi failed to return to pick us up, Schofield and I discovered a tunnel entrance, through which we could see a pinpoint of daylight at the other end. It went straight through the Rock from side to side, so we walked back carrying our diving gear. It was an unexpected shortcut.

* * *

In the mornings we usually met for breakfast at a bar called Tarik, while lunch was sandwiches in the boat, and on most evenings the younger divers dined at Smokey Joe's. This amazing eating house was a gloomy, smoke-filled rectangular den, with a central alleyway between two groups of trestle tables and bench seats. The cuisine was composed of steak (very small), fried egg, chips, baked beans, fried tomato, sausages, bacon, peas, spaghetti, and mushrooms. The menu displayed every conceivable mathematical permutation of every item, in pairs, triplets, quintuples etc., each given a different price in pennies or half-pennies, and across the top was written "Special Today! Steak, egg..." through to "and mushrooms" as a complete set. Each menu was stuck onto the sort of crinkly cardboard used to make packing boxes, with, stuck on the reverse, an enticing and healthy-looking pin-up. At the bottom of this gastronomic work of art was the signature "M. F. Martinez", and then in brackets, just to be sure, "(S. Joe)". We always ate the "Special Today".

In exposed locations on the coast of Gibraltar, the wind, waves, underwater visibility, and current determined each day whether we could dive or not. Scientists cannot be choosers. We do not have the option always to dive in sun-drenched places with clear blue calm water. The data accumulated steadily, filling notebooks and boxes of Ferrograph printouts showing the profiles of the seabed and the flat terraces on it. On the east side of the Rock, at a point where the cliff descended straight into the sea, the modern notches incised by the seawater eating into the limestone cliff had formed at two levels, about 1 m (3 ft) apart, instead of directly at mean sea level. Since the tide at Gibraltar is driven from the Atlantic and rises and falls by about 1 m, I wanted to measure the notches and see if they were exactly 1 m (39 in) apart (Fig. 4.8).

It was another breezy day from the east, with the waves smashing up against the rock.

"Give me the ranging pole," I said to Greer as Waechter brought the *Rev. James* to within 20 m (66 ft) of the cliff. "I'll swim over there and measure the notches." I plunged over the side and hung onto a rope strop.

"You fucking idiot!" shouted Greer, "It's far too rough. You'll get scrambled into little pieces."

"I can fend off from the cliff with the pole. It'll be OK. Come on!"

Figure 4.8. Double solution notch at sea level 1 m apart and the author with ranging pole. Greer thought I would be smashed against the cliff. The broad scalloped indentation at a height of 6 m is the solution notch from the last interglacial high sea level 125,000 years ago.

"OK, here it is." He handed me the measuring pole. "An accident always adds prestige and class to an expedition," he added. "I promise to take superb pictures of your body parts plastered onto the cliff. Good luck."

I was able to swim right up to the cliff quite safely. I measured the notches, which were just 1 m (39 in) apart, showing that the seawater dissolved the rock so quickly that it was able to dissolve one notch at high water and one notch at low water on the tidal cycle, when the surface of the sea was almost stationary at each turning point.

* * *

When a NATO fleet sailed into town the British, Dutch, French, and Canadian sailors stormed into the high street searching for souvenirs and entertainment. The shopkeepers were ready for them. There was a French aircraft carrier and a British cruiser and several other ships. I sat that evening listening to the music from a bar playing the songs of Nina & Frederick, *Maladie d'Amour* and *Yellow Bird*, watching four Russians in identical badly fitting suits drinking coffee outside a fancy restaurant on the other side of the road. Perhaps these stereotypical, almost "joke", Russians were just decoys, and perhaps the real spies were completely inconspicuous. Gangs of moustachioed French sailors rollicked down Main Street singing. A nondescript sailor staggered up to the door of the restaurant, lurching drunkenly, but was blocked by an alert waiter. The sailor shouted through the door "I hope you all die of food poisoning!" There was some fighting further down Main Street. I retired to bed.

Towards the end of a project the priorities are often to fill gaps that have occurred due to bad weather or bad luck earlier in the schedule. It can be a grinding routine. I

Figure 4.9. The author on top of O'Hara's Mount, Gibraltar. The vertical drop behind me is about 300 m or 1000 ft.

remember Waechter stopping halfway down the diving ladder as the *Rev. James* rolled in the murky swell, taking his mouthpiece out of his mouth, clutching the ladder with his other hand, and saying with a grin, "It's this quiet academic life that I like so much!" Then he disappeared below the surface. Given that he had only recently learned to dive, he had adapted to it remarkably well. On one of our traverses around Europa Point, the echo sounder picked up a shallow rocky outcrop just below the surface, flanked by a precipitous cliff dropping to more than 60 m (198 ft). Sceptical as we were about the currents round Europa Point, we asked fishermen if there was ever a slack water there. We dived when the fishermen recommended, and Schofield and I descended the cliff to plot the undercut erosion at its base as best we could, although we were in almost total darkness. I discovered years later that the shoal is called Los Picos, and has been responsible for a catastrophic recent shipwreck.

A visiting film crew wanted to have pictures of Waechter in Gorham's Cave, and me climbing a cliff. They were introduced to us by Tony the Cliff-climber, who was acting as their guide. Tony arranged for pictures of me on the cliff next to O'Hara's Battery (Fig. 4.9), and the next day to film Waechter on the beach. It was the film crew's last day on the Rock, but the surf was pounding on Governor's Beach so that we could not land there with our boats.

"It's no problem," said Tony. "We can abseil down the cliff from Arow Street."

"Most of us will have no problem with abseiling," I agreed. "But how do we get up again?"

Figure 4.10. Boathoist Cave, photographed in 2005. In stormy weather the waves break right inside the cave eroding and polishing the rocks so that no unconsolidated prehistoric deposits could survive. This cave can be reached by a tunnel from inside the Rock (Photo Clive Finlayson).

"Easy, just use Prusik loops." We all looked baffled. Tony added, "Don't worry, I'll show you when we get there." This was a bit optimistic.

After the filming on the windswept beach, Tony showed us how to knot three loops of different length onto a vertical rope suspended down the cliff. You can ratchet the loops up, and they don't slip down again, so that with your foot in one loop, a second attached to your waist-belt, and the third higher up for your hands, you can climb steadily up a thin rope. This is the Prusik loop. All went well, except for one of the film men who, though strong, was 6 ft (1.8 m) tall, broad-shouldered, and so heavy that he could not work the sliding loops, partly because the main vertical rope was synthetic fibre, which just stretched under his weight. After several false starts, we abandoned him alone on the beach while Tony assured him that he had a plan for his rescue.

At the top of the cliff, the ever-resourceful Tony explained that there was a secret passage from inside Arow Street Tunnel down to Boathoist Cave, just beyond the end of the Governor's Beach. He obtained the key from a nearby military guard-post, found the thick wooden door on the landward side inside the tunnel, then, accompanied by Greer, Waechter, and Julius, our taxi driver, we followed him down the slimy rock-cut steps groping for balance in the dim torchlight. The passage turned and twisted, doubling back under Arow Street Tunnel, until it opened into the top of Boathoist Cave where there was a constant roar of crashing surf (Fig. 4.10). A rusty ladder with several rungs missing led down into the sloshing water that swirled between the boulders. From the entrance to the cave, I was able to swim out by plunging under the surf, then swim around the bluff of the cliff, to land on the beach near the disconsolate cameraman. His clothes were soaked through by the time we got him back into the cave, safely, but we soon had him dried out, restored with a hot meal, and in the taxi in time to catch his flight to London.

* * *

Because of the expense, the divers dined together at the fancy Ascot Restaurant only a couple of times. On those occasions we were thanking our Gibraltarian benefactors, with Waechter present to pay the bill. But apart from this, since the younger divers ate very basic food, usually at Smokey Joe's, we did pine for more sophisticated fodder than we could afford. The opportunity arose just before the end of the trip. Our dives on Vladi's Reef had revealed that most of the caves were occupied by large langoustes, weighing several pounds each. We collected a good many of these, went around the town giving live langoustes to our helpers and friends, and kept just three for ourselves. I negotiated with the manager of the Ascot. He knew that we could not afford to eat there under usual circumstances, so I proposed that if we delivered the langoustes and his chef prepared them in the most festive way possible, we would pay for aperitifs, wine, hors d'oeuvres, vegetables, and desert, but have the langoustes for free. He agreed. I handed the langoustes to the chef who was delighted at the challenge.

That evening Schofield, Greer, Illing, and I put on our best or only clean shirts, scrubbed our calloused, rock-scarred hands, cleaned our fingernails, shined our shoes, and trooped into the Ascot at the most fashionable hour. The other expensive guests noted our arrival surreptitiously and pretended not to. You did not go to a smart restaurant in casual dress in those days. A prominent table had been reserved for us. After ordering aperitifs and wine, and consuming our hors d'oeuvres, we waited for the great moment. We were not disappointed.

The chef and assistant chef in full professional rig entered the dining room carrying two huge silver platters, while the manager held the door open for them so he could watch for the impact. One platter bore two large langoustes, and the other bore the gigantic orange body and carapace of the largest langouste of them all. Each langouste had been baked with spices and herbs, doused with hot butter, and then surrounded by a bed of prawns, shellfish, and a variety of molluscs, while the spines on the backs of the langoustes were decorated with dozens of impaled olives and anchovies. The other diners gazed in astonished admiration. For a few seconds knives and forks were almost motionless as people tried not to stare. The manager watched the effect with satisfaction.

The platters were deposited with a gracious flourish on our table, dishes of vegetables arrived, more wine, and the chefs departed. Pretending not to notice the sensation that this performance had created, we quietly started to break off legs, scooped up scallops and mussels, and hacked huge chunks off the great tail muscles. (Langoustes do not have big front claws, unlike lobsters). The noise of conversation in the rest of the room regained its previous level, while we happily drank a toast to the completion of a successful project. After the pudding followed by cheese and port, we thanked the manager and the chefs, paid with a generous tip, and retired happily, feeling that we had made our mark.

* * *

My work at Gibraltar in 1962 was undertaken before the theory of plate tectonics, and before the dating of multiple ice age fluctuations based on deep ocean cores, which provided a timescale of over one million years. In the 1960s we thought that, on evidence

Figure 4.11. Divers kitting up for a dive on Vladi's Reef in 2005. The cliff on the left is close to Europa Point, the southern tip of Gibraltar, and the coast of North Africa is visible in the background. The Gibraltar Strait was never dry even during the lowest sea level during the glacial maximum, but it was a bit narrower than it is now (Photo Clive Finlayson).

from the glaciers in the Alps, there had only been four glaciations. Also, because so many fossil beaches around the Mediterranean are found high up on cliffs above sea level, we thought that the global sea level had been falling for many hundreds of thousands of years. This was a mistake, and the uplifted terraces on many Mediterranean cliffs were caused by the plate tectonic convergence of Africa moving towards Europe. The observations of submerged erosion features that we made at Gibraltar were accurate, highly original, and innovative. Recent studies by divers have confirmed caves at similar depths around the Mediterranean. But we have had to revise our understanding of the timing of glacial low sea levels over a longer period.

Clive Finlayson, Director of the Gibraltar National Museum, has organized a steady programme of scientific diving around the Rock in recent decades, and in 2005 Geoff Bailey and I worked with an international team, including Abdullah Alsharekh from Saudi Arabia, to update our knowledge of the submerged caves there (Fig. 4.11). This work has continued, with many dives by Darren Fa and Geraldine Finlayson, with especial success in defining the extensive plain on the east side of Gibraltar, which was the primary hunting and foraging area for the Neanderthal families when they lived in Gorham's Cave.

The narrowness of the Strait of Gibraltar, where you can see the other side clearly, and any smoke from a fire is visible, suggests immediately that prehistoric peoples might have crossed, either by swimming or with simple rafts or canoes. Swimming the Strait is a standard achievement for ambitious open-water long-distance swimmers, with the usual crossing time about five to six hours. These are not the fastest. Studies of the Neanderthal stone tool technologies on both sides, and DNA research, neither confirm nor refute the possibility that people crossed the Strait. Earlier hominin species such as Homo *heidelbergensis* lived in the Spanish peninsula one million years ago, and the first hominins to arrive certainly travelled along the north Mediterranean coast from the Middle East. The question remains whether the Neanderthal occupants of Gorham's Cave could travel directly across the Strait. My guess, and it is only a guess, is that the Neanderthal and Homo sapiens prehistoric people living close to the shore on both sides of the channel were skilled seafarers, and they crossed the Strait routinely when they wanted to. It was not a big deal. But nor was it a migration in the modern sense of a purposeful mass movement of people in one direction. Both Morocco and southern Spain were pleasant places to live throughout the ice ages, and there was no motive or compulsion for a mass migration that would show up in modern analysis.

* * *

If all my observations of caves in the Mediterranean had been in limestone cliffs, there could have been an unidentified bias due to the geological regions in which limestone most frequently occurs. With Natalka as my co-diver, in early summer 1962, I conducted several weeks of diving based on the Observatoire Océanologique de Banyuls-sur-Mer in southern France, known as the Laboratoire Arago. Banyuls is on granitic and metamorphic rock just north of the Pyrenees. From Banyuls we made excursions to Sant Feliu de Guíxols in Spain, just south of the mountains, similarly on granite, and then to the half-built Club Mediterannée village at Cap de Creus, near Cadaqués. This resulted in a fabulous dive straight down a rocky slope "into the blue", but no caves.

The Laboratoire Arago gave us free accommodation in a hostel attached to the laboratory, the occasional use of a powerful diving boat, and delightful food for lunch in a cafeteria overlooking the sea. The marine station was situated at one end of a gently curving bay, and in the evenings a number of dance bands performed on the seaside promenade, playing the instruments known collectively as a *cobla*. The *cobla* band includes several oboe-like double reed instruments of different pitches, producing a strident harmony, supported by one or two brass instruments, drum, and double bass. Circles of people dancing in the nearby streets grew larger as more people joined the fun, and the music rippled across the bay.

This idyllic peacefulness was disturbed in summer 1962. President de Gaulle was dismantling the French Foreign Legion bases in Algeria, almost disbanding the Legion completely. He recalled troops to mainland France and abolished many elite units, contrary to all tradition. Everybody speculated as to what trouble would result when these rugged, war-battered soldiers of many nationalities who had seldom known a productive civilian life would be demobilized and wander onto the streets. There must have been a holding barracks near Banyuls. During our stay the squads of legionnaires marched slowly through the town on most days, tightly disciplined, wearing rubber-soled boots, so that in spite of being in squads of 100 men or more, they moved in eerie whispering silence. Their facial expressions were depressed, sad, staring grimly and passively, as if they knew that their time was past. They knew that they were an extinct species.

The dives that we made on the rocks of the Pyrenees revealed a few caves, a few isolated chunks of upstanding rock that had obviously been eroded around the base, but always the shape and depth of the feature were controlled by pronounced weakness in the rock, not by consistent long-duration erosion at a consistent level. The exercise was worth doing, but it confirmed that detailed and accurate data on a large number of eroded forms would be limited principally to limestone coasts.

* * *

Our base for diving throughout August was the Station Marine d'Endoume, on the outskirts of Marseille. John Woods, who had recently conducted a successful mapping and excavation of a wreck full of amphorae off Xlendi Reef, Gozo, at a depth of 60 m (198 ft), joined us at Banyuls for the last few days, before the train journey to Marseille. Natalka left as soon as we got to Marseille to return to her work at the Natural History Museum in London.

John, who was an active diver in the Imperial College Diving Club, was determined to build a new discipline of underwater observational science, a view that coincided with mine. We had often met to discuss plans back in London. John's career has criss-crossed mine throughout our lives and he was at times my immediate boss when we worked together in a team that designed the Global Ocean Observing System (Chapter 9). He is, and was, a fiercely intellectual man, totally focused on his research in the physics of the ocean and atmosphere, always with far distant plans and goals. He has contributed greatly to theoretical oceanography and global modelling of the ocean. During his career he worked as a research fellow at the Meteorological Office, held professorships at Southampton University, Kiel University, and Imperial College, and was Director of Marine and Atmospheric Research at the Natural Environment Research Council (NERC) from 1986 to 1994, during which time he created the National Oceanography Centre at Southampton (Chapter 9).

The coast of Marseille is famous for its amazing deep caves. After mistakenly trying to enter a demobilization barracks of the Foreign Legion, John and I presented ourselves at the Station Marine. Next morning we arranged our equipment in the large tiled laboratory on the ground floor. Professeur Pérès, the director, explained the local arrangements, stressing the need to shut all windows securely when the mistral wind was blowing. This became a standing joke, as he would tour the building on windy days shouting "Fermez les fenêtres! Fermez les fenêtres!" At midday we descended the cliff by a flight of steps to the sandy beach of the Anse de Maldormé, the name of a *calanque* (a narrow inlet) just below the laboratory. A minute, rocky promontory to our right supported one precarious villa. To our left the houses were piled upon one another from the water's edge in random tiers that sprawled across the hillside.

The lowest houses consisted of only one or two ramshackle rooms over a ground-floor boathouse which opened straight into the sea. Here holidaying families ate on the balconies while the children splashed in the water below. A narrow, paved path ran around the *calanque*, level with the roofs of these houses, where, on the sun-baked stones, the boys and girls of Endoume sunned themselves in groups, laughing, shouting, eating ices, drinking, talking, and making cautious overtures of love. Lush gardens with palms and bougainvillea bordered the path, which was overhung with creepers of all kinds suspended from painted pergolas, while behind these were tall houses of many styles plastered and painted in every hue of yellow, white, pink, and pale brown. The whole effect was topped by a sea of red roofs, behind which still higher waves of orange tiles rose up against the blue sky. In the intervals between diving excursions, I painted a large ink and watercolour view of the bay, always watched attentively by clusters of teenagers who made helpful comments, suggestions, or enquiries. They tried to recognize their friends in the picture.

Beyond the *calanque* the bare limestone coast extended with numerous headlands, finally rising to the far peaks of the mountains of the Marseilleveyre. From these mountains, and from the high valleys of the Rhône, Marseille draws its freshwater so that, even in midsummer, streams of water gush from hydrants to flow in the streets. Flowers and trees bloom in every square.

The month that I spent in the suburb of Endoume, with the necessary visits to the Vieux Port and the centre of Marseille, developed in me a great love of that city. For all its problems, Marseille has always been my favourite French city on the south coast. It is a serious, large city that could not care a damn about tourists. Its attitude is: if you are happy in Marseille, that is fine. If not, please go somewhere else. We won't change to please you. Marseille is, or was, confident in its own character.

We first explored the small caves and rocky tunnels immediately under the promontory with the little villa on top. Using full scuba gear we made detailed drawings of the tunnels, although it took a few minutes to become accustomed to the darkness. From within the tunnel, far away, I could see the green glow of daylight. For the next few days, the mistral wind howled down from the mountains so that diving was difficult, not to say cold. Nevertheless, we persuaded three of the young French scientists, Pierre, Paul, and Nardo, to help us launch the aluminium dinghy, which was kept in the private *calanque* belonging to the Station Marine d'Endoume. In this John and I rowed hazardously to the Îles d'Endoume, about 200 m (660 ft) offshore, where we made several dives and discovered another cave and a submerged rock arch.

Pierre then suggested that we should drive to the north side of Marseille, away from the wind, and make a large group dive in the Grotte de Niolon, where there were many connected caves and tunnels. I agreed with Pierre that after exploring the caves we should swim 100 m (330 ft) seawards to see what happened further offshore.

We plunged into the water and followed single file down a long rocky gorge. At a depth of 14 m (45 ft) Pierre swung left into the first cave. I swam up to him, he nodded, and I took the lead. Shining the powerful beam of the Crystal brand lamp in front, I finned slowly forward until I could see where the sandy floor sloped up to meet the roof. The rock was dripping with sponges and coral, blood-red and mottled with white polyps (Fig. 4.12).[1] The floor sloped away to the right where I forced myself through the narrow gap below the roof, emerging into a huge arched chamber with several exits to the open water. I turned and hovered as Pierre slid out of the darkness from the other cave, and I signalled to him that I was going to look around for several minutes. While I worked, the others swam in and out of the caves trailing their necklaces of bubbles so that the silver columns of air were woven amongst the arches in a maypole dance.

---

1 To bring the beauty of the Marseille caves to a modern audience I have been helped by a French colleague from the old days, Jean-George Harmelin. Jean-George, known as "Jo", is a biologist who has studied the creatures and plants living in caves, and he has generously provided the pictures shown in Figs 4.12-4.15, together with some extra comments for the captions. Thank you for the wonderful pictures, Jo.

Figure 4.12. The Grotte Niolon, west of Marseille, where John Woods and I made our first dive on a cave with divers from the Station Marine d'Endoume in 1962. The cave was eroded by low sea levels during the Ice Age. Its complex topography, with a broad overhang, a blind tunnel, and arches, results in a wide range of ecological conditions and assemblages of sciaphilic (shade-loving) organisms, including yellow sea-fans (Eunicella cavolinii), yellow encrusting anemone (Parazoanthus axinellae), red coral (Corallium rubrum), and many bryozoans and sponges, first described by Laborel and Vacelet (1958) in a pioneer work (Photo Jean-Georges Harmelin).

In preparation for the bottom traverse at Niolon, I set a compass bearing on which we swam out to sea in file again, until we had reached a depth of 27 m (90 ft) on the gravel. Then I felt a tap on my shoulder. Pierre was beside me reaching for my drawing board. I watched as he wrote in straggly letters "*Nous devons retourner à la falaise. C'est la deuxième plongée seulement pour Paul*". I turned the compass though 180° to guide the swim back. Luckily, as there was no current, we arrived exactly at the caves. Pierre showed us some more underwater caverns before we surfaced. It was a fascinating place. I had covered my writing board in sketches and depths.

When the mistral gale finally blew itself out we drove early in the morning to the Vieux Port to board the *Antedon*, a fishing caique converted to serve as a research boat. The port was already brimming with activity. The cafés were in full swing, the pleasure boats were advertising trips to the Château d'If, and fishermen and yachtsmen alike were messing about in their boats in the traditional way. There was a stink of fish and oil. To the west we could see the clouds of smoke from forest fires in the mountains. That summer was very hot so that the perpetual risk of fire had produced a worse blaze than usual, fanned by the mistral. The progress of the fire was reported in the press each day as *Le Sinistre*.

Antedon nosed out between the castellated jaws of the port into the great bay dotted with islands. We passed the Château d'If, around which the pleasure boats nestled like piglets around a sow, and dropped anchor off Cap Cavaux, the southern tip of Île Pomègues. Nardo offered to show us the best diving spot, and we kitted-up. It was dead calm, the sun beat back from the blistering hot rock illuminating the dead whiteness even below the water. At depth, the white rock darkened to a nacreous Grünewald green, but further away from the cliff, the water was a pungent Dali blue. We dropped down a sequence of giant rocky steps plunging 30 m (100 ft) into a sloping bed of sand where countless necks and handles of broken amphorae projected from the seabed. It was the last resting place of a Roman shipwreck, one of many that we saw in the course of a few weeks. John and I measured the profile of the cliff using tapes, a ranging pole, depth gauges, and writing boards, while the French divers collected biological specimens. We made several dives in the area, then crossed in the boat to Île Maïre, eating lunch on the way. In those days it was acceptable to drink wine diluted with water between dives. We warmed up in the sun, enjoying our lunch of cheese and garlic sausages on crunchy baguettes.

Île Maïre is the last outpost of the Marseilleveyre: a last statement of white cliffs and spires before the land disappears into the ocean, to emerge momentarily at Île de Planier, and then to plunge into the abyss forever. We skirted the southern cliffs of the island to anchor by the twin pinnacles known as Les Pharillons. These precipitous blades of limestone rose sheer from deep water.

John and I dived, with a standby diver on the surface. We swam into the shadow of the cliff and down the rock face until a bench sloped out at 18m (60 ft). We floated outwards and down again to 40 m (130 ft), where we settled onto another bed of amphorae. In spite of the shadow cast by the island, we could see the arc of the cliff face curving away from us into the deep sea. We swam slowly along the foot of the precipice at a depth of 46 m (150 ft), where the rock was heavily encrusted with enormous growths of coralline algae reminiscent of the fungi which grow on trees. On the rocky ledges grew the ebullient fronds of gorgonians, bloated and black. As we ascended, we found

Figure 4.13. This drowned arch of rock, close to Boulégeade Cave between Riou Island and Moyade islet, has a complex underwater topography with steep walls, overhangs, arches, and cavities. In this narrow arch, which crosses the summit of a drop-off, the water circulation is channelled and strengthened, leading to a particular development of sessile suspension-feeders, such as sponges, sea-fans, and this yellow encrusting anemone (Parazoanthus axinellae), which covers large areas (Photo Jean-George Harmelin).

a cavern that pierced the cliff at 9 m (30 ft) (Fig. 4.13) so that the slender flying buttress, brightly lit in the shallows and sheathed in pastel-coloured algae, arched like a rainbow into the blue. We climbed back on board as the engine roared and echoed from the cliff. The captain steered for the Vieux Port.

* * *

We dined often at Le Leman Bar, where for five new francs each we could eat well including a glass of beer with the meal. We were waited upon by the patron's wife, whom we called Madame Twanky, while her husband played interminable card games with the legionnaires awaiting demobilization. As always, our budget was very limited, so we tried to keep the cost of meals below ten francs, at a time when the exchange rate was about 14 francs to the pound. The restaurants we visited quickly spotted that we were counting the centimes and trying to get the best bargains, but Madame Twanky excelled them all for generosity and friendly spirit. Her bouillabaisse was stuffed with great chunks of fish, depending on the landings that day in the Vieux Port, rascasse, bream, or dorade, piled with mussels and crab meat, in a broth so rich in oil, pepper, and garlic that you could smell it from 20 yards away. She gave us huge portions. In the mornings we bought the cheapest 8%-alcohol wine, which was decanted straight from the barrel into the bottle that we brought for the purpose, to be consumed at midday on the boat.

John's friend Charles Carmichael arrived from Germany a few days later to help us conduct a series of deep dives. He was a regular army officer who combined business with pleasure by working with us during his leave. I had arranged with Professeur Pérès that the three of us could dive most days from a small vessel called the *Sainte Marie*, which had its own boatman. The three of us also planned to hire a car to drive along the coast looking for the remains of ancient classical cities, but that excursion will be postponed to Chapter 6.

Next morning, when we carried our gear down to the jetty in Calanque Maldormé, we saw across the calm sea the approaching shape of the *Sainte Marie*, cleaving a straight wake from La Madrague, the village on the south side of the bay where she was kept.

"*Bonjour Messieurs*!" shouted René, the boatman, as he threw us a line. "Now where exactly do you want to go?" he asked as we finished loading.

"Grotte de l'Ile Plane," I said. "And if there's time we'd like to see any other caves you know about."

"I think I can find plenty for you. *Je suis à votre service, messieurs*!" He cranked the motor and we cruised across the bay towards Cap Croisette. René crouched over the tiller as he lit a cigarette. He was short, lean, tanned a ruddy brown, and his thin face appeared almost emaciated as he drew in the smoke.

"What time do you have to get back?" I asked

"It doesn't matter to me really."

"That's most unlike a Frenchman, if you don't mind me saying so. What about your lunch?"

"Ah, my wife is away staying with her sister. I am going to enjoy as much time as I can in the boat, sailing and fishing. I used to be a sailor you know."

"For many years?"

"No – only my National Service, I was a navigator, a very responsible job, and we saw a lot of the world."

We rounded Cap Croisette and passed through the shallow strait separating it from Île Maïre. Before us, away to the southeast, stretched a chain of white, skeletal islands, Île Jaire low and spiky, Île Plane flat and smooth, the craggy ramparts and gothic peaks of Île de Riou, and beyond that the stubs of Grand Congloué and Petit Congloué.

"That headland on the horizon is the Bec de l'Aigle," said René. "There is no road on this part of the coast and the only way to reach the shore of the Marseilleveyre is by boat."

We chugged into the shadow of Île Plane. Near the farthermost point, René cut the engine and jumped ashore onto the *trottoir* of calcareous algae which fringed the cliff.

"The cave is just below here," he said. "How long will you be?"

"About twenty minutes I expect. Why?"

"I would like to collect some mussels for my dinner this evening. They grow very well on the cliff here."

"Well," I hesitated, "as long as you are ready to give us a hand as soon as we surface, that's all right."

"Of course, *Je suis à votre service.*"

"Charles," I said, "can you do snorkel cover? Just swim on the surface and watch where we go."

"Right, Nic."

John and I swam deep into the shadows, pausing at 30 m (100 ft), then swooped low over a sandy ledge and stopped to look around. Beside us the cavern gaped dark and blind. We swam under the brow of rock where we waited until the retinas of our eyes became sensitive to the darkness. The vault of the roof was fringed with black gorgonians streaked with yellow, and the inner walls were adorned with macabre fans of the purest blue-black, funereal and sombre. Towards the back wall, yellow, white, and red gorgonians spread fine fingers and tiny polyps interspersed with precious coral (Fig. 4.14), protected forever from the sunlight. We rose slowly to the roof until the walls drew close around us and we were enclosed in a vertical tunnel from which smaller tunnels radiated. As so often, solution channels of groundwater had penetrated the rock from above and gushed out into the cave when it was high above the sea during the ice ages.

The red coral, polyp-spotted like fly agaric, clothed the rock in scarlet and white. Still higher we floated, turning slowly in the water to see any sign of an exit. Above us the daylight suddenly glared between jagged spears of rock, which guarded the mouth of the tunnel like teeth. We slid between the bars and rolled free into the open sea only 4.6 m (15 ft) below the surface.

As we rose beside the boat, René jumped to help us, lifting our equipment over the gunwale and stowing it carefully. Charles climbed back on board to help us in.

"My God!" he said. "I didn't know what to do when you vanished into the cliff like that. Then a few minutes later the bubbles began to come out through those cracks in the cliff near the top, so I knew you were all right."

"We'll take you down there tomorrow," I said. "We must look into the side tunnels, but we shouldn't do another deep dive just now. René, where are those other caves you were talking about? Are there any shallow ones?"

Figure 4.14. The Boulégeade Cave on the west side of the island of Riou, just southwest of Marseille, where we dived in 1962. You can see the brittle fronds of red coral on the walls of the cave. The multiple levels of the cave confirm its origin on a fossil shoreline. The cave is composed of a small tunnel which leads at 23 m depth to a chamber with two large entrances, beautifully adorned with red coral (Corallium rubrum). This coral, used for jewellery for thousands of years, occurs at shallow depth in the Marseille area due to wind conditions that limit excessive warming during summer. Thanks to recent protection measures, the status of the red coral population of this site, now included in the National Park of Calanques, will improve (Photo Jean-Georges Harmelin).

"It's too late to go to the Triperie, M'sieur, but there is another cave on this island I believe. Nobody has ever been in it. Do you want to see it?"

"Please."

We cruised back to the midpoint of the island, where John and I were able to swim through a narrow wave-smoothed tunnel only 3 m (10 ft) deep which cut 30 m through a headland.

The next day John and I dived again at the Grotte de l'Île Plane, but this time we checked the profile of the cliff below the cave. We ended up at a depth of 64 m (210 ft), where the sea was so still that a layer of dense silt-laden water lay like fog in a valley. We surfaced slowly, decompressing for ten minutes at 3 m (10 ft). After a suitable rest, I took Charles into the cave. I had the big torch with me, and while he waited patiently, not to say bravely, in the throat of the main tunnel, I explored the side ones. Luckily none of them were more than 6 m (20 ft) long, so that we were scarcely separated, and we were able to surface together.

From that day on we all dived frequently, though seldom below 61 m (200 ft), exploring the cliffs and caves along the south side of Île Maïre, Île de Riou, and Grand

Congloué. We also dived on the wreck excavated by Cousteau on the north side of Grand Congloué, where the seabed was still littered with broken Campanian black pottery.

I should comment on the depth to which we often dived during this month. The official maximum diving depth for a diver with the most elementary scuba qualification breathing compressed air in the open sea is now 20 m (about 66 ft), with 35 m (115 ft) for an experienced diver, and 50 m (164 ft) the limit for an advanced diver with lots of experience (Chapter 7). These limits are very reasonable where large numbers of divers are concerned, especially since sports divers who wish to go deeper nowadays can use trimix and rebreather equipment. Now there is no need to dive deeper than 50 m (164 ft) on air. As I described in Chapter 2, the nitrogen in air becomes increasingly toxic as one goes deeper, which our experiments in 1965 confirmed conclusively, proving that it was an advantage to change to helium-rich mixtures to maintain mental alertness. But in 1962 there were no such options, and John and I were hardened divers, familiar with our own responses to nitrogen narcosis, while the conditions at Marseille for deep diving in clear warm water were exceptional. As a result, we frequently dived below 61 m (200 ft) to depths that would now be regarded as excessively dangerous, as we shall see.

* * *

One morning René called for us in the *Sainte Marie* when the sea was exquisitely calm, blending at the invisible horizon into a soft mist.

"Will the weather hold?" I asked.

"Until midday at least," said René. "But there will be a strong westerly wind after that."

"Can we go to Grand Congloué, or is there time to get to the Triperie?"

"The weather will break, I am sure of it. There is not time."

"Well, can you suggest somewhere we could find really deep water then?"

We discussed the possibilities as we rounded Cap Croisette, where the flotilla of white islands sailed into view.

"You might dive on the Empereurs, M'sieur, but the current is often bad there, and we must certainly get into the shelter of Riou before midday because it is terribly exposed."

"How deep is it there?"

"Oh, I don't know exactly, but well over 60 m, 200 feet."

We sailed beneath the white towers of Riou until the twin coronets of the Empereurs rose before us, pinnacles of rock rising from the sea in gigantic imitation of the Pharillons.

"Let's do a little dive first," said John. "I want to test this new mouthpiece on the Normalair set."

"What's the idea?" asked Charles

"This valve," John explained, "is usually attached to a full-face mask, so that you can breathe through your mouth and nose together. It's good to use with underwater telephones and so on. But the big disadvantage is that it is difficult to clear your ears, and difficult to economize on air. The basic valve is so good that I've stuck an ordinary mouthpiece straight on it, and I'm going to use it with an ordinary mask."

"Do you want me to do snorkel cover?" asked Charles, as René secured the boat to a projecting rock on the outer pinnacle.

"No, I think you'd better kit yourself up with a full tank and regulator," I said. "Stay swimming on the surface and watch us, but don't go below 50 feet [15 m] unless there's trouble. René, no musseling today I'm afraid. In case we come up far from the boat I want you to be ready to cast off and pick us up quickly."

"Je suis à votre service M'sieur."

"Don't panic yet," said John. "I've still got to test my valve".

René and Charles helped us to get dressed. We checked the pressure in our cylinders, 150 atmospheres, and checked our valves doubly carefully. Then John dived in and swam to 9 m (30 ft). I followed him and watched as he rolled and twisted, testing the valve in every attitude. When he gave the OK sign to Charles we turned downwards.

Below the turbulent surface layer, the water was fantastically clear, and beneath us, the cliff dropped into infinitely darkening blue. The wall was coated in a glittering array of plants and sessile animals illuminated by the shallow sunlight. At 24 m (80 ft) the cliff bulged like the buttress of a Norman keep, but we coasted swiftly over the slope and again the wall dropped sheer. At 52 m (170 ft), where the light was dim and blue, the rocks were encased in brittle interpenetrating shelves of coralline algae. Still, there was no sign of the bottom.

When I looked over my shoulder to see that John was still with me, he gave the OK sign. I checked the pressure in my cylinder and continued down. At 61 m (200 ft) we landed amongst huge boulders with heaps of rubble, beyond which a slope of sand and stones fell away steeply. I kept my pressure gauge, watch, and depth gauge, in a row before my eyes, and scanned them like an aircraft instrument panel as we finned steadily down the submarine scree. We had only been underwater for a few minutes, so still had plenty of air reserve.

The rocks were now encrusted with phosphorescent algae which glowed vermilion, turquoise, and yellow in the dusk. The stones were spread more thinly and I could see white waves of shell shingle undulating down into the absolute midnight darkness. I checked the depth gauge – 83 m (272 ft) – that's enough.

Suddenly I felt light-headed as the instruments blurred before my eyes. I was possessed by a jolly feeling rather like when one has a pint of bitter before lunch after forgetting to have any breakfast. It was nitrogen narcosis, rapture of the deep.

I had been watching my depth gauge intently because I was afraid of this, so I immediately whipped round to see if John was affected. I signalled violently to him as I kicked upwards, gripping the mouthpiece between my teeth with the studied relaxation of a drunk pretending to be sober. I bumped into John as I soared upwards. I watched the depth gauge unwind, and at 61 m (200 ft) the mental fog cleared. To my right I saw the cliff, not ten yards away. I stopped my meteoric ascent.

John rose beside me in a cloud of bubbles, turning towards the cliff. We ascended slowly from 61 m (200 ft), checking all the notches and ledges which might be traces of old sea levels, noting the depths on my writing board until we reached 6 m (20 ft). I signalled to stop. With our hands hooked onto the rock to anchor ourselves against the current, we waited, fluttering our fins slowly to keep warm. Above us the bubbles spiralled and burst into the surface, and Charles swam down, hovering around like a guardian angel. After five minutes I signalled to rise to 3 m (10 ft), which puzzled John, so I wrote on my board "270 ft" and underlined it. We were getting very cold. John kept making query signs, pointing upwards, but I replied stay-put. After 15 minutes at ten feet, I gave the signal to go, and we floated very, very slowly to the surface.

The brilliant sunlight, the glare off the rocks, and the warm colours of the *Sainte Marie* seemed as unfamiliar as if we had been submerged for a week, not just half an hour. René and Charles helped us into the boat, where we stripped off our wetsuits to warm in the sun. The smell of the hot paintwork and wood of the boat seemed to come from a new planet.

"Christ!" shivered John. "Why did we spend so long decompressing?"

"We had to," I said. "We hit 270 feet on the bottom."

"What! I didn't understand why you wrote that on the board. I thought you meant that because the bottom was at 270, we had to turn back before we got there! Oh lord, you mean we really went that deep?" He looked astounded.

"Cheer up," said Charles as he rubbed John's back with a towel. "At least you've proved that the Normalair valve works at that depth."

"Why did you come up so fast?" John turned to me again. "You nearly knocked the mouthpiece out of my mouth."

"I'm sorry, I had the narks."

"That's funny, I never felt a thing. I had no idea we were that deep. Thought we were only about 200 feet. I wondered why it was so dark. You know, your narcosis may have been self-induced just because you knew how deep we were and were expecting it."

"Self-induced or not," I said, "it's real enough, and when it hits you there's only one thing to do. Go up."

René steered the *Sainte Marie* into a sheltered cove where we basked in the sun for half an hour.

"I may have had an advantage by having the valve so close to my mouth," said John, "because this narcosis business is caused as much by increased resistance to breathing as by nitrogen. The gas viscosity increases with pressure and density."

"You're probably right," I said. "By the way, have you checked how much air you've used?"

"There were 75 atmospheres in mine when we finished decompressing."

"Exactly the same as mine".

"You mean you only used half the air?" asked Charles.

"Yes," I said, "but if we'd stayed another minute or two on the bottom our decompression time would have been nearly an hour instead of 20 minutes, so it was only a reasonable safety margin."

When we were warmed and rested René sailed us round to the Moyadons, a group of rocks off the west end of Île de Riou, where Charles did a small survey while I covered him from the surface.

René took us next day to the Triperie, a long journey which terminated in a bay surrounded by high cliffs. We made several shallow dives on the way, so there was not much air left.

"I'll do this one if you like," said John. "René, how deep is it?" "Only 15 m [50 ft], I think, M'sieur. The entrance to the cave is directly in front of the boat."

John kitted-up and went over the side. I stood by fully kitted-up, while Charles snorkelled. I watched John's grey figure snake towards the dark space at the foot of the

Figure 4.15. Diver in submerged cave at La Triperie where John Woods retrieved a piece of stalactite in 1962. Stalactites and stalagmites on submerged sea caves show that they were once exposed to fresh air, and that water seeped through the rock above, becoming saturated with limestone. This karstic cave is located in the Triperie cove, on the west side of Cap Morgiou, in the Sormiou Calanque (Photo Jean-Georges Harmelin).

cliff. Then we waited. Nearly quarter of an hour later John glided out into the sunlight, rising towards the boat, bearing a chunk of white rock in his hand.

"It's full of stalactites!" he shouted and passed the rock to René (Fig. 4.15).

"How far does it go in?" I asked.

"About 50 yards [45 m]. The entrance is quite deep but the floor slopes steadily so that there's only about six feet [2 m] of water when you're right in there. There are hundreds of stalactites, and I stopped only when I couldn't squeeze between them. Right at the other end I thought I saw daylight, but it might have been a reflection of the entrance on an air pocket."

"Let's have a look at that rock," said Charles.

"I broke it off a stalactite," said John.

We all looked at it. The outer layer of marine encrustations was only a fraction of an inch thick, but on the broken face, we could see the concentric circles of deposition which must have been formed when the cave had served as a channel for freshwater flowing down to an ancient sea below the present surface (Fig. 4.16). On our way back we explored several half-submerged caves, which we entered above sea level and could then see the sunlight glowing through the water from a submerged entrance. In the next few days, John and I explored other caves with stalagmites and stalactites, which provided similar evidence to the Triperie.

Figure 4.16. Section of a stalactite found in the submerged cave at La Triperie by John Woods in 1962. Concentric rings of growth, some marine, some freshwater subaerial deposition, show the periods when the cave was submerged, and when it was high above the sea level, which had dropped during the Ice Age.

* * *

We breakfasted early at our usual café on the crossroads at the bottom of the Impasse de la Batterie des Lions. Five roads met at this point with the café occupying one corner of the junction so that people wishing to use that road had to walk between the tables.

The local events which could be viewed while we had our croissants and coffee were so varied that we came to regard them as a free cabaret. Sometimes it was two cats fighting up a telegraph pole; sometimes naked children fighting with plastic swords and space-guns, or a little girl wearing her mother's high-heeled shoes.

This morning we had a group of scoundrelly-looking ex-legionnaires who had been demobbed the night before. They all wore the same style of dark city suit with open-necked shirt. In the far corner sat a young Nordic legionnaire with button nose and floppy mouth, carefully drinking from a bottle of beer. Suddenly he threw his useless white kepi on the floor, stood up, shouting gutturally, "Last night I had a wife! Oh, how I had a wife!"

When I walked into town to collect the hired car, I met everywhere the wandering soldiers in their useless black suits. One asked me pathetically the way to the frontier. "What frontier?" I asked. "Any one," he said. I suppose he had an inborn urge to get away from places. He could not speak more than two consecutive words of any language I could recognize, so I put him on a bus to the railway station. The British newspapers were portraying the situation in Marseille as if the town was in a state of siege or military rebellion caused by the ex-legionnaires. Brief and local episodes of violence did occur, but we never experienced them, and the demobbed legionnaires were more disoriented and sorrowful than threatening.

* * *

We drove along the coast to Villefranche, where French divers had reported large caves. The deep bay of Villefranche is surrounded by forested mountains that stretch out in prominent capes on either side of the deepwater anchorage. The town, which is piled steeply into one corner of the bay, rises from a crowded yacht basin, near which we found the Observatoire Océanologique de Villefranche-sur-Mer housed in converted galley sheds. It used to be the Czarist Russian research laboratory with territorial diplomatic immunity until the director died in very suspicious circumstances in the early 1900s. The French then took control of it. We introduced ourselves to the present Sous-Directeur, Dr De Bougis, who kindly arranged for us to dive from the laboratory's boat at ten o'clock the next morning.

We always searched for the cheapest good meal but, as the days had passed, we became thinner and more obviously muscled and bony, as well as more darkly suntanned. As luck would have it, the restaurant which we chose for dinner in Villefranche was part of a small hotel, where many residents had not turned up for dinner. The manager, who must have been expecting more guests, encouraged us to eat as much as possible. We made up for weeks of minimum diet by consuming two or three servings of soup, several platters of cold saucisson slices, roast meat courses, followed by huge piles of cheese, grapes, and peaches. The patron and the waitresses watched in amazement but insisted on refilling our plates again and again until we could genuinely eat no more. The last fruit and coffee were consumed on the terrace beneath a dense canopy of vines where we unanimously voted the restaurant the best in France.

We slept in sleeping bags on the shore. I woke at seven in the morning when the bay was still deep in the shadow of the mountains. I went for a quick swim, finding that the water felt warm after the bite of the morning air. The sky clouded over after breakfast, so that, by the time we had boarded the boat and sailed out to the point where there was supposed to be a cave, it was almost raining. We dived all three together and quickly found the cliff which started at 12 m (40 ft) and dropped to 24 m (80 ft). For 46 m (150 ft) along its face, an immense overhanging cave was cut deep into the rock and the whole ceiling of this great chamber was thick with small branches of coral. After a dive to 52 m (170 ft) off Cap de Nice, we thanked the crew of the boat for their assistance, and by midday we were on the road back to Marseille.

* * *

On the last day at Marseille, we conducted a recreational dive. This is a proverbially dangerous occasion for working divers. Accidents tend to happen when you are not focused. As we swam lazily at 60 m (198 ft) near the Grand Congloué, I recalled the hit of intense narcosis at Les Empereurs at 83 m (273 ft), and I made an experiment to check how serious the narcosis was at 60 m (198 ft). I relaxed completely and did not try to reject or control it. John and I had been working at 60 m (198 ft) day after day and considered that we had partially adapted to the stress of narcosis as a result of the repeated exposure. The effect of my experiment was shocking. Within seconds I was overwhelmed by dizziness and vertigo as if I was spinning rapidly and falling from a great height. My jaw relaxed and the mouthpiece began to drift out of my mouth. Instantly I tried to regain control. I clenched my jaws and put my hand to my mouth to secure my breathing. It was harder to defeat the fuzziness in my brain. For a minute or more I struggled to regain control of my thoughts and decisions as the mist cleared. At last I restored my normal controlled thinking so as to complete the dive safely.

This experiment convinced me that narcosis was a serious impediment, even if you think you have adapted and have learned to live with it. Recent physiological research does not show changes with repeat dives indicating adaptation or acclimatization. On the other hand, experienced divers are convinced that they perform more efficiently and safely as the season progresses with deep dives. This suggests psychological habituation, especially when conducting repeated tasks or learned tasks of a routine nature. You learn how to cope, at least partially, with the limited speed of thinking and reduced reaction times, but the deficiencies are still there. Simply put, you get used to it. My frightening experience of nitrogen narcosis and its limiting effect on research under water provided the motivation for my later research on diving with mixed gases, including helium (see Chapters 2 and 8).

From Marseille, Charles took the train back to Germany, and, for John and me, the next port of call was Naples.

* * *

Bill Hemmings was studying fish shoaling behaviour at the Stazione Zoologica Anton Dohrn Napoli. He advised us that the sea close to the town was so filthy that diving was dangerous. He suggested that we should arrange with the director, Andrew Packard, to dive from the Stazione's villa on the island of Ischia. This island was one of my prime targets to see if we could measure ancient shorelines on volcanic lava, and I asked Andrew to make the arrangements.

We rose at six the next day to catch the Pozzuoli bus, which roared out of the city past smoking steelworks, an oil refinery, and magnificent new buildings in levitated concrete. White and red ferry steamers huddled against the quay at Pozzuoli where the fish market was in progress. Boxes of silver, blue, black, and red, fish were stacked everywhere, while fisherman shouted prices or shovelled ice and salt. The quays ran with crushed fragments of fish, and the paving blocks tipped as one walked so that black stinking water belched up and flowed over the stones. Little boys sneaked through the throng balancing trays of drinks, while the barbers shaved their customers seated on deck chairs.

The research station's lorry arrived with the filled diving cylinders, which we loaded onto the boat. We steamed out of the little harbour where the once famous Bridge of Caligula is now concealed beneath mounds of rubble and concrete. The petite cone of Monte Nuovo, which in 1538 poured ash and lava onto the town of Pozzuoli, rose green with trees at the sea's edge, beyond which the random crests of the dead and dying volcanoes of Solfatara loomed above the mist.

The island of Ischia is the accumulated wreckage of a million years of activity by a hundred volcanoes. While the ice age glaciers and the oceans waxed and waned, the volcanoes poured forth lava unceasingly, as one vent cooled and died, another burst out to take its place, piling up the mountains of lava and ash to create the peak of Monte Epomeo. The oldest volcanoes were eroded by the sea until only the solid plugs of magma that sealed their vents were left standing in the water. One of these sank below sea level, creating the shallow bank of Secca d'Ischia. As a literary aside, it is curious that commentators on the humorous novel *South Wind* by Norman Douglas always assume that the setting is Capri, although the geology, which is described in detail, is clearly Ischia. The novel is unique, so far as I know, in that one of the main characters is a geologist.

As the post-glacial sea rose for the last time it flooded the lower volcanoes to create conical bays with steep sides. The Greeks who landed on Ischia in the 8th century BC were forced to abandon the island because of the incessant flow of lava, but after the eruption of 400 BC, described by Strabo, there was a volcanic silence. Throughout the period of the Greek occupation of the Bay of Naples, there was no eruption of Vesuvius, although Strabo recognized from the shape of the mountain and the nature of the rock that it had been a volcano. Complete peace reigned on Ischia and Vesuvius until AD 63, when tremors shook Pompeii and continued intermittently until the final cataclysmic eruption in AD 79. Ischia was quiet for another thousand years but erupted once more in 1302, the last flickering spark of the embers that seem finally to have been snuffed out.

Leaving to the south the Castello Aragonese d'Ischia perched on its volcanic plug, we sailed into a green, tree-lined bowl, the flooded crater in which lies the little town of Porto d'Ischia. Ahead lay Cleopatra's barge in all its glory, moored majestically above its own golden image, abandoned only a few days earlier by Elizabeth Taylor. The film company scrapped the luxurious ship after the filming of *Cleopatra* with Richard Burton. Above the village on the eastern spur of the bay stood the square red villa belonging to the Stazione Zoologica Anton Dohrn Napoli. We disembarked, piling our kit on the quayside, where a man with a small motorbike and trailer immediately offered his services.

"How much to the Villa Aquario?" I asked.

"Two thousand lire, signor."

"Don't make me laugh! We're not tourists. We've come here to work." I turned my empty pockets inside-out to prove my point.

"All right, signor. For you, two hundred lire."

The tiny vehicle whined up the road out of the town, stopping after a few minutes in the yard outside a small hotel, behind which was the Villa Aquario. I soon found a tubby man basking in a deck chair on the seaward side of the terrace. He stood up as I approached.

"*Buongiorno, signore*. My name is Antonio and I am the caretaker here. What can I do for you?"

"*Buongiorno, Antonio*. My name is Flemming. I have arranged with Dr Packard at the Stazione Zoologica that we could stay here until Sunday."

"He has said nothing to me. Ah well, he seldom does. Never mind, there is plenty of space. My wife will prepare a room for you." He suddenly beamed with hospitality and I took an immense liking to him.

"I am sorry that we cannot cook meals for you, but you will find that there are many cheap cafés in town. I suppose you will want to go out in the boat to dive?"

"Every day, if possible," I said. "We want to dive at each point where the lava forms cliffs in the sea. I have a map here." I showed him the beautiful geological map compiled by Rittman in 1930. As so often with a man whose profession it is to know a piece of coast or land through direct experience, he looked at the map like a toy. He was obviously trying to see if they – the scientists – had got his island right.

Antonio agreed to take us out in the diving boat to dive at Punta Molino at four o'clock. I spent the early afternoon plotting the measurements of the caves at Marseille to see if they fitted into a consistent pattern. John, speaking as a physicist, was pleasantly surprised to note that the measurements, recorded with haste, and with no conscious bias when under water, gradually lined up on my huge sheet of graph paper. He conceded that my methods lacked some of the precision of the physics lab, but that it was reassuring to see that we had not been wasting our time. We were relieved that the data from Marseille provided rather good correlations between the depths of erosion in different locations.

* * *

At four Antonio helped us to carry two cylinders and weight belts down the twisted steps to the quay. We passed the artisanal boatyards, the barrel makers, the ice factory, the wine-bottling plant, and the mineral-water factory, which clung together at our end of the harbour away from the fashionable souvenir shops and cafés. Antonio's boat was a stubby rowing boat with a small inboard engine fitted amidships. We set out of the harbour and in ten minutes reached Punta Molino, which turned out to flank a plush bathing beach decorated with coloured umbrellas sheltering expensive tourists. A swathe of unforested land sloped back from the point, showing where the lava flow from the Arso volcano in 1302 had not yet been weathered to fertile soil. On the beach itself there were a few outcrops of lava, and more black rocks projecting from the water. I wanted to see if the lava flow retained any sign of its contact with the sea level at the time of eruption.

"What do you want to do?" asked Antonio.

"Just get a sample of that rock," I said, and jumped into the water with a geological hammer. As I swam to the nearest rock and began to chip a piece off it I heard Antonio asking John, "Why do you want pieces of stone? I have often taken biologists in my boat, but this is mad."

"We want to take the rocks back to England and have them dated in a laboratory by radioactivity."

"But why do you have to dive for the rock?"

"We want to find out how deep the old lava flows went into the water, and whether the sea eroded them deep down."

"Ah – I see, but it's still mad if you ask me."

I splashed back to the boat and handed the contorted grey lump of stone to Antonio.

"Now you know all about our work – eh. Come on, let's get over to the Castello."

Antonio cut the motor as we slid into the shadow below the castle on its colossal cliff, the remnants of one of the earliest volcanoes on the island. The rock was so weathered that twisted blades and sloping shelves of punctured stone thrust 2 m (6 ft) and more from the pitted mass. I jumped onto a low ledge and struck a sample off the tip with one blow of the hammer, then swam back to the boat.

To obtain a deep rock sample I slung my battered old yellow 80 cu ft (2200 litre) cylinder on my back and rolled into the water again. I eventually surfaced with a sample of rock from a depth of 23 m (75 ft) over 270 m (300 yards) from the cliff. We continued southwards past powdery, banded cliffs as colourful as a jar of sweets. The Grotta di Terra was a cave that had been cut into an intrusive flow of magenta rock, which rose some 15 m (50 ft) in the cliff face and then mushroomed out into foliated coloured layers. I dived in the cave and was nearly mown down by an amorous couple in a speedboat searching for privacy.

We cruised on past the crumbling yellow cliffs to Punta San Pancrazio, the southeastern corner of the island, formed by a lava cliff dating from a later period than that on which the Castello stood.

I took a sample from the surface rock before John dived. He sank rapidly into the darkness. Antonio stood up and rowed slowly, efficiently, keeping the boat over the bubbles with short fluttering movements of the oars. The sea was hypnotically calm. As the sun sank lower the pleasure boats and launches returned to the port to be replaced by the fishing boats chugging off to the horizon for a night at sea. John surfaced slowly, reporting that the cliff dropped sheer to 50 m (165 ft), below which was fallen rubble. He handed to me a rock covered in orange and purple algae.

Antonio turned the boat for Porto d'Ischia, cruising back past the procession of fisherman filing out into the dusk. The next day we dived on the north coast of the island, at Monte Vico and Punta Caruso the lava flow of 400 BC. On the journey out we met Antonio's brother's fishing boat, which we boarded to be introduced to him as an essential gesture of etiquette. There was a crew of six who looked extremely fit and well fed, engaged in sorting out lines. They proudly held up their night's catch of steely swordfish for us to admire.

We hove-to off Monte Vico, another very ancient volcano. I dived to 37 m (120 ft) and found the lava so hard that I could not break a sample off. I eventually found a small boulder, but could not break it in half, and so tried to swim with it to the surface. Because my rate of ascent was very slow, I began to breathe harder and harder until the cylinder was almost empty. Only 6 m (20 ft) from the surface I had to drop the rock and swim up on a single breath to the open air.

We sailed past the fleshy beaches of Punta Molino, past the fairy castle and the deserted cliffs of San Pancrazio, into the huge, semi-circular bay of Scarrupata, rimmed with dark cliffs which once formed the walls of an immense crater, now lost in the sea. At the far end of the beach, a spit of sand curved out towards another towering volcanic plug.

As soon as Antonio cut the engine I plunged overboard and swam to the cliff. With all our dives at Ischia we were forced to dive singly in order to conserve air, though we always had a full cylinder in the boat for the second man who was kitted-up in case of emergency. I descended cautiously to 61 m (200 ft) where the shadow of the cliff and the late hour of the day produced total darkness. At this point discretion got the better part of valour. I chipped a rock off the cliff face and returned slowly to the boat.

John dived next off Capo Grosso. The sun was very low in the sky as he left the boat, and already a bright star gleamed over Vesuvius in the east. Capri sailed away like a high-pooped galleon into the darkness, and the only sound was the creak of the rope toggles on the thole pins as Antonio paddled slowly after the bubbles rising from John's aqualung.

We took it in turns to sit near the hot engine exhaust pipe during the long cold ride back in the semi-darkness. At Molino, the neon signs on the roofs of the new hotels flashed their messages across the sea. The lighthouse at the harbour entrance was already beaming, and from Capri to Gaeta a hundred others answered. At the foot of the Ischia light was a statue of the Virgin Mary in an illuminated niche. It was an old tradition.

We dived once more the next day, on the foundered volcano which forms the shallow of Secca d'Ischia, before preparing to return to the mainland. Up at five in the morning on Sunday, we crept downstairs to find Antonio waiting for us in the hall, smiling gently as ever. We carried the gear down the hundred steps and loaded the boat for the last time. Antonio rowed us across the silent harbour and helped us transfer straight onto the ferry. We shook hands sincerely and parted. Antonio, like René at Marseille, was far more than just a boatman. He performed his job with such devotion to our safety and efficiency that it is unthinkable to have done the work without him. For every spoken word with people like that there are a thousand unspoken but understood.

* * *

The late Peter Gill was the first Englishman, maybe the only one ever, to serve as Chef de Plongée in a Club Mediterannée diving centre, in this case, Palinuro, in southern Italy in 1963. It was a great achievement at a time when superiority in all things to do with diving was supposed to be a French monopoly by right. The French reputation and domination of diving spread out from the sensational movies of Cousteau, combined with the unique Club Med formula of sunshine, good food, and diving in simple village accommodation. The Chef de Plongée who could manage boats, the boat crews, instructors, training of novices, and adventurous diving for some of the guests, all conducted with an ineffable quiet authority, was a man to be reckoned with. I knew that the Italian archaeologist

Figure 4.17. The upper and lower entrances to the cave at Palinuro, Italy, where Natalka and I dived in 1964. The cave goes right through the headland. Emerging at a level of 50 m depth (Photo Fabrizio Antonioli).

Alberto Blanc, one of the founders of prehistoric archaeology and the study of the ice ages, had, in the late 1930s, found bird bones plastered in cemented deposits on the walls of the caves at Palinuro where they could only be entered now by boat. He had reasoned that, during ice age low sea levels, the caves might have been occupied by humans. Since the caves were advertised in 1963 as tourist attractions for divers visiting the Club Med nearby, I arranged with Peter Gill that Natalka and I could stay at the village in the usual way, using some of the diving time and boat time to explore underwater caves.

The Club Med at Palinuro accommodated several hundred *gentils membres* who were housed in stone chalets with running water, not straw huts. Every day Natalka and I enlisted on one of the diving boats that was exploring the deeper caves (Fig. 4.17). Capo Palinuro is a beetling crag of melodramatic proportions extending into the Tyrrhenian Sea and plunging deep into dark blue shadowy water. At the surface on the side of the headland facing the village, there was a small cave entrance into which one could swim or row a small boat. This was the entrance that Alberto Blanc had used.

The roof of the cave descended at the back so that it seemed to end suddenly. But a glance downwards revealed something quite different. A turquoise blue glow glimmered up from the depths. Furthermore, the glow seemed to come from two directions, one immediately below the initial entrance, and one from the opposite side to the entrance, as if there was a tunnel right through the mountain. Over several days we made many dives on this convoluted cave. Equipped with measuring tapes and ranging poles we gradually pieced together its shape, and its possible origins hundreds of thousands of years ago.

Starting from the side where we usually dived into the cave we dropped quickly to 20 m (66 ft), depth where a rocky floor extended into the distance towards the pale glow of a huge submerged arch on the other side. Finning slowly over the rock floor we reached a steep slope of fallen stones and cobbles that descended to a sandy floor at 35 m (115 ft). This rubble scree fascinated me since I imagined that it might conceal artefacts such as stone tools or carved bones.

The far arch was completely submerged so that it could not be entered by boat from that side. It was the most colossal arch of blue that I have ever seen beneath the sea. But the submerged cliff concealed many more secrets. Outside the arch the headland seemed to continue with three successive pinnacles of rock, *secca*, that rose from 50 m (164 ft) depth and were cut off flat at the top at 20 m (66 ft) depth. One of the three *secca* was penetrated by another arch, where part of the roof had fallen in. The whole assemblage was the perfect example of fossil cliff shoreline features, like those in England showing rock pinnacles or stacks at Old Harry's Rocks and arches like Durdle Door. The Palinuro cave was eroded and shaped by waves on a sequence of fossil sea levels, at 50 m (164 ft), 30 m (98 ft) and 20 m (66 ft) below present sea level.

* * *

The repeated deep dives that we made in remote locations during 1962-63 were the last exuberant flourish of a style of diving that quickly became extinct. Day after day we descended vertically, spearing down like swordfish into the deep-blue gloaming of the wild ocean, then looking up to 61 m (200 ft) of water over our heads. We trusted our reflexes. We trusted each other and we trusted ourselves. We trusted the boatman. The sensation was an escape from the world of land. Liberating in the sense of being alone and free on a new planet.

With modern knowledge of the physiology of bends and the nucleation of bubbles in the blood when the diver starts to ascend, our conduct was dangerous. But it was permitted by the official decompression tables at that date. Modern divers and diving physicians will be shocked at what we did. But we survived, and we enjoyed it. In Chapter 8, I describe my role in promoting the improved regulation of safe diving by scientists in the 1970s and '80s.

During the years 1965 to 1987 I made a few dives in other useful places such as Malta, Mallorca, Ibiza, and Corfu, during which I measured a few more terraces and cave floor levels. These have all added some confirmatory details to the original measurements that I made for my thesis in 1961-63. Comparison of all the caves and rock-cut terraces that we dived on between 1961 and 1963, which included observations from Gibraltar (14 locations), Sant Feliu de Guíxols (four locations), Banyuls (three locations), Marseille (20 locations), Ischia (nine locations), and Palinuro (three locations), showed concentrations of cave floor depths at 10-15 m (33-49 ft), 24-30 m (79-98 ft), 50-60 m (164-197 ft), 80 m (262 ft), and deeper than 96 m (315 ft). This summary of the findings gives a feel for the volume of data obtained but ignores numerous qualifications and get-out clauses explaining local variations due to geology and tectonics. The results showed that submerged coastal caves and erosion features on ancient shorelines revealed unique facts about the changes of sea level and climate during the ice ages. The technique was proven. Anyone seriously interested should consult the original publications or the work by modern authors in this field.

In recent years a fresh generation of divers with more modern equipment and measuring and recording devices, as well as video cameras, has succeeded in adding many more details to the succession of rising and falling sea levels in the Mediterranean. The modern interpretation of multiple sea levels has become more complicated, but also explicable, as we have learned about the multiple glaciations and the so-called interstadials of sea level change when the ice caps half-melted and then froze again.

My early proposal that submerged caves sometimes contain prehistoric remains was not entirely wrong: some of them do. But I was wrong about the frequency with which large caves below present sea level might have been occupied, and the chances that deposits within the caves might survive the subsequent rise of sea level. While it is obvious that waves will break violently into the mouths of some caves, it is not obvious that this should be so destructive in all cases that no archaeological deposits ever survive. The caves on Vladi's Reef, Gibraltar, and one of the large caves at Villefranche, face away from the open sea, being eroded into the landward side of rocky spurs which protect the caves. Similarly, some of the caves at Marseille are between rocky islands so that one would expect them to be protected from the largest storm waves. But almost nothing survives. (For prehistoric sites on the seafloor in open water see Chapter 10.)

Since the many dives by John Woods and me into the caves of the Marseilleveyre coast, an astonishing discovery was announced in 1991. A local diving instructor, Georges Cosquer, and his comrades had explored one of the many tunnels near the Bec de Sormiou and the Calanque de la Triperie and found that it penetrated 140 m (460 ft) into the cliff, slowly rising, until the divers broke the surface inside a huge chamber that was isolated within the rock. Although the chamber was partly above sea level, there was no entrance from inland. The tunnel through which they had entered at a depth of 40 m (131 ft) was the only access.

As if this was not sufficiently astonishing, the divers found the walls of the cavern above the water were covered in patterns of hands imprinted from charcoal, and more rock paintings of birds and horses. For several years professional archaeologists thought that the reported paintings were a hoax but, gradually, experienced archaeologists who were also divers entered the cave and confirmed their genuine nature. Samples of the charcoal that were taken to laboratories for carbon-14 radioactive dating confirmed two active periods of occupation, at around 27,000 years ago, and 19,000 years ago, when the sea level was more than 100 m (330 ft) lower than now.

The topography of the Grotte Cosquer is so utterly unusual that no second example is ever likely to have existed. Tantalizing hints have been found by some French and Italian divers: a deposit of charcoal here, a single worked tool there, the bone of a deer in the back of a cave in Sicily, a probable burial just below sea level in Corsica. The findings are so sparse, that there is clearly something wrong with the supposition that many of the submerged caves were occupied. I have discovered since the 1960s that Stone Age occupation sites survive on the flatter areas of the seafloor, but not in caves (Chapter 10).

5

Round the World to an Octopus Farm

"Global Co-ordinator Two", said the sign on the door, "Dr Eric Swindlehurst".

This was in the UK head office of the global company Unilever, where anything could happen, located in the vast marble-columned curved edifice at 100 Victoria Embankment, London. The opposite door read only "John Worcester", with no job description or title. As I was five minutes early for the appointment with Swindlehurst I walked up and down the corridor reading door labels to see if "Global Co-ordinator One" existed, but apparently he (or she) did not.

At ten o'clock I knocked on the door.

"Enter!" snapped an authoritative voice.

I went into a magnificent office overlooking the Thames. Although it was sparsely furnished with grimly efficient office-type shelving, desks, lights, and a plain carpet, the panorama through the vast windows facing the Thames and Blackfriars Bridge was worthy of an impressionist painting. Behind the glass-topped desk, partly silhouetted *contre jour*, sat a lean dark-suited man with a long, narrow, intelligent face, neatly cropped hair, and deep-set questioning eyes.

"Good morning. Please be seated. You are here to explain how you would design and run research for the Commercial Oceanology Study Group, if we appoint you. Is that correct?"

"That's right. John Wren-Lewis from Imperial Chemical Industries suggested I should develop some preliminary ideas, and then present them to you at an interview. I have had a few weeks to think about it."

A secretary brought tea which I sipped while I prepared my pitch. I was about to take a big gamble. Late in the summer of 1965, I had completed two months of research diving with oxy-helium in Malta and Plymouth (Chapter 2), but my Department of Scientific and Industrial Research (DSIR) Research Fellowship at Cambridge had only a year to run, so that I would soon need new employment. Wren-Lewis, a neighbour in Chiswick who worked on research policy at the company Imperial Chemical Industries (ICI), had mentioned that he and colleagues in different large companies were puzzled by the spate of publicity being given in the serious news media, as well as in popular and technical magazines, to new exploitation of the seas including the deep ocean floor.

Were they missing something?

In: Flemming, N.C., 2021, *Apollonia on my Mind. The memoir of a paraplegic ocean scientist*, 181-220. Leiden: Sidestone Press.

Was there a bonanza that their own advisers had failed to identify? For more than a year there had been books, TV programmes, and dramatically illustrated features in magazines like the *National Geographic*, *Life*, *Time*, *New Scientist*, and *Scientific American* promoting optimistic visions of deep-sea diving, humans living in houses on the seafloor, humans farming fish and even farming whales while living under water, cultivating seaweeds, and mining manganese nodules at depths of 8 km (5 mi) in the central Pacific. John Mero had published his book on deep-sea manganese mining to which he added diamond mining for good luck. Jacques-Yves Cousteau, Ed Link, Robert Stenuit, and the US Navy were building underwater houses in which aquanauts lived for weeks at a time at progressively greater depths. Hans Keller had dived unexpectedly to 200 m (656 ft) in a Swiss lake breathing a secret gas mixture, while commercial companies and national navies were attempting more and more ambitious dives to depths in the sea that seemed quite frightening. Exploration companies had recently discovered oil and gas under the North Sea, while many other countries were prospecting for oil offshore. In South Africa they were already mining for diamonds in the shallow coastal waters.

The smell of money and frontier-style risk-taking were in the air. The bathyscaphe *Trieste* had dived to a depth of nearly 11 km (6.8 mi) in the Pacific in 1960 with a two-man crew, then searched for the lost nuclear submarine USS *Thresher* in 1963. General Dynamics, Westinghouse, Lockheed, and General Electric seemed ready to venture millions on civilian submarines and seabed dwellings. There was talk of cargo submarines that could take the shortest route through the Arctic regardless of weather or ice. General Electric, in their Bottom Fix project, planned for hundreds of people to live in spherical glasshouses at a depth of 3600 m (12,000 ft) in the central Atlantic Ocean (Figs 5.1, 5.2). Russian and Japanese factory ships were hauling unprecedented tons of fish out of the sea, the Peruvian anchovy fishery was booming, and pundits predicted that the world fish catch at sea could be trebled (or more), even while fisheries experts predicted that fish stocks would be destroyed by industrial overfishing.

The sense of excitement about exploiting the sea was a mixture of high adrenalin all-conquering adventure to obtain new resources that would help to save the world from starvation and poverty, combined with the beginnings of an environmental awareness of uncontrollable pollution and resource depletion. Rachel Carson had published *Silent Spring* and *The Sea Around Us*, alerting the public to the dangers of pollution. Some TV programmes predicting environmental disaster had already become so formulaic and ritualistic that they were grouped under the heading of "Eco-doom". There was a rumble of band-wagons rolling on every side. Behind it all, the great navies at loggerheads in the Cold War were conducting dangerous reconnaissance patrols in nuclear-powered spy submarines. There had been sensational accidents. The *Thresher* had disappeared in the deep ocean in 1963. The British public still remembered the catastrophic floods of 1953 in the North Sea and the loss of a de Havilland Comet airliner in the deep sea off Elba in 1954. It did not surprise me that responsible scientists and policy-makers in global companies wanted to sort out the truth from the dreaming and scare-mongering.

The NIO that I had visited in 1958 to enquire about changes of sea level at Apollonia (Chapter 1) had been formed out of teams of civilian scientists working in the Royal Navy at the end of the war. The Institute conducted research in the oceans all over the world, having been set up as a separate establishment in 1949, but it had no teaching role, and there were no university courses in oceanography in Britain in 1965. National

Figure 5.1. Futuristic concept of an underwater science city at several miles depth on the mid-Atlantic ridge proposed by American aerospace companies in 1965. Supplies would be delivered by submarines, and products taken back to the surface (Artwork © Ed Valigursky, reproduced with permission from the Estate of Edward I. Valigursky; credit: Popular Mechanics, February 1968).

Figure 5.2. More details of the ocean bottom city (Artwork © Ed Valigursky, reproduced with permission from the Estate of Edward I. Valigursky; credit: Popular Mechanics, February 1968).

marine research laboratories of a similar scope existed in the major countries such as the USA, France, Germany, Japan, and Russia. But, apart from more pragmatic research in direct support of fisheries in dedicated laboratories, most of this civilian work was curiosity-driven and open-ocean, with the lurking military requirement always in the background. Understanding how sonar sound-waves propagated in the deep sea would determine whether nuclear submarines could hide indefinitely or could be found by surface tracking ships, aircraft, or sonar listening stations. There was a wide gap of information and expertise between the civilian laboratories and the world of commerce, mining, shipping, and offshore oil and gas represented by my potential employer, the Commercial Oceanology Study Group (COSG).

Wren-Lewis suggested that I could use my practical marine experience to separate genuine economic and commercial opportunities from those popular obsessions that had no possibility of success. The appointment would be for one year only. It was an appealing chance to break out of the narrow cloisters of Cambridge, which had become for me rather stifling. But it was truly a big gamble.

I began to explain my ideas to Swindlehurst, while he sat relaxed and impassive behind his wide desk, on which there were only one or two thin documents. I suggested visits to many companies and laboratories. Today one might spend a few weeks on the web to obtain much of the information needed by the six companies collaborating in COSG, but even now it would have to be followed up with targeted visits to carry out discreet enquiries on tricky problems. Commercial confidentiality would always create obstacles. Big companies with successful developments underway do not publicize their achievements, at least until critical innovations have been patented. Even today a serious investigation would require seeing some of the machinery, the fish-farms, and equipment live in operation, since web data can be notoriously unreliable and self-indulgent. Back then all information was on paper or on microfilm in libraries. Telex was the fastest way of communication, and letters were dictated to and typed by secretaries. Since electronic databases did not exist, the only way to store and retrieve large quantities of correlated facts was on punched cards. Library catalogues were apt to be leather-bound books, sometimes on chains.

I rambled on: "It is important not to pre-judge or bias the outcome of this study. We must not start by collecting facts under headings that we expect to be profitable. The classification system must be neutral. So potential mineral resources should be indexed under the terms of the periodic table of elements. Organic and food subjects can be indexed at high taxonomic levels of marine phyla or genera. They can be correlated with geographic regions, and with each other, by using superimposed see-through punched coincidence cards. Environmental impact and economic issues can be grouped under wide headings defined by different markets and their external side-effects."

Swindlehurst leant forward, steepled his hands on the desk, lowering his head slightly as if peering over a pair of spectacles.

"Tell me. What is the environment?"

I groped in all directions in my mind to make a logical answer.

"Well ... when a company or government agency starts a new project, launching a new detergent or building a harbour, you calculate the costs of the raw materials, transport, machinery, wages, and so on, the transport of the product to the retail outlets, the price to charge, the profit margin. If you get the sums right the project is a success.

But there have been many other side-effects as you run your business that don't appear on your accounts or balance sheet. Destruction of the habitat where the materials were excavated or farmed, waste products from the factories polluting the surrounding countryside, noise and fumes from transport, packaging discarded after use. All these things might cause damage and costs to someone else, or they damage wildlife and possibly cause public health risks. These negative effects occur in the "environment" surrounding the project. They produce individual costs or social costs that are not seen as the responsibility of the commercial company. There is a lot of publicity now about these risks. The media are interested." I stumbled to a halt.

"Fashions like this come and go. It doesn't sound as if the facts have been measured accurately."

"You may be right," I said, not wanting an argument. I knew that the facts were much debated. "But instruments for the detection of biologically active chemicals are becoming much more sensitive and accurate, and the Minamata case of methylmercury poisoning on the coast of Japan has become a symbol of environmental dangers, as well as insecticides like DDT on land. I think that discoveries of unfortunate side-effects from industrial processes and food production are going to get more worrying. Scientists will pick them up first, then the media. It may not die away."

Swindlehurst looked sceptical. I tried to relax by drawing calm comfort from the beautiful view of the sunlit Thames behind him.

"If we offer you this job there would probably be a lot of travel. Is that acceptable to you?"

"I have visited many European countries and went to Libya twice as an undergraduate, and I have been to the USA once. Both coasts. I like travel."

"What about Russia, or Japan?"

"I would find that interesting."

Reaching to the papers on his desk, Swindlehurst picked up a document which he held delicately by one corner between finger and thumb.

"The *Journal of Sedimentary Petrology*," he read out. He was holding a copy of my first and only solo academic publication, which was just ten pages long.

"Yes."

"'The Sorting by Shape of Beach Particles during Cusp Formation'". He read out the title with heavy emphasis. It was the result of the experiments I had done in the wave tanks in the Geography Department at Cambridge.

"Yes."

"To what extent do you think that the analysis of the shape of beach sand and gravel particles qualifies you to make snap judgements on the world fish stock, or the use of submersibles in mining manganese nodules?"

The expression on his face was not unkind, there was even a hint of a smile. As if to say: let's see if you can dig yourself out of that one.

It did not seem wise to quote Blake's "To see a World in a Grain of Sand/And a Heaven in a Wild Flower/Hold Infinity in the palm of your hand/And Eternity in an hour." I tried a more mundane approach.

I explained that studying the shape of sand grains had taken me to the frontier of knowledge, albeit in a very restricted field, but that I thought that this experience would help me to sort truth from fiction. It would help me to separate plausible possibilities from absurd dreams.

"I will consult with my colleagues," said Swindlehurst, "and we will let you know within a week whether we wish to employ you in this job."

The next week I received a letter offering me a one-year contract to work for the COSG at a salary of £1800 for the year. Since my DSIR grant was £500 per year, albeit tax free, the offer promised great riches. I accepted promptly and resigned my DSIR Research Fellowship at Cambridge. The companies clubbing together to support the COSG were Unilever, ICI, British Petroleum, Richard Costain Construction, Hawker Siddeley Dynamics, and Rio Tinto Zinc. My office would be in the Costain main building near Westminster Bridge. Collectively they wanted me to find out the future economic and commercial benefits from the world oceans with a focus on unconventional technologies. What was boondoggle, what was sound development?

* * *

I started work for COSG in September 1965. Costain had a computer (it filled a large air-conditioned room), although I do not know its capabilities. Its performance was probably laughable by modern standards, but they used it for computations on commercial contracts like building a dam or a motorway. Geoffrey Rice, my boss at Costain, was very proud of it. For my own work, I relied on the superimposed punched card system that I had described to Swindlehurst.

When I moved into my office on the third floor with a view level with the rooftops across to the dome of St Pauls, I was proud to be in the real world, wearing a smart city suit. I had lunch each day on the top floor in the 111 Club for the Directors, where, within a few days, I found myself by chance sitting next to the Finance Director. Looking at me suspiciously he asked,

"When do you pay off?"

Meaning, when would my value to the company produce enough cash to pay my salary and overheads? With optimism I replied, "I think I will save you enough money to pay my salary in the first year, and the information I gather will enable you to make profits many times my salary within a year after I have left."

He looked happy.

* * *

Once in office I subscribed to a variety of new magazines and newsletters, some quite short-lived amidst the hectic novelty. These included *Geo-Marine Technology*, *Ocean Technology*, *Hydrospace*, *Sea Technology*, *Marine Systems*, and the journal of the new United States Marine Technology Society, and so on. These magazines were not responsible for the misguided and exaggerated stories that had infiltrated the popular press, since they focused on up-to-the-minute technical facts, accurate engineering drawings, recent commercial contracts, and government or military statements. They were read by experts who would detect errors immediately. More well-established magazines such as *Fishing News International* and *Oil and Gas Journal*, together with a stack of academic reference books, completed my reading list.

Notwithstanding my assertion to Swindlehurst that we should avoid assuming in advance what resources or services would be profitable, and what were impractical fancies, there were several activities that the sensational press had trumpeted so blatantly and repeatedly that I had to resolve whether they were feasible or not. If I failed to do so, COSG was a failure. While planning my deliberately wide and unbiased lists of contacts, proposed visits, enquiries, and surveys, four themes were often in the back of my mind.

The first big idea that the popular press and magazines were promoting was the proposition that divers could swim in the open sea and survive under pressure breathing special gases at depths of hundreds of metres under water, where they would live in underwater houses or habitats for weeks or months at a time. It was true that research and field trials were going on in shallow water, which were accurately reported, but there was no attempt to estimate the limits of this physiological exposure, or the costs or risks. In 1964 the US Navy SEALAB I trial was completed with four men living at a depth of 58 m (192 ft) off Bermuda for 11 days. The futuristic dreams of people living for months underwater, even in underwater cities, envisaged prolonged productive work on the seabed, often fish farming, operating oil wells, or mineral extraction. The fantasies were going too far. But where was the real limit?

The second concern was the mining of manganese nodules. These polymetallic nodules, first discovered on the seafloor in the 19th century, contain a rich ore of manganese, cobalt, copper, nickel, with other trace metals. The nodules look like potatoes or golf balls, and in some parts of the deep ocean, at depths of several thousand metres, they are scattered on the seafloor rather as if a lorry had tipped over and they had rolled all over the road. The thought that "they are just there to be picked up" was woefully alluring and deceptive. Experiments had already started with attempts to dredge up tons of nodules, but everything depended on the cost of the dredging and the ore-refining processes. A book by John Mero suggested that the recovery could be economic within a few years.

Next was fish farming and even whale farming. People in Europe, East Asia, and other regions have been farming freshwater fish for centuries. But not rearing sea-fish from larvae or small fish up to adult size. There were rumours that the Japanese were already cultivating many sea-fish species in cages, and discussion had started in Europe about the feasibility of farming salmon. These modest facts were extrapolated by the popular magazines to predict a sort of wild-west farming in the open ocean, with shoals being farmed in vast cages, supervised by divers on motorized underwater scooters. The most fanciful depictions were clearly absurd, but where was the real scientific and economic limit? What species could best be farmed where and how? Could fish farming only be applied to luxury species, or could cheap food types be produced in quantities that could change the global supply of protein? What food would farmed sea-fish consume?

Finally there was the possible usefulness, whether scientific or commercial, of small deep-sea submersibles capable of diving to many hundreds of metres with a crew of one or two pilots and a further one to five passengers or scientists. Cousteau's "Plunging Saucer" had already hit the headlines, and in the USA many large engineering companies had adapted part of their aerospace competence to manufacture submersibles. Fitted with large glass portholes, external remotely operated manipulator arms, industrial

Figure 5.3. The Aluminaut submersible built by Reynolds Metals Company in 1964, weighed 80 tons, and was used to find a lost hydrogen bomb in the Mediterranean (Virginia Historical Society: Reynolds Metals Company Records, 1919-2000, Series 6.1, Aluminaut, Mss3 R3395).

power tools, cameras, and other instruments, they could clearly do some useful work. But the question was: how much, in what sea conditions, and at what cost? What surface support ships did they need? Were they really useful, or just expensive toys?

These four sensationalized topics had to be on my "to do" list, but none of them exhibited the economic scale and significance of the offshore oil and gas industry, cargo shipping, port construction, subsea cables, or possibly even coastal protection and flood prevention. I had to analyse such well-established industries as well as measure their rates of change and technical progress or profitability. Were there unexpected technologies that could revolutionize these established industries? Other curiosities included proposals such as the Atomic Weapons Research Establishment scheme to extract uranium from seawater to support electricity generation, and there were also projects to extract energy from waves, tides, and ocean currents. Environmental monitoring and pollution prevention were only just creeping onto the agenda, but this was a growth sector in itself.

In late December 1965, while I was refining my year's plan for my new job, the first jack-up oil platform to work in the North Sea, the *Sea Gem*, collapsed with a loss of 13 lives. The horrifying risks, both for human safety and for large financial or environmental disasters, hit the headlines. The first big offshore technology in British waters had sent a warning of danger. Then in January 1966 the US Air Force lost a hydrogen bomb off the coast of southern Spain when a bomber broke up in a mid-air refuelling collision and crashed. The daily newspaper reports of the desperate search for the bomb in deep water drove home again the drama and risks of high technology at sea and the frightening limitations of even the best available equipment from the richest country in the world. The bomb was finally recovered from 870 m (2850 ft) depth using the 80-ton aluminium-hulled submersible Reynolds *Aluminaut* (Fig. 5.3). The search was led by John Craven, whom I met several times over the following decade, but I never began to guess his real technical genius, which was largely hidden by military secrecy, until the end of the century. He even wrote a chapter on submersibles for me in a book I would later edit on marine technology. Perhaps I contributed to the innocent cover he needed for his clandestine activities.

* * *

During autumn and winter of 1965-66, I visited the research laboratories of the member companies of COSG, including windswept fisheries labs in Scotland, oil and gas engineering research at BP's laboratory, the storm-lashed ICI marine lab at Brixham, as well as a range of industrial companies that I judged to be developing successful new technologies in coastal defences against erosion, anti-fouling paints, diving, submarines, and offshore engineering. I built up a list of government laboratories and companies across Europe that I wanted to visit, while making long-range plans for the Soviet Union, the USA, and Japan for late in 1966. I had to assess the competence of countries that would compete with the UK. The travel department at Costain used telex for foreign reservations and hotel bookings. Airline tickets could be exchanged for a different route or date at any stage of a journey, without any penalty, so that to own a ticket gave one a princely feeling of autonomy and freedom. A licence to go.

I visited Southampton University, where there was a department of marine biology, and a similar department at Bangor. In both cases there was an awareness that the emphasis on biology and fisheries needed more support from physical oceanography, marine chemistry, the study of currents, tides, waves, and mixing of the seas in coastal waters, combined with the geochemistry, and interactions of the water with the seabed. The traditional laboratories tended to divide their effort between exploratory "blue skies" research in the deep ocean, like the NIO, or fisheries biology in coastal seas, as at Lowestoft and Bangor. At the NIO the early research in the physics of waves, both in deep water and as they broke on the shore, had already begun to produce economic benefits in coastal engineering. Later, the wave research produced benefits of hundreds of millions of pounds for the North Sea oil industry. Dr Deacon, Director of the NIO, and several of the senior staff were keen to improve relations with industry further, and they supported me strongly in 1967 when I was helping to launch the new Society for Underwater Technology (Chapters 8 and 9). In 1965 the gravel and dredging industry was dominated by the Dutch, and Costain had several joint ventures with them, so that there was a very strong COSG interest in seabed processes in coastal seas.

My first European tour was early in 1965, with snow and frost still in the air. I flew to Paris from Heathrow, departing in a scudding wind of drizzle that ruffled the sheets of water on the tarmac. My targets were the Institut Français du Pétrole offshore oil drilling research lab in the suburbs of Paris, the coastal engineering hydraulic laboratories of the Société Grenobloise d'Études et d'Applications Hydrauliques (Sogréah) in Grenoble, and then down to the coast at Marseille. There I sat in the cold air beside the Mediterranean Vieux Port, falling in love with the city all over again, but now in a very different mood. The colour of the Mediterranean is so different in winter, with the wind scattering patches of indigo on slate and flecks of silver beneath a grey sky. Cold blasts cut through my tweed jacket while I sat in cafés muffled up, clutching cups of coffee, watching the turbulent sea. Is this the weather that previous generations sought in winter to get away from British gloom and snow? They were hardier than I had imagined.

I worked eastward by train along the coast, visiting companies and laboratories developing deep-diving systems, saturation diving support chambers and diving bells, new gas mixtures, and submersibles, with my journey ending at the Oceanographic Museum in Monaco where Jacques-Yves Cousteau, rather controversially, was Director.

The Groupe d'Exploration et Recherche Sousmarine in Toulon and the COMEX company nearby in Marseille had pioneered deep commercial diving. COMEX, founded by Henri Delauze in 1961, had many contracts to support operations on offshore oil rigs. It has been a world leader ever since. The two laboratories both competed and exchanged some data. They led the world in deep dives, with COMEX making working dives to 152 m (500 ft) for 24 hours in 1958, experimental sea dives to 1000 ft (304 m) in 1966, and later working dives to 152 m (500 ft) with saturation for 100 hours in 1968. As scientists pushed forward the frontier of depth and duration in terms of engineering, the fundamental physiology of how the gases affected the human body created one problem after another. Physiologists and doctors studied these not only to avoid serious accidents but also to minimize long-term health effects. In the laboratories there were huge pressure vessels, tanks, control panels, banks of hundreds of cylinders storing gases at high pressure, with piping and cables running along every wall and ceiling. The seriousness of French investment in deep ocean technology and the competence to work underwater was impressive.

Jacques-Yves Cousteau was, at this date, director of the famous Musée Océanographique, established by Prince Albert of Monaco. The Musée was a sombre, almost Victorian style, museum building with musty exhibition galleries branching off long dim corridors. Although I had met Cousteau on several occasions to exchange a few words at conferences, on this visit I met his chief engineer, M. Alinat, and then various scientists, particularly Günther Giermann, who had made many dives in the "Plunging Saucer". Giermann's dives to study the detailed geology of the seafloor in the Strait of Gibraltar revealed very clearly how a scientist in control of a small submersible could use on-the-site judgement to follow traits and clues that would help to solve scientific questions. At depths of 100-200 m (330-660 ft), he could pause over unexpected outcrops of rock, back-track to check or compare observations, take photographs of key features, and apply his expertise continuously as if he were on land. This flexibility in decision-making remained the strong point justifying the use of piloted submersibles throughout the 1970s and '80s, finally giving way to sophisticated remote operated vehicles and robotics in the 1990s.

I returned northward through Switzerland, where I spent an afternoon with Jacques Piccard, one of the pilots of the bathyscaphe *Trieste*, which had made the world-record-breaking descent to a depth of 10,916 m (35,813 ft) in the Challenger Deep of the Pacific Mariana Trench in 1960. Sitting together in his warm drawing-room in Geneva, while the winter winds howled outside, we discussed his plans for large tourist submarines that would transport paying customers to explore the ocean depths. These plans never came to fruition. Land-dwellers are seldom prepared for the devastating limitations of poor visibility underwater, even in the most ideal conditions, and this effect has doomed many attempts to exploit underwater tourism. Divers are used to it, but the first-time tourist in a submarine with portholes is always disappointed. I did not express this sceptical opinion. One more cup of coffee, and thence back to London to submit my reports to the COSG companies.

* * *

Through the spring I visited companies and laboratories in Germany, the Netherlands, Poland, Denmark, Sweden, and Norway. Each country provided examples of excellence. In Germany it was instrumentation and the study of manganese nodules; in the Netherlands, the design and operation of dredgers working into deeper waters – more than 33 m (100 ft); in Poland, the shipyards and research laboratories at Gdansk, Sopot, and Gdynia, close together on the Baltic coast, specialized in the design and operation of factory fishing fleets and their skirmishing trawlers. Most vessels in the so-called Russian fishing fleets were built in Poland. The modular design, economics, and logistics of the fleets were carefully calculated, but not the environmental impact. Efficient fishing fleets compensated for the inefficiency of Soviet agriculture. Easter was approaching, but the wind off the Baltic bore the sharp flavour of ice. In the Polish villages, groups of schoolgirls wore decorated peasant skirts and were escorted by nuns. Most of the traffic was horse-drawn in the countryside, with the plodding heavy horses walking on dirt roads either side of a narrow tarmac strip for the few cars and lorries.

My travels in Norway, after visits to the impressive laboratories in Bergen, culminated with an excursion to an offshore island near Bergen where I saw what must have been one of the first salmon farms in Europe. An enterprising fisherman, with no scientific credentials, had dammed off sections of rocky creeks and bays, separated by metal or netting grilles, where he raised the salmon progressively, stage by stage, until they were adult fish swimming in small shoals in the largest enclosure. He had worked out his own methods of feeding them, ensuring a flowing water supply, and preventing diseases.

I stood with the proud fisherman on the banks of one of his pools, gazing in astonishment at what he had achieved. He towered above me, radiating the strength of his broad Nordic frame, hardened by many years at sea. But today he was nervous and apprehensive.

"Who do you really represent?" he asked.

"I'm working for a group of companies, just gathering information. We have no immediate plans for investment. The company most interested in fish cultivation is Unilever."

He shifted his footing nervously, twisted his hands together, and looked longingly over the crowded pools all jumping and alive with fish.

"I was afraid that would be the case."

"Why afraid?"

"Look what I have done all by myself." He gestured with a wide sweep of his arms. "It has cost me so much time and money. My wife is furious with me. But I cannot patent it. Anybody can copy what I have done. I have no clever lawyer. I cannot raise any outside investment. They will steal everything, I am sure. I am really afraid of that."

The spring sunshine glowed on the wet rocks at the edge of the pools, and the new shoots of leaves and grass made the whole place seem like heaven. The fisherman was torn between a wish to boast, quite legitimately, about what he had achieved, and a fear that all his skill and personal knowledge would count for nothing if a big company chose to buy him out. He was terrified that he would be cheated for a pittance.

"Have you thought of raising the fish in cages moored in the open water?" I asked,

"The cost of the cages would wipe out all the profit. I have thought about it."

I tried to reassure him as we walked back over the hilly ground to the dock where the ferry came in from the mainland, but I had no authority to commit or promise anything. In my next report to COSG I tried to explain his fears using calm bureaucratic language, but I do not know if he was ever approached by Unilever, or whether his own career in salmon farming prospered. I hope it did.

* * *

It is difficult to recall now the awed respect and fear inspired by the Soviet Union's engineering and science in the 1960s. There was the so-called "missile gap" with more powerful and greater numbers of Soviet nuclear-tipped missiles aimed at the West. The first Soviet Union space satellite was launched in 1957, followed by the first manned space flight by Yuri Gagarin in 1961. After several more triumphs, the first human spacewalk outside a spacecraft took place in March 1965. It was only decades after the death of the USSR's chief rocket designer, Sergei Korolev, in 1966 that anyone in the West understood the magnitude of his achievements. Every time the Americans tried to catch up in space, the Russians seemed to make another jump further into the lead. At sea the Russian factory trawler fleets were reputed to be extracting unimaginable quantities of fish from the open seas on a global scale, including in the Southern Ocean, leaving competing coastal trawlers with nothing to catch. The USSR had exploded its first megaton hydrogen bomb in 1955, much earlier than the USA expected. Soviet nuclear submarines were a military threat, though I was in no position to comment on that. The USSR was reputed to have built small submersibles, and an underwater house in the Black Sea, while Soviet hydrofoil ships provided routine high-speed services in the Baltic and Black seas and on major rivers. The hydrofoil design was quite different from the designs being developed in the West. These technologies, combined with formidable achievements in theoretical oceanography, meant that I must make an investigative trip to the USSR.

There was no point in my visiting the USSR alone, since I did not speak a word of Russian, and the state tourist agency, Intourist, was well known to be a branch of the KGB dedicated to preventing tourists from wandering around unsupervised. I had a friend at Cambridge University, Joseph Uscinski, who spoke fluent Polish and Russian, who was a physicist specializing in fluid dynamics, and who was delighted to accompany me as interpreter.

Joseph, who had a fascinating background, was a lively and reliable companion, brimming with initiative and ideas. He had been born in Australia, into a Polish émigré family that had originally fled eastwards to China from the Russian Revolution in 1917, and then fled again southwards from China to Australia in 1949 when China became Communist. He had studied mathematics and physics in Australia, then won a place to study high-altitude atmospheric physics for a PhD at Cambridge. He was keen to meet his opposite numbers in atmospheric research in Moscow.

It would have been absurdly dangerous to plan the trip in a way that made my enquiries seem to be clandestine or subversive, let alone spying. I had to be completely open about my intentions. I would rely on the accepted conventions of free exchange between scientists to visit their laboratories. If officials said no, then I would retreat immediately and never push my luck. If they said yes, then Joseph and I would visit the labs and try, by normal, friendly enquiry (plus, perhaps, some tactful flattery), to learn from the Russian scientists' experience. I did not assume that the Soviet agencies would play by the unspoken rules, but it was worth trying.

Joseph and I registered for the Second International Oceanographic Congress in Moscow that was organized by the International Scientific Committee for Oceanographic Research from 30th May to 9th June 1966. The advance programme showed over 1500 scientists attending from 57 countries, with nearly 500 papers to be presented in numerous parallel sessions. I searched all available catalogues and international directories to identify the marine laboratories and research institutes that existed in European Russia, comparing them with the technical problems that most interested COSG. There were a dozen labs ranging from the Black Sea coast to Leningrad on the Baltic coast, and to Kaliningrad for the big Atlantic fisheries fleet.

We obtained USSR visas for the total time of the stay including excursions after the conference. The visas were small bits of densely printed paper clipped into the passport, not stamped on the page itself. They could easily be confiscated by the police, so I made several copies. Intourist compelled Costain to purchase expensive first-class coupons in advance for all accommodation and every meal, including a first-class car and driver for tours at weekends. The coupons were only exchangeable for lodging or food at establishments operated by Intourist itself. I protested to the Costain travel department that, with Joseph as guide, we would spend much of our time eating cheaply in cafés or restaurants off the usual circuit, but Intourist insisted. They also said that if I returned the unused coupons in complete unbroken sheets for each day at the end of the trip, the Intourist Office in Russia would refund the cost of the unused coupons on the last day.

Joseph and I flew to Moscow two days in advance to acclimatize during the weekend. We stayed the first days like tourists at the old-fashioned 19th-century Centralny Metropol Hotel near Red Square, before moving to the colossal Stalinist skyscraper Ukraine Hotel where most of the foreign conference participants were accommodated. Three grandiose pyramidal skyscrapers from the Stalin era still dominated the city like George Orwell's Ministry of Truth that dealt in lies; the Ministry of Peace that dealt in war; and the Ministry of Plenty that dealt in famine. Joseph, who was familiar with the émigré Russian community in Australia, blossomed in his first experience of 100% Russian life. But his old-fashioned classical Russian accent caused some amusement. In the heat of a Moscow summer, I developed a taste for *kvass*, the cheap beer sold from tankers at street corners, and a further taste for *okroshka*, the cold vegetable soup flavoured with dill and kefir, a sort of weak, rather liquid, yoghurt. I remember *okroshka* as also having some *kvass* in it, but modern recipes do not show that as an ingredient.

The opening conference session took place in the Great Hall of Moscow State University, all endless rows of white marble columns and gold curlicues. Then we scattered to Lenin University, on the Lenin Hills, where Joseph and I tried to identify and chat up the speakers I had listed as important contacts. We found it difficult to break the ice, as the Russian scientists were reserved and wary.

"We've got to find some divers," I told Joseph. "I'm sure we can make friendly contact with them."

We had the street address for the Russian State Sports Diving Club, the equivalent of the British Sub Aqua Club, whose members, I hoped, would know scientific divers. We found the club offices in a suburban building. We were soon gossiping and chatting in Joseph's perfect Russian interspersed with broken English by the club staff for my benefit. Their hospitality was almost overwhelming. Coffee, vodka, and small cakes appeared as if by magic. We discussed diving training, different kinds of breathing regulators, and the problems of motivating sports divers when you live a thousand miles from the sea with very cold winters.

"We are attending the International Oceanographic Congress at the Lenin University," I explained, "and I want to meet scientists who dive in their research, or who do research on diving physiology. Are there any people in the club who are scientists, and who might be attending the conference?"

After a hurried discussion in Russian, they gave us two names, Pavel Borovikov and Slava Yastrebov, who both worked at the Institute of Oceanology in Moscow and were attending the conference. They made some phone calls to the Institute and arranged that Borovikov would meet us in the foyer of the University the next day.

Our meeting with both Borovikov and Yastrebov began a friendship which lasted for a decade, including their visiting London for a conference on diving science in 1969, after which they stayed as guests in my house. In Moscow, after a few mutual introductions and discussions, everybody became helpful and welcoming. I am sure that all invitations, visits, and discussions were reported to the watchful security officials but, for some reason, our intrusions and enquiries were tolerated. A great mailed Soviet fist might have grabbed us and crushed us like helpless worms at any time, but it did not.

The narrow path of permitted excursions was revealed a few days later when a man who had invited us to his fisheries laboratory in Kaliningrad met us the following day in a state bordering on mortal terror. Beads of sweat ran down his face, which had turned a ghastly pale grey-green. He trembled. He had been told very bluntly that he should not have invited us. One could scarcely imagine the threats that had been made. Joseph spent a long time helping him to calm down, after which we went to the conference president, Academician Vinogradov, to explain that the invitation had only been issued at our request. The scientist had not done this spontaneously. If there was any fault, it was entirely ours. We saw the Kaliningrad scientist in the corridors several times later, without renewing any conversation, but he seemed relaxed and calm again. Episodes like this occurred every now and then in the margins of our stay in Russia, reminding us constantly of the penalties of life in a totalitarian state. In every hotel so-called "dragon ladies" sat in windowed cubicles on every corridor noting who went in and out of every room, and presumably informing the KGB when they had time to search your luggage.

At the weekend halfway through the conference, Borovikov invited us to go to a Moscow park that was famous for its gardens, where there were a ruined basilica and some open-air cafés. The basilica had probably been bombed during the war. The day was beautiful, and we felt at ease talking about the conference, diving, experiments with helium gas mixtures, and the importance of trying to maintain the intelligence of scientists or engineers when working in deep water. I am sure that neither of us mentioned any technical details that could not have been obtained by careful reading of open literature, but the spirit of mutual understanding, constrained inevitably by cautious security, recognized both ways, was exactly what I had hoped for. As we entered the deep shadows of the bombed basilica, Borovikov pointed up to the half shell of the shattered dome, where some bright mosaic pictures and icons had survived the explosions.

"Look," he said. "There is a picture of the first diver!" he was pointing to an elaborate cartoon-like image of Jonah inside the whale, dramatically surviving under storm-tossed waves, flanked by other pictures in a biblical series.

We all scrutinized the mosaic for a minute or so. Its near-perfect preservation was extraordinary.

"I don't think Jonah was the first diver," I said gently. "Gilgamesh was the first diver."

Borovikov turned to me in surprise.

"Who was Gilgamesh?"

"It's from Sumerian legends, about 2000 BC. Much older than the bible, but there is a beautiful story there of Gilgamesh of Uruk diving into a lake to retrieve a magic rose that gave eternal life. He tied a stone around his feet so as to get deeper."

"They don't teach us things like that," said Borovikov. "We are not allowed to learn about religion."

"You don't have to believe in a religion," I said. "I'm an atheist myself. But you can study and learn about these wonderful stories from the past as legends, myths that helped to build societies in the way they were. We don't have to believe them. They are just beautiful stories from the past."

Borovikov looked at me long and hard, straight in the eyes. Then he spoke very slowly and clearly, emphasizing every syllable.

"You know that." Pause. "I know that." Pause. "But they don't know that." We all understood what he meant.

We turned back to the gardens, relaxed, found a nice shady place to drink beer, and, I hope, felt that we understood one another at a level that the authorities might not have approved of.

* * *

We connected together the schedule of our visits, which included a flight to Simferopol in Crimea, several days at Yalta with visits to laboratories and research ships in the great military port of Sevastopol, a flight across the Black Sea to Sotchi, then a train journey to Gelendjik, where there was an important marine geophysics institute. Back to Sotchi for a flight right across Russia to Leningrad, several Arctic institutes there, and then back to Moscow, and via a stopover in Warsaw, to London.

Joseph flagrantly flirted each day with the same beautiful clerk at the long Intourist counter in the Ukraine Hotel. Irritable queues shuffled and jostled on both sides. One link of the journey after another seemed to have all seats sold on the required flights, and we could only get on board because she knew someone in another office who could release seats that were pre-allocated for a trades union or a politician. Her persuasiveness on the telephone was triumphant. As the last part of the trip clicked into place, we bought her an enormous box of chocolates at the "tourist only" shop, for which she was very grateful. Many other conferees gave up trying to arrange tours, or got stranded, or in one case ended up in a prison cell only to be rescued by his ambassador.

One evening we bought tickets for a performance of Rimsky-Korsakov's opera *The Legend of the Invisible City of Kitezh*, which is seldom produced outside Russia. There are the usual plots with a maiden from the forest, a love-struck prince, and a deceitful traitor, but a turning point at the centre of the opera is the sack and looting of the town of Little Kitezh by Mongols before their savage assault on the capital city. For a good five minutes the vast proscenium arch of the Bolshoi Theatre was filled with action from top to bottom. Cannons boomed; columns collapsed; city walls and gates foundered; swords flashed; pistols cracked; flames roared to the sky and panicked horses stampeded. Smoke drifted into the auditorium. The looters carried off gold and silver and vases as they ran away with women thrown over their shoulders. Then the looters came back and rolled up the carpets from the ruined houses and took those too. The audience went wild. You always know where you are with the Mongols: they steal the carpets. The orchestral crescendo crashed with a last drum roll of terror. I won't give away the secret of the invisible city in the plot.

The final conference banquet was in the graceful new marble Palace of Congresses in the Kremlin. Long tables seating hundreds of people on each side were placed in parallel rows so that over 1000 could sit for dinner. Every table was decorated along its length with vases of flowers interspersed with clusters of bottles containing various wines, amongst which, as we soon found out, one of the bottles that looked like white wine contained a strongly fortified liquor, almost like a cheap cognac. This ensured that the party accelerated rather more rapidly than might have been expected.

I sat with a group of friends and we calculated that each setting of four bottles of liquor was intended to serve six guests. The meal was sumptuous and, as the wine flowed, people began to go up to the top table, taking the microphone to propose toasts of varying solemnity or absurdity. Hundreds of white-coated waiters placed bottles of vodka and shot glasses by the empty wine bottles on every table. For unknown reasons, one of my neighbours bet me that I would not eat the flowers in the vase in front of us, which I promptly did. I was informed later that tiger lilies, which formed a central part of the floral bouquet, are very poisonous, but I experienced no ill effects, perhaps due to the previous consumption of cheap cognac and vodka.

The next morning, we took off for Simferopol, the capital of Crimea, in a half-empty aircraft which climbed at an alarmingly steep angle, seeming to have been only recently converted from an earlier life as a four-engine bomber. The pilot banked to the south with an abruptness that suggested he was avoiding an attack by fighters. The canvas seats were primitive, but we were served huge quantities of iced caviar followed by strawberries and champagne, which helped to pass the time. The absence of fresh vegetables or fruit in Moscow, even in mid-summer, had been depressing. I developed a passion for large glasses full of kefir every morning at breakfast, only discovering later that it contains as much vitamin C as fruit. I felt a constant pang of hunger for green vegetables or fruit.

Several of the other passengers on the aircraft were conference participants who had also managed to obtain permission to visit Sevastopol, which was, in Russian terms, a "closed city". That meant that you had to go through the equivalent of a passport checkpoint to enter or leave the city, and non-residents had to leave the city at night.

Our party was staying in a small seaside hotel in Yalta, on the Crimean coast 40 miles east of Sevastopol. In spite of the historic importance of the Yalta meeting in 1945 between Churchill, Stalin, and Roosevelt, this was seldom referred to, and Yalta was just a cheerful holiday town, where the adjacent tsarist-era palaces, scattered along the mountainous coast, were converted into rest homes or holiday camps for mine workers and other deserving folk. It was a good use for them.

We had a Sunday in Yalta before the bus journey to Sevastopol. The Intourist coupons thus provided us with a large car, accompanied by a beautiful and lively guide, to visit the famous Tartar palace inland at Bakhchysarai. Tanya, who spoke excellent English, had freckles and a glint of Scottish copper in her hair. The three of us struck up a lively conversation with quick back-chat and jokes, which we enjoyed at a café table while planning our 40 km (25 mi) journey to the palace of the Tartar Kahns. The rapport was so immediate that if Intourist or the KGB were trying a honey-trap they could have taken any number of suspiciously intimate looking shots of our heads close together smiling and laughing. But why would they bother? Tanya's conduct was scrupulously correct throughout the day, which just happened to be the day of the national elections. Voting was compulsory, she said.

On the journey through the mountains, sitting in our luxurious Zil car, driven by a peak-capped chauffeur, we discussed life, the universe, and everything. Tanya was convinced that the world was really run by a secret cabal of rich Jews and defied us to say that we had ever met a poor Jew. Joseph and I both said that we had. When we asked her why Sevastopol was a closed city, she refused to believe that there were no closed cities in England, or indeed anywhere in Western Europe. "Does your government have absolutely no secrets that it wants to hide?" she protested. We tried to explain that secrets could be hidden without such a show of force as to have roadblocks counting people in and out of a whole city. We may not have sounded very convincing, given the recent defections of Burgess, MacLean, and Philby to the USSR.

As we drove into the small town of Bakhchysarai the asphalt road turned to dust and gravel between overhanging timber houses rich with balconies and awnings. The Tartars were warlike and violent but the old palace expressed isolation, peace, beauty, and intimate calm, with shaded courtyards and Islamic geometric screens in perfectly balanced proportions. We gazed at the Fountain of Tears, dedicated by the Khan Qirim Giray to his dead beloved, the weeping drops of water falling slowly from marble rose-petal to marble rose-petal. It is the most understated and most moving monument to human passion. When I saw the fountain, I had not yet read Pushkin's dramatic and very long poem about it, written in 1824. The fountain speaks for itself in pure silence. Solitude reigned everywhere. During our wanderings through the cool rooms and tranquil courtyards, we did not see another soul.

After lunch in a nearby café, supplemented by a mug of *kvass* for me from a tanker in the street, we took a leisurely walk back to our car. The driver, who had been drinking something much stronger than *kvass*, set off like a rocket, swerving around corners at maximum speed. We were stopped twice by speed cops, who were very unfriendly, especially the second time, but Tanya charmed them, and we escaped without trouble. The next morning during breakfast Joseph translated the newspaper, from which we learned that the Communist Party of the USSR had been voted back into office with a majority of 99.7%, and there had been no voting irregularities. They counted fast in those days.

We boarded the bus for Sevastopol. The coast road west from Yalta curved through the foothills, overhung on the landward side by vertical cliffs, but sloping to seaward through sub-tropical woodlands and gardens, down to the Black Sea. Improbable tsarist palaces perched on headlands and rocky pinnacles. Most improbable of all was the so-called Swallow's Nest, at Haspra. The crenellated towers looked as if they would collapse into the sea in the first storm. The bus was stopped by the roadblock and military checkpoint outside Sevastopol where we all showed our passports. The security men took the passports into their bunker for half an hour before waving us through.

As the bus descended the hills into Sevastopol we saw spread out below us acres of dockyards dotted with grey warships, interspersed with store sheds, warehouses, cranes, and a few civilian cargo ships. Anyone with good naval experience and a notebook could have recorded the movements of every warship, its state of repair and maintenance, when weapons were being loaded or removed, and so on. Perhaps this is so true of all military ports that the information is effectively common parlance and thus useless to both sides. Our party was escorted around a large research ship, with lectures on their research cruises in the Atlantic and Indian Oceans. While the overall picture confirmed their global interests and capability, there was no mention of research on sound transmission through the ocean, which everybody knew to be a key issue in the Cold War confrontation between fleets of nuclear submarines.

The instruments on board the research ship, both on the bridge and in the laboratories, were more or less standard, though bulkier than the equivalents in the West. In several laboratories there were some benches and counters draped in sheets that were carefully pinned down. We were told apologetically that these concealed instruments that were confidential. Many years later a Russian colleague told me that there was nothing secret under the sheets and that this was just done to confuse the wicked Westerners. After visits to several institutes in the city, including the Hydro-meteorological Institute and the Oceanology Institute, we wearily boarded the bus for Yalta and were counted out of the city by the KGB.

* * *

Joseph and I flew to Sotchi the next day, a famous resort from the tsarist era, rather like Cannes or Nice. The town is on the shore at the extreme southern limit of Russia proper, close to the border with Georgia. The weather deteriorated as we flew eastwards over the Black Sea, with white horses breaking and streaming foam, but we were totally unprepared for the shocking scene which confronted us at the airport where we somehow managed to land, buffeted by cross-winds. Bedraggled people were sleeping on bundles on the floor everywhere. Children cried and called out for help. Stranded workers and travellers sat wherever they could, jammed close together, trying to eat or sleep or just keep warm. There were some caged live chickens somewhere in the mêlée. The chaos had obviously been getting worse for several days, but there had been no advance news or warning in Yalta. Rain was falling slowly but steadily outside the terminal, and sheets of standing water lay on the runways and on the roads where we searched for a taxi. But we still did not begin to realize the scale of the disaster caused by a storm.

The taxi drove a few miles north along the coast to the Grand Hotel, which had probably sported a casino in the 19th century. Joseph spoke rapidly with the taxi driver, gradually getting the story of what had happened. Nobody has been able to fly out of Sotchi airport for three days, and there had been landslides everywhere. The thin rain fell remorselessly, and I gaped in disbelief at the sea where millions of trees rolled in the surf. Whole trunks of pine trees, complete with green branches, bobbed and heaved and spun in the waves, covering the sea surface out to a distance of a hundred yards from the beach, and continuing towards the northern horizon until they were lost to view miles away in the mist and rain. Entire mountainsides covered in forests must have collapsed into the waves in colossal landslides to have created such a bizarre sight.

We were supposed to spend one night at the hotel before taking the train to Gelendjik for two days, then returning to Sotchi. Joseph explained our situation at the check-in desk, where we agreed to extend our stay at the empty hotel until the feasibility of our train journey could be checked.

"The taxi driver and the check-in staff both tell me that the railway line has been washed into the sea," said Joseph. "It doesn't look good. I'll get more information tomorrow."

Joseph tried to telephone the Gelendjik Geophysical Institute after lunch, but the phone line was broken. Muffled up, we ventured onto the terrace in front of the hotel, from which a ceremonial white marble staircase reminiscent of the Odessa stairs scene in *Battleship Potemkin* descended to the seafront. The rain had washed numerous well-stocked flowerbeds, complete with bright tulips, onto the great zig-zag staircase, where the red-brown mud slithered, oozed, and slurped its way, carrying the red and yellow flowers with it, dripping step by step down onto the parade.

We descended and turned north. The fallen forest rolled and crashed in the surf, where waves of one or two metres (3-6 ft) were still breaking. The hillside inland was dotted with a few more bleak hotels, and then nothing. In the distance, through the rain, we could just distinguish a dark sloping mass cutting diagonally across the parade. This was probably the first of many landslides. There had been no talk of fatalities back at the Grand Hotel, but the emergency services must have been fully stretched to find villages and houses cut off by the disaster. Very much sobered, we returned to the hotel.

The next day the front desk reported that the railway line was still broken. We worked in the morning and decided to stretch our legs in a walk to the south in the afternoon. The rain was now just a soggy mist, but the sea was still lumpy with scattered white horses. A hydrofoil craft appeared in the distance motoring steadily northwards about half a mile offshore. So-called "surface-effect" craft were the subject of many experiments in both the West and East during the 1960s. These included hovercraft, catamarans, and hydrofoils of many different designs. The research goal was to achieve high speed and large load-carrying ability combined with smooth sea-keeping in bad weather, all at an economic cost. The Russians had developed hydrofoils that worked on a unique principle, and they were rumoured also to be experimenting with colossal flying boats that flew very low over the sea, getting extra lift from the interaction with the air pressure between their huge wings and the sea surface – the so-called Wing In Ground effect.

Unlike the Italian hydrofoils of the kind still operating between the Italian islands today, that use very large and deep U-shaped foils, the Russian craft used much flatter

Figure 5.4. Meteor hydrofoil vessel first designed and launched in 1959 for high-speed travel on the major lakes and Rivers in the Soviet Union. Note the shallow depth of the hydrofoils (CC Licence: LxAndrew).

straight-across foils mounted quite shallowly below the hull fore and aft, and sometimes with two or three separate foils projecting from the supporting legs. This meant that they could and did travel at very high speed on the great rivers and lakes which abound in Russia, while having a shallow draft for entry into small harbours. Based on Western principles of hydrofoil design, it was difficult to understand how the foils worked safely, and how the craft would perform in rough seas. I watched fascinated by the hydrofoil in very choppy water.

The Russian foils worked because, as the ship accelerated, the straight foils rose towards the underside of the water surface, and eventually began to lose extra lift as the volume of water passing over the top of the foil became less. At this point the elevation of the hull above the water became stable, though the design of the foils was very sophisticated to achieve this effect. The physical process that generates this stability is now known as the Alekseyev Effect after the designer of these foils. From the Western point of view, the question was: if the stability of the ship depends on such a subtle balance created by the thickness of the flow over the foil, what happens when there are big waves, or even a low swell? Maybe the slightest bad weather would force the vessel to pitch wildly or to slow down and rest on its hull? For COSG I had identified seven different makes of Russian hydrofoils, ranging in tonnage from 25 to 117 tons, each with a romantic brand name like Meteor (Fig. 5.4) or Kometa. The hydrofoil that I saw at Sotchi was a Kometa of 58 tons, designed to carry 150 passengers.

The use of hydrofoils in the Soviet Union decreased after 1980, and many of these powerful monsters are now wrecks or abandoned on land. They have recently attracted fan clubs who appreciate the unique combination of power and sleek beauty that the Russian hydrofoils displayed, somewhat strangely for a tyrannical and otherwise unimaginative regime. From pictures on the web, you can see the slightly angled foils, most of which stayed completely below the water surface.

Abandoning the attempt to reach Gelendjik, we returned next day to the airport, where the chaos had died down a bit. We boarded our flight to Moscow and thence to Leningrad. The people who had found themselves trapped in the airport were presumably passengers intending to make short-haul flights to provincial towns inland, which the storm must have cut off. We had no trouble getting our scheduled flight to the north.

At Leningrad we stayed at some distance from the city centre in the Cosmos Hotel, which was aligned as a long rectangular building beside a busy railway track so that all the customers could hear the shunting of trucks. The curtains did not quite meet across the windows so that the midnight sun disturbed our sleep throughout the night. The building was about eight storeys high and adjoined a low rectangular block that was a large post office.

Joseph and I went down to breakfast at 7:30 in preparation for meetings in town at 9:00. These were at an Arctic research institute, a Baltic Sea institute, and a ship design laboratory where they were interested in submersible design. Our grumpy sleepless mood was not improved when the waiter announced, "We do not serve individuals before nine o'clock. Only trade union groups and official Intourist tour parties. No individuals!"

I asked for a mug of kefir while we waited, trying to work out a new plan of campaign. The waiter said there was no kefir. I pointed to a nearby table where there were obviously several mugs of kefir, but the waiter just walked away. With a calm fury that I have never experienced before or since, I got up, walked into the kitchen, found a large glass jug of kefir, poured myself a glass, picked up a plate of food, and went back to the table. The kitchen staff smiled, showing no surprise or resentment, perhaps aware of the way the waiter had treated us. Joseph followed me a few minutes later. He confirmed that the kitchen staff, who were at war with the waiters, had no objection to our behaviour.

Over the next three days, we chatted with some of the guests at neighbouring breakfast tables, mostly American tourists, from whom we gathered that many had experienced rude behaviour from different members of the hotel staff. This was not a matter of cultural misunderstanding, since we all understood the relative poverty of the USSR and the pervasive security controls. The hotel was simply worse than usual, with many rude unhelpful staff. Our meetings in town went well, during which discussion of the possible Arctic route by sea from Arkhangelsk to Vladivostok was a lively topic. The sea ice in summer seemed to be getting thinner, so ice-breakers could force routes through more frequently. This could eventually open up access to currently isolated towns on the Arctic coast with potential exploitation of minerals and oil. The director of one of the institutes confided to us that he and most of his colleagues still referred to Leningrad as St Petersburg, regarding Moscow as an upstart village.

The next day, after meetings at the Museum of Arctic Exploration, we took a trip on a hydrofoil for 24 km (15 mi) along the coast to the tsarist summer palace of Peterhof. Since the waters of the Baltic, or more strictly, the Gulf of Finland, were flat calm, we experienced the sensation of the hydrofoil going at full speed. It was very impressive, apart from a slightly unpleasant vibration from the huge diesel engines. Peter Wyke, at Hawker Siddeley Dynamics, later confirmed that Russian diesels were notorious for vibration. The sun was brilliant and, from the landing stage, we gazed along the artificial lake where dozens of larger-than-life gold-painted statues coruscated with showers of water falling from colossal fountains. The palace itself is beyond the lake, raised on a grassy slope, embellished with many marble steps.

On the last morning in Leningrad Joseph and I packed our cases, deposited them in the porters' lodge, and went to the Intourist counter to claim the refund on the unused coupons. There were frustrated queues as usual, but while we waited we could see, behind the counter staff, through a door that opened directly into the public area of the post office. When I reached the counter, I displayed unbroken sheaves of coupons and asked the clerk for a refund. It was several hundred pounds. She looked puzzled and asked, "Where are the nights?"

"We have spent the nights in Intourist hotels," I said, "But I was told I could reclaim all the money on the days if they were in complete, unbroken sheets for all meals."

"No, no," said the clerk. "A day means 24 hours."

"But the night coupons are in a different booklet entirely," I objected. "If they are the same thing, why aren't the meal coupons and night coupons all bound in together?"

Joseph saw that I was about to raise my voice, if not lose my temper. He gestured to me to keep quiet. He and the clerk talked rapidly, until a fierce female manager came over from a booth at the opposite end of the counter to see what the fuss was about. Eventually Joseph turned to me with resignation.

"They insist," he said, "that a day means 24 hours, and in Russian it is much more likely that one would see that immediately. It should have been explained to you more carefully in England. The manager says that you can spend the coupons on any food or vodka you want in any Intourist shop or hotel, and she says you should just get drunk and enjoy yourself, and buy tons of caviar and vodka to take home."

Joseph gave the manager an insincere smile, and we left to conduct a morning's work in town. Our last visit was to an oceanographic institute specializing in bathymetric surveying of the Arctic Ocean. During lunch, we hatched a plot. We bought half a dozen bottles of expensive champagne with some of the coupons, and took a taxi back through the sun-bright streets to the Cosmos Hotel, arriving at about four o'clock. With the champagne concealed in plastic bags, we sat in the hotel lounge adjacent to the foyer, making conversation with tourists of several nationalities, chatting with people we had met at breakfast. We told everybody that at five o'clock there would be a big party in the Intourist Office, but they were recommended to bring tooth mugs from their bathroom, or wine glasses from the restaurant, as there would not be enough glasses. Some believed us, and some didn't but were too polite to say so.

At five minutes to five, carrying the champagne bottles still in the plastic bags, we went out of the main hotel door and walked the 20 m (65 ft) to the post office entrance. Once inside we turned back across the postal queuing area towards the hotel and the door opening into the back of the Intourist Office. We waited until the big clock on the wall showed exactly five. Then we threw open the door into the Intourist Office, appearing behind the counter clerks, and popped the first two champagne bottles. The corks flew to the ceiling. The queues on the other side of the Intourist counter broke into disorder as people came to the counter with glasses at the ready. We filled their glasses rapidly, passing some to the bemused counter staff. The more sceptical tourists ran back to the restaurant to grab glasses and returned eagerly. We popped two more corks and everybody started to drink toasts, gossiping merrily with each other. The manager came storming out of her booth, but Joseph intercepted her, and elegantly passed her a glass of champagne. He spoke rapidly in Russian, pointing several times to me.

I stood motionless holding the last two bottles. More empty glasses appeared on the counter, and we filled them as rapidly as possible. Even the manager seemed to relax slightly. I popped the last corks ceiling-wards, filled a couple more glasses, left the almost full bottles on the counter, and together we slipped out through the rear door into the post office. We ran like hell round to the hotel foyer, grabbed our cases, and jumped into a taxi, telling the driver to go to the international airport.

"What on earth did you say to the Intourist manager?" I asked Joseph when there was a safe distance between us and the Cosmos Hotel.

"Ah," he said, "I told her that you had obeyed her orders, and had been drinking vodka all day. I said that you were very drunk and that you were very dangerous when you were drunk. I warned her not to upset you, or that you would then do something very damaging and people might get hurt. I advised her to enjoy herself, and that nobody would mind."

At the Moscow Sheremetyevo airport hotel, we spent another uncomfortable night, during which the bathroom washbasin fell off the wall, cascading very hot water into my bare toes through the dislocated drain pipe. In the airport departure lounge, we bumped into a haggard Canadian mathematician whom we had met during the oceanography conference. He had spent ten days in prison near the airport because he had had the wrong visa to give a lecture in Warsaw, and had been deported back to Moscow. We boarded our flight for Warsaw, where I was due to say farewell to Joseph because he wanted to try to find the village where his father was born.

* * *

My final grand tour for COSG was designed to take in both coasts of the USA, Canada, Japan, and Hong Kong, returning round the world via Bangkok and Beirut to England. There were many minor appointments, which I will not list here, but the main targets in North America were advanced submersible design, deep-diving programmes of the US Navy, professional fisheries assessment of maximum sea-fish catches, and manganese nodule mining economics. After this I planned to study fish farming in Japan. Clutching a thick bundle of airline tickets clipped together, I boarded the flight from Heathrow to New York in late July.

The significance of my enquiries in the USA was greatly enhanced when, on 17th June 1966, the US Congress appointed Vice President Hubert Humphrey to chair a new National Council on Marine Resources and Engineering Development, which President Johnson had established. I had been monitoring the scientific and technical lobbying in the USA, especially by the Marine Technology Society, since the start of the COSG exercise, but the political endorsement shown by this appointment was unthinkable in the UK or any other European country. The Council would not produce results before I submitted my final report to COSG, but its existence emphasized the serious commitment to marine technology at the highest national level. The Stratton Commission, appointed by the president and reporting to the Council, spent more than two years compiling a book-sized report entitled *Our Nation and the Sea*.

Figure 5.5. Asherah submersible, built by General Dynamics for archaeological research conducted by George Bass in 1964, could dive to a depth of 182 m (600 ft) (Photo Chuck Nicklin).

Figure 5.6. The submersible Deepstar being lifted from the sea by a hydraulic crane onto a flat barge deck. Deepstar was built in 1965, and had a working depth of 1050 m (3500 ft).

Edward Wenk, who had been the first science adviser to Congress, became executive secretary of the Council. The work of the Stratton Commission led directly to the creation of the National Oceanic and Atmospheric Administration (NOAA) in 1969 and the Coastal Zone Management Act in 1972. I visited Ed Wenk several times during my work on the establishment of the Society for Underwater Technology (Chapter 9) in the UK from 1967 onwards, after which he co-edited with me an illustrated popular book on marine resources and technology that we published in 1977. The new Council on Marine Resources and Engineering Development showed how seriously the US government regarded the need to ensure rapid advancement in marine technology, but it could not discuss the unacknowledged confidential military issues that arose from the Cold War. I could only report to COSG what I saw during my visits to the USA, and then add my comments as to whether the investment under discussion appeared to be commercially viable or not. Some investments seemed to have no reasonable commercial application or benefit.

I drove from Kennedy Airport to General Dynamics, Groton, Connecticut, across the river from New London, where GD Electric Boat produced the Polaris submarines under intensive security protection. I was welcomed at a modest office responsible for the design and operation of the small submersible *Asherah*, launched in 1964 to support the work of George Bass in his underwater archaeological research and excavations in Turkey (Fig. 5.5).

Asherah, rather than Cousteau's *Soucoupe (plunging saucer)*, established the geometry of all subsequent submersibles apart from the very few weighing more than 40-50 tons. For the small vehicles up to about 10 tons, the universal design became a sphere with the air inside at one atmosphere pressure, providing living accommodation for the pilot and crew or passenger, and fitted with portholes or a large acrylic spherical window to give visibility forward and downwards (Fig. 5.5). Behind the occupied sphere, the batteries, motors in orientable pods, electronic packages, buoyancy control, gas cylinders, permanent buoyancy, variable buoyancy, emergency ballast, and other workaday features were packed into a teardrop-shaped metal supporting frame that was exposed to, and flooded by, the surrounding water pressure. The whole package was then encased in a free-flooding teardrop-shaped outer hull, providing streamlining, with just the forward segment of the occupied sphere exposed. A manipulator claw was usually fitted in front, plus an overhanging frame to support lights and tools, with landing skids below the main hull. Since these vehicles did not have a large air-filled hull like a military submarine, which would have given permanent buoyancy, they also needed additional material lighter than water to counter-balance the weight of the machinery. This had to be a plastic foam, which also had to be incompressible under high pressure, so-called syntactic foam. Improvement in the design of these foams was a valuable component in increasing their depth range, ultimately to more than 1000 m.

The three vehicles in the heavyweight class were the Lockheed *Deep Quest*, the Reynolds *Aluminaut* (Fig. 5.3), and the General Dynamics *NR1* (launched in 1969), all over 50 tons. At this size, the crew have more space to move around, and much of the machinery can be brought inside the hull, though the details of design varied greatly. The very secret 400-ton *NR1* was nuclear powered.

The arrangement of pressurized sphere plus a free-flooding teardrop-shaped engineering package was re-worked in detail by numerous companies later, including a collaboration between Westinghouse and Cousteau to manufacture the *Deepstar* (Fig. 5.6). Other companies produced the *Star 1* and *Star 2*, *Pisces*, *Sea Cliff*, *Alvin*, and so on. Success went to those companies who thought through the design in every detail, working out how to reduce costs of surface support, ensure rapid launch and recovery even in choppy seas, and provide easy maintenance, quick battery recharging, and accurate navigation. *Pisces Canada* was one of the most commercially successful later, and I chartered their submersible twice, operated by Vickers Oceanics, in 1970 and 1971 (see Chapter 7).

I had identified 42 different submersibles in use around the world in 1966, of which 29 were in the USA. Most had been constructed for military use, or for scientific research, with the oft-stated hope that they would be of great commercial value soon. But, as I noted at the time, "clear-cut arguments for their commercial use are extremely rare". The need to rescue the crews of crashed military submarines was one of the driving motives for the military. Over half of the submersibles could not dive below 2000 ft (600 m). Already in 1966, I reported that remotely operated vehicles (ROVs) controlled by cable from the surface, but without any crew on board the vehicle, were beginning to compete with submersibles.

* * *

At Woods Hole Oceanographic Institution (WHOI), near Cape Cod, I discussed ocean physics with Bill von Arx and seabed geology with Ken Emery. From Boston I flew to Halifax, Canada, to spend a day at the famed Bedford Institute of Oceanography (BIO), overlooking the grand harbour. The fisheries experts there were scathing about the predictions of the wilder mavericks who were predicting possible world fish catches many times the present catch. At BIO they were adamant that stocks in some places were already over-fished, and that disastrous collapse of some stocks was inevitable in future. Given the extraordinary collapse of the Grand Banks fisheries more than two decades later, these views were prescient. In fact, the Grand Banks fisheries stock, for which the catch peaked in the 1970s, collapsed completely in 1992 to a mere 1% of its earlier value, in spite of ill-judged attempts to limit total catches and restrict non-Canadian fishing vessels. The stock has never recovered.

I flew coast-to-coast to Vancouver where I visited the big fisheries lab at Nanaimo on Vancouver Island. The fisheries scientists echoed their East Coast colleagues' horror at the irresponsible proposals that factory fishing could be expanded by a multiple of two or three without damaging fish stocks.

* * *

Don Rosencrantz, whom I had previously met when he was providing submersible support for George Bass, took me to see the *Deep Quest* under construction in the "Zero Defects Zone" of the Lockheed factory at Sunnyvale, in the southern outskirts of San Francisco. The shimmering outer hull of welded aluminium was suspended at head height, with engineers in dust-proof white suits working on frames above and below it. The twin-sphere steel pressure vessels were designed to work at 8000 ft (2450 m) depth, and the motors with directional propellers were not mounted on external pods, but in ducts inside the hull frame. Thrusters gave power up and down, or sideways, with a fixed propeller at the stern for straight ahead. Computer controls selected the amount of power needed on each thruster to achieve the motion that the pilot wanted. This was an application of aircraft and helicopter control technology, not shipbuilding.

Weighing 50 tons, *Deep Quest* was too large and expensive to be competitive commercially but was used in the engineering research needed to design the US Navy Deep Submergence Rescue Vehicle, intended for crew rescue from crashed nuclear submarines. Between its launch in 1967 and 1980, when it was decommissioned, *Deep Quest* found and recovered several items of military hardware, crashed fighter aircraft, and the flight recorders from crashed civilian airliners. *Deep Quest* was supported by a custom-designed 425-ton mother vessel called *Trans Quest*. In 1966 I was impressed by the quality of precision engineering that had gone into the half-finished *Deep Quest*, but I saw that this was not the future of hardworking commercial vehicles.

I crossed the Golden Gate Bridge one bright morning to visit the office, on the north shore of the harbour, of John Mero, the author of *The Mineral Resources of the Sea*, that had caused so much sensational interpretation by the media. His book explained in elaborate detail the composition of oceanic manganese nodules in different deep ocean basins, suggesting that the costs of recovery and refining the ores were, or soon would be, economic, in comparison with land-based mining. By the time I met him he seemed almost exhausted, perhaps traumatized, by the reaction to his book. He was somewhat reticent in supporting the claims he had made, and I could only guess that he had been shaken by the sensational publicity it received, as well as chastised by the criticism and reserve from mining companies. I recorded my negative impression for COSG.

My last appointment in the USA was at a US Navy engineering laboratory in San Diego where they were responsible for the design and construction of the SEALAB III habitat (Fig. 5.7). SEALAB I had been launched in 1964 to a depth of 58 m (190 ft) off the coast of Bermuda, with four divers spending 11 days submerged. This was followed by SEALAB II that was already in place in La Jolla Canyon at a depth of 62 m (203 ft), close to the Scripps Institution of Oceanography. A few weeks after my visit three teams of divers were due to spend 15 days each working in the habitat, conducting scientific experiments both on their own physiology and the surrounding marine biology. One of the aquanauts was the astronaut Scott Carpenter (see Chapter 8).

SEALAB III was planned to work at a depth of 312 m (1025 ft), with successive crews of nine aquanauts testing salvage techniques. While I was discussing the future plans with the US Navy engineer, he was briefly called out of the office. I read upside down the first page of the thick document lying on his desk. It was a request for tender quotations inviting companies to bid for the contract to provide major parts of SEALAB III. Since the project was planned for the near future, soon after SEALAB II, I remember being surprised that such components were still not ready. In the event, construction of SEALAB III, which was adapted from SEALAB II, took a further three years, and there were several setbacks and accidents in its early days of operation in 1969.

I was baffled by the huge US Navy investment in seabed habitats for divers. There was always a façade of scientific investigation of the seafloor environment around the habitats, and the study of human physiology under pressure for long durations, which was certainly genuine, but I still could not understand the overall purpose of the programmes. Overt publicity stressed the potential economic benefits of working on the seafloor, which the media quickly embroidered with idealistic images of humans sharing the ocean with the fish. There was no conceivable military use for a massive

Figure 5.7. SEALAB 3 under construction (Naval History and Heritage Command).

Figure 5.8. Hitachi seabed tracked crawler vehicle for underwater foundation engineering. It was powered by hydraulic pressure from the surface control, and the buoyancy tanks could be adjusted to provide the best contact pressure with the seabed (Photo 1972).

underwater house that could not be camouflaged, could not move, and only allowed the divers to explore to a few hundred yards' range around it. The true explanation would come only in 1999.

* * *

I departed the US West Coast on a Continental Airlines flight for Tokyo, with a short stopover in Hawaii. The flight was probably one of the last to offer all its passengers a lounge situated amidships, with wicker chairs, a library, and bar. It made a long journey positively enjoyable, encouraging people to walk about and chat to each other. Continental Airlines went bankrupt a few years later. On that flight, I briefly recaptured the sense of occasion that used to surround air travel in the 1930s, when people actually put on their smartest clothes for the flight. I half expected to see Hercule Poirot or Humphrey Bogart in the next chair.

In Tokyo, since I stayed at a large European-style hotel, I did not have time to explore the more interesting traditional hotels and guest houses, nor the local food. Small technical things were striking, such as the taxis having broadcast TV programmes running on miniature screens for the passengers, the diversity of neon strip-lighting and advertising at night, which amounted almost to a new art form, and the early examples of handheld calculators and computers that were already small, presaging the desktop. Shops and family businesses still used the abacus for rapid calculations, while professional abacus experts maintained that they were just as fast as electronic calculators. Competitions were organized to demonstrate the truth of this claim. The professionals could do square roots and other extraordinary calculations so fast that you could hardly see their fingers moving.

ICI, Hawker Siddeley, and Unilever had arranged for me to meet a number of contacts in their subsidiaries and partner companies, especially in MITI, the notorious Japanese Ministry of Technology and Industry. MITI seemed to have found the magic formula for assisting private industry to develop new technology without letting them relax on the comfort of repeated government subsidies. Investment in futuristic seabed technology was being heavily subsidized on a scale that was difficult to understand. The hardware resulting from these investments was not yet visible in 1966, but would be much on display at conferences and trade exhibitions when I visited Japan again in 1972 (Figs 5.8, 5.9).

The Unilever office in Tokyo put me in contact with Mr Hachiro Hirata in Takamatsu, a small town on the north shore of the island of Shikoku, facing the Seto Inland Sea (Fig. 5.10). Hirata was an expert in fish farming. I flew south one morning on a Japanese internal domestic aircraft in which all the other passengers were Japanese.

Hachiro Hirata met me at Takamatsu Airport, which was dry and dusty, surrounded by low deciduous trees. The buildings were typical of small provincial airports of the day, low huts and outbuildings, a small concrete control tower, and a café-bar/check-in desk. As the saying goes, "The fun went out of flying when the terminal buildings got bigger than the aircraft."

Hirata beamed welcome. He was a chunky, broad-shouldered man, speaking fluent English which he had perfected visiting Unilever in England. He brimmed with cheerful explanation of the plans he had made for my visit.

Figure 5.9. Komatsu seabed tracked bulldozer. This machine weighed 43 tons and could work in a water depth of 7 m (23 ft).

Figure 5.10. Fish farming on the island of Shikoku, facing the Inland Sea, similar to the fish-farm that I stayed at for several days in 1965 (Aquaculture farms near Kumamoto in Shimabara bay Stock Photo -Image ID: M6W9GA Alamy).

"We'll drive over to the fish-farm research college, where you can stay in the student hostel. As you are here for three days, we'll work along the coast looking at different types of farm and then visit a prawn farm in concrete tanks, not in the sea, on the last day. Just follow me."

We boarded a small pickup and I threw my case into the back. Dirt roads between patches of dense woodland led past scattered houses to the coast, where Hirata turned along the shore for a few miles. Looking out to the Inland Sea, I saw islands everywhere, some connected by long suspension bridges or causeways. We drove into a yard amongst a collection of two-storey wooden buildings. Beyond the yard, I could see acres of floating platforms and trestles rocking in the small waves of the sheltered bay. Dark tanks and pumps stood in rows between the buildings and the shore, creating a constant hum of electric motors accompanied by the gurgling slosh of flowing water.

Hirata escorted me upstairs to a sparsely furnished room with a bed, a chair, and washbasin, overlooking a sandy inner yard.

"Make yourself at home," he said. "Lunch is downstairs in the cafeteria with the students at one o'clock, and I'll pick you up at two."

He left. I opened the big window into the yard only to discover a wasps' nest the size of a small melon hanging from the eaves. Wasps were busily flying in and out doing whatever wasps need to do. The air outside was fresh and breezy in spite of the heat, but the room was stale and musty, having not been occupied for some time. I wanted to leave the window open if possible. In the corridor outside my door, I found a long-handled broom, which I used to break off the wasps' nest, which crashed into the yard amidst a cloud of buzzing insects. They did not come back towards my window, so I left it open.

At lunch the students were queuing at counters to collect bowls of rice and stew. I joined the queue, picked up chopsticks, and settled down at one of the long tables where two or three students sat already. They nodded to me but showed no surprise. I ate clumsily with my chopsticks, but made quite a good fist of it, as the students watched with furtive amusement. I was in no risk of going hungry, just a bit untidy, and I hoped that they awarded me about 6 out of 10. Nobody complained that they had found a wasps' nest outside their room.

After lunch Hirata walked with me out onto the first floating pontoon, from where we turned back looking towards the shore so that we could view the whole establishment.

"The buildings on the right are where we are experimenting with cultivation of planktonic algae as fish food, and also growing small shrimps and other larvae as fish food. You can see all the cultivation tanks and the seawater supply. We are trying to avoid needing a subsidiary fishing industry to catch the food for the fish that we have in these tanks we are standing on."

He gestured downwards.

"Of course, we are not cultivating edible algae here. That is a completely different business, which is done in other parts of Japan."

"It's a very impressive complex of cultivation industries all in one place. How long has it taken to develop all these skills and special buildings and cage designs?"

"The attempts to farm fish in this region in a modern technical way go back to the 1920s, the first in Japan, but the last decade has seen an enormous improvement in the science and management, and especially the control of fish diseases and parasites."

"It's fascinating," I said. And it really was.

"Come on. Let's look at more of the tanks and different types of fish."

We were standing on a mesh of interconnected bamboo poles and narrow wooden planks that connected on all sides to plastic buoyancy floats and submerged nets and containers of varying sizes. As we moved unsteadily from tank to tank, I looked down at the swirling, gyrating, endlessly turning shapes of yellowtail amberjack, mullet, salmon, and sea trout, ever-restless and searching for food. The open seawater could flow freely through the mesh of the tanks, while pipes from the shore distributed food that was pumped in quite low down in the tanks, not on the surface. We walked balancing on the boards, finding smaller tanks with octopus climbing around in some and pufferfish in yet more. The variety was extraordinary. Hirata explained that many of the tank designs, nets, and feeding routines were experimental. In every case, the aim was to increase the tank size so as to operate in deeper open water with more oceanic clean flow, but this would obviously increase costs, and the best design had to be achieved by experience, step by step. The more open the water area the greater the risk of storm damage from waves, and hence the stronger the whole structure needed to be, with bigger anchors.

The multiple dilemmas in much sea-fish farming had already begun to emerge in Europe as researchers considered the conflicting demands of finding a food source that did not result in depletion of other marine fish stocks, the need to make the cages sufficiently small to pipe the food efficiently, the need to make the cages large so that the fish did not breed contagious parasites, the need for a good natural current of clean seawater through the cages, and the danger of spreading pollution from unconsumed rotting food debris. The Japanese were many years ahead of us, both in identifying problems, and in solving them. For two days I accompanied Hirata to different outstations along the coast, seeing more and more arrangements and designs. My skill with chopsticks improved. On the third day, we drove several miles to the factory – there is no other word for it – of the Takamatsu Prawn Company.

The prawn cultivation factory consisted of a series of workshops and low sheds, with many concrete enclosures. Hirata explained to the manager, to whom he was obviously well known, the purpose of our visit. Once inside the main enclosure, I saw row after row of enormous concrete-walled tanks, each the size of a very long narrow swimming pool, perhaps 6 x 50m (20 x 160 ft). Most of the tanks contained prawns already selected for commercial exploitation, being grown for processing and selling either as fresh or frozen food to markets. Other tanks contained different species on which the biologists were experimenting with feeding and growth rates.

We returned to the airport, where Hirata told me how much he had enjoyed his time in England and Scotland, and how much he liked British beer. The barman had one dusty bottle of Guinness, so we drank very sincere toasts to each other's health, and to fruitful collaboration between Japan and Britain. I had enjoyed intensely my days in Takamatsu, and I flew back to Tokyo with genuine regret.

My conclusions from my visit to Japan, which were scarcely original even in the 1960s, were that this was an economy that was going to boom and boom, with one new burst of technology after another. We were about to enter the Pacific Rim decades.

* * *

The journey back to the UK involved stopovers in Hong Kong, where I had one interview with some engineers on tunnel building for COSG, then Bangkok with an overnight stay, Delhi to change aircraft, Beirut with an overnight stay, then Heathrow. Flights were not allowed over the USSR, and, in any case, aircraft did not have the range to overfly without stopping inside Russia.

A few days after the completion of my global circumnavigation I was due to travel to France to spend a short break with Natalka, who was staying for a holiday with cousins who lived at Montrésor in the Loire valley. At the start of the flight from London to Paris, a blonde young woman of quite disturbing beauty threw herself on my mercy as soon as I had found a seat in the aircraft. She sat next to me, asking me to arrange her accommodation at a hotel in Paris. Where would I recommend? Where should she go in France after Paris? She talked rapidly and poured out her heart, relating the dangers and risks for a young woman travelling alone in Europe. My first response was to assume again that she was a Communist spy who was conducting a honey-trap, but I could not imagine what information anyone should want to obtain from me. My COSG endeavours were of long-range commercial value, not to say obscure, and would not be of much value to the Russians. She told me that her name was Barbara Guggenheim. I promised to help her.

In Paris, we walked together across the bridge over the River Seine between the airport bus station and the railway. At the hotel near the Gare Austerlitz, where I found a room for her, I assured the *madame* that we wanted a single room for the young lady only. My French was quite fluent by this date. *Madame* repeated again and again that surely, I wanted a double room, and she was baffled by my insistence. I convinced her that there was no misunderstanding. After a delightful meal with Barbara in a nearby restaurant, I left her to her fate and then caught the late train to Montrésor. We corresponded subsequently, and Barbara became a life-long friend whom I often visited during the following two decades when work took me to the USA. Having obtained her PhD, she built up a thriving business in art consultancy and then moved from Manhattan to Los Angeles. She is now a long-distance e-mail correspondent in California, where she married and has continued to build a very successful career.

Natalka's cousins owned a hilltop fortress originally purchased by the Polish Branickis in the mid-19th century. The family lived in this outrageously romantic and decorative castle, which was partly in ruins. The magnificent structure rose at the top of a rocky pinnacle in the middle of the village of Montrésor, and the towers and battlements from several centuries provided living quarters known as the new castle and the old castle. Not all of it was habitable. A twisting flight of rock-cut steps led down on the south side to water meadows and a small river with banks golden with flowers. My grand tour was over.

* * *

I submitted my report to COSG in late January 1967. It was commercially confidential, but I think BP conveyed a copy to the Cabinet Office, and I remember a brief meeting with Sir Solly Zuckerman, the government's chief scientific adviser. Fifty years after submitting my report attempting to forecast the progress of marine technology it would be cowardly if I failed to summarize my successes and mistakes. Let us start with the four hot topics that I identified earlier in this chapter.

Diving and underwater living

During the 1960s, and through into the '70s and '80s, companies and navies conducted research on progressively deeper diving with more complex breathing gas mixtures. The objective was to obtain greater working depth and longer duration at depth, combined with greater efficiency, comfort, and safety for the divers. The wilder publicity about "living underwater" envisaged divers swimming freely away from their underwater houses at depths of tens to hundreds of metres (about 100 to 1000 ft), riding on diver transport vehicles, and living underwater for weeks or months at a time. There was a world of difference between the reality and the sensational dream.

The longer a diver spends at depth, and the greater that depth, the more diluent gas, usually nitrogen or helium, dissolves in the diver's tissues. This means that in order to return to the surface the diver must stop for several minutes or hours, even days, to allow the gas to diffuse from the tissues into the blood slowly, without forming bubbles, and thence to the lungs to be exhaled. If bubbles do form they can block blood circulation, and in extreme cases cause death. For any given depth the tissues will eventually become saturated after a certain time and not absorb any more gas. This means that the diver may be able to stay submerged for several hours or several days without requiring an increase in decompression time or stops during ascent. This characteristic of physiology can be exploited by sending the diver to the working depth in a submersible "bell" chamber and, when the immediate task is completed, returning to the bell chamber, maintaining the chamber at high pressure and, at the surface, transferring the diver into a deck decompression chamber, always at the same saturated pressure.

The diver can then relax, sleep, shave, have hot meals, read today's newspaper, and repeat the same dive the next day without doing decompression. While in the deck chamber the diver is in close contact with support staff and doctors. At the end of several days, or a week's diving work, the diver would have to decompress very slowly, taking many hours, or even days, of gradually reducing pressure, until the doors of the chamber can be opened, allowing the diver to come out into the one-atmosphere world again. This working technique, which is known as saturation diving, has been used for many decades as a standard method to do underwater engineering and repairs in support of the offshore oil and gas industry (see Chapter 8).

Living underwater in a habitat for weeks or months exploits the safe methods of saturation diving, but instead of coming to the surface in a pressurized bell at the end of a working task, the diver swims into a special underwater house, or habitat, on the work site at depth under the sea. The diver stays physically under water. This looks neat, even romantic, and seems to save money by avoiding the expensive support ship, and eliminating the submersible transport bell chamber which requires massive cranes and surface equipment every day to transfer divers up and down.

Experiments in the 1960s and '70s (Conshelf I and II, SEALAB I, II, and III, Tektite, Helgoland, Cerno Mor, projects by Cousteau, Link, Stennuit, *et al.*) showed that habitats were feasible, and that divers could indeed live in them for weeks at a time at depths of 10-100 m (33-330 ft), and even greater depths for a limited time. There was also a certain kind of long-term scientific biological observation which could be conducted from habitats, but the question I had to consider for COSG was whether they were useful commercially, and whether they would become a standard method in offshore industries.

There is a problem. A diver who is saturated and at depth is in a very dangerous situation physiologically. In the event of emergency, they cannot surface. They are trapped under pressure, needing complex technology to allow them to decompress slowly. If the diver is swimming away from the submersible bell or habitat, their only hope of safety is to return to the bell or habitat, not to the surface of the sea. Surfacing could be fatal.

In a working context this means that a bell diver with support ship or platform can move from site to site when there is a new job. But a habitat is, by definition, in a fixed location for months or years. A diver in a habitat can work in the water, either on an umbilical hose or free scuba, but, in either case, their range of excursion is very limited. If the diver swims too far from the habitat, more than 100 m (330 ft) or so, the slightest accident would become very serious, probably fatal.

The predicted cost savings through elimination of the on-site surface support turned out to be an illusion. If there is no surface ship or platform above a habitat, and no cable connection with power and communications from shore close by, the design and engineering of the habitat must achieve a degree of reliability and fail-safe emergency backups comparable with prolonged manned space flight. The habitat batteries must never fail; the gas supply must never fail; the maintenance of oxygen partial pressure and low CO_2 must never fail; and so on. Such is the potential cost of this calibre of engineering that in practice no research group or navy ever attempted it. The so-called cost advantage of habitats therefore disappeared. Surface ship support, or shore connection, was always needed.

I recommended to COSG that habitats would have little commercial application, and no British company or government agency invested in them, much to the anguish of some enthusiasts and the media. A small number of shallow habitats have survived to this day, mostly used for scientific experiments, and in some cases tourism. But they have never proven useful to industry.

Regarding very deep diving, the research laboratories that did experiments on divers found that there was an unavoidable problem known as high-pressure nervous syndrome, which effectively prevented working below 600-700 m. Dives to this depth were very seldom undertaken even in the 1970s and '80s.

I recommended to COSG that deep diving would be increasingly associated with machines and robots in an integrated way, but I did not foresee that, after more than 20 years, the machines would greatly reduce the number of tasks for which diving was needed. On a timescale of five to ten years, my advice was correct. Oil and gas are now produced from water depths of thousands of feet, so that the seabed engineering is conducted completely by robotics or ROVs.

I said earlier that the reason that the US Navy invested so much in habitats in the 1960s was only revealed in 1999, and that it was a surprise. The 1999 book *Blind Man's Bluff*, by Sherry Sontag and Christopher Drew, explained that in the 1970s the US Navy had been mounting a habitat on the outside hull of a nuclear spy submarine, then sailing into USSR waters, where they released divers who lived in the habitat for a week or so as they attached recording devices to Soviet military subsea communications cables. It was a staggeringly successful and ambitious piece of clandestine technology. The project had been managed by John Craven, whom I had met several times without the slightest suspicion of the scale of secret work that he did. Craven wrote his own version of events in a book called *The Silent War* in 2002. He confessed that he felt guilty that the cover stories for some of his projects had encouraged poorer countries to invest in manganese nodule mining and underwater living, when it was not in their interest to do so. When I read about the real reason for the US habitat program I thought of writing to Craven to explain that I had not been taken in by the assumed economic need for habitats, and I had reported to COSG that I could not understand the US investment in habitats. But I never did write to him, and Craven died in 2015. A truly extraordinary man.

Manganese nodules

So-called manganese nodules contain also nickel, copper, and cobalt. A successful programme to raise the nodules from ocean depths of 4 km (3 mi) or more would be profitable only if the nickel could be produced more cheaply than from mines on land. During the 1960s and '70s, almost 500 million US dollars was invested in experimental technology to raise nodules and refine them. Hundreds of tons of nodules were raised. Companies from many countries were involved. But the cost could never be reduced to a level that would make the offshore mining process profitable.

The rate of progress was devastatingly slow and, by comparison with the offshore oil and gas technology, where there was a guaranteed market for the product, the development needed to mine manganese nodules would take decades. In my recommendations to COSG, I wrote, "The difficulties of mining manganese nodules have been severely underestimated by enthusiasts for this form of mining."

Fish farming

The wilder predictions of the popular press for divers farming or ranching fish in the 1960s were quickly shown to be absurd, while the economic and effective cultivation of numerous species in controlled tanks and cages, as pioneered by the Japanese, gradually spread all over the world. My report to COSG showed that research and experimental testing was needed at every stage, and that success would be neither instant nor easy. I added that world sea-fish catch would never grow to the upper limits proposed by the maverick forecasters, and thus the world demand for fish would never be satisfied by sea catching. In 2016 the tonnage of fish caught at sea and the tonnage of farmed sea-fish were the same globally. The side-effects of intensive fish farming – pollution, genetic alteration, fish diseases, and other problems – have not been solved, but progress is slow and persistent. A substantial part of the food fed to farmed fish still comes from the sea and is not cultivated separately, which still undermines a part of the potential benefit.

Although Japan was and is the global focus of culinary taste for high-quality tuna, and pioneered the cultivation of tuna, 50 years later bluefin tuna are now caught in the Mediterranean and fattened up in hundreds of floating fish-farms there for export to Japan. My anticipation of farming fish in the open sea has come to pass as a ship-shaped floating farm 430 m (1400 ft) long launched by a Norwegian company, Havfarm, in 2019. The process is not cheap, and the floating farm can contain 10,000 tons of salmon at a time.

Submersibles

The loss in mid-Atlantic of the nuclear submarine *Thresher* in 1963 had already persuaded the US Navy to invest in new deep ocean technology, including the small research nuclear submarine known as NR1. Many small submersibles were built, particularly in France, Germany, the USA, and Russia, most of them using the pressure sphere for crew, and the teardrop-shaped, streamlined, free-flooding machinery compartment already described. Military rescue submersibles must be bigger as they need to accommodate 15-20 crew from the boat being abandoned on the seafloor. Submersibles had an essential role during the 1960s and '70s before robotics and ROVs became efficient, and those surviving today are mostly for military rescue. Of the ten *Pisces* vehicles built, only two are still in use. Modern materials have stimulated a new type of small submersible with a completely transparent personnel sphere, allowing all-round visibility. These are very attractive for tourists and for some kinds of research, but the numbers in use are still small.

As late as 1983 European agencies were considering the possible construction of a deep ocean submersible comparable to the Russian *Mir*, granting a contract to Stewart Rusby and myself at the NIO to study the potential use and applications of such a machine. Stewart and I flew around Europe in a Piper single-engine aircraft visiting labs to study the technologies available. I piloted the flight from Hamburg to Lubeck just to prove to myself that I could still remember the controls. Although we concluded that Europe had all the necessary skills and materials to build a vehicle that could work at depths of several thousand metres (many thousands of feet), the case for doing so was not strong for either scientific or economic applications.

* * *

Working for the COSG was an extraordinary and unique experience. It gave me an understanding of the interface and two-way traffic between business and engineering on one side, with scientific research, whether basic or applied, on the other. This was valuable during the foundation of the Society for Underwater Technology in 1966-67, the Underwater Association for Scientific Research with John Woods and Bill Hemmings in 1966 (both described in Chapter 8), for evaluating the potential profit and value from oceanographic digital data in 1976 for the Marine Information and Advisory Service (Chapter 9), and ultimately the assessment of the economic payoff from British Marine Science and Technology (CCMST/IACMST) in 1990 (Chapter 9) and the Global Ocean Observing System and EuroGOOS in 1994-96 (Chapter 9).

The assessments, recommendations, and forecasts that I made in the COSG final report, which I submitted in January 1967, included many descriptive observations, comments and judgements that were short term. I assumed my projections would be relevant for about five to ten years at most. Many of my recommendations and predictions were simple projections of trends that I had identified and then extended for two to five years. The four "hot topics" that excited the popular press I have already described, and I got the judgements just about right. I confess to having retained a secret enthusiasm for submersibles a bit longer than was rational, but I could see their limitations.

Some of my other assessments related to large-scale industries and long time horizons, which can be compared with what really happened in the following years and decades. What did I get right, and where were my major mistakes? Did I help the member companies to earn money, avoid losses, and help to protect the environment? I did foresee that environmental assessment for large projects such as new ports or manganese nodule mining could take years before a final and correct decision would be reached and agreed by all parties. This insight came as a shock to me in 1966 but was confirmed within a few years by the endless debates on the location of airports, oil terminals, and coastal barrages. I definitely did not foresee that such debates would become so confrontational and adversarial that the conflicts would generate complete stagnation with no decisions for decades.

I recommended areas of basic and applied research which came about pretty much as I suggested. That was because I had detected trends that ought to grow and develop, and other people soon saw the same benefits that I had spotted. I probably gave the COSG a year or two of advance strategic-planning lead on these trends.

Here are some of the biggest and best correct predictions and worst errors:

Correct

- Satellite observation of marine meteorology as a scientific tool, and marine weather forecasting for safety at sea. (I did not anticipate remote sensing of the properties of the seawater itself, such as current velocity, salinity or sediment load.)
- Potential for satellite navigation systems at sea.
- Mining of manganese nodules will not be economic for many decades.
- UK industry was a leader in desalination and had a good commercial future.
- There is no commercial economic case for divers or other people to live on the seafloor. Habitats will not be economic.
- Predicted that global fish catch from the sea could not be more than doubled.
- Predicted steady increase in depth of offshore hydrocarbon production.
- Predicted that fish farming would grow rapidly for expensive and luxury species, but not for cheap trawled open-ocean fish.
- Predicted growing concern about all aspects of pollution and environmental damage.
- Described multibeam acoustic experiments as potential future for bathymetry.
- Foresaw need for increased instrumentation for environmental monitoring, navigation, and control of operations in hostile and varying conditions.
- Foresaw need for computers to process oceanographic measurements.
- Stressed need for seabed mapping of sediments and bedrock. Showed that many industries needed measurements of the properties of the seabed regarding the ability to support the foundations of structures or suitability for pipe burial.
- Predicted that generation of electricity from wave energy would not be economic at a large scale.

Incorrect

- Overestimated the growth in offshore production of tin, sulphur, diamonds, and gold, although diamond mining has developed continuously off southern Africa. I expected them to continue as minor industries rather than to die away, which is what happened.
- Expected maximum ship size to grow beyond 500,000 deadweight tons, which did not happen.

Failed to anticipate

- Robotics and automated untethered vehicles that only became practical in use by 1990.
- Miniaturization of motors and components. Many devices were clumsy in the 1960s, and we kept trying to achieve things decades before the components became small enough and reliable with low power consumption. I saw the problem, but not the solution. It was slow and gradual.
- Satellite measurement of the sea itself, which only started in 1978 with SEASAT.
- Loss of distant water Atlantic fisheries, which meant that European fishing fleets could no longer fish the Grand Banks and around Iceland.
- Fibre-optic trans-oceanic cables. The first trans-ocean fibre-optic cable was laid in 1988. That was beyond my prediction horizon. Communications satellites in 1965 seemed to be on the point of making submarine telephone cables redundant.

6

Cities in the Sea

Underwater cities became part of my life for several decades after the mapping of Apollonia in 1958-1959. To discover tens and even hundreds of ruins beneath the waves, more each year, became a quiet obsession. Whenever I had the chance to visit the Mediterranean coast, whether on a scientific project or on holiday, I would search for local ruins in the sea. Everything that I found under the sea I sketched, mapped accurately if there was time, photographed if possible, and reported to the national archaeological museums and authorities. I began to search for ruins of the Bronze Age and earlier, even back into the Stone Age when the sea was still rising after the ice ages (Chapter 10).

The French diver Philippe Diolé wrote a delightful book, now forgotten, called *4000 Years Under the Sea*. It was published in France in 1954 and I read the English edition in 1959. He used simple diving equipment, usually just snorkelling, sometimes scuba diving, often working on his own, as he travelled around the western Mediterranean just after World War II, diving on ancient ports and ruins in the sea. Algeria was still French in administrative and literary culture, and many of the sites he visited were in North Africa. He swam over ancient ruins in Tunisia and Algeria, as well as studying the literature on archaeological diving in the eastern Mediterranean.

Jacques-Yves Cousteau promoted a personality cult that dramatized deep-diving exploits and he soon overshadowed the work of Diolé, but Diolé's descriptions made a great impression on me. Instead of focusing on the fragile bits and pieces of wrecks from the ancient world, and the endless repetitive cliché of salvaging amphorae by the thousand, Diolé painted a more subtle picture of cities, trading ports, buildings, breakwaters, garum fish-paste factories, city defensive fortifications and permanent structures that supported the network of sea trade and military might of the ancient civilisations. Maybe these remains were often only in shallow water, precluding claims to heroic diving in the savage dangers of the ocean, but they told a story that was more tangible and richer than that of the shipwrecks.

Diolé was entranced as much by single villas, fragments of a wall, one house, a small mosaic floor in the shallow sea, or by one fish cultivation pool, as by a great city such as Algerian Caesarea (Cherchel), or Carthage. He reminded me of the little islands to the west of Apollonia where we had found tanks and storage pools. He used these details of wealth, comfort, and luxury at the small scale to conjure up the delights of life and the security enjoyed by the Roman landowners and merchants in North Africa during the flourishing of the Empire. The idyllic life of the great villas by the sea, such as Selin

In: Flemming, N.C., 2021, *Apollonia on my Mind. The memoir of a paraplegic ocean scientist*, 221-270. Leiden: Sidestone Press.

in western Libya (Chapter 11), continued even while the barbarians in the north were crossing the Rhine, then surging through Spain, until finally they crossed at Gibraltar to devastate the African foundations of the imperial economy.

Between major cities there were many minor provincial buildings, along with quarries, tanks, rock-cut silos, dye works, and artificial pools. Every one of them could give me a measurement of the change of sea level since the days it was in use. Lying in the shallow waters, every construction from the ancient world, however prosaic, would give me a message.

I read Diolé's book and was hooked. I bought atlases of the Greek and Roman civilisations, and, though I was reading for a degree in science, I started to make lists of ancient ports, comparing the official reports of archaeologists with the adventures of Diolé as he swam over each coastal ruin. The engineering skills of the ancient world were astonishing, and I read all I could about Roman hydraulic concrete, ancient lighthouses, and the essential support given by harbours to the economy of the Roman Empire, which depended on a colossal transport of grain from North Africa, the military domination of the Mediterranean, and the suppression of piracy. My list soon included more than 1000 ports over 2000 years old from Gibraltar to Beirut and the Black Sea.

The scattered remains of ancient Greek and Roman ports, of which there were so many, would provide a consistent measuring rod for changes of sea level on a timescale of 2000 years. The 19th-century British Admiralty hydrographic chart-makers had observed several areas of ruins on the shores of the Mediterranean, which they had recorded accurately as hazards to inshore navigation. Beechey's work at Apollonia was by far the most detailed, but other captains and surveyors like Beaufort and Smyth had correctly noted ruins in the water at other cities in Italy, Tunisia, and Turkey. I dived on a few ruins on the coasts of the western Mediterranean during my PhD research at Cambridge, and joined the NIO in 1967, where I extended my research to the eastern Mediterranean.

* * *

Researchers before me, both archaeologists and geophysicists, had found about 30 submerged ancient ports scattered irregularly through the east and west basins of the Mediterranean. The majority appeared to be in the east, especially in the Aegean, but there was also a concentration around Naples, probably influenced by earthquakes from Vesuvius. With so many "sunken cities" as evidence, most researchers concluded that the world sea level must have risen about 2 m (6 ft) in the past 2000 years

You can approach this argument the other way up. Since there had been over 1000 ancient ports in the Mediterranean, which was manifestly the case, the observation that only 30 or so were submerged did not prove anything. What about the other 970?

If they were not submerged, why not? If the sea level globally had risen 2 m (6 ft) in the past 2000 years, or the whole Mediterranean basin had sunk geologically, every ancient coastal city should be deep in the water, not just 30 of them. Furthermore, if the submergence had been gradual and uniform, all ports of the same age should be submerged by the same amount if global sea level change was the main cause. It was obvious that the Bay of Naples was prone to repeated earthquakes, with eruptions both of Vesuvius and at Solfatara, and the Aegean area is proverbially unstable with frequent devastating earthquake shocks, so it would be surprising if 30 cities were not partly submerged as a result of earthquakes alone. Nobody seemed to have been interested in finding stable, let alone uplifted ancient ports, but perhaps they existed.

Journalists repeatedly retold Strabo's story of the sunken city of Helike, lost in the Gulf of Corinth, adding the legend of Atlantis for extra colour. Various well-publicized expeditions tried, but failed, to find the ruins of Helike under the sea. The 1954 Walt Disney film of *20,000 Leagues under the Sea* included a brilliant scene of Atlantis with standing columns deep in the ocean which stimulated imaginations further. The impression of the Mediterranean being subject to both earthquakes and sea level rise was reinforced repeatedly without any clear logic. It was almost deliberately muddled. By 1960, Ed Link was using acoustics to map underwater ruins at Caesarea in Israel, which added to the mix, as well as promoting a further journalistic outpouring.

The confused deductions about the Mediterranean sea level were resolvable by taking a comprehensive approach. Measurements of the relative change of sea level at all ancient ports in the Mediterranean, or at least a few hundred, would reveal the truth. If all ancient ports were submerged by the same amount, then either the sea level globally had risen, or the whole Mediterranean area had sunk geologically as a flat rigid plate, which was unlikely, though not impossible. If some cities were submerged, and some uplifted, and if the displacement was varied in magnitude, and if the displaced cities were concentrated in areas of known earthquake instability, then their change of level was mostly due to local earthquakes. If most cities did not show any displacement at all, then the global sea level had not changed in the past 2000 years.

The logic was clear. But I would need to visit many tens of coastal ruins, preferably hundreds, to obtain the right answer.

The ancient Greek, Roman, and Phoenician ports, such as Apollonia, contained specialized buildings right on the edge of the sea. These included slipways for warships, fish tanks, docks and jetties, breakwaters, perhaps a lighthouse, and other structures with the foundations and cut rocks extending well below sea level. As a consequence, for a diver simply to note that there were cut blocks or rectilinear rock cuttings in the water did not prove that the city has sunk into the sea, nor that the sea has risen. The purpose of each building must be worked out from its shape, profile, construction, and location relative to other buildings, to check whether it was deliberately built in the sea or not. In contrast, roads, pavements, steps, closed rooms with a tiled floor and a doorway, mosaics, graves and tombs, ceremonial structures such as churches, or public features such as market places, race tracks, or forums all indicate dry land. The borderline between these two types of construction would show the original position of the shoreline and the local sea level 2000 years ago.

Previous observations by divers and chart-makers went back centuries. Reports from the early 20th century cited articles from the 1850s, which in turn quoted scholars from the 18th century, and so on. In those early days, the major European countries had trading stations or "factories" in the Mediterranean, including on the coasts of North Africa and the Levant. In the British case, there was usually a chaplain to His Majesty's Factory, and such men were fascinated by the local customs and history. They and their local colleagues and scholars published reports, sometimes luridly exaggerated or plain wrong, about ruins underwater that they had found in their neighbourhood. These publications were in obscure journals from Algeria, Spain, Sardinia, Sicily, and Greece.

The University Library at Cambridge stored such journals in seldom-visited wings under the heading of "Defunct Periodicals". Clutching my notes, copied from the vast handwritten catalogue to the library, I entered dark galleries and wings that were so dusty that I learned to stand back as clouds of dust tumbled down from the switch cover when I turned the lights on. Doors creaked with unoiled hinges, and the leather-bound books had probably not been opened or consulted for a hundred years or more. In these gloomy galleries I never met another soul.

For my thesis work in 1960-1963, I analysed mostly archaeological reports for the western Mediterranean and found time to visit and dive on a dozen or so sites in Spain, the south of France, and around Naples. The combination of literature and my own observations provided good data on 79 ancient ports, from which it was obvious that most of them had not been displaced at all, other than those in the Bay of Naples. It looked as if the prevailing academic image of "universal rising sea level for 2000 years" was misleading, researchers having arrived at this conclusion by the simple process of only looking for submerged sites and ignoring the rest. I published this conclusion in 1969. I stated that the global sea level had not changed very much in the past 2000 years, less than +/- 50 cm (20 in), and that the submerged towns in the western Mediterranean were all due to earthquakes. This was quite sensational for the small community of specialists who worried about such things.

* * *

By 1967, when I joined the NIO, the picture of the geology and geophysics of the Mediterranean had advanced in a way that was revolutionary. Geologists accepted plate tectonics as the "Big Picture" for global earthquake generation. The complex collision of Africa and Eurasia with lots of local "micro-plates" crushed between them was beginning to emerge. If I continued to observe and measure vertical changes at ports in the eastern Mediterranean, I could make a truly novel contribution to the understanding of vertical earth movements in the framework of plate tectonics, as well as conducting a double-check on my conclusions from the western Mediterranean. It might also be possible to check on the evidence for global change of sea level amongst the statistical confusion of geological movement and earthquakes. I also wanted to reduce the uncertainty of my conclusions from 50 cm (20 in) to 25 cm (10 in).

Between 1959 and 2003 I visited and snorkelled or dived over hundreds of coastal archaeological sites in Italy, France, Spain, Tunisia, Greece, Turkey, Cyprus, Morocco, and Israel, but most between 1962 and 1983. To be specific: Sicily 1959; Spain 1961; southern France and Bay of Naples in 1962; Tunisia 1966; the southeast Peloponnese 1967; southwest Peloponnese 1968; southwest Turkey 1969; southern Turkey 1971; Crete 1972; Cyprus and Rhodes 1973; Karpathos 1974; Israel 1975-76; Mallorca 1980; Tel Qatif, Gaza,1983; Morocco 1985; and the whole Libyan coast in 2003.

Each time that I swam over a new site there was a heart-jolting moment when the first straight wall or street, the first rectangle of a tomb, or even a single cut block appeared through the blue haze, followed by a mass of questions and a struggle for answers. What did it mean? How old was it? Exactly how deep? Who built it and why? Some discoveries emerged as major towns underwater, like Salamis in Cyprus, and Plitra or Pavlopetri in Greece, but I enjoyed finding single fish tanks in remote locations, abandoned grain silos cut into the rock and now full of fishes swimming in seawater, or miniature slipways where a fisherman launched a single boat. The entire coast of the Mediterranean was alive with these relics.

Because most of my academic writing was published in the journals dedicated to geology and geomorphology, I rarely wrote articles about my archaeological discoveries. But I alerted colleagues in archaeology to my finds for further exploration and research.

Year by year, diving archaeologists studying shipwrecks discovered more about the cargoes transported across the Mediterranean 2000 years ago and even found a few warships. I and a growing band of diving archaeologists who studied ports were able to show how this network of trade and warfare was managed when it reached the shore. Ships need continuous repair, security, and re-supply every time they put into port. These needs provided skilled jobs for many specialized trades. The trades needed buildings.

In the mid-1960s when I embarked seriously on this project, marine archaeologists were still uncertain as to when the first artificial harbours protected by breakwaters had been built. They certainly existed at some sites 2500 years ago, with just a few about 2800 years ago, in the Iron Age. But there was no evidence yet for constructed harbours in the preceding Bronze Age from 3000 to 1000 BC, although we know from Homer and the wall paintings dated to about 1700 BC on the Aegean island of Santorini that seafaring was a major aspect of the culture. Trade by sea and naval campaigns to secure territory or protect trade were continuous features of Bronze Age history. Coastal islands off Alexandria in Egypt and at Arvad in Syria bore rock cuttings that suggested that they had been exploited in the Bronze Age to shelter ships from the waves of the open sea, but that is a simpler technology than transporting large blocks into the sea to build a breakwater in a depth of several metres.

Ehud Galili and his colleagues at the Israel Antiquities Authority discovered in 2019 that the submerged Neolithic settlement of Tel Hreiz near Haifa, Israel, dating to 5000 BCE, was protected on the seaward side by a long wall of boulders that the villagers had built to prevent inundation by the post-glacial rise in sea level (see Chapter 10). This is probably the world's oldest coastal defence structure, but the boulders were placed on the ancient shore, not in deep water. The technical ability to build artificial harbours came later.

It is proverbially difficult to prove a negative, but, as the years passed, with various research groups mapping increasing numbers of ancient Mediterranean ports, we had to suspect that artificially protected harbours with built breakwaters and moles did not exist there in the Bronze Age, in spite of the fact that ships had been used for long voyages for thousands of years before.

Each year I researched in advance the coast that we would be visiting, collecting maps, references, charts, published articles and books about the antiquities and ruins close to the shore. I prepared small bound booklets containing the key information, including photocopied sections of maps, so that it would not be necessary to unroll large charts or maps when we were on site. This is an advantage because it avoids trying to handle flapping sheets of paper in wind and spray, but it also avoids spreading charts on café tables in a way that might attract too much attention from the suspicious police or military.

* * *

When John Woods and Charles Carmichael were working with me at the Station Marine d'Endoume (Chapter 4), I spent a few days at the Musée Borely while John and Charles dived around the islands plotting the depth of caves. Professor Benoit, grey-haired, short, and bouncing with energy and helpfulness, supplied me with a steady stream of books and references which he thought I would find useful. And they were useful. At the same time, he kept exclaiming, "I am sure there has been no recent change in sea level. Absolutely sure of it, sunken cities are fairy tales." He went on finding books for me.

When I had compiled a list of the exact sites that I wanted to visit, I thanked him for his kindness and took my leave to re-join John and Charles. I hired a Renault 4L car in which we picked our way slowly eastwards from one coastal village to the next, searching for the archaeological remains. In 1962, places like Cassis, Bandol, La Ciotat, St Maxime, St Tropez, and St Raphael were still pretty fishing villages or small towns that attracted tourists, and they knew how to do it, but they were not overrun by hordes of tourists, let alone celebrities.

We trekked along the shore, noting bits of ancient walls, occasional rock cuttings and rectangular pools incised into the shore, some on the coastal islands, but nothing indicated a sea level change of more than a few centimetres. Professor Benoit was right. The south coast of France is stable. This excursion is memorable because at one village we drove the car onto the hard ground near the beach, leapt out of the car in our swimming trunks, picked up ranging poles, fins and survey gear, and strode down to the beach past a large open-air café, packed with holiday-makers. We must have looked unusually purposeful and exotic because a bikini-clad girl at one of the café tables stood up and hailed us so that she could take our photograph. Within a few seconds, every woman in the café was on her feet or standing on a chair trying to get a better look at us, shouting, squealing, and waving as if we were pop stars. The men remained seated staring at their drinks. We pretended not to notice as we sauntered down to the sea to do our work.

* * *

Later that summer, after diving in caves (Chapter 4), John Woods and I departed from the volcanic island of Ischia at dawn, taking the ferry back to Pozzuoli. The mainland was still invisible as we left the harbour, and the first wisps of cloud were being drawn by the sun into the blue space above Monte Epomeo. We sailed past the sleeping Isle of Procida and rounded Capo Miseno into the bay of Baiae. Pozzuoli and Posillipo were shrouded in smoke and fog. We drew alongside the quay to feast our eyes again on the riot of colour and activity on the dockside. Bright dinghies darted about the water, heaps of fish were scattered on the quay, and everywhere there was running water, ice, salt, tar, ropes, shouting, smells, and colour.

At Pozzuoli we had a short time to visit the so-called Temple of Serapis – more accurately a Roman marketplace, with its tall marble columns standing in 30 cm (1 ft) of seawater (Fig. 6.1) – before our pre-arranged rendezvous with a diver from the NATO centre in Naples. Geologists and scholars for the past two centuries have noted that the exposed height of columns of the temple, and the depth of water around them, vary

Figure 6.1. The so-called Temple of Serapis, really a Roman marketplace, close to the shore in the town of Pozzuoli, was, in 1962, below sea level, and the seawater soaked through the beach to flood the ruins. A shepherd found the tops of the columns projecting from the ground in the 18th century. The borings of marine molluscs in the large columns show that the structure was once much deeper in seawater and has been uplifted by previous earthquakes.

from year to year as the volcanic forces raise and lower the land in the neighbourhood of the town. Our friend drove us around the northern curve of the Bay of Naples to the village of Miseno, where the Roman navy had had its main base in a fortified harbour 2000 years ago. During this drive, I consulted my maps and notes to remind myself of the route of the submerged Roman road from Naples which runs parallel to the modern shore, passing through several submerged cities on the way. These were first detected by the Oxford don R. T. Günther, who, in the early years of the 20th century, employed octopus fishermen with glass-bottomed buckets to search for the ruins. As we neared our destination, we crossed a low bridge that spanned the channel between the Laco Miseno and the old Roman port, then swung left past a crowded bathing beach into the village of Miseno.

The cluster of houses around the church was overshadowed by the headland of Cap Miseno, and flourishing trees grew between the buildings right down to the edge of the water. We took drawing boards and snorkel gear as we climbed down the iron ladder from the jetty into the shallow water between rocking dinghies. From there we swam on the surface towards the mouth of the bay. The water was a transparent yellowy-green within the harbour and graded into oceanic blue as we approached open sea. In the deeper water, we immediately came across the first pier of the defences erected by Emperor Agrippa across the harbour entrance. These had originally consisted of two rows of huge square concrete towers disposed alternately so that the waves that passed between the first row were broken by the second. The water is now 6 m (20 ft) deep between the towers, which rise to within 1 m (3 ft) of the surface in some cases, though many of them are broken. In 1835 Guiseppe de Fazio, an Italian diver, found a bollard in one of the blocks farthest from the shore, but though I dived around it for half an hour, I could find no sign of a bollard. To strengthen the port defences, chains may have stretched between the towers. Before the land subsided, the towers would have extended well above the sea level.

I cruised back between the crumbling columns of weed-covered concrete. In the cliff to my left, I suddenly noticed a vertical rectangular slot, rather like a doorway, almost completely submerged, which penetrated the headland so that a pale-blue light shone through, silhouetting the clouds of fish within. I swam cautiously into the tunnel, which sloped downwards until the roof was under water. I floated slowly through the shoals of fish, holding my breath and finning gently, until I emerged in a peculiar half-light on the far side. Below me was a deep, square basin surrounded by a narrow pavement – a perfect Roman piscina for cultivating fish.

I surfaced to get my bearings. I was astounded to realize that I was in an artificial cave whose vertical walls descended below the sea, where they ended at a pavement on the brink of the drowned pool. The piscina was more compact but more elaborate than that at Apollonia (Chapter 1). I sank into the water and swam around the walls where small votive niches were cut into the rock. The pavement was at a depth of 2.75 m (9 ft), and the rubble filling in the tank itself was at 4.5 m(15 ft), so that the real floor must have been even deeper. The top of the seaward wall was level with the pavement, and through it, two tunnels were cut which were partially blocked by fallen stones. I eased my way down one of these channels and found myself outside the cave in a shallow trough, which sloped into the open sea. This was a perfect fish-tank in every detail. I swam back into the cave and hovered at the ancient water level, looking around at the green walls with their little recesses, in which I imagined decorative sculptures glowing white in the darkness. Perhaps the slaves used to walk through the flooded tunnel to adjust the sluice gates and catch fish for their master's dinner.

We spent the rest of the week checking the maps published by Günther. In the murky waters near Posillipo, we found flights of steps that began and ended below the sea with many other signs of submersion such as rooms with tiled floors. Near the Casa degli Spiriti, a multi-storey Roman house with the first floor under water, we found a flooded tunnel that could be entered on the shore, but which opened out well below the surface. The whole coast on the northern shore of the Bay of Naples, and the surrounding land, has been submerged much more deeply at times within the past 2000 years but has then been partially uplifted again to the present position. The marks of the sea at different heights on the columns of the Temple of Serapis provide a record of these strange events.

In 1971, I worked with TV producer Philip Daly of the BBC and science writer Nigel Calder, to make a section of their series *The Restless Earth* on the yo-yo movements of Pozzuoli and the shore round to Miseno (Fig. 6.2). We obtained some dramatic footage, but an earthquake in the intervening years (reported quite widely in the international press) had uplifted the region again, so that the tunnel through Cap Miseno out to the submerged piscina was half out of the water, and not so dramatic as it had been nine years earlier. Plate tectonics and associated regional volcanism do not stop and start for the convenience of filmmakers (Fig. 6.3).

Figure 6.2. In 1971 the author revisited the Temple of Serapis with a camera team from the BBC to make a TV programme on earthquakes, written by Nigel Calder. The floor of the area, really a market place, was still below sea level, but not as much as in 1962.

Figure 6.3. The Roman market place at Pozzuoli in 1996 showing that uplift continued slowly through the 1980s until the floor was dry and above sea level (Photograph John Dvorak).

* * *

My first season of coastal surveying at the NIO was summer 1967, for which I had a permit from the Greek Department of Antiquities, led by Nikos Yialouris, to survey the east and southern coasts of the Peloponnese. I had adapted a Ford Transit van as a diving wagon, with racks for cylinders, a bookshelf, hanging storage for clothes and diving gear, and an air mattress to sleep on. Eastern Europe, the Balkans, and the more remote parts of Greece were pretty rough in those days, so I also carried large water cans that I could fill from streams and fountains.

Greece was in a peculiar state politically, since the "Colonels" had taken over the government on the 21st April, with the excuse that they were preventing a Communist coup. The king tried to accommodate the dictatorship for most of the summer and then, after an abortive counter-coup, fled the country in December. Many of the friends that I made in Greece that summer eventually despaired of the excesses of the right-wing dictatorship and fled to France, Germany, or the USA in the next few years. Since I spent most of the time in very remote locations, the politics in Athens did not impinge on my thoughts, until, during a farewell party at a big hotel at the end of the trip, we were told to stop playing songs by Mikis Theodorakis on the piano, or we would have to leave the hotel. His music was banned by the Colonels.

On my arrival in Athens, I made the obligatory visit to the British School. The problem of Early Bronze Age harbours, and the absence of Bronze Age settlements on the southern coast of the Peloponnese, had made a strong impression on me. I took the risky step of writing in the Visitors' Book that I was searching for Bronze Age harbours, which was somewhat over-ambitious.

I started by working my way southwards on the Peloponnese through my checklist of 18 coastal ruins to measure their submergence, uplift, or evidence for lack of vertical displacement. Then onwards to check the extreme southern bays and headlands for early settlements or ports that would have provided shelter, food, and freshwater for Bronze Age shipping between the highly developed coasts on the east and west of the Peloponnese. Homer's *Odyssey* shows, at least allegorically, how important it was for warships to round the southern headlands safely. There must have been safe havens for replenishment of stores or refuge from storms.

Leaving Athens for the south in 1967, I drove from site to site through mountainous country, donning fins when I arrived at the ruins, plunging into the cold blue water, sketching and measuring for an hour or so, peering into the shadows and ripples of light transmitted by the waves, and then drying off in the sun as I wrote my notes. I always travelled with a typewriter, so, in addition to sketches in my notebooks, there was a tidy script to go with it. Since there was seldom a road along the coast because the mountains dropped steeply into the sea, I had to drive for most of the time inland through the hills, find a local dirt road down to the coastal site, and then return by the same road back inland after the dive. There were no surfaced roads south of Sparta, and only the main inter-town roads were surfaced south of Tripoli. A serious problem was the absence of petrol south of Sparta, since nobody in the region had a car. The only fuel they needed was paraffin (kerosene) for their tractors and lamps. I saved fuel by coasting down hills with the engine switched off.

The food in the villages was simple and incredibly good, at least for a hungry diver. Fresh homemade yoghurt, honey, dry bread, and fried little fishes for breakfast; kebabs, stuffed peppers, stuffed tomatoes, or stuffed vine leaves with stewed beans at lunch; and chicken stew and rice in the evenings. The villagers cooked in wide aluminium dishes about 60 cm (2 ft) across, with five or ten dishes lined up over simmering charcoal, supported on two rows of bricks. Often the food was on display in the street where it was being cooked, but if not, you could go inside to see what was cooking and pick the dishes that looked best.

The locals drank only retsina wine, and their hospitality was shame-making to a rich northerner. As soon as I sat down at a café table the landlord placed a copper jug of retsina before me, already paid for as a gift by someone at another table. When I stopped at night to sleep, often in a forest, the smell of pine resin was overwhelming, as every tree had a cut incised deeply into its bark above a tin cup inserted to catch the oozing resin. This was used to flavour the wine. A salad in those days was chunky feta cheese, sliced tomatoes, cucumbers, and lots of raw onions. No lettuce.

I always drove with the sliding doors of my van open, waving to passers-by, and hailing them with the regional greeting "Yassou". In the south I knew that I could park the van, leaving the doors open to keep it cool and come back after swimming for an hour, during which time nobody would have stolen a spanner or my binoculars.

I found ruins in shallow water at many sites but at varying depths. An American university was already mapping the submerged houses and sensational athletic race track underwater at Portocheli, and the other sites produced scattered small foundations which were all useful. From Lechaeum, to Asine, Zarax, Monemvasia, Epidaurus, and so on, my journey continued. All the ruins that I found under the sea at these sites were from the period around 2000 years ago, when most building was done with large rectangular well-cut blocks known as ashlar. There was no cement between the blocks, but the fit was so good that friction held them in place. Some of the later sites from the Roman and Byzantine periods had buildings made of bricks and concrete.

My main discovery that summer resulted from both careful planning and pure accident. On the shore in the Bay of Vatika, which was one of my "hot" sites for Bronze Age remains, I found that a German family was building a house that would also serve as the summer base for a boys' school near Hamburg (Fig. 6.4). Their camp on the beach near Aghios Georghios became my local home for a few days. When I explained what I was looking for they told me that there were rock cuttings and rock-cut tombs on the beach near the island of Pavlopetri, a few miles away. I snorkelled there the next day, accompanied by Solvey Engelke, one of the family, and after initially drawing a blank, we swam back towards the mainland from the island of Pavlopetri. Suddenly I was staring down through 2 m (6 ft) of crystal-clear water at a tangled mesh of walls, rooms, and tombs, all made of irregular, unshaped stones. The tombs were the most obvious artificial feature, since they were made of four slabs of stone standing vertically, with a fifth slab over the top. The past occupants of the town had built groups of rooms aligned so that one façade fronted onto a street, with a row of stones making a kerb. This was a town, with rows of adjoining houses, not a village or scattered prehistoric huts.

The strange pattern of walls, none of which was straight, mystified me. I sensed that the outlines of the buildings were purposeful. But to what purpose? The arrangement of the stones could not be chance or a natural phenomenon. But who would build in this

Figure 6.4. Members of Dr Koetschau's school group watch while the author attaches a twin-hose reguator to the aqualung cylinder on the beach at Pavlopetri in 1967. The ranging pole was used to provide scale in underwater photographs.

Figure 6.5. Channel cut through solid rock on the beach at Pavlopetri, leading to a small bridge, and slots in the channel for a sluice gate. This channel is adjusted to present sea level, but is a few hundred years old, probably leading to a salt pan. This picture is from 1967, and the bridge was destroyed a few years later.

Figure 6.6. A small tomb made by standing four slabs of stone on edge, known as a cist grave. The stone slab that would have covered the grave is missing. The water depth is about 2 m, and the scale bars on the ranging pole are 25 cm. This picture was taken in 1967. In 1968 the divers found that there were 35 such graves distributed throughout the underwater city.

curving, almost fluid way? One building stood out having the bordering stones standing on edge and a curved apse at each end. Human purpose was undeniable.

Extraordinary claims require exceptional proof. In order to publish that this was the remains of a lost town from the Bronze Age before the use of ashlar more than 3000 years ago, I needed the most complete evidence.

I spent the next two days measuring and sketching these strange walls, also taking photographs under water before moving on to the other sites on my list (Figs 6.5-6.7). The walls at Pavlopetri were not what I had expected a Bronze Age town to look like, so I had to do careful research back in London at the Institute of Archaeology before I could convince myself what they were. I learned that the box-like tombs were called cist graves.

In September, the *Sunday Times* published my discovery of the first underwater Bronze Age town, and in 1968 I published a scientific paper in the journal *Nature*. It was, and still is, the oldest town ever found under the sea. My urgent next task was to obtain a proper permit for a full team of diving archaeologists to spend a month on the underwater town in 1968. Nikolaos Yialouris again granted the permit, so the team led by David Harding, backed up by divers from CUUEG, could work there with scuba gear to make an accurate map. I continued my tour around the Peloponnese in 1968, sketching the map of the submerged ruins at Plitra, just north of Pavlopetri, then making a short visit to the team working at Pavlopetri on the way. Their discoveries exceeded all expectations, with roads, houses, public buildings, and graves plotted on huge sheets of paper pinned up on a wall, plus a steady haul of ceramics that confirmed the date for those ancient citizens. The pottery dated from at least the Middle Bronze Age, 4000 years ago. Delighted with the work of the real archaeologists, I left them to continue their studies. Their sensational analysis of the submerged town of Pavlopetri was published in 1969 by the British School at Athens, and in the *Illustrated London News*, an influential weekly magazine.

My geological and oceanographic calculations from the sites that I measured in the Peloponnese and on the island of Antikythera showed that there was a widespread variability of uplift and subsidence which could only be explained by earthquakes and tectonic processes. The residual change of global sea level was less than +/- 50 cm (20 in) in the past 2000 years.

For forty years, Pavlopetri remained undisturbed by prying scholars, but economic progress did not leave the coast untouched. As the decades passed, tarmac roads soon brought coach-loads of tourists to sunbathe on the sweeping sandy beaches of Elaphonisos Island; local farmers used more irrigation and fertilizers, washing soil into the bay; new hotels had to install water purification to treat sewage. All these signs of progress were conducted responsibly, and the village mayors did not ignore the archaeology, but it was obvious that modernization would eventually create a risk of damage. Pavlopetri needed protection.

Having seen the destruction caused by modern development at other coastal sites in the Mediterranean when I had revisited them, I revisited Pavlopetri in 2007 to see how much it had changed. I contacted Solvey in Hamburg, with whom I had kept in touch over the years, to see if she would like to make an anniversary visit. Her three children were grown up, and she did not need to stay in Hamburg. Solvey drove our hired car from Athens to Neapolis, and there I contacted the local coast guard to see if there were restrictions on diving over the underwater town at Pavlopetri. He assured me that diving was restricted, but that he could organize a boat so that Solvey and I could examine the site to study the state of the ruins. The next day we boarded the supply boat that shuttled between Neapolis and Elaphonisos with stores for the restaurants and hotels. After delivering the stores, the boat was ours to dive on the underwater town.

We anchored by the ruins. I clambered from my wheelchair onto the wooden handrail of the boat, already wearing my mask and snorkel, and asked the sailors to push me into the sea. As I splashed down and hit the cold water, I saw the captain crossing himself with a devout look on his face. Solvey and I snorkelled amongst the ruins for about half an hour, finding many buildings that were still recognizable. My first impression was surprise at the continued existence of any intact and complete walls, as such fragile structures could have been scattered irretrievably by winter storms. I wrote a report for the British School at Athens, suggesting that in view of the intense local commercial development, the underwater city needed to be resurveyed using modern electronic techniques, with, if possible, permission to excavate. This survey should be followed with a preservation programme to protect the site from encroachment from the land, to restrict boats from anchoring over the ruins, and to prevent tourists from collecting pottery from the seafloor.

A year of negotiations and correspondence resulted in an Associate Professor of Archaeology at Nottingham University, Jon Henderson, being appointed as co-director

Figure 6.7. Kirsten Flemming diving on the new walls discovered by Jon Henderson's team at Pavlopetri in 2010 (Photo Jon Henderson).

Figure 6.8. Diver at Pavlopetri excavating a large jar or pithos on the seabed. Note rigid frame of fixed tubes so the diver does not have to rest on the seafloor and disturb deposits. The broad pipe on the left is the airlift dredge that is sucking up the sediment to reveal the artefacts (Photo Jon Henderson, Pavlopetri Underwater Archaeology Project).

Figure 6.9. AUV working at night to avoid sunlight ripple effect on the seabed. The algorithm that matches adjacent frames of the automatic survey would be confused by the recording of the flickering highlights of sunshine through the waves, so it was more accurate at night (Photo Ariell Friedman, Pavlopetri Underwater Archaeology Project).

of a joint project with the Ephorate of Underwater Archaeology in Athens. Henderson's team worked from 2009 to 2012, making a superb new map (Figs 6.7-6.10), and conducted excavations with analysis of the ceramics showing that the earliest phases of the town's foundation were more than 5000 years old, in the Neolithic. This was a fascinating discovery but, due to inadequate legal protection, the ruins are still threatened by pollution and casual damage by boats anchoring and souvenir collectors.

Henderson's international team of divers, which included my daughter Kirsten, discovered many new buildings. A huge foundation must have performed an important public function. Oscar Pizzaro and his team from Sydney University, Australia, programmed an autonomous underwater vehicle (AUV) to conduct a grid of close-spaced tracks over the drowned city, making a complete photogrammetric survey. Excavation pits revealed a varied range of pottery, which Chrysanthi Gallou recorded and interpreted.

The modern survey and excavation of Pavlopetri was a sensational success, with the BBC producing an excellent film, *City Beneath the Waves,* that has been re-broadcast several times. The ancient town grew from a more primitive Neolithic settlement more than 5000 years ago, when the sea level at the site was about 5 m (16 ft) lower than at present. The town grew in the Early Bronze Age, exhibiting the transition from local fishing and artisanal seafaring towards a culture of organized trade and long-range shipping. This transition took place at many coastal settlements throughout the Aegean and the Levant coast, but we still do not understand the details. It was a profound shift in the culture of seafaring and marine skills.

A combination of earthquakes and sea level rise due to melting of ice caps resulted in Pavlopetri being flooded in the Late Bronze Age, about 3000 years ago. There are no

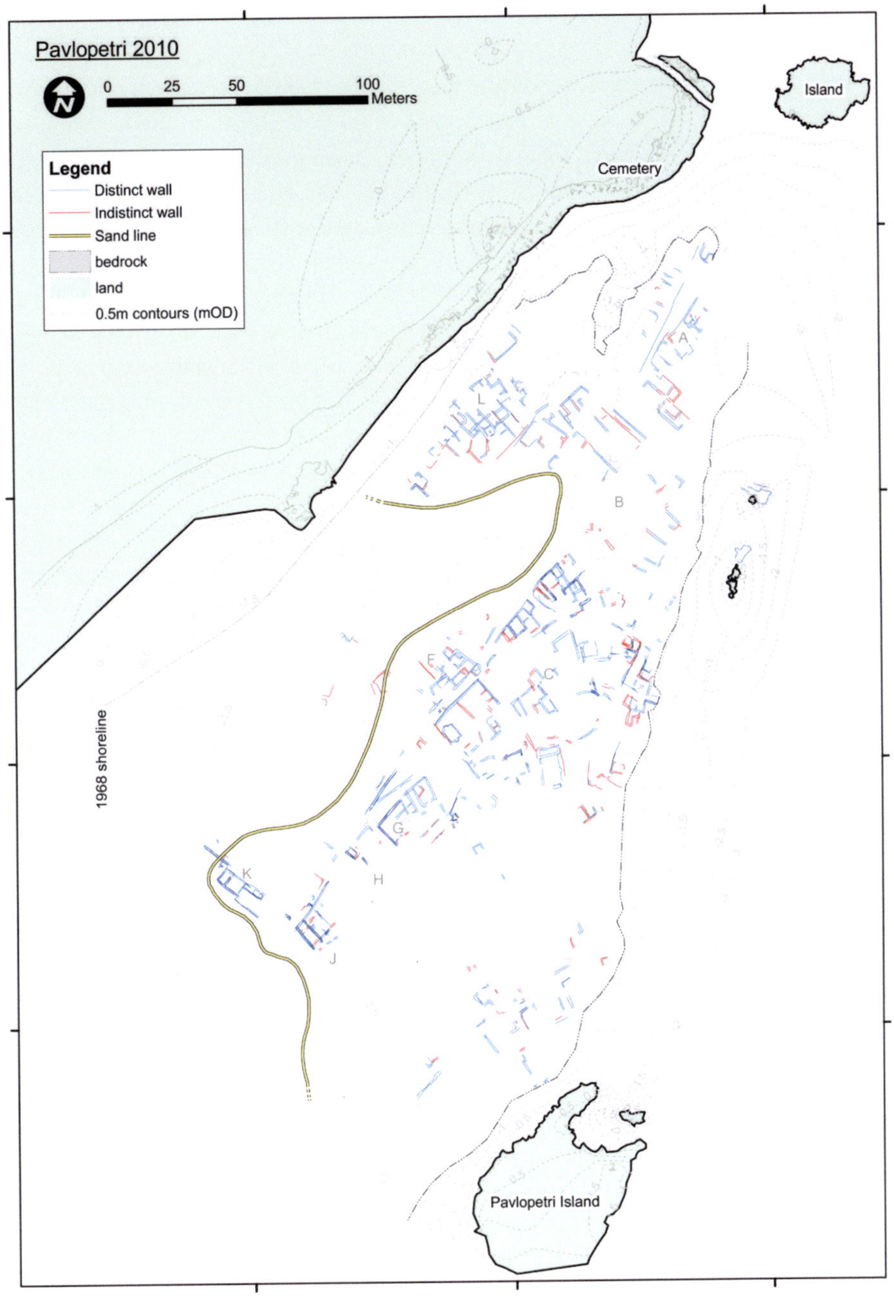

Figure 6.10. Map of Pavlopetri (Jon Henderson, Pavlopetri Underwater Archaeology Project).

buildings of later date and very little pottery. The ruins of the Bronze Age town remain isolated, with nothing later on top of them. In the nearby villages interest and pride in the underwater town of Pavlopetri are strong.

* * *

In 1969 and 1971 I continued my study of local land and sea level change in the eastern Mediterranean by diving on 37 archaeological ruins from Izmir, on the Aegean coast of Turkey, to Kaş, in the south, then 31 sites from Phaselis (modern Tekirova) along the south Turkish coast to Iskenderun and Seleucia Pieria close to the Syrian border. For the 1969 trip, I worked with Natalka Czartoryska, my then-wife. We drove across Europe, through the Balkans, across Bulgaria, into Turkey, and on to Izmir.

From Izmir we drove westwards along the Cesme peninsula, stopping to swim on ruins at Clazomenae and make sketches while snorkelling. As usual, I had a set of maps photocopied and stapled as small books. The archaeological guide books written for the general reader by George Bean provided an excellent introduction to this fascinating coast. In ancient times the cities and city-states on both sides of the Aegean were Greek, while Greek and Roman influence continued, with its later Christian culture, until the Arab and later Turkish invasions of the 12th century, and the eventual sack of Constantinople in 1453.

We spent the next two days snorkelling and diving over ruins at Erythrae, the modern Ildir. Erythrae was one of the 12 cities of the Ionian League, and at the battle of Lade in 494 BC it contributed eight ships to the fleet, compared with 100 from Chios. It was a prosperous but minor city. The remains include parts of the city wall 4-5 m (13-16 ft) thick extending round a hill, at the top of which is an acropolis 70 m (230 ft) high with, on the northern flank, a theatre. The city was famous for its thermal springs. The acropolis hill drops sheer on the west side to the modern beach.

As we approached Ildir there were many houses not more than 100 years old, but derelict, roofless, and in some cases falling into the sea. What I had not anticipated was a Greek city from the 1920s that had been abandoned at the time of the great massacre in Smyrna in 1922 when the army of Atatürk conquered that city and burned it to the ground, with many of the occupants. At Ildir, after the departure of the Greeks, Turkish peasants from Europe who were uprooted for transport into Turkey from Greece, were given the abandoned lands. By 1969 a few houses were occupied by Turkish farmers, but most were empty. During my travels in Turkey in 1969 and 1971, we met other groups of Turks who felt that they or their parents really belonged in Europe. In mirror image, we met a few Greek fishermen who somehow avoided the great transportations of whole populations in the chaos after 1918, and whose families remained in Turkey when the country collapsed. In spite of the brightness of the late afternoon sunshine as we approached Ildir and the green cultivated fields, there was an air of melancholy desolation and poverty. Nature was reclaiming its own.

We planned to spend the night at Ildir, but there was no çay ev (tea house), or café, and certainly no restaurant. As we paused uncertainly by a wrecked street of derelict roofless houses a man came out of the nearest house that was still occupied.

"Günaydın. Can I help you?" He explained that he had a few words of English, and more German.

"Günaydın." I replied. "Thank you. Teşekkür ederim. We were wondering if we would be able to buy food for a meal later this evening. We plan to stay nearby overnight."

"You are most welcome," he said. "I will see what food we can find, and you can have dinner here in a couple of hours."

We thanked him and drove cautiously down through fallen stones and dusty ruts to a flat area at the other end of the abandoned town, near the beach, with the acropolis hill rising steeply beside us. Astonishingly, the derelict 20th-century houses continued into the sea, where tiled floors, flights of steps, and doorways were a few feet deep in the lapping water. There were islands in the sea a half-mile offshore, protecting the town from the wind and waves, so that the sea was very calm.

In the past 50 years, the modern Greek town had sunk into the sea like an ancient ruin. The whole area must be sinking from earthquakes at an almost unbelievable rate (from a geologist's point of view). I walked over the shallow remains of houses from the 1920s, into water 1 m (3 ft) deep before putting on my fins. Swimming out a little further I could see fallen columns from 2000 years earlier, half rolling down the slope into deep water. It was an extraordinary palimpsest of different periods all collapsed and tumbling into the sea, one layer on top of another. The detailed work would have to be done the next day.

We drove back to the edge of the derelict town to find our Turkish friend who had offered to arrange dinner. He had placed a table outdoors in anticipation. As we approached he picked up the table, moving it to the other side of the track, where we were on the brink of the slope down to the sea.

"You will have a better view of the sunset from here," he said.

We thanked him and sat on the chairs that he placed for us.

"I am sorry," he went on, "but we do not have much to give you. I have asked my neighbours, and we have put together some eggs and tomatoes with herbs in a dish that we call 'menemen'. I think you will like it."

With self-tortured distress and guilt, I realized the extreme poverty of our host, and the sacrifice that he and his neighbours had made. He served the food and sat with us while we ate. We conversed in German and English. He had worked for several years as a machine tool operator in factories in Germany, returning finally to his home region. He did not comment on the bizarre state of the empty ruined town, nor refer to the few poor Turks living in it.

When the meal was finished, I offered to pay him, but he refused so vehemently that clearly no persuasion would overcome his sense of hospitality and honour that were at stake. I went to the car, found a large box of cigarettes, and offered it to him.

"If that is instead of payment," he said, "I cannot accept it."

In desperation I searched in our little dictionary, where I found the Turkish word "dost" as a translation for the English word "friendship". Suddenly his heart seemed to melt. He accepted the cigarettes with a big smile, and we spoke happily together for a little while, as the sun set into the mirror of the sea. Then Natalka and I set out to find somewhere to sleep. I did not smoke myself at the time and never have smoked.

Later on the trip, I noticed that children frequently called out "Arkadash", which I looked up in the dictionary, finding that this word was also translated as "friendship". Through further enquiries I found that *arkadash* means something more like chum, buddy, or pal; while *dost* means soul-mate, commitment, faith, and trust. There is a whole website now discussing these two meanings. Luckily, way back in 1969, I found by chance the one word that enabled our generous and hospitable acquaintance to accept my gift with honour.

Before returning to the main subject of archaeology, I want to relate another incident, a few weeks later, where the folk-obligation of generosity almost broke my heart. There were many other lesser events of this type, which we learned to navigate tactfully with honour on all sides, but this one sticks in my mind. Near Karatoprak, after diving with George Bass at Yassi Ada island, I was driving a few miles north intending to find some small ruins which were supposed to be on the shore. I was on my own as Natalka was shopping in the town. The dirt road climbed slowly into the hills a short way back from the sea, becoming narrower with each mile, sandier, then grassier, until it petered out in pure grass and thorn bushes, surrounded by a few rolling fields. The weather was stormy with low clouds and a few raindrops. There was nothing for it but to do a classic three-point turn and abandon the search.

As I was turning, I looked across the roughly tilled field to a distant white house nestled into the trees at the foot of a hill. There was no other building in sight. A small figure was walking from the house towards me, unsteadily, picking careful footsteps through the sparse crops. Soon I saw that it was a child, a little girl, and that she was carrying something in both hands, which she clasped together. I waited for her.

When she reached the car, she walked around the front to where I sat by the open door on the right-hand side. She was perhaps five years old, dressed in the simplest and poorest clothes in dull colours, but very neat and practical. She had no shoes. Her face was serious with that premature weight of responsibility that comes to children who know that one mistake looking after the goats or chickens will lead to weeks of hunger for the whole family. The wind blew her hair back from her solemn face.

She unfolded her hands. Reaching forward, she offered me an egg.

What could I do? I knew that in the house a grown-up, probably her mother, had seen me driving the van into a dead-end, and she knew I was lost. A lost stranger, and probably rather stupid as well. The grown-up had told the little girl to take me a gift and to help me return safely to the town where I belonged. The child brought me that gift, which I could not refuse.

I opened my case, found my wallet, and gave her quite a lot of money. She knew that she should, by the Turkish etiquette, refuse it, but, as I had hoped, she was not brave enough to object. I smiled at her. “Teşekkür ederim,” I said, that is “thank you”. Her face relaxed slightly. She watched as I placed the egg carefully in my food box. She smiled for the first time, before walking back through the wind and misty rain to her house on the edge of the forest. I hope that her parents were glad to receive the money and did not criticize her for accepting it.

* * *

Back at Ildir, Natalka and I mapped the underwater buildings the next morning, working in the cool shadow of the acropolis hill, noticing at the same time that the farmers using a track along the beach, avoiding the hill and the acropolis, had raised the footpath with large stepping stones. The shore was still sinking so rapidly that they had had to do this in order to preserve the quickest path to their fields.

We drove around the western and southwestern shore of Turkey that summer, with many beautiful archaeological ruins, extraordinary examples of generosity, threats of danger, and unexpected discoveries. Turkey produces a full gamut of emotions very quickly. We saw bulldozers flattening ancient buildings at the coastal ruins of Lebedus, pushing marble columns and capitals into the sea; at Loryma castle, near Bozburun, a man who had been bombing ancient tombs to collect the ceramics inside appeared out of the bushes and hitched a ride on our hired fishing boat to save the walk back to town carrying his sack of stolen loot; when we were searching for the track through the forest to Cedreae, two soldiers appeared waving their rifles and shouting "Fotograf! Fotograf!" I thought we would be arrested for taking pictures of ruins, perhaps in a security area, but they just wanted photographs of themselves to show their friends. In the region of Bozburun, in addition to an unexpected discovery of ruins at Saranda, with a great scree of pottery descending into the deep, we saw dozens of underwater foundations on small submerged rocks or islands in the bay, which nobody had identified previously, and, as far as I know, they have not been mapped since. I did not have time to explore them properly. Perhaps the strongest abiding sense is of the collective beauty of so many of the ruins we visited, with their haunting ghosts of past greatness: towers crumbling into the sea where the grass came down to the water in a bay so sheltered that the waves never broke across the beach; vast trees growing close to the shore in remote ruins, where one or two men sat in the shade drinking tea; herds of cows standing in calm seawater a few inches deep where the fields and the sea met in such a gentle gradient that it was difficult to tell the difference between them.

Once again, many archaeological remains of the same age were submerged to different depths, or had not been submerged at all, so that the only explanation had to be that the cause of most of the submergence must be earthquakes.

On the way back from this season's work my van was forced off the road near Belgrade by a careless driver, and I describe the consequences of the crash in Chapter 7. I was paralysed from the chest down and have used a wheelchair since then. In summer 1971 I was back in the active world of diving with a plan to continue the research along the whole south coast of Turkey, around the bay of Iskenderun, to the Syrian border. My vehicle now was an Australian Chrysler Valiant with hand controls, right-hand drive and automatic gearbox, a feature very unusual in England at that time. The engine was a lovely straight-6-cylinder that ran very smoothly. To provide extra muscle and swimming assistance, Natalka and I were accompanied by Peter Hunter, my assistant from the NIO, and James Ward, a diving friend from CUUEG days.

We drove into Antalya on the south coast of Turkey in the late afternoon, where we did some shopping before leaving town to the south towards ancient Phaselis. We found a place to sleep, near a running stream, with enough firewood to cook a meal before bedding down for the night. The drive onwards next day to Phaselis, close to modern Tekirova, was not easy. The gravel and dirt track crossed many small stony rivers, some still flowing weakly in mid-summer, but there was never any attempt to create a ford or shallow strip. Presumably, the other vehicles using the road were farm tractors, which were very high for clearing obstacles.

The only way we could cross these streams was for my three companion divers to move large stones and pebbles so that there were two raised shallow ridges that I could manoeuvre across inch by inch, as if on railway lines, with some stones being moved into the best place in front of the wheels as I progressed. Laboriously we crossed several rivers like this until we reached a forested area close to Tekirova. Farmers pulled my wheelchair across the last stream, which was quite deep, and we had a hospitable glass of tea in their house on the bank. The forest trees hung with vines seemed to consume the houses by the river. It would be impossible to drive any further south.

After a suitable polite period of gossip, the farmers and their families pointed us to the ruined city on a small forest-covered headland. We drove past a marsh onto a grassy isthmus, several hundred metres wide, with the calm sea like a mirror before us. A huge stone sarcophagus, carved with human figures, lay broken open on the beach, tilted towards the water. Small oleander bushes and flowers grew out of the sand. A section of an aqueduct, with many arches still intact, led towards the raised headland, with more ruins visible through the trees. The city was overgrown, with fronds and vines tangled between the trees and the ruined buildings.

Phaselis was founded in 690 BC by settlers from Rhodes, rapidly becoming the main trading port on this coast. The harbour and the main city buildings were enlarged and expanded over the years, culminating in a series of major streets and a ceremonial arch to honour the visit of the Emperor Hadrian. Substantial buildings and water cisterns were added in the Byzantine period.

We explored the three harbour basins, especially the central one that still has a quay with stone mooring bollards projecting from it, but rather low in the water. A paved avenue crosses the isthmus, bordered by a raised sidewalk, linking the harbours, and in all the harbour basins we found walls and foundations indicating slight submergence of the city, about 20 cm (8 in). Swimming over the rubble breakwater of the southern harbour we could see a long sandy beach to the south, behind which the mountains climbed to the snowy peak of the Ak Dag. Somewhere in those mountains was the eternal volcanic flame of Yanartas above the ruins of Olympos. The ruins of Phaselis are now flanked by two well-developed small towns, accessible by a wide tarmac road, and there is a cable car up to the flame above Olympos. So it goes.

* * *

East of Antalya there was a fairly wide road along the south coast in those days, some of it tarmac surfaced, some just gravel, so the coast road itself was not a problem. Getting off the road to visit the ruins we needed to see was usually a problem. Whenever I asked the way to a village or archaeological site off the main road the reply was always, "Yok! Yol bozug," meaning "No way! Bad road". Usually we managed to get to the sites, even if it meant filling in potholes with rocks.

We spent several weeks driving steadily eastwards, sleeping every night either on the beach or in the hills behind the road, cooking our evening meal over a fire of twigs and driftwood collected each day for the purpose. The Taurus mountains came closer and closer to the shore, but occasional small roads twisted inland up the precipitous valleys to the Anatolian Plateau that occupies the centre of Turkey.

We examined several minor sites before reaching Alanya, the ancient Coracesium, which was of special importance to me, because the main buildings on the shore dated from 1226-1227 AD, thus providing a fix about midway through the 2000 year span that was the purpose of my study. The original Greek city, with an acropolis on the hill 250 m (800 ft) above the sea, experienced the cycle of local piracy, Roman suppression, and then overbuilding in the Byzantine and later periods. The headland with steep cliffs is surrounded by crenellated fortifications built on top of the earlier Hellenistic foundations by the Turks, who captured it in the 13th century. The fortifications of the hill were then joined to massive walls protecting the harbour, with a focal tower known as the Red Tower providing formidable defensive strength on the shore. Close to the Red Tower are five sloping ship sheds built in 1227 AD. The ships, when pulled out of the water, would have been completely enclosed by the thick walls and a roof of stone. The sheds are each 34.6 m (114 ft) long and 7.3 m (24 ft) wide. In the dark shadows, the seawater laps gently on the sloping slipways about one-quarter to one-third of the way up the slope, so that a vessel 20-30 m (66-100 ft) long and 5-6 m (16-20 ft) wide could use the shelter perfectly to this day. The ship sheds provide elegant proof that the sea level in 1227 AD was the same as today, to within about +/- 20 cm (8 in)

Within a few kilometres either side of most large ancient towns, we found numerous quarries cut into the coastal cliffs. Two thousand years ago transport by sea was the easiest way to move big rocks to the town. Transport by cart would have been very laborious. None of the quarries showed submergence. The incised grooves, like in a chocolate bar, showed where the ancient quarry-masters had been cutting rectangular blocks, but the signs of quarrying did not go below sea level.

We arrived at Antiochia ad Cragum near Gazipasha just after midday under a cloudy sky. The Chrysler Valiant purred nicely as we curled through the low hills up towards a group of rugged headlands that projected into the sea. Although the city is on the coast, it is dominated by a temple several hundred metres (about 1000 ft) above sea level. The city was founded by Antiochus IV Epiphanes in approximately 170 BC, after a series of political betrayals and murders that helped him secure the throne. A farm track curved upwards through fields, and then past rough scrub and grazing. The clouds were so low and the air so damp that the summit was almost in the mist. I drove carefully along the last section of grassy track which terminated in a flat area the size of one or two tennis courts, with a few scattered boulders. We were now on a bluff from which I gazed down at the storm-dark water pock-marked with white horses far below us, stirred by the brisk wind from the southwest. I could see scattered walls and remnants of towers on the slopes. Reports of the archaeological remains claimed that there was a small harbour, a huge natural rock arch, and a sheltered small bay with some caves that could be used to shelter vessels at the base of the mountain.

"You had better climb down the slope to the shore," I said to Ward. "You can all go. Take some fins and a notebook. Here are the maps from previous studies. Give it a couple of hours. I'll be OK here reading and catching up on the reports. Take some drinking water."

The descent to the sea was steep, but by no means a vertical cliff, so that the party of three was soon out of sight amongst the grassy gullies below me. I turned to my notebooks and maps, writing up and clarifying records from the past few days, adding notes to sketch-maps, planning the next stops along the coast. I felt uneasy after a while, as if somebody was watching me. I looked around, but there was nobody there. The hilltop was deserted. The wind sighed in the bushes on the slope. Fixing my eyes on the notebook, I again felt observed and then heard marching footsteps and the muted rattle of buckles, leather trappings, and clanging of metallic objects.

I had been sitting with my back to the wind, looking inland, so that in order to see where the noise was coming from I had to turn half into the wind. I do not believe in the objective existence of ghosts, but my mind on that day manifestly experienced a ghostly occurrence. A body of armed men was marching obliquely across the flat grassy space, not aggressively, but carrying their weapons “at ease”, convenient for a long march across country. They looked like a Roman legion on the march. This was factually absurd, since the entire flat area was no bigger than a few tennis courts. The marching men were transparent, and I could see the sky and sea beyond them, through them, but that suggests some sort of filmy gossamer structure, which was not there at all. I sensed the existence of the troops in every visual detail, filling my field of view, even though my rational mind told me that there was nothing visible.

For a moment, since I felt intrigued by the experience, I was inclined to let it roll by, noting what happened, the calm observer. Then I was alarmed to see that one of the approaching soldiers in the rank closest to me, a rank that was marching past just a few feet in front of me, was carrying a long-handled lance with a sharp blade at its tip which was projecting sideways so that it would strike my head, or even cause serious injury. I could see their eyes, but the soldiers obviously could not see me, and they just stared straight ahead. Since I knew that what I was observing was a non-existent psychological illusion I determined not to duck, and to let the lance-blade pass over or through me as a fiction. Fear mounted rapidly as the illusory shaft of the lance swung closer. The hair on the back of my neck prickled uncontrollably. I struggled not to duck my head.

But I did.

I ducked as the lance-blade swept over the top of my crouching head.

A few seconds later I raised my head in self-disgust and looked inland as the marching ghosts disappeared into space, vanishing high over the sloping pastures and scrubs where goats were grazing.

I can only report what I experienced. Psychologists may wish to explain this event or suggest why I conjured up this illusion at this time. Neither then, nor at any time since, have I been able to find any precursor, experience, stress, or thought that could have stimulated sensitivity to such an illusion. I did not tell my friends what had happened in their absence. There was work to do.

* * *

The ancient city of Anemurium, which occupies hills at the southernmost tip of Turkey, retains huge areas of ruins on the hillsides, a necropolis, two aqueducts, and many buildings close to the shore. There is evidence of a stable sea level, and the same at the nearby Armenian castle of Anamur. From there we drove on to search for the ruins of Holmus. The maps and guide books showed that it was on the prominent headland at Ovacik, but there was no obvious road to it. Local enquiries revealed a stony track with the usual warning of “Yol bozug!” We started up the track nevertheless, with Ward and Hunter hopping out to place stones in the frequent potholes. Trees hid the view of the sea until we suddenly broke out of the cover on the crest of a ridge and saw the fantastic sight in front of us. It was so beautiful that I stopped the car. We all gazed in wonder.

The mountain sloped down below us until the track snaked out onto an isthmus of such pure brilliant green grass that it was almost unbelievable. Beyond this, the peninsula sloped up onto a hill, and then further small peaks formed the mass of a rocky headland. The trees growing on the far hills were sparse but equally bright in colour. The location provided the classic two harbours for an ancient city, with the blue sea glinting on both sides, and an island on the eastern side giving extra shelter. My cynical mind could not accept the simple unspoilt beauty of it.

"If I was a millionaire," I said, "I would buy this land, build a big hotel on that isthmus and without building any roads, I would just fly rich tourists in by helicopter."

"Don't be so horrible," shouted my companions.

I started the car again, and we coasted down onto the green grass that was as smooth as a cricket pitch or bowling green. There was a small derelict brick hut to one side, where for a second I thought I saw a person moving, but it seemed unlikely. We explored the shoreline on both sides of the isthmus but, apart from a few ancient walls in shallow water, there was nothing significant. We were preparing to swim out to the island when a man did appear from the brick hut. He looked like Groucho Marx, with a solid black moustache, heavy rimmed glasses, and a large nose, though incongruously dressed in baggy Bermuda shorts and a decorative Hawaiian shirt. He looked suspiciously at us, but after brief explanations of our interests, he took us to a place on the beach where he kicked away the sand and earth to reveal a beautiful mosaic, several metres across. We thanked him.

He then explained that he represented a millionaire in Ankara who had bought the land to build a large hotel and fly guests in by helicopter. He showed us some plans on the table in the hut. We wished him success and drove back up the hill.

We drove on for several days, examining many fascinating ruins on the beaches, before we crossed the bridge over the Göksu River in which Emperor Frederick Barbarossa drowned in 1190, thus weakening the military forces of the Third Crusade, which failed to reconquer Jerusalem. At Soli Pompeiopolis, modern Viranşehir, I measured the elevation of the large breakwaters that had originally enclosed an almost elliptical harbour basin. The Greek colony of Soli was built on an earlier Bronze Age settlement, but by the time that Alexander the Great occupied the city, the Greek spoken there was so bad that Alexander punished them, which is the origin of the word "solecism". The city was rebuilt by Pompey during the 1st century BC, with a splendid avenue of Corinthian columns, and a sophisticated harbour basin was added in the 1st century AD.

The harbour basin, sheltered from the open sea by the intact, curved, western breakwater, was mostly infilled with sand, driven by wind and waves. The basin was almost 200 m (650 ft) wide and 300 m (1000 ft) in the elongated axis of the ellipse. The breakwaters are 20 m (65 ft) thick, constructed of large ashlar blocks clamped together with dovetailed keys of lead or iron on every surface. The strength of this interlocked construction is extraordinary. The interlocked ashlar massif created hollow inner caissons that were filled with poured concrete. The upper surface of the breakwaters is at present 1.5 m (5 ft) above the sea, and there may have been an additional road or wall on top of that, which was subsequently pillaged for materials. This magnificent testimony to the skill of Roman harbour engineers is protected from modern development, although Google Earth shows that the built-up town is now very close on all sides. Apart from the sand infill, the harbour and its walls are in the correct relation to present sea level, to within an error of +/- 30 cm (12 in).

We snorkelled over several archaeological ruins in the area of the Sehyhan Delta and were ahead of schedule. The next targets were a reported minor ruin just south of the small town of Arsuz, 55 km (34 mi) from the Syrian border, and finally Seleucia Pieria, only 20 km (12 mi) from the border and just north of the mouth of the Orontes River. We drove around the corner at the northeast limit of the Mediterranean to head south. The main road south from Iskenderun curved inland to avoid the Amanus Mountains, but Arsuz can only be reached by the minor coastal road that crosses a small river and terminates at the foot of the mountains. There is a Turkish Navy base at Arsuz. We hired a small fishing boat that morning, leaving the car parked on the road beside the river, close to a row of cafés and shops. The sun was extremely hot, and by the time we returned, after finding only small walls and cuttings set back from the shore, we were desperately thirsty and hungry.

It was after midday when our boat motored slowly past the navy base, under the bridge that was the only access to the town, past our parked car on the road, and then nosed into the shade of an overhanging plane tree where several fishing boats were moored. We unloaded our gear quickly, helped by other fishermen. As I paid the boatman I noticed a plump unsmiling man standing behind the tree, clicking his worry beads and staring at us. He wore a clean white shirt that was unusual in the circumstances. I should have been warned, but we were so tired and hungry that we dumped the gear in the car on our way to the nearest restaurant. We ordered beer and tons of food.

The beer was on the table, and the first dishes had appeared, when the doors of the restaurant burst open to admit two uniformed military policemen carrying small machine guns and crashing their boots unceremoniously as they marched up to our table.

"Who is the driver of that big grey car parked outside?"

"I am. What's the matter?"

"Come with us now. Immediately!"

I turned to Ward and Hunter, talking very quickly and quietly.

"Pay the patron for the whole meal. Keep smiling. Walk slowly. Leave the restaurant very quietly and slowly. Whatever you do keep smiling and chatting to each other. Then follow me to the car. For chrissake don't look anxious or guilty."

I bounced down the steps out of the café, pushing my wheelchair as if I was quite weak so that the policemen had to walk slowly beside me. Their guns were cradled across their arms each with a hand on the trigger-guard as if ready for use at any moment. Glancing back, I saw the others beginning to leave the restaurant.

At the car, the policemen stopped on either side of me, and the leader barked at me.

"Where is your passport?"

I opened the door, took out a box of cigarettes, and offered one to each of the military police. Fumbling with their guns they accepted one each, and tucked them behind their ears, looking a bit disconcerted. I deliberately wasted a few extra minutes as I searched for my passport in my case, then handed it to them. I noticed that Mr Plump-with-White-Shirt was standing behind the corner of the nearest building peering round the wall at me as he clicked his worry beads. He was the town police-informer: I should have got out of town as soon as I saw him by the plane tree. I had no wish to discover what it was like in a Turkish prison, and the thought of the police finding all my notebooks and maps was not reassuring. Pleading to be an archaeologist or biologist was the oldest cover-story on earth for low-key spying. The other three had arrived by the car and started to get in.

"Oh, you are British," said the policeman, leafing slowly through my passport. "You are not Israeli?"

"That's right, I am not Israeli."

The policemen saluted, they both slung their machine guns over their shoulders, and together they turned to march efficiently in step, boots thudding, back down the road towards the navy barracks by the river mouth. When Mr Plump-with-White-Shirt saw that his plot had failed, he started to hurry after them.

"Keep chatting and smiling," I said to the others as I switched on the ignition and did a slow three-point turn. The car had been facing towards the boat mooring, away from the bridge. People were peering from every window in every house along the road back to the bridge. The café owner had spread the message. Were we guilty or not? What was going to happen?

I drove slowly. To rush would be an admission of guilt. There was a soldier with a rifle by his side guarding the bridge which was the only exit from the town. Mr Plump-with-White-Shirt could not run very fast, but he had already overtaken the two military policemen in his frantic rush towards the navy barracks. I turned slowly towards the soldier on the bridge. I smiled and waved as the car rumbled onto the wooden planks, and I drove slowly across. The solder smiled back.

The road was straight for a short distance beyond the bridge so that we were still visible from the town. As soon as there was a bend with trees that concealed us I slammed down the accelerator so that we could get away as far and as fast as possible. We were all silent but thinking the same things. A small road junction appeared on the right, which would probably lead over the hills to the main road from Iskenderun that passed by inland. I swung onto it and drove as quickly as possible along the winding road that luckily had a good surface.

As the minutes ticked by it seemed less likely that we would be followed or chased by the military police at Arsuz. There was no sound of a helicopter.

The minor road finally joined the big north-south road to Antioch as I expected. We turned south, taking a further turning beyond the coastal Amanus Mountains, leading us down the Orontes Valley onto the beach close to Samandag. This was the site of the colossal ancient Roman military port of Seleucia Pieria. A long sandy beach promised good swimming, and the hilly country in front of us framed the far side of the basin in which the ancient port had been built. There were low buildings behind a wall on the inland side of the road.

As we were getting out of the car a man in high-ranking army uniform came out of the gate in the wall. I watched with horror. What if the Navy Police at Arsuz had alerted other military bases in the area to look out for suspicious foreign spies with maps?

"Good afternoon," said the officer. "This is a Turkish Army Rest and Recreation Centre. Military families come here for a holiday. If there is anything you need, please inform me."

"Thank you. We are planning to swim and snorkel here for a day or two, and to look at the archaeological tunnels in the mountain. I hope that is acceptable."

"Yes, of course it is. Please make yourselves at home. I see that your wife is travelling with you. If she wishes to meet the wives of our soldiers, they will be very pleased to entertain her."

Natalka took the opportunity the next day to spend several hours in the base as the guest of the military wives, which was both a pleasant experience and good diplomacy.

The Roman tunnels were extraordinary constructions on the slopes behind the harbour, which gave us a good reason for roaming far and wide. The Hellenistic foundation of Seleucia Pieria was expanded by the Roman emperors Vespasian and Titus late in the 8th decade AD, creating a grand harbour, protected on the seaward side with colossal ashlar breakwaters, but unfortunately prone to infill and blockage by storm floods from the mountains behind. The main city at Seleucia Pieria was on the hill above the port, where the rivers flowed through the city itself. These rivers tended to flood, causing damage to the city in the spring rains. It was logical, having built the harbour, to try and divert the floodwater away from the city and prevent the influx of stones and mud that washed down from the Amanus Mountains. Over the course of several years of quarrying and excavation, carried out on imperial instructions, the Roman engineers and slave workers cut a network of huge tunnels and canals through solid rock to divert the river flow north of the city, away from the harbour, direct into the sea. The tunnels seem to have functioned well for at least two centuries, since the port at Seleucia Pieria was enlarged several times, continuing to be a major base for the Roman navy as late as the reigns of the co-emperors Diocletian and Constantine (284-305 AD; 305-306 AD). The tunnels are 1400 m (4500 ft) long, and, although rectangular in cross-section, bear comparison with modern railway tunnels, albeit spectacularly embellished with trees and vines.

Notwithstanding the tunnels, the harbour eventually filled up with stones and soil, and it is now a patchwork of green fields. On the seaward side, close to the modern beach, we saw a wall of well-fitted ashlar blocks of such gigantic size – 7 x 2 x 1.5 m (23 ft x 6 ft 6 in x 5 ft), each weighing about 50 tons – that it is truly difficult to understand how they were moved and lifted into position. A vertical wall of masonry when attacked by large breaking waves has to absorb all the energy without any scattering of the water motion between blocks of rubble. The wave energy must be reflected back into the sea, which is not good engineering. The breakwaters at Seleucia Pieria seem to have tried to solve this problem by making the wall of stones weighing many tens of tons, fitted closely together. From the position of the breakwaters and the levelled infilling of the harbour basin, the relative sea level at Seleucia Pieria had been stable, with no change in the past 2000 years, to within an error of +/- 30 cm (1 ft).

* * *

We were still ahead of my working schedule when someone suggested that it would be interesting to drive into Syria and return through Aleppo. We did that and met many delightful and hospitable people on the way. However, to describe the pleasure of that journey now would be a senseless insult to the modern people living in northern Syria, especially in Aleppo, where they have suffered unimaginable degradation, deprivation, torture, bombing, and poisonous gas during the conflicts of the past few years. I truly wish peace, happiness, and recovery to those tragically hurt people, and I will never forget the kindness that we received there 50 years ago.

From Gaziantep, just inside Turkey from the Syrian border, we drove rapidly back westwards across the Anatolian Plateau. There was time to fit in one more site before departure for England. We drove down to Demre where we could hire a boat to map ruins at Kekova, a village to which there was no road at all. There were reports that ruins were deep in the water near that village.

The boat had a wide sun awning, so the shady cruise down the Demre River provided a relaxing start to our trip. As the boat navigated the sinuous meanders between the marshes, we could see the tops of dozens of stone doors and windows projecting just above the water in the reeds. There were acres of ancient ruins drowned under the estuary, but nobody seemed to wonder why. We reached open water, where we turned westwards, past Myra, the birthplace of St Nicholas. The mountains began to close in on all sides, with further mountainous pinnacles on bare white islands to our seaward side. We were motoring into a closed and sheltered sea basin surrounded by protecting islands and headlands. The trade along this coast, protected by the chain of islands, must have been immensely profitable 2000 years ago.

The approach to the little harbour of Kekova itself, now known as Kale-Üçağız, was between two sharp headlands, beyond which the bay opened out to reveal a modest wooden jetty, leading directly onto the dirt road at the centre of the village. A crowd of villagers consumed by curiosity flocked onto the jetty as we approached. A short, broad-shouldered man with bulging muscles and a big moustache, ordered them about, taking command, to the obvious resentment of some of those standing by.

On all sides, domed Lycian tombs and bits of wall rose from the seabed with the waves lapping around them. Everywhere there were collapsed buildings in the sea with the top stone lintels of doors and windows just above the surface. Overshadowing us on the skyline, high above the village, was the crenellated fortress known simply as Kale, "The Castle". Tall Lycian tombs on dry land (Fig 6.11) gave us a sense of the depth of water covering similar ones in the sea (Fig. 6.12). They populated the slopes of the hillside, high on their plinths, rising above the vegetation like a forest of stone. Windows and doors projected above the sea close by to our right (Fig. 6.13). I gestured to the others to wander about aimlessly like tourists, to see what they could find. A little group of villagers stayed with me as I wheeled myself into the village, until we arrived at the focal meeting place, under a large tree, where two dirt tracks crossed in front of a café or çay ev.

Most of the crowd filtered into the background discretely at this point, sitting on benches, banks, trees, or whatever was handy, leaving the moustachioed muscle man and two or three others sitting with me at the largest table in front of the çay ev.

"I am the school teacher," explained one of my hosts. "This gentleman," he gestured to the muscle man, "is the *muhtar* of Kekova, the mayor."

"Thank you. I am honoured to meet you." I nodded to the *muhtar*. He made a speech in Turkish which seemed to impress the surrounding villagers in the shadows. By this time numerous little boys were sitting in the trees above us, twittering.

The school teacher and I had a friendly halting conversation on the usual subjects, while tea was served, which we sipped as we wished health and prosperity to each other. The *muhtar*, through the teacher as translator, asked why we had come to Kekova. I explained my interest in the ruins, without making it into a scientific project. Just curious tourism. They were happy with my explanation, genuinely proud of the strange sunken city.

"The *muhtar* would like to arm wrestle with you," said the school teacher.

Figure 6.11. Lycian domed sarcophagus close to the shore at Kekova. From the ground to the base of the domed lid is 2 m, so that tombs in the sea must be submerged to about this depth.

Figure 6.12. Lycian domed tomb in the sea at Kekova. I have been snorkelling round the base of the sarcophagus. It was broken open many hundreds of years ago.

Figure 6.13. The lintels of doors and windows in a massive submerged building at Kekova. The masonry of the walls, and probably parts of the roof have fallen inside the main building to create a raised artificial island. The domed lid of the tomb shows that the ruins have been submerged by at least 2 m.

I protested politely, turning the subject back to the beauties and uniqueness of Kekova. This ritual cycle was repeated a couple more times.

"The *muhtar* insists," said the teacher. "He really wants to arm wrestle. You should accept his challenge."

I guessed that this was what everybody had been waiting for. I tried not to gape at the *muhtar*'s upper arms which were two or three times thicker than mine.

We adjusted our sitting positions, jockeyed for a good elbow angle, and grasped hands. I waited for the inevitable humiliation, accompanied by glory for the *muhtar*, followed by generous glasses of tea all round to show his magnanimity in victory. I braced my arm and waited for the coming thrust which would flatten my arm back onto the table. There was some pressure from the *muhtar*, but not more than I could resist quite easily.

I waited like this for what seemed like minutes, assuming the *muhtar* was testing my strength, before it occurred to me that maybe he was not as strong as he looked.

Instead of maintaining a bracing force that I could assert for a long time, I increased my force to the maximum to see what happened. To my astonishment, the *muhtar*'s arm just twisted over quickly so that I rammed the back of his knuckles onto the table.

This was a diplomatic blunder. If I had realized how weak he was I would have checked myself. But it was too late. He gave way in seconds. The little boys in the trees burst out laughing. The *muhtar* scowled like thunder and glared at me with true hatred. The men sitting further back hid their smiles behind their hands as they slipped away into the shadows before anyone tried to talk to them. Apart from the giggling boys, there was then silence.

"Would you like another glass of tea?" asked the teacher in an attempt to rescue the situation.

"No thank you," I said very politely.

The *muhtar* looked at his watch, muttered something to the teacher, pushed back his chair, and departed quickly. Everybody else avoided him or pretended not to have seen his humiliation. I looked at the teacher as apologetically as I could, gesturing helplessly with my hands, but could not think of anything intelligent to say. Then the teacher smiled but said nothing. I think that he, and everybody else, was secretly glad that the *muhtar* had received his comeuppance. I thanked the teacher for his kindness and made my way back to the jetty.

When my companions gathered round after their explorations, I donned snorkel gear, and we searched amongst the colossal flooded tombs, measuring the water depth at the base of each one. Close by in the bay, in real deep, dark, blue water, were pinnacles of rock rising from much deeper in the sea, but flattened out on top with the foundations of houses, rooms, flights of steps, and other signs of domestic life, all more than 2 m (6 ft 6 in) under the sea. We boarded our boat again, sailing out of the bay, searching everywhere for these submerged islands which supported just a few houses on each pinnacle. We saw fish swimming through the doors and windows where people once lived and breathed. The ancient town had been distributed over many such small islands, and one could imagine neighbours rowing by boat a short distance to drop in for lunch on the next island, before rowing over to the next rock for tea. A sort of Venice on bedrock instead of mud.

The drowned city at Kekova had sunk by 2.2 m (7 ft 3 in), with an error of +/- 30 cm (1 ft). The complete scope of these many submerged towns in the 200 km (120 mi) between Tekirova and Bozburun has still not been fully assessed or mapped. For many centuries, the earthquake shocks had been concentrated on this short section of coast, but there is no record of when they happened or how people escaped or were drowned in the disasters. The modern tourist trails appreciate the commercial value of the ruins in the sea as a curiosity and attraction, a gimmick, and the ruins are protected, but the application of serious scholarship by Turkish scientists is just beginning.

In geological contrast, the coastal ruins along the south coast of Turkey further east, 600 km (370 mi) from Antalya to the Syrian border, show that that coast is stable to within an error of about +/- 30 cm (1 ft).

* * *

For the next three years, I recruited small teams of divers to work on Crete, Cyprus, Rhodes, and Karpathos.

The routine crossing of Europe had now been reduced to a swift combination of trains and ferries, with the car loaded on the special transport carriage, so that the journey to Crete in 1972 was achieved in a very few days. The last stage was the ferry from Athens to Heraklion. The observation of changes of sea level around Crete has a long history, with the first scientific reports by Captain T. A. B. Spratt, RN, published in 1865. He noticed that the sea dissolved notches deep into the limestone cliffs at present sea level, and that similar notches could be seen in many parts of western Crete as high as 5-10 m (16-33 ft) above sea level. From this, he concluded that western Crete had been uplifted by earthquakes. Since no such notches were observed in eastern Crete, Captain Spratt proposed that the whole island had tilted, up in the west, down in the east, like a see-saw.

By the 1960s we knew that this model of simple tilt was incorrect, since archaeologists had found Bronze Age towns on the shore of eastern Crete. There must be a hinge-line somewhere in central Crete, with the island distinctly bent at this point. Accordingly, I asked for a permit to visit archaeological ruins in the western and eastern thirds of the island but not in the middle. A request to visit all sites on the coast of Crete would be rejected.

Our assistants in 1972 were Dr John Gifford, a distinguished American archaeologist who was expert on sea levels, and Bernard Potocki, Natalka's half-brother, who lived in Canada. I will focus mainly on three sites out of the dozens that we visited and swam over: Rethymnon and Phalasarna in the west, and Mochlos-Siteia in the east.

At Rethymnon the coast road had been newly cut, or at least widened, so that it skirted around the base of the turreted fortress on a headland projecting into the sea. I was driving along this road when I suddenly saw slipways cut into the bedrock and actually buried where the new road had been laid across them. I slammed on the brakes so hard that it was lucky there was no car just behind us.

The three slipways, one of which had a central groove runner like those at Apollonia, were well preserved, apart from being partly covered by asphalt (Fig. 6.14), sloping down to the sea level at just the right depth. This example on the north coast of Crete shows a point that has definitely not moved up or down in the past 2000 years. We must be close to the hinge-line, I reasoned. We parked, then measured and photographed the slipways in detail. Many thousands of people had passed by this place but never realized what these strangely precise cuttings were for.

Immediately to the west of Rethymnon the present sea level has cut a neat and deeply incised notch into the limestone cliff. As we drove westwards, we saw that a second notch had been dissolved into the cliff, occurring higher and higher towards the west. This upper notch was pre-Roman, the result of erosion and solution into the cliff before it was uplifted when western Crete tilted in a terrifying earthquake. This was the notch seen by Spratt. There was no road to the ruins of Diktynnaeo on the northern tip of the Rodopou Peninsula, but I needed to see the ruins there, to measure the height of the notch. The US Navy Seals based near Khania gave us assistance in reaching Diktynnaeo. During the trip on their landing craft, we measured the notch at a height of 5 m (16 ft) (Fig. 6.15). Greek geophysicists calculate that the extraordinary uplift of western Crete was caused by a magnitude 8.5 earthquake in 365 AD, that destroyed buildings in cities over a wide area, as well as creating tsunamis in North Africa.

We continued on to the west coast of Crete, arriving at the harbour of Phalasarna, which needed several days of research. We found a sheltered corner for cooking and sleeping and settled down to work.

Phalasarna harbour, if it was a harbour, was uplifted by over 6 m (20 ft) and was so overgrown with grass and scrub that some writers doubted that it was a harbour at all. The acropolis at Phalasarna was a hill rising to 100 m (330 ft). Closer to the shore, the signs of raised sea level notches, tower foundations flanking the presumed harbour entrance, a channel for boats to approach the harbour, and other signs of the original

Figure 6.14. Ancient Greek slipway at Rethymnon, emerging from under the modern roadway. The central groove probably had a wooden beam in it originally. The lower end of the slipway is adjusted to the present sea level, to within an acceptable margin of error.

Figure 6.15. Pre-Roman Sea level notch uplifted to a height of 5 m on the cliffs of the Rhodopou peninsula, 74 km west of Rethymnon. Note snorkeller in the sea for scale, and the modern solution notch cutting into the rock.

Figure 6.16. The author with Bernard Potocki on the edge of the uplifted fish-tank at Phalasarna, Crete. The area of rock lifted out of the sea by the earthquake of 365 AD still has no vegetation on it.

sea level were so convincing that I had no doubt at all about the harbour's existence. This information was greatly appreciated by Elpida Hadjidaki a few years later when she excavated the site to confirm all the constructions of a major port, with fortified walls, entrance channel and towers, quays, mooring bollards, and city roads. It had been a very grand port in the Roman Empire.

The most dramatic proof of uplift was a small piscina that had steps down into it (Fig. 6.16). This piscina is a different design from that at Apollonia, and is more a storage tank for fish, with internal dividing walls, and some flushing by wave action. The combination of many clues together at Phalasarna showed that, since the city was occupied, it had been uplifted by 6.6 m +/- 20 cm (21 ft 8 in +/- 8 in). The earthquake destroyed it and obliterated its function as a harbour.

To check that the eastern half of Crete had not been substantially submerged, I wished to see the reported slipway at Siteia, which was quite close to the islands of Mochlos and Psira. Mochlos itself was covered in well-preserved buildings, houses, roads, and ruins of the Bronze Age. The island might have been joined to the shore by an artificial isthmus or causeway.

We spent the day swimming and snorkelling around Mochlos before hiring an inflatable boat which we drove the 4 km (2.5 mi) to Psira, where my companions spent several hours climbing on the steep island. Psira, which was occupied from the Neolithic to the Late Bronze Age, had, in its most prosperous period, 60 buildings ranged around an open square. The sides of the island are steep, but we could find no buildings close to sea level. On return to Mochlos, we refreshed ourselves with a meal of lamb and beans accompanied by excellent strong red wine, which prepared us for a good night's sleep. The restaurant claimed to make the best baklava in Crete. So we ate some to make sure.

Sites in eastern Crete showed very little submergence, although Itanos, in the extreme northeast, close to the only grove of wild palms in Europe, has ashlar walls a few centimetres under the sea.

* * *

The ferry from Athens pulled into Limassol harbour, on the south coast of Cyprus. Once ashore, we drove 10 km (6 mi) eastwards to the ancient site of Amathus (Fig. 6.17). Here there was a large mound of ruins and ancient walls on the landward side of the road, partially protected by some fencing and a few dusty signs. On the seaward side, down a low rubble slope, we could see more walls extending to the water level. Our assistants this year were Jean-Francois Aubineau, an ex-fighter pilot who lived in Paris, and Michael Leitner, an American archaeology student. Within minutes of snorkelling into the warm sea, we had found a massive submerged breakwater of large ashlar blocks enclosing a rectangular basin about 100 x 250m (330 x 820 ft). When Amathus was a booming city 2000 years ago, this port supplied ships trading with Anatolia, Greece, and the Levant coast. The main export was copper.

The breakwater is totally submerged now, providing no protection from storms. There are many fallen blocks, and the upper courses of ashlar may have been removed at some date to build structures on land. But if we imagine that the sea level was much lower, then the basin within the breakwater becomes too shallow for ships of any size. The relative sea level had risen by 0.75 m (30 in), most probably caused by tectonic land subsidence.

Figure 6.17. Jean-Francois Aubineau with a 2 m ranging pole, measuring the header blocks which make the submerged breakwater at Amathus, a very early Phoenician or Greek harbour on the south cast of Cyprus, in 1973.

We were accompanied by a Greek archaeological guide, who made sure that we always had access to antiquities in those cases where there was an on-site guardian. We spent a few days exploring the submerged walls and streets at Salamis, on the east coast. Then, after a long drive out to Cape St Andreas and back along the northern shore, swimming on many small archaeological sites, with the grey sun-baked pinnacles of the Kyrenia range looming above us, we arrived at Lambousa, 11 km (7 mi) west of Kyrenia. The air temperature reached 40° C (104° F) and smoke from forest fires in the foothills drifted over the mountains

The Bronze Age town at Lapethos also flourished again in the Roman and Byzantine periods with the name Lambousa, until it was sacked in 650 AD by Arabs led by Muawiyah I with a fleet based in Alexandria. In 688 AD the emperor Justinian II reached an agreement with the caliph Abd al-Malik to share the rule of Cyprus. The discovery of a Byzantine hoard of silver in the early 20th century resulted in predictable bouts of looting. In the Graeco-Roman period, the citizens constructed a complex of three sets of multiple fish tanks extending for 200 m (650 ft) along the rocky shore. We spent two days measuring the array of channels and tanks at Lambousa which, given the high temperatures, required ample supplies of cold beer to keep us going. The multiple fish tanks here show perfectly the ways in which the ancient engineers used waves and currents to ensure a constant through-flow of fresh seawater.

As at Apollonia, this amazing network of tanks was probably decorated with statues and provided places to relax in the shade of awnings or screens, where the gentry could enjoy the calm contemplation of the swimming fish. In the background, slaves would have been surreptitiously busy opening and closing the sluices and gates at the right times, and throwing food into the pools to feed the fish.

The eastern complex contains a tank which is 30 x 13.25m (98 x 43 ft) by 1 m (3 ft 3 in) deep (Fig. 6.18), fed from the sea by three oblique channels, each tens of metres long, equipped with small intermediate settling tanks, presumably so that sand is not washed directly into the fish pool. Ordinary wave action from any direction would feed water into the channels on one side, refreshing the pool, which would spill over and drain on the opposite side. The central group of tanks contains pools typically 4-5 m (13-16 ft) across, connected to each other by channels controlled by perforated stone sluice gates (Fig. 6.19), one of which is still in position. The western group consists of two interconnected tanks, each about 7 m by 2 m (23 ft by 6 ft 6 in), refreshed from the sea by two channels about 5 m (16 ft) long. There are grooves for sluice gates, but the gates are missing. Some of the smaller tanks cut into high-standing rock had rectangular slots in the walls as if to support the beams of a roof or sunscreen to keep the fish cool.

You can see at once that the whole system of tanks and channels is just about perfectly adjusted to the present sea level. Solution notches inside the big tank line up with the present high tide and low tide to within a very few centimetres, providing an indication of stability to within an error of 5-10 cm (2-4 in). Like the southern Turkish coast just 120 km (70 mi) away, the north coast of Cyprus is stable. The island arc structure of plate tectonics swings to the south of Cyprus, causing the subsidence of Amathus, but leaving the north coast stable. Seeing a whole piscina like this at water level, you can see how, once submerged (or uplifted for that matter), it is possible to measure the change of relative sea level to within 10-20 cm (4-8 in).

Figure 6.18. At Lambousa, Cyprus, there is a large Roman fish-tank 30 m long, with rectangular side tanks, visible on the right of the picture. The main tank is fed with fresh seawater by a network of long channels, so that waves or tidal fluctuations pump water into the tank. There are many smaller tanks nearby.

Figure 6.19. One of the smaller Roman fish-tanks at Lambousa is fed by a channel that is controlled by a perforated sluice gate, that is still in place. The cut indentation in the rock above the sluice suggests that some of the tanks may have been roofed in.

As we circumnavigated Cyprus and came south again several weeks later by Paphos and Lara, we ended up at the Akrotiri military base, where, by some very careful diplomacy, we had permission to dive within the base. A helicopter pilot who knew of our interest reported immediately that he had seen a straight wall sticking out into the sea, but all underwater, at the foot of the cliff by a place known as Dreamers' Bay. I gritted my teeth to refrain from telling him that this was a ridiculous place for an ancient breakwater but instead thanked him for his observations.

The base loaned us a boat used by the military sports diving club, from which we made several scuba dives as we worked our way around the eastern headland of Akrotiri, and then steamed slowly west along the cliffed south coast. As we approached Dreamers' Bay, exactly in the spot described by the helicopter pilot, there was the shadow of a straight breakwater deep below the surface. When I had mapped this after an hour snorkelling, I found that it was a perfect ashlar masonry breakwater or jetty sticking straight out into the stormy sea, with no protection from the ocean waves. Thank goodness I had not been rude to the helicopter pilot. I thanked him profusely when we were back at the base.

Nevertheless, the function of the breakwater at Dreamers' Bay is not easy to understand. It is exposed to the open sea storms and, even after allowing for subsidence of 1 m (3 ft) or more in the region, so close to Amathus, it does not seem to perform an important function, as it is not easily accessible. Presumably, it was designed to be used only in calm weather, when it would allow ships to land stores directly onto the high land that backed the cliff.

Since 2016 a team of archaeologists from the University of Leicester has discovered warehouses and storage buildings under the soil at the top of the cliff. Many of the buildings have been eroded, falling in fragments into the sea. So Dreamer's Bay may have been a significant port after all.

* * *

In 1975, Professor Elisha Linder, with whom I had worked training disabled divers the previous year (Chapter 7), invited me to visit Israel to give a course on submerged archaeological sites as a visiting Professor at Haifa University. Linder was one of the founders of marine archaeology, and the invitation was for two terms during the winter of 1975-1976. I drove to Marseille where the car was loaded onto a small cargo ship that carried a few passengers direct to Haifa. My hand-controlled vehicle at this date was a Chrysler Valiant V8, which was very powerful, but fuel-hungry. From October to February inclusive I lived in Haifa, where Linder arranged a beautiful apartment for me in a house in Denia, high on Mount Carmel. It had a full-width bow window that gave a view of the sky-scraping tower of the Crusader castle of the Templars on the coast at Atlit, ten miles to the south, with the sea beyond glinting in the silvery winter light. I sat there often at the end of a day's work to watch the sunset. Mist and storm clouds over the sea sometimes provided added drama behind the ghostly black silhouette of the tower.

By this time I had many friends in Israel, including Dr Yehuda Melamed in the Israeli Navy, Avner Raban at the University, and Catherine Goetschel, who lived in Jerusalem, but whose family came from Switzerland. My responsibilities required me to give lectures to the students and to participate in field courses arranged by other staff for the students, as well as conducting my own research on sea level change. During the five months I was there, I visited, snorkelled, or scuba dived on every identifiable archaeological ruin on the coast from Rosh ha Niqra on the Lebanese border to Tel Rafa close to the Egyptian border of Gaza.

Notwithstanding the many beautiful stretches of the coast of Israel and Gaza, and the many discoveries and adventures during my stay in Haifa, the most important issue from a research point of view was the ongoing intense dispute about the tectonic stability of the Israeli coastline. A senior geologist in the Geological Survey of Israel had proposed that the major part of the southern coast of Israel on the Mediterranean side had been submerged and uplifted again by as much as 5-10 m (16-33 ft) in the past 2000 years. Not only was this earthquake oscillation a greater tectonic event, if it had occurred, than any other earthquake or earth movement ever observed or recorded in the history of the Mediterranean, but it seriously jeopardized a plan to build a nuclear power station on the coast to produce electricity. It was technically feasible to build a nuclear power station to withstand the statistically normal Mediterranean seismicity, but not the colossal and rapid submergence and uplift proposed by the geologists.

I examined coastal archaeological remains at Rosh ha Niqra on the Lebanese border, and worked steadily southward, month after month, checking known sites, but also finding previously undocumented quarries, cuttings, silos, and salt pans to add to my list. For various reasons, I visited some sites south of Gaza quite early in the study.

I had documented about half the coast when the senior Israeli geologist came to visit me to discuss my observations and deductions. Although I found his proposed massive earth movements frankly incredible, and although my observations already made on the Israeli coast showed that very little vertical movement had happened, I pleaded that my research was incomplete. He told me very firmly what he expected me to discover, and what I should conclude. He did not want to see any questioning of his theories. I said again that my work was not finished, and that I could not anticipate my final conclusions. He departed in a state of frustration.

During the five winter months that I worked in Israel, I visited and surveyed 35 archaeological ruins on the coast, obtaining written data on a further 25, making a total of 60 sites with good data.

A second research goal was to prove the existence, or non-existence, of Bronze Age artificial harbours. The submerged town of Pavlopetri in Greece was very complex as a town, but the naturally protected bay in front of the town seems to have served adequately as a harbour with no artificial improvements. The same was true of the known Bronze Age towns on the coast of Crete. But "absence of evidence is not evidence of absence". Did we have enough no-shows to be convincing? The coast of Israel exhibited 26 Bronze Age settlements, and eight sites in the earlier Neolithic and Palaeolithic. Linder was determined to focus my techniques on as many of these sites as possible to see if we could find any remains of artificial breakwaters, docks, quays, or jetties. Quite rightly, he saw that if this huge concentration of Bronze Age coastal towns, often showing marine trade from the evidence of excavations on land, did not contain one example of an artificial harbour, then they probably did not exist at all. I agreed with this approach.

* * *

As I worked along the coast, I saw obvious signs of submerged walls and arches in the defensive structures on the seafront at Akko, and the submerged Roman harbour walls at Caesarea were already famous. But apart from these two regions, there were no signs of submergence, with most sites providing evidence that they had not been displaced relative to sea level in the past 2500 years, to within an error of observation of +/- 20-50 cm (8-20 in).

Some ancient settlements provided evidence from more than one archaeological period, each one showing that the buildings from that period were at the correct level relative to the sea. This confirmed that, at least at that place, the land had been continuously stable. A prime example is Atlit, which is so romantic in appearance, with its vast towers on a headland covered in ruins. It is easy to forget that what you are seeing is only the last and latest layer of buildings, the Crusader castle piled on top of remnants from every preceding archaeological period back to the Neolithic. There are Bronze Age walls on the beach; a Phoenician breakwater dated by cedarwood wedges between the blocks of stone to about 800 BC; and then the Crusader walls on the bedrock close to the sea. All the archaeological periods, back to the Bronze Age, exhibit ruins that seem to line up with present sea level to within an error of the order of 50 cm (20 in) or less. Ehud Galili has, since the early 1980s, shown that there are several well-preserved Neolithic and Mesolithic (also known as Pre-pottery Neolithic) villages under the sea close to Atlit, with others submerged along the coast for 10 km (6 mi) to the north. These villages were submerged by the global rise of sea level at the end of the last ice age, and date from about 8000 years ago, while the post-glacial sea level was still rising. (See Chapter 10.)

Newe Yam is a small Neolithic settlement with several rectangular, stone-walled rooms in shallow water. In 1975 it was my first sight of intact buildings on the seafloor older than 5000 years, and it gave me courage to believe that eventually we would find older and older prehistoric sites. I sat by the beach on a windy winter afternoon, discussing with Dr Avraham Ronen the chances of finding remains from the Palaeolithic, perhaps 10,000 years old or more, under the sea, somewhere out there. We gestured optimistically with a wave of our arms. We both dreamt that discoveries were near, in spite of the storm waves crashing on the beach in front of us.

During my winter at Haifa University, Linder and Raban conducted many weeks of work at Tel Nami, a Bronze Age town on a headland, connected to the mainland by a wide sandy isthmus. Linder argued that the sand had probably built up since the Bronze Age, and that the settlement could originally have been on an island. If that was the case, then harbour works could exist under the sand, and could in fact be the cause of the sand accumulating on such a large scale. We visited Tel Nami week after week, where the staff and students excavated Bronze Age walls, tracked buildings all around the headland, but not extending down into the sea, and drilled numerous cores into the sand isthmus. There is plenty of evidence that Bronze Age trade was conducted from Tel Nami, with later discoveries of cedarwood and bronze objects, but we could not find evidence of artificial improvements to the natural shelter.

Next down the coast was Dor, a major city that was very instructional for the students. There were rambling networks of buildings, cuttings, tanks, channels, temples, and slipways, with the sea lapping against a wide shelf of cut rock and coral *trottoir*.

Figure 6.20. At Dor on the Mediterranean coast of Israel there is a large ancient town with temples so close to the sea that they could easily be eroded. This picture shows how a ridge of rock has been deliberately left standing, with a trough excavated between it and the temple. This serves as a wave-trap to prevent erosion. Since there has been no change of sea level, it is easier to see how the trap worked than at Apollonia in Libya (Photo Dorit Sivan).

Figure 6.21. Caesarea from the air. The dark curve under the sea is the submerged Herodian breakwater (Photo courtesy Avner Raban).

Several trips to this large settlement, where structures date from the Bronze Age to post-Roman times, provided valuable data. The temples which are very close to the sea on high ground are protected from wave attack by rock-cut ridges or wave-traps, just as at Apollonia in Libya. Interestingly, a rock-cut wave-trap still worked perfectly since there had been no change of sea level (Fig. 6.20).

Figure 6.22. The fish-tank at Caesarea is comparable to that at Lambousa in Cyprus. It has not been submerged, and the seawater flows into and out of the tank nowadays at the correct level.

The famous city of Caesarea, built by Herod the Great in the 1st century BC, had been studied under water for many years. It was well known that a huge arc of rubble and masonry in the water was the remains of the breakwater that the Romano-Jewish historian Flavius Josephus described as being as large as the harbour at Piraeus. Since Piraeus exploits several natural semi-enclosed bays, and Caesarea was built on a straight coastline, with an artificial barrier against the sea, this is a bold testimony to Roman engineering. Diving excavation over several decades has shown that the great breakwater was constructed using wooden caissons filled with concrete, then protected on the seaward side by loose rubble. The technology of the Roman engineers, combined with Herod's ambitions, resulted in a prosperous trading port and a rapidly growing city. In spite of the usual Crusader looting of masonry to build their own castles and ports, much of the Herodian harbour at Caesarea remains intact under the sea (Fig. 6.21).

The anomaly at Caesarea was that the large fish pool a few hundred metres south of the breakwater (Fig. 6.22), the massive aqueduct just to the north, and various drains from the ancient city debouching onto the beach all showed that they were at the correct level in relation to the sea. No displacement up or down. How could the shoreline itself be stable for 2000 years, when the great breakwaters built by Herod had subsided into the sea by several metres (about 10 ft)? The facts of the case were clear by the time I finished my work in Israel in early 1976, but the mechanism for the local offset was not known with any certainty. It could have been a local fault parallel to the coast, or an earthquake that caused the sediments under the breakwater to liquefy and compact, or simply that the weight of the rubble and concrete compacted the loose sediments underneath. Years of work by the late Avner Raban, and then by Bob Hohlfelder, and others have revealed the sophistication of the timber and concrete caisson construction. But the cause of the local subsidence is still open to doubt.

When I wanted to visit coastal ruins for my own research, not as part of the university course-work, I was usually accompanied by Yehuda Melamed, Catherine Goetschel, or both of them. Most of the ruins could only be approached across sandy beaches, or over rocks and boulders, so that I often needed help. Catherine and Yehuda both became expert at digging my wheelchair out of deep sand. Thank you.

* * *

As the years passed, and my research and bureaucratic obligations moved on to other topics (Chapter 9), the rate of discovery of ancient ruins on the coast became more episodic and erratic. A holiday in 1980 with Vane Ivanovic at his exquisite villa clinging to the cliffs at Formentor, on the north coast of Mallorca, produced unexpected results. Vane was one of the great pioneers of fin swimming and spearfishing even before 1940. In the post-war years, he designed one of the first dedicated diving boats built with a low cut-out stern and easy access to a diving ladder. Our dives together were mostly for fun, always followed by excellent food and champagne served from concealed rotating lockers hidden behind the diving gear (a sort of James Bond-villain system in reverse). Dives with Vane, and with Billy McLean, his brother-in-law, were enlivened by Billy's hatred of moray eels, which he felt compelled to shoot with a spear-gun on all occasions. Jay and I drove one day to Son Real, a megalithic necropolis 20 km (12 mi) to the east, just beyond Can Picafort, to conduct an archaeological study. The necropolis was close to the shore, but there was no sign of tombs on the beach or in the water.

Jay swam to a little island just offshore, where she found an amazing concentration of ancient buildings, although we were unable to date them. Returning to Can Picafort, where we planned to take a break and get some food, I was astonished to see, right against the tarmac coast road, a colossal rock cutting that sloped down into the sea. It was a huge ancient slipway, comparable with that at Matala in Crete. It was partly filled with rubbish, and a derelict hut had collapsed onto the floor of the slipway halfway down. Having quickly consumed a tourist snack we clambered across the rocks to take a better look at the archaeological remains (Fig. 6.23).

Like the slipways at Rethymnon on Crete, here again were structures that nobody had recognized in the middle of a modern town. The incised rock cutting was 34 m (111 ft) long and 3.75 m (12 ft) wide, which could have housed a sizeable galley. We tracked down the slipway to its deepest end, where the walls cut into the rock were 2.70 m (9 ft) high, but the slipway surface only just touched the water. It did not go below the sea level, making it difficult to haul a boat out onto it. It seemed that the land had risen about 1 m (3 ft) out of the sea. As if to confirm this, the centre of the slipway had been down-cut for the last 10 m (33 ft), making it steeper, in an attempt to reach closer to the water. At the left-hand side of the slip, at the bottom, a large rectangular niche had been cut into the rock wall, as if to take a religious statue (Fig. 6.24). Or perhaps it was to store equipment for the galley. Jay and I measured and photographed everything we could to provide complete documentation, and reported our discovery to the museum of Alcudia the next day.

By chance, in 2011, I attended, with Jay, a conference on satellite altimetry and sea level change in Mallorca. To give the scientists a bit of outdoor experience of real sea level change, I suggested to the organizers that we take an excursion to the slipway

Figure 6.23. Can Picafort slipway in 1980. The slipway is partially blocked by modern derelict buildings and rubbish. Note that the floor of the slip has been deepened at the lower end, probably after the land had been uplifted slightly (Photo Jay Kleinberg).

at Can Picafort. When we arrived at the site, I saw that, after a lapse of 30 years, the slipway had been labelled with a large but faded sign, and that the rubbish had been removed. Although the tourist parade had been widened, and a yacht marina built very close by, the ancient slipway was undamaged. Our busload of scientists climbed all over it with fascination, quickly provoking the interest of nearby tourists and holiday-makers clutching ice creams, followed by children with their water wings. Soon there were dozens of people debating the construction, discussing the size of boats that could have used it, why it had been cut deeper, and at what date, and how old it might be. With a few extra display signs, this seemingly inert cutting in the bedrock could have been a considerable tourist attraction. Furthermore, everybody could climb over it without doing any damage.

Observation of coastal cities and ruins of the Greek and Roman periods along the whole coast of Libya (Chapter 11) provided data across a big gap on the coast of North Africa when we took a family holiday there in 2003. All the observations of ruins on that coast, except in the stretch for 100 km on either side of Apollonia, showed that there had been no change of sea level to within +/- 0.5 m (+/- 20 in) in the past 2000 years.

Figure 6.24. Can Picafort, Mallorca. At the lower end of the slipway which we found in 1980 there is a large votive niche carved in the rock. This would probably have held a religious statue that would bring good luck to the sailors (Photo Jay Kleinberg).

* * *

By the year 2000, my discoveries and deductions resulting from measurements of coastal and underwater ancient cities had contributed to the solution of several problems. I had found many previously unknown ancient ports and ruins in the sea, especially Pavlopetri in Greece, but also at Plitra, just north of Pavlopetri, Bozburun, Gemili Ada, and at Aghios Petros in the Sporadhes (Chapter 10). The vertical displacement or constant relative sea level measured at hundreds of sites correlated well with a model of active centres of volcanism and plate boundaries, leaving large, almost stable areas in between. Using statistical methods, I was able to separate the earth movements from regional and global sea level change, which was less than 0.5 m (20 in) for the past 5000 years in the Mediterranean; and we had shown that there were no built breakwaters or artificial harbours in the Bronze Age on the Mediterranean coast.

Each year I published the maps of coastal sites visited, and presented them to the authorities who had provided my permits. Construction of tourist hotels and coastal roads was already beginning to encroach on some of the coastal ruins that I investigated during the 1960s to 1980s. Many of them now have modern construction up to the edge of the ruins, while the growth of yacht marinas, and commercial boating trips, result in boats being moored or anchored over the drowned ruins. In many cases there is no announcement or warning that there are antiquities under the sea. In the 1990s, Lucy Blue at the University College London Institute of Archaeology compiled an integrated database of all the sites we could document on the Mediterranean coast, and this was merged later with the 'Ancient Ports' web-based catalogue compiled by Arthur de Graauw.

When I visited the NIO in the winter of 1958-1959 to find out whether the land or the sea had moved at Apollonia (Chapter 1), I could not have envisaged that the uncertainty about the answer would still be a problem 60 years later. Each time that I and other researchers on coastal ruins improved the accuracy of our observations, the oceanographers and geophysicists detected additional subtle processes going on that needed explanation at a new level of accuracy.

My measurements from Greece, the Aegean, Turkey, and Israel improved the accuracy of determination of vertical movement at archaeological sites to within an error of +/- 25 cm (10 in), or even 10 cm (4 in) at best. This analysis revealed subtle regions of uplift, subsidence, and stability which correlated well with the models of plate tectonics in the Mediterranean, including the motion of so-called micro-plates, and the subduction of the crust of the ocean floor under island arcs. The individual lateral movements of Italy, Greece, the Aegean, and Turkey are a response to their being crushed and squeezed sideways by the convergence of Africa and Europe. A satisfying discovery was that the place where the Cretan island arc and the Cyprus island arc intersect in southern Turkey coincides exactly with the 200 km (125 mi) of coast, from Bozburun to Cape Gelidonya, which has the most extensive and dramatic submergence in the whole of the eastern Mediterranean.

By the 1990s to 2000s geophysicists skilled in applied mathematics had worked out that the melting of the huge ice caps on the continents at the end of the last ice age, 20,000 years ago, had not only caused the crust of the earth to be distorted within a few hundred kilometres (a hundred miles or so) of the original ice cap, but that, additionally, the weight of water transferred back into the oceans caused the ocean floor itself to subside slightly everywhere on the globe. This process is known as Glacial Hydro-Isostatic Adjustment or GIA.

As the geophysicists used ever-more closely spaced grids, needing bigger computers, they showed that the earth's crust in the Mediterranean basin responded to the added water from the ice caps by distorting like a rubber bucket. Instead of the whole region descending like the platform of a weighing machine with a weight added to it, the basin deformed to subside by 5 m (16 ft) or more in the centre of the deep sub-basins, while the shoreline remained more stable, with very slight uplift or subsidence from the GIA factor, varying slightly from region to region. This factor, which continues still, can now be picked up by increasingly accurate analysis of the archaeological data.

These earth movements, which were identified by statistical methods, left a residual change through time that was common to the whole Mediterranean region. This common change had to be attributed to global sea level change resulting from the melting of ice caps more than 5000 years ago. For the past 5000 years this curve showed that the fluctuation was less than 0.5 m (20 in), but for earlier prehistoric sites going back to 10,000 years ago, the sea level was 30-50 m (100-160 ft) lower.

Since the 1980s, people have been increasingly concerned about climate change which can produce a modern rise of sea level, both through the heating and expansion of the upper layers of the ocean and from melting of glaciers and ice caps. Can we learn from the past so as to understand the mechanism?

The answer is yes. An international group of interdisciplinary scientists is now joining together the information that we have on past climate, variation in the size of ice caps, variation in temperature of the ocean, models of the volume of seawater on the globe, and the GIA, together with measurements from tide gauges and ocean-observing satellites (Chapter 9). It is not easy but, in order to learn from what happened in the past, we must be able to separate indicators of global sea level change from the effects of local earthquakes and tectonic events. The study of ancient cities on the coast has, unexpectedly, proved very useful in this effort. The demonstration that the supposed 2 m (7 ft) rise of sea level in the past 5000 years did not occur is equivalent to saying that 722,000 cubic kilometres (173,000 cubic mi) of ice, or 722 trillion tonnes (710 trillion imperial tons; 796 trillion US tons), did not melt during that period. This makes a significant difference to how we understand the relation between climate and the volume of ice caps.

Throughout the 1960s and 1970s marine archaeologists hoped to find older and older harbours going back into the Bronze Age and earlier. We knew that people used boats in these early days, and probably there were sailing or rowing boats, or navigable rafts, tens of thousands of years ago. But what harbours did they use? Pavlopetri shows a beautiful example of a busy town with facilities for fishermen and traders but without – to my sorrow – any sign of a developed waterfront: no quays or breakwater. Even large cargo boats must have been pulled up on the beach, or offloaded onto donkeys and carts standing in the water, or onto lighters.

My work in Israel and the survey of coastal Crete confirmed that there had been no artificial waterfront structures to protect and support ships against storms and currents in Bronze Age harbours. They exploited the natural terrain. There is one group of rectangular foundations at Kommos, in Crete, that may have been sheds to store boats, but they are back from the waterfront, and are horizontal, not sloping. There is no structure to create a shelter from waves or storms. Since the sea level had been rising from the end of the last ice age, 20,000 years ago, and levelled off about 5000 years ago, this all seemed to fit together, though the argument is circumstantial. Before 5000 years ago, at least in the Mediterranean, seafarers used the countless creeks, waddis, and estuaries that were being infiltrated by the rising sea. When the sea level stopped rising, then sand and mud started to block these natural harbours, and some time in the Iron Age or early Classical period, after 1000 BC, people started to build artificial structures to improve the shelter for ships, and to make loading and unloading easier for larger vessels.

The transition from a culture of coastal villages and towns without artificial harbours to one where every major town had a complex of massive protective breakwaters and docks was motivated, I think, both by environmental change and cultural priorities that we would now call capital investment in infrastructure. Before 5000 years ago, long-distance shipping was probably sporadic, seasonal, with infrequent trips for special cargoes like obsidian; or, at a local level, for fisheries. After 3000 years ago only ports with good protection from adverse winds or storms could attract trade. The pilots knew which harbours had good breakwaters. At some point, at a given level of prosperity and civic organization, it became a good investment to attract more ships and trade. This latter transition seems to have occurred at about 800 BC, that is 2800 years ago. A well-designed harbour with artificial walls could also be fortified to provide a base for military vessels and troops.

The apparent delay in technical progression towards artificial harbours during the Late Bronze Age probably has an additional cultural cause. The two centuries after 1200 BC saw the disintegration of Bronze Age civilisations due to the chaotic influx and migrations of tribes into the eastern Mediterranean lands and outbreaks of plague. The collapse was more devastating than the fall of the Roman Empire. Even the art of writing was lost. After 1000 BC the newly settled populations gradually adopted technical skills and writing from the Middle East, and slowly rebuilt the indigenous organization needed for towns, colonies, and seafaring trade throughout the Mediterranean. Around 800-700 BC the renewed growth in ship size and cargoes, combined with military rivalry, and the infilling of creeks and estuaries already described, forced the coastal cities to build artificial breakwaters and docks.

7

Disability: A Statistical Sample of One

"Merhaba Ay!" screamed the headlines of the Turkish newspapers in huge letters that filled the page on the 20th July, 1969.

"Hello Moon!" was their reaction to the NASA Apollo rocket landing, and the first man walking on the moon.

Natalka Czartoryska and I, after diving and snorkelling on coastal ruins and mapping submerged classical buildings for five weeks in southwest Turkey, were due to drive back to England a few days later. Sitting in a seaside café in Fethiye we translated as much of the sensational reports as we could, using our dust-stained Turkish dictionary. With the survey of the coast from Izmir to Kaş completed (Chapter 6), we drove back through the forested mountains of northern Turkey, crossing the Dardanelles strait at Canakkale. We made excursions through Chalcidice and Thessaloniki, then swung north to cross the Yugoslav border.

Late on the 28th July, we had already eaten our evening meal and the clouds were darkening the dusk. About 40 km (25 mi) south of Belgrade I was looking for somewhere to pull off the *autoput* to spend the night. We needed somewhere quiet. A small side road looked enticing. I drove a short way up the hill, twisting around bends into a sheltered corner surrounded by trees that dripped gently in the drizzling rain.

The next morning it was still damp and misty. There was a depressing café-bar for farmers a few hundred metres down the road where we went into a dense dark fug of smoke and alcohol fumes. Although it was early in the morning the soggy customers were drinking slivovitz and other spirits as they crouched looking miserable in the gloom. We managed to purchase some dry bread and coffee, then scuttled out as quickly as possible. Natalka was driving when we re-joined the *autoput*, heading north through intermittent light showers of rain. The gauze of mist drifted over the hills, draining the colour out of the landscape. The *autoput* was engineered with gentle gradients and curves like a superhighway so that it was easy to drive very fast and there was no speed limit. But there was no central divide or reservation, the total width was only just sufficient for three lanes of cars, and there was no restriction on access from side roads and farm tracks. It was not unusual to find a farm tractor or even a wandering cow or sheep obstructing the *autoput*. Predictably the accident rate was horrific. The numerous burned-out wrecks of cars and lorries were left on the verge to act as a caution to drivers.

In: Flemming, N.C., 2021, *Apollonia on my Mind. The memoir of a paraplegic ocean scientist*, 271-306. Leiden: Sidestone Press.

A steady rumble of traffic headed south, mostly heavy lorries, and a moderate density of vehicles on our side of the road, with no need to overtake or dodge other drivers. The steep terrain dropped away to our right with the lumpy texture of rough moorland. The lorries threw up spray from the wet road, but the visibility was otherwise quite good. As a passenger in a right-hand drive vehicle driving on the right side of the road, it was my job to monitor oncoming traffic, judging if it was safe to overtake, while keeping a general watch out for overtaking traffic visible in our wing mirror. I saw with concern that a car was coming up very fast behind us, and it was obviously going to overtake without any attention to approaching vehicles. As it shot by in a cloud of spray, I was even more alarmed when a second car followed immediately behind it apparently keeping station in convoy. The second car roared past, cutting sharply in front of us and narrowly missing a large oncoming truck. There was a third car speeding behind that one. The three cars were driving together in close convoy. It was stupid and dangerous beyond belief.

The reckless driver of the third car was staring death in the face. To avoid the radiator and bumper of the lorry driving straight at him, an onrushing wall of steel, he swung his car violently to the right. He crashed into the side of our van, pushing us onto the gravel verge still driving at about 80 km/h (50 mph). Natalka wrestled with the steering. With a rending clash of metal our nearside front wheel hit a bollard marking the edge of the *autoput,* forcing us to swing even more to the right, hanging over the hill-slope. The next bollard flipped the van sideways so that it tipped onto its front bumper and somersaulted down the hill. The door on my side flew off, the cab was half-crushed, and I shot into the air spinning and flailing like a Catherine wheel. Sky and earth alternated in flickering images like a juddering old movie. At the summit of the soaring trajectory, my twisting body must have displaced several vertebrae and ripped the spinal cord into an irreversible break. I sensed that something strange had happened, but was staring downwards and anticipating my impact on the ground with foreboding.

When the impact came it was not painful. A soft landing, I thought. No crunch of broken bones. No concussion.

I tried to sit up, but could not move. I sensed the damp soil and stones under my shoulders and head, but not under the rest of my body. Sparse misty raindrops were cold on my face. There was no sensation below my chest. At first I thought I had been cut in half, but it did not seem that I could be alive if that was the case. *Cogito, ergo sum.* I had never expected those words to be so comforting. I was not dead. My hands explored around me, discovering that my body was still there, but without sensation. I was paralysed. "Thank goodness I'm not a footballer," I thought.

I tried to drag myself over the rough ground and hummocks of boggy grass, perhaps moving a few metres before Natalka came to find me.

I rolled over on my back again, lying still while people came running down from the road. Out of the corner of my eye, I could see a queue of traffic beginning to build up. I could also see our van nearby, lying on its side. Natalka, who had had nursing training, knew pretty well what had happened, although she did not let on to me at the time. She had received head wounds and cuts, and a number of severe blows to her shoulder and body, but she quickly managed to prevent any of the local farmers or motorists from moving me, while she arranged coats to be eased gently under my body and over me so that I could keep warm.

Time did not seem to matter much. My head was clear but time drifted. I thought about insurance, getting the van out of Yugoslavia, sending a message to the Foreign Office to say I would be a day or two late travelling to a United Nations meeting in New York. Dark birds circled high in the mist. I thought that in this year of 1969, with jet aircraft and electronic communication, everything could be sorted out somehow. More and more people gathered gazing at me with curiosity but Natalka kept them away from me.

There was complete deadness now below my chest. Just nothing. And a bad pain every time I breathed. I wondered if the paralysis would spread. There seemed to be such a small island that was still me, with waves of paralysis lapping round it. The proverbial sandcastle in a rising tide. I moved my shoulders and head now and again to reassure myself. *Cogito, ergo sum.* Not only was I not dead, but apparently not dying.

A man loomed high over me looking down from the grey sky, holding my Calypso Phot underwater camera in his hands, which puzzled me. It was very expensive with a huge, domed wide-angle lens. Then he walked away, and it only occurred to me minutes later that he was stealing it. Even now I cannot understand the look on his face as he examined me. Did he think I was going to die and therefore it didn't count as stealing?

Natalka was able to talk to people in a mixture of Polish and Serbian, and somebody went to get an ambulance. I think there was a policeman in uniform, but he was not helpful. I asked Natalka to take out of the car all the cameras and photographic cassettes, notebooks and maps with all our findings, and the car papers, logbook, insurance certificate, and also to note the mileage on the speedometer so that I could make a correct claim of expenses to my employers. She was reluctant to leave me for long to do all these things but managed to run backward and forward, looking after me and rescuing the gear.

Arguments were breaking out in the crowd, so that Natalka had periodically to dissuade enthusiastic helpers who thought that with a little encouragement I could be made to stand up and walk, or at least should be given a stiff cognac and carried to a car. A lean Swiss man pushed his way through and explained that he was a doctor.

"Je suis paralysé," I said in my best French, which seemed to sum up the situation.

He prodded me a bit and went back to his car to get his medical bag. Two more doctors quickly appeared, one French and one German. An argument broke out as to who was superior and they faded away again.

The Swiss returned first and gave me an injection of digitalis to keep my heart going. The other two doctors returned, one now wearing a surgical coat and stethoscope and looking very impressive. They began arguing again as to who was in charge in various languages. The Swiss explained that he had already given me digitalis and they calmed down.

Eventually, an ambulance arrived. Before moving me one of the doctors administered an injection of morphia, but it did not make any difference to the pain. They rolled me onto a flimsy stretcher, on which they carried me up to the road. The ambulance was not a large truck, but an ordinary saloon car painted white with a red cross. There was only one bed inside which already had an old lady lying on it. There was more arguing and shouting and the policeman joined in, apparently saying that I could not be taken away until I had properly accounted for the wreckage of the car, etc., and destroying three of the People's State bollards.

Then everybody got mad. Natalka told the policeman what he could do to himself, and he slunk to the back of the crowd. They decided that I would have to lie on the floor of the car beside the bed. The stretcher, which was only canvas with two poles, was put on the floor, and I immediately felt a strange protuberance on my back, as if I was lying on a golf ball. But it didn't hurt any more or less than before.

There was no room in the ambulance for Natalka, so a little procession of cars was sorted out, including the Swiss doctor. Natalka went with them. We drove slowly along the *autoput* for the 30 km into Belgrade, then over the cobbled suburban streets to the Belgrade Traumatological Clinic. During the journey, I could see nothing but the roof of the ambulance. A team of men carried the stretcher into a receiving room where I was put on a trolley, and the kind Swiss doctor said goodbye after giving us his card.

Natalka cut my clothes off very carefully. The trolley was wheeled along dark corridors, in and out of a lift, and finally into a small ward, where there were two beds. There was a big window through which I could see the tops of trees and hear street noises. That felt good.

After a minimum of fuss, the nurses arranged that Natalka could share the ward with me, and two nuns tucked me into bed. I remained in Belgrade for five days, during which time Natalka organized the rescue party. David Harrison from the British Embassy was extremely helpful. Much of our luggage, after salvage from the car, was stored in the Embassy for transport home by rail. The Embassy informed my office at the NIO in England. Since the airlines would not accept a passenger in my condition without an accompanying doctor, the Embassy sent a cable to my uncle, Dr Cecil Flemming, to see if he could come out to travel back with me. He was a senior orthopaedic surgeon at University College Hospital London (UCH). The first flight we could get would be in four days, when Cecil could accompany me for the return journey.

For most of the first day, I was left in peace, while the nuns brought me occasional warm drinks or a little soup. It was very soothing, since everybody seemed to be doing all they could. I felt that I was amongst friends and was not worried. I did not really know what had happened, or why I was paralysed. I had no idea how long it would last, or whether I would ever walk again or whether I would die. Because everybody was so calm and reassuring I again felt that with electricity and telephones and machines everywhere these days and jet aircraft and computers, my feet did not really matter very much.

In the afternoon Dr Mitic and his staff of surgeons came in, in a group. One of them spoke German, and, as usual, Natalka could help with various Slav languages. They explained that I had broken my back, that the X-rays showed several broken vertebrae, with cracked ribs, so that the offset of the vertebrae was pinching the spinal cord. They were very gentle and factual, talking with the evident intention of sounding out my fears and needs, and to reassure me in every way they could.

They discussed the kind of food I would like, and they wanted me to know that they would make porridge and scrambled eggs, fried fish and other outlandish dishes if that would please me. Although I was beginning to get hungry, the thought of anything with a strong taste or smell was frightening. We agreed on meaty soup with soft rice, followed by those stewed fruit, which they call "compote". Over the few days my appetite, which increased rapidly, was a source of wonder and excitement to everybody. I drank lots of local herbal teas.

On the second day, Dr Mitic came back with his colleagues to explain that it would be dangerous for me to travel in my present condition unless they first made an attempt to straighten my spine. They would then encase my back in rigid plaster. He said that the decision was entirely mine, whether to undergo the treatment which might help to reverse the paralysis or whether to travel as I was and to risk the spine dislocating further. After we discussed this all together like a group of friends, I agreed to have the treatment.

They wheeled me to the bright glare of a theatre where I was to experience spinal traction. Lying face down I held on to a shiny steel bar with my hands while my feet were attached to a pulley, so that my body was stretched out and hanging freely in space. The pain was considerable: the effort of holding on to the bar was so great that it absorbed all my attention. They kept taking X-rays and developing them immediately to see how the vertebrae were settling down. They injected a powerful relaxant into the vast knot of muscles that had bunched up around the damaged bones. Suddenly someone delivered a sharp blow to the injured spot.

By this time I was half dazed and anyway could not see what was going on because I was looking down at the chequered-tiled floor. There was a steady hum of voices, then a doctor was wrapping me round with bandages while I hung there, encasing me with plaster of Paris. Soon I was back in the little ward wrapped up like an Egyptian mummy. The Embassy supplied me with back numbers of *The Economist* and *Playboy*, and I settled down again. Natalka bought a fiendish mathematical puzzle from the marketplace, which diverted my attention for hours.

My uncle arrived, calm, wise, elegant as always, with commanding presence. Uncle Cecil sported a monocle. He had a long discussion with Dr Mitic and the other doctors as we prepared to leave. We drove to the airport with an escort from the Embassy, which ensured that we were quickly through customs and emigration. Pan American Airways looked after us well. In London Natalka's brother Andrzej came to meet us. He made sure that I was comfortably installed in UCH before he took Natalka away to rest at his house.

At the hospital, where I was in an open ward with about thirty accident patients, one of the doctors quickly decided to remove my plaster. This resulted in incredible pain which lasted for a whole day. My parents came to see me, bringing some books, which were a great comfort. One was translations of Chinese poetry and another of Tolstoy short stories.

The next day I telephoned my employers at the NIO to speak to the Secretary, Mr Ray Williams. I told him that I was back and safe but wanted to know what the administrative position was. After a few minutes the timing pips went on the telephone, forcing me to grope for more sixpences to pay for more time.

"If you are having to pay for this, I will send you a telephone credit card immediately," said Ray. The card which arrived the next day enabled me to start getting in touch with all the various people with whom I wanted to maintain professional contact.

This was typical of the way the Institute reacted to my accident, contributing enormously to my peace of mind and ability to carry on with my work. They were perfect employers. Mr Cunliffe, chief personnel officer of the NERC, the Whitehall Department responsible for the NIO, came to see me, which I found very impressive. It had not occurred to me that I warranted such attention, but I suppose he and other people knew more about the seriousness of my condition than I did. He also had to consider the Institute's financial liability as my employer.

Armed with a telephone I started finding things to do. My secretary sent me some files up from the office, which, combined with the papers that I had had in the van, enabled me to sort out the aftermath of insurance claims, transport of baggage, etc. The Foreign and Commonwealth Office representative who was to take my place in New York wanted to call to talk things over, and when he came it was quite a party. Natalka and her brother arrived with a friend, Alan Munro, who produced a bottle of champagne and bowls of strawberries so that we celebrated something. I have no idea what.

The urgency now was to find me a bed space at Stoke Mandeville, the hospital where the legendary Sir Ludwig Guttmann had first devised the treatment and rehabilitation of people with spinal injuries. My level of spinal break is called paraplegia. Paralysis was total from the chest down, but my arms were fine. Stoke Mandeville was so busy that I had to stay in UCH for the next four weeks. The doctors moved me to a single-bed ward. Another telephone call to the Institute of Oceanography resulted in whole boxes of files being sent up for me to work on, and all sorts of people dropped in so that at times it was like a permanent committee session. Captain Kenneth Goudge from Vickers Oceanics came to talk about submersible charters for research diving in the English Channel; David Stott of *Hydrospace* magazine to talk about special scientific features and a trade index; a representative from Iliffe Press for a book chapter; Dr Ambraseys of University College London to talk about earthquakes in Turkey, and so on.

A book that I often carried with me when travelling was T.S. Eliot's *Four Quartets*, which my mother had given me when I was in my teens. I read the poems many times, and would sometimes open the book at random just to drift through a few lines or familiar pages as if visiting a personal but secret world. In the summer of 1969, Harry Blamires published his commentary and analysis of the *Quartets* entitled *Word Unheard*. I don't know how I obtained a copy, though I had perhaps seen it reviewed in the daily paper. Maybe I asked my mother to buy it for me. I was now able to explore a deeper world of multiple reflections and kaleidoscopic references and cross-references in the poems that kept me entranced with their unexpected richness: "Garlic and sapphires in the mud / Clot the bedded axle-tree. / The trilling wire in the blood / Sings below inveterate scars / Appeasing long forgotten wars." Unforgettable lines that I had always loved, and now even more so. "The Dry Salvages" with its more reflective themes of rivers, the sea, and death from shipwreck struck home more profoundly than ever.

They say that in time of adversity you find out who your real friends are, implying that many whom you trusted will desert you. I found to my shame that many people whom I had thought of as only professional acquaintances wrote obviously heartfelt letters expressing their concern, offering to help in any way they could. These letters meant a great deal to me as they helped to prevent me ever feeling cut off and forgotten. Some of them expressed the thought that I would probably cope with disability so completely that it would make no difference to my life.

With that encouragement, I knew I could keep going.

A letter from the western division of the Institution of Mechanical Engineers asked me to give a lecture on the "Future of Oceanography", to be delivered in Salisbury in early December. I explained to them that I was injured in hospital, but hoped to be mobile before then, and they confirmed the booking. That gave me a date to aim for.

My uncle had told me in Belgrade that the doctors would soon have me walking on crutches, so it was a bit of a shock to find that "soon" meant twelve weeks at the earliest.

That took me well into November. Also, I was told that it would take all that time, and possibly longer, before I would know whether I would ever be able to move my legs again. My physiotherapist at UCH, Miss Szilasi, was charming and optimistic, and I did the mandatory exercises with a will.

When the day came to move, the ambulance drove me with Natalka to ward 1-X at Stoke Mandeville Spinal Injuries Centre, near Aylesbury. The redoubtable Sister Mac, who was in charge of the ward, took one look at Natalka and recognized the symptoms of untreated concussion. She just ordered her to bed. Soon our wonderful friends Monica and Bill Edwards took Natalka under their wing, inviting her to rest all through the autumn and winter at their beautiful farm in the Punchbowl in Surrey.

At Stoke I found myself in the care of a cheerful outgoing doctor named Carla Gardner-Brown. Natalka had arranged for boxes of papers to be kept provisionally in the almoner's office, but as the days passed more and more files accumulated beside my bed. My locker never had the naked Spartan tidiness expected of it, while boxes of papers, books, tape-recorders, and a typewriter piled up under my bed. I was working against a deadline to complete the proof corrections for my paper on earth movements and coastal archaeology for the Geological Society of America, to be published later that year. It was 125 pages of closely printed text with dozens of maps.

Everybody predicted that sooner or later I would incur the unbelievable wrath of Sister Mac, compared with which thunderbolts, fire, and brimstone, were like a summer breeze scented by flowers. But the blast never came, and indeed Sister Mac seemed to approve of my industry. Angelo, the six-foot-tall West Indian orderly who could lift the heaviest patients as gently as if they were babies, smiled and said that I was very lucky. I arranged for a local secretarial agency to send in a shorthand-typist, while other letters and papers were dictated onto tape to send to my secretary at the Institute.

The weeks went by. My bed was moved further up the ward as the new cases moved in at the bottom and the advanced patients moved out at the top. It was cheering to see that those who had been in hospital several months were already bowling about in their wheelchairs, talking about the mysterious "Circus" where they stood up on crutches or walked holding on to parallel rails. At the same time, our hearts went out to those critically injured with broken necks, lying with tubes inserted in their throats attached to lung machines because even their lungs and diaphragm were paralysed. I was still never allowed even to sit up in bed, so that many of the voices that I heard every day in the ward belonged to ghostly people I had never seen. We could all see only the ceiling and our immediate neighbours on each side when the electrically-powered bed tilted to change the pressure on the skin. Although I could see the sky through the windows, I developed an obsessive desire to see close-up a tree with leaves on, hoping that I would surely be moving about before they fell to the ground leaving bare branches.

There was a military spirit of barrack-room camaraderie in the ward, but everybody's character and individuality was respected and appreciated for whatever they had to offer. Nobody, however badly injured, seemed to begrudge the luck and happiness of those better off. But the one thing that got no sympathy – and I mean from the patients, not staff – was malingering or skiving or playing for sympathy. Ninety-nine per cent of the men there – and the women too, I am sure, in the female wards – appeared to triumph over their adversity and adapt to it by their own strength of mind.

The miracle depended on the fact that we trusted the system and the staff of untiring physiotherapists, doctors, orderlies, nurses, cleaners, and cooks who looked after us.

Attitudes have changed in the past 50 years. In 1969 nobody in the ward had heard the phrase post-traumatic stress disorder, though some of us would have known about soldiers after violent campaigns experiencing "shell shock" or "battle fatigue". I have no doubt that some of my fellow patients concealed much greater emotional distress than they felt they could admit to. People's response to permanent disability, as to other extreme shocks, varies greatly. The staff at Stoke Mandeville were probably more aware of this than the patients. Each patient needs help and support in a different way, and the modern emphasis on mental health, openness, and counselling has improved the ability to be flexible and adaptable to each case.

When Uncle Cecil came to see how I was progressing, he gave me reassurance.

"Don't worry about the broken bones," he said. "Speaking as a professional I can tell you a secret about orthopaedics. If you put two pieces of bone within half an inch of each other, they will join together. The problem is entirely with your spinal cord." He was right.

Whenever I could I talked to passers-by about the future. Since there was no return of feeling or movement to my legs, I accepted that I would never walk again. I wanted to know about driving cars, getting into a bath, swimming, and travelling. How could I get into an aeroplane without a medical drama? Everybody was full of stories about devices and tricks which make things easier for the disabled. The representatives of the Institution of Mechanical Engineers visited Stoke to arrange publicity for the Salisbury lecture, which I found disconcertingly unreal. They talked about the afternoon lecture, then the evening lecture, the dinner and reception with the mayor, and did not seem to realize that, as I had not yet been outside the hospital, I had precious little access to lecture material. In fact I had not yet even sat up in bed. I telephoned the photographer at NIO, Arnold Madgwick, to explain a list of items for which I needed photographic slides. He prepared them excellently. This attitude of "business as usual" seemed perfectly natural to everybody, and the only doubts I ever encountered were my own. Sometimes a little voice inside would mutter: "For Chrissake, don't they realize I am paralysed and might not be able to do all they expect?"

Ten weeks after the accident, the director of the hospital, Dr Walsh, said that the bones of the vertebrae and ribs had mended firmly enough for me to sit up. For two or three days the bed-head was tilted up a bit, for longer each day, then I was in a chair for a few minutes, before finally I was allowed to sit in a wheelchair all day. Then started a hectic round of balancing exercises, weightlifting, learning to get in and out of bed alone, learning to dress myself, to go to the lavatory, etc. In the absence of waist muscles, even sitting still requires a delicate sense of balance.

After a few days, I ventured outdoors through the French windows at the end of the ward so that I could see the trees with their leaves on. I breathed the fresh air of mid October. My Dylan Thomas month: "Through the parables / Of sunlight / And the legends of the green chapels". When I turned to go back into the ward, I found that it was very difficult to ascend the slight slope of the ramp up to the door. Each time I let go of the wheels to make a new push the chair rolled backward. By working very hard I just made it to the top, breathless and in a very thoughtful frame of mind. Another cause for deep thinking arose when I looked in the mirror for the first time and saw that my face constantly betrayed a mixture of pain and stress. I looked worried and anxious. Like most paraplegics, I sensed a continuous stream of burning pain from my lower body caused by the damaged ends of the spinal nerves. For some people it is overwhelming, requiring special treatment. For me, the pain was bearable, so I ignored it most of the time, but I was frustrated to see the expression on my face that did not represent the way I felt. I tried raising my brows a bit to remove the lines around my eyes, but that caused a furrowed forehead. I tried to force a bit more of a smile at the corners of my mouth, but it looked like a grimace. After ten minutes of careful practice, I could remove all the lines on my face, and look relaxed. That sounds as if it would be a conscious and artificial effort all the time to maintain the unworried look, but it settled down, with my usual range of facial expressions and responses remaining unaltered.

For my birthday on 10th November, Natalka, Andrzej, and my brother, John, and sisters, Mandy and Sara, laid on a splendid party at a fine hotel nearby. This was my first car journey. I was excited beyond belief. Every tree, every leaf, every person walking, was a joy to behold. We blew up balloons, drank wine, and I smoked a cigar and felt wonderful.

A week or so later my father drove me up to London, where I changed into my dinner jacket at Andrzej's house so that Natalka and I could go to a dinner at the Mansion House in my capacity as Honorary Secretary of the Society for Underwater Technology. My chest and neck had expanded so dramatically from the recent exercises that the collars and the shirt itself had to be slit down the back so that I could be buttoned up down the front.

On the day of the Salisbury lecture, they sent a car to the hospital, in which Natalka and I were driven to Salisbury arriving about midday. I was wheeled up the ramp on to the stage of the lecture hall, where I ate lunch in the wings to save time. The first audience, about nine hundred students and schoolchildren, asked a lot of excellent awkward questions and were very lively. There was an exhibition of offshore engineering machines in the car park, with vast chunks of yellow-painted buoys, measuring instruments, rock drills, and diving suits and helmets. The children were overawed and happy. My old friend Harry Wardle of Strongwork Diving had provided a marvellous contribution in the form of a huge yellow submersible decompression chamber. These machines are so strong that even the children knew that they could not break them. Then there was a little rest, cocktails and snacks with the mayor of Salisbury, and the lecture repeated with longer words to another audience of engineers and the general public.

The evening lecture was packed-out, but I had a twinge of guilty conscience about one episode. During the question period a knowledgeable gentleman took up some point that I had made about the hull alloys for deep-sea submersibles, embroidering it to show off his vast knowledge of all the things going on in the world of submersibles before very slowly getting back to his original point, to ask if I agreed. I said "Yes," just like that. There was a long pause while he waited for something more, and then the whole audience burst out laughing.

My next outside engagement was the Annual General Meeting of the Society for Underwater Technology. I had originally offered to resign from the position of honorary secretary, but Ian Richardson, then vice president of SUT, insisted that I stay on, and indeed I had had much time to think about the plans and affairs of the society. With great excitement I was allowed at Christmas to go home, where Natalka and I had a happy few days visiting friends, going to pubs, and receiving visitors. During a drinking session at the Red Lion in Milford, I first learned to balance my chair perfectly on its back wheels, holding the balance for several minutes.

Back at Stoke Mandeville hospital for a wild New Year's Eve party in my new ward, Number 2, I found that after a few beers I could dance balanced on my back wheels and establish a rhythmic rapport with the nurses and physios who enjoyed dancing close in front of me. Then back to the routine of training. By now I was learning to stand up with plaster casts strapped to my legs to prevent my knees from folding up. Liz Kerr, my ebullient Australian physiotherapist, was endlessly patient as I staggered backward and forward between parallel bars before I finally made the first hop leaning on crutches. Liz and I made an expedition into Aylesbury to practice shopping. There were also archery sessions run by Peter Ayton, and woodwork and art classes, so that I was never bored. I made my first experiments with oil painting.

Years later, in 1977, I was visiting Australia to attend a diving research conference in Melbourne, where I met Liz Kerr for a nostalgic meal and a bottle of dark-red wine in a Lebanese restaurant. She told me that I had been the first patient for whom she had been fully responsible, reporting each week to a panel of doctors and senior physiotherapists. Whenever she described my activities to them, she was warned that my situation was psychologically dangerous since obviously I had not yet realized that I would never recover the use of my legs. The experts assured her that I was suffering a delusion of denial and that one day I would wake up screaming when I realized that I was permanently paralysed. I was in a very dangerous state, they said. Liz endured this scepticism for several weeks, but then plucked up courage to tell her bosses that I was fully aware of my medical condition, and was genuinely happy, full of zest for life, and optimistic.

* * *

Still, the papers, books, and files were piling up around my bed, so that eventually, early in 1970, the Sister in charge of my ward got upset. Carla Gardner-Brown came to the rescue. Now that I was in a new ward, no longer under her professional care, she said it would be quite all right ethically if I wanted to use a spare room next to her accommodation in the doctors' wing. I went along to look at the room one afternoon when she was not there, to see where I could put my boxes, but when I saw how tiny her rooms were, and how little space she had for herself, I felt so ashamed that I decided to go away and say no more about it. I almost wept at her generosity. I did not see how she could possibly put up with me crashing in and out and typing on a clacking clattering manual typewriter and generally making a noise when she had so little time to herself and so little privacy. But I hesitated, changed my mind, and decided to try it for a few days.

Now was the great moment when I learned to swim again. In the heated therapeutic pool, I made the first tentative arm strokes under the guidance of Peter Ayton and Liz Kerr. It turned out to be much easier than I had thought, although my feet tended to float up and my head to fall face forward in the water. After a few lessons, I obtained the Disabled Swimming Advanced Award, which gave me the ambition to try snorkelling and even scuba diving.

Mike Davis, a friend from CUUEG and subsequently a consultant anaesthetist in New Zealand, agreed to bring a complete set of weights, snorkel, several different masks to try, regulator valve, and a full air cylinder. Mike had been one of the diving team in Malta in 1965 when we tested the effects of helium-oxygen breathing on the cognitive function of divers at depth (Chapters 2, 8). I practised removing and replacing the mask while in the hydrotherapy pool, discovering that I had to accept that I would sink as I raised the mask to my face, so that the technique was always to don the mask under water. This quickly revealed the basic problem of diving with no leg or waist movement, namely, that your hands can be used either for propulsion or for doing some technical work, but not both at the same time. If you want to work with your hands underwater you have to accept that you will roll over, lose balance, or sink in an uncontrolled attitude.

Then I tried the aqualung, a single 60 cubic foot (1700 litre) set, again finding no serious problem. The pool was very small and warm as bathwater, but it was a fabulous moment to be bubbling about again, turning somersaults, generally being a diver. I tried to remove and replace my mask underwater and found that it was all right provided I did not mind falling sideways and controlling my breathing in response to whether it was above or below the demand valve, which changes the ease of inhalation.

By now the steel callipers for my legs had been delivered, on which I was able to do several hours of standing and walking each day. The callipers had probes that fitted into sockets in the heels of my shoes, and metal bars on each side of my legs, with hinges at the knees. Leather straps secured them around my shins, knees, and thighs, so that I could stand up with the knee hinges locked straight. To sit down, I lowered myself backward into my wheelchair and then unlocked the knees to put my feet on the pedals.

While standing in the parallel-bar walking frame in the exercise area known as the Circus, I noticed an impressive young man, probably in his mid twenties, come slowly into the room balanced on his back wheels. He flipped his wrists lazily and smoothly to keep rolling forward, while he leant back haughtily in his tilted chair. He was dressed in close-fitting black clothes and had a slightly sinister, hawk-like face. He turned his head slowly from side to side, looking for someone, failed to find them, spun around easily to wheel out of the Circus still balanced on his back wheels, and departed as silently as he had arrived. I thought "That's the way to do it."

On the same day, the representatives of a film production company came round talking to people to get ideas, and possibly to recruit "roll-on" part-time actors and extras for a feature film with a paralysed main character. Their spokesman asked me whether I was anxious about going back into the real world.

"I'm looking forward to it," I said.

"But surely it is worrying that you will be slower than other people, and jobs will take more time?"

"I have a good employer, thank you. And I have been practising my work while in hospital. I'm sure it will not be easy, but I'm going to give it a good try."

"But you might lose time through being sick, and on sick leave. That tends to make disabled people less employable, surely? We keep hearing that complaint."

"It is a risk, yes. You're probably right. But Stoke Mandeville training, like this Circus, gets us very fit, and disabled people just have to try a bit harder. Reliability will be a problem. Life is not fair. I will just have to accept that. But I am sure it is possible to keep working. I will try."

They looked disappointed and drifted off looking for someone who fitted better into their preconceptions and prejudices. I hope that they failed.

At one time I had hoped that I would be able to walk with callipers all day and dispense with the wheelchair, but eventually I realized that this was almost impossible, since a person with no waist muscles, standing on callipers, is effectively standing on their hands, so they cannot take one crutch off the ground without falling over. In this position it is all but impossible to open a door, turn a key, open a drawer, or pick up a book from a table, making it almost impossible to do a day's work.

I have found that in any given hour of the day I can either work at my job to earn my living, or do exercises trying to walk better, and the first has always taken priority. However, at weekends and evenings for the first few years after I returned home, I found that it was always a pleasure to put on the callipers to walk about the garden, or even to go to the pub or a party without the wheelchair. After a few years, the effort to do this, the lack of general safety, and a few nasty falls convinced me that the callipers were only practical for occasionally standing up to stretch and to improve health generally. But after developing a pressure ulcer on my left heel, I gave up using callipers altogether after 1999.

My programme at Stoke Mandeville was getting very full, with writing, working, swimming, walking, and driving lessons. I had also reached an agreement with Carla to work in her spare room at times that caused her the least disturbance. Carla was usually working in the wards or away in London, so that I was able to do a lot of work without our getting in each other's way. Late in the evenings, when she came back from town, we would sometimes drink coffee together, and I would enjoy every minute of not being exposed to continuous television in the ward. Her radio was permanently tuned to Radio 3. Carla was engaged to be married, and we were careful not to develop any emotional dependence, but we were both widely read and I realized after a while that we were communicating at a secondary level by using references to literature and music as a proxy for our feelings and indulging in a sort of intimate shadow boxing. As I left her rooms, the sound of a Radio 3 string quartet would die away in the distance behind me, and the continuous rattle of gunfire from a dozen TV sets playing loudly in the wards would gain in bellicose volume.

Amongst my companions in the ward was Michael Ebokpomweh, a Nigerian who also spent most of his time in working on files of technical papers. I discovered that he was a qualified accountant studying for further exams. He had a great sense of humour and remained a close friend for several years before he returned to Nigeria. It is pleasant to be able to look back on six months in hospital and remember that socially and intellectually it was as bright and exciting a time as any.

Liz Kerr and I did some serious diving in the big swimming pool in the sports stadium administered by Sir Ludwig Guttman. I went through the whole British Sub-Aqua Club pool training schedule but had a bit of trouble putting all the equipment on under water. All the rest turned out to be surprisingly easy. Liz, after careful practice, had a go with the aqualung and proved to be a natural diver. After that I borrowed a Calypso Phot camera, which I taught Liz to use in the pool, so that we could take underwater pictures, partly for amusement, and partly because Sir Ludwig Guttman wanted documentation for the book he was writing on sport for the disabled.

In February 1970 I was "demobbed" from Stoke and returned to the outside world. Natalka welcomed me back to the only room in the house which I could enter, where the Welfare Authorities, in the person of the excellent Mr Titcombe, had installed a hospital bed of the right kind. Mr Titcombe had also fixed a ramp at the front door so that my chair could be manoeuvred easily. My GP told me to take life easy, but I was determined to return to work at once, as I sensed acutely that only if I maintained a regular appearance at work would I be able to keep my career going to obtain a steady income. I needed enough to pay for the additional expenses which I already saw looming ahead.

It would take some time to find a suitable car, even to decide what kind of car was best for a disabled driver who wanted to carry aqualungs and heavy equipment.

Since I hoped to continue travelling and diving, I wanted a hand-controlled car big enough to carry diving gear. I tried to hire a self-drive hand-control car but was told that this was not possible by Hertz, Avis, and Godfrey Davis in turn. Then I rang Godfrey Davis again, offering to hire an automatic Cortina which I would convert at my own cost. They said all right if I insured it at my own cost also. I telephoned my car insurers, Harrison Graham Pattison, and put the idea to them. They said OK and we were off. Godfrey Davis took the car to Reselco Hammersmith who fitted the hand controls, and the next day the car was delivered to my house by a Godfrey Davis driver.

Armed with a car, I returned to work four days after leaving hospital. For a few days people tried to open doors for me or carry my tray at lunchtime but, after being nearly hacked in the ankle by the pedals of my silent wheelchair coming fast around corners, they soon decided they had better look after themselves, not me.

My employment and salary gave me security. Nevertheless, there were unnerving moments. While I was at work one day in spring 1970, an occupational therapist visited Natalka at home, assuring her that I would be all right, and that they would teach me to make baskets, which, they said, would give me a sense of independence, thus reducing her to tears. A year after I had returned to work, I received a letter from the Income Tax Inland Revenue saying that they had discovered that I had been working while in hospital and that therefore I had not been eligible for some small sick benefits that I had received (Fig. 7.1). Please would I send the money back?

The period defined in the letter was roughly from the time I first sat in a wheelchair to the time I left Stoke, but this still made me annoyed. I explained vigorously that, since lying in bed for three months with nothing to do was debilitating, I had decided to work while flat on my back rather than just sleeping or watching television. I declined to return the money. They wrote a generous letter back stressing that I was now entitled to various state benefits that I had never heard of, such as Industrial Injuries Benefit and Disability Living Allowance, which they invited me to apply for. They offered to pay the backlog. I applied and have found the benefits very helpful. Paraplegia can be expensive. Clothes wear out more quickly. One may have to use taxis more frequently because much public transport is inaccessible. Adaptations to the home and car, adaptive accessories, and wheelchair gadgets all cost money.

At the NIO, two major projects were coming up. Before discharge from hospital, I had already attended meetings at the Foreign and Commonwealth Office preparing for the next session of the Standing Committee on Peaceful Uses of the Sea Bed at the United Nations in New York, and I had to go there to work with Alan Archer of the IGS (Chapter 9). Together we were assisting the staff of the United Kingdom Mission to the United Nations in presenting the UK policy. Everything was so hectic that I did not have time to worry. Hertz in London arranged for a self-drive converted car to be available in Manhattan. At Kennedy Airport, Natalka and I were met by the UK Mission car which

xxxxxxxxxxxxxxxxxxxxxxxxxxxxxxxx
Haydon Place, Guildford, Surrey.

Dr N.C Flemming,
Flitrae,
Milford,
Godalming,
Surrey.

Benefit Section

23 July 1971

Dear Sir,

I am writing to let you know that a doubt has arisen whether you were entitled to sickness benefit amounting to £236.35 which you received for the period 15.9.69 to 10.2.70 because you were engaged on official duties for your employer.

It is necessary to refer the question of your entitlement to this benefit to the Insurance Officer, who is the independent authority appointed under the National Insurance Acts to decide such matters. If he decides that you were not, in fact, entitled to the benefit in question, the legal provisions are such that his decision must include a requirement to repay any money overpaid unless it is shown to his satisfaction that in the obtaining and receipt of the benefit you have throughout used due care and diligence to avoid overpayment.

Before the Insurance Officer decides whether you were entitled to this benefit and, if not, whether you must repay it, you may wish to explain why you did not advise this office that you were carrying out official duties whilst in hospital. If so, will you please write your explanation on the back of this letter, add your signature and the date, and return it to me within seven days. A copy of this letter is enclosed for you to keep if you wish.

An addressed label is enclosed.

Yours faithfully,

Manager.

Figure 7.1. Letter received from the Inland Revenue.

delivered us to our hotel. The next morning, I collected the Hertz car on 47th Street. The conversion for hand-control was fitted while I waited.

Besides the routine work at the UK Mission and the daily meetings at the United Nations Secretariat buildings, there was a round of after-hours negotiations, parties, and visits to old friends. Natalka managed to visit numerous galleries, museums, plays, the zoo, and the aquarium, while I was immured in air-conditioned offices. Alfred Poor very kindly arranged for me to have access to the swimming pool at the Racquets Club on Park Avenue, and as this was the first time I had swum outside the hospital, I was very grateful. We had a picnic by the sea with Barbara Guggenheim at a beach on Long Island. Because of a threatened airline strike, we decided to drive north to visit friends, and to fly direct to London from Boston. Eddie and Anita Firestone took us to the theatre, drove us down to the Woods Hole Oceanographic Institution, and looked after us royally.

In England again, I began to slip into the old life of laboratory work, library reading, and committees. Natalka fought with architects and planners to obtain permission for us to build a single-storey extension to one side of the house so that I could have a bath.

At the Institute, John Wilson and I were putting finishing touches to our plans to charter the research submersible *Pisces II*, operated by Vickers Oceanics. This was the project that I had been discussing with Captain Goudge in UCH the previous summer. For this charter, we designed a practical series of geological research dives in the area southwest of Eddystone lighthouse in a depth of about 60 m (200 ft). Part of the objective was to map, in collaboration with Arthur Stride, the boundaries between the transport paths of sand

Figure 7.2. The author demonstrates diving into the pool without any equipment, before putting on the aqualung that has been placed on the bottom, with the cylinder turned off. The weightbelt is put on first. Then the valve turned on to start breathing from the regulator. The other items of kit can then be donned at leisure. Watched by Roger Clement, one of the NIO divers in training. This is a standard BSAC test of competence.

and gravel ribbons on the seafloor, extending the work that we had attempted previously, both with air diving and with heliox, in 1964-65 (Chapters 2, 8). Sidescan sonar mapping around the British Isles had shown ribbons of high and low reflectivity, which Arthur interpreted as ribbons of sand hundreds of metres wide being transported by waves and currents over underlying gravel. The ribbons transported huge quantities of sand to the edge of the continental shelf. This sand was partly coming from rivers, and partly being washed out of the ancient landscape inundated at the end of the last ice age. Arthur wanted to examine close-up the mechanism of movement of the sand, how often and how fast it was moved by different storm conditions, and what processes stabilized the edge of each ribbon. This required very close-up viewing and sampling at known distances from the edge of the ribbons. We scheduled the work for early June.

A problem arose when scientists discovered that the huge ocean-going acoustic survey machine GLORIA (Geological Long Range Inclined Asdic), designed, built, and operated by the National Institute of Oceanography, could only be launched and recovered with the assistance of swimmers and divers. GLORIA contained a massive array of acoustic transmission-receiving equipment inside a cylinder that was 10 m long and 1.8 m in diameter (33 ft x 6 ft), weighing several tons, all of which had to be lowered from the ship and connected to towing cables with electrical connections (Chapter 8). During the first trial in 1969, two members of the Institute staff, Vince Lawford and Stewart Willis, had become volunteer technical support divers by force of circumstance, but now the NIO needed a properly trained diving team. We began regular training sessions for four budding divers at the Guildford Lido, where I ran a full instructional course like the early days at Cambridge University (Figs 7.2). Of the newcomers, Ray

Peters was an absolutely natural diver, and he, together with Stewart Willis, became the mainstay of the Institute Diving Team. Those divers who attended the 1971 meeting of the Underwater Association were awestruck by Stewart's description of the skill and complex routines involved in the handling and rigging of the cables and power supply for GLORIA while diving from a research ship underway in the open Atlantic a thousand miles from shore.

In May 1970 I bought a 1966 second-hand Chrysler Valiant Safari for £700, with which, at the end of the month, I drove down to Fowey in Cornwall to join the *Vickers Venturer*, the support ship for the submersible *Pisces II*. We joined up with John Wilson and his assistant Colin Pelton, and a Cambridge University diver, David Cooper. The weather was bad, with a nasty wind from the southwest. During the week we only managed eight out of the fourteen planned dives.

Bob Easthough, the *Pisces II* team leader, rigged up a rope gantry over the submersible hatch so that I could be hoisted in a parachute harness high in the air and lowered through the hatch into the spherical hull (Figs 7.3-7.6). When the hatch was sealed shut the huge hydraulic rams pushed the swinging A-frame out over the stern of the ship, and the winch then lowered the submersible into the sea. A diver unhooked the lifting cable and the pilot adjusted the ballast so that we descended slowly to the seafloor.

Figure 7.3. The stern deck of Vickers Venturer, 1970, showing the submersible Pisces II docked under the A-frame. The winch in the foreground lifts the submersible up to the swinging frame hanging from the A-fame, and then hydraulic rams push the A-frame out over the sea, so that the submersible can be lowered smoothly into the water.

Figure 7.4. The author being hauled up on a block and tackle before lowering into the hatch of the submersible, which is visible on the left. I am wearing warm padded clothing mostly just for protection in case I bump into things.

Figure 7.5. Pisces II being recovered after a dive. The divers have attached the towing line and the lifting cable so that the submersible is already partly out of the seawater.

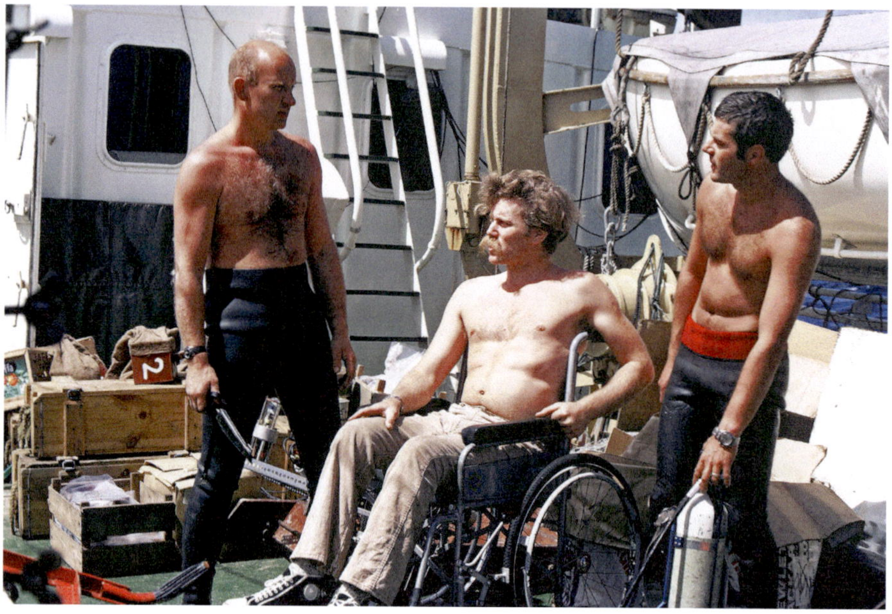

Figure 7.6. Bob Easthough, with the diving knife, the author, and one of the Vickers diving team discussing plans for the next dive on board Vickers Venturer, support ship for Pisces II.

Cruising close to the bottom, each dive lasted for several hours, during which the professional pilot steered the submersible while watching the technical controls, battery power, the manipulator arm, the sampling grab, and air purity. We sat close together in the 2 m-diameter (6 ft) sphere, looking out through the two small portholes. I made notes, triggered the underwater camera, and decided when we needed to take sediment samples. Fish hovered and swam slowly over the muddy gravel, and scallops and sea urchins were dotted around. From time to time the pilot let the submersible settle on the seabed, so that we would take a break, sipping tea or eating sandwiches. Being paralysed was no disadvantage, since you cannot walk far in a 2 m (6 ft) diameter sphere.

Bringing the submersible to the surface at the end of the dive, the pilot had to be sure that we were well clear of the *Vickers Venturer* support ship. Close was convenient, but not too close. When we reached the surface, *Pisces II* wallowed and rolled in the waves with the water splashing and plunging over the portholes. While we waited for the divers to come with an inflatable boat to tow us under the A-frame, which would lift us back on deck, the pilot would always put a music tape into a player. The only tune he had was Acker Bilk's *Stranger on the Shore*, which is a tune with some originality, but if I hear it now, I am immediately back in that little sphere, rolling and lurching with the waves lapping over the portholes.

I had been apprehensive about moving around the cramped quarters of a 400-ton research ship, but it worked out well. My wheelchair could be manoeuvred around in the dining saloon or on the open upper deck, but everywhere else, including the companionway, was too narrow. On the other hand, my rubber wheels with brakes meant that during a storm I was less likely to slide around than people sitting on normal chairs. For getting on deck the sailors rigged a small chair on two broom poles and would carry me in this as in a sedan chair. One of them said, "Don't let the directors of Vickers see this or they will all want one!"

Later we decided it was easier for Bob or one of the bigger sailors to pick me up on his shoulders. I know some disabled people dislike being manhandled, which I fully understand, but in a really tight situation, like a small ship, exploring a cave, or getting into a small aircraft, I would rather be picked up by an honest-to-goodness strong-armed person than be fussed over by dozens of nurses or professional ambulance people worrying about my dignity and self-respect while they try to manoeuvre an electrical hoist into position, or wait hours for the one person who is officially qualified to lift me.

After a very successful week at sea, though with much trouble from the weather, we returned to Plymouth on a Saturday night. The next night we were due in Malta for the start of the *Pacem in Maribus* Conference sponsored by the Center for the Study of Democratic Institutions. Natalka discovered that the entire strawberry crop in the garden had ripened in our absence, so we nearly made ourselves sick trying to eat them all in the one hour's drive to Heathrow.

In Malta, one of my old stamping grounds, I met my UN friends again, including Tom Busha and Peter Thacher. Tom drove us to Wied iż-Żurrieq, where we met John Woods and Bill Hemmings, divers who feature elsewhere in this book (Chapters 1, 4, 8). They helped me on with an aqualung in preparation for my first sea dive. There was a small concrete quay providing easy access to the fjord-like bay. I rolled off the warm concrete into the water, which felt icy, clutching my mask and mouthpiece. Although there was a little chop on the surface that knocked my mask about, I felt fine as soon as I was under water, and within minutes we were swimming at a depth of 25 m (82 ft). Seeing the seabed with its grey rocks decorated with weeds and starfish was like entering a well-remembered room and rejoicing to see cushions and furniture and even old books still in their usual places. With a firm sculling motion of my arms, I could keep up a steady speed without the disadvantage of the stop-start of normal breaststroke. Much exhilarated, I returned to the surface to sunbathe on the warm rocks by the jetty.

In the autumn there was more sea diving when John Woods and I were members of the BSAC delegation to the First Symposium of the Scientific Committee of the Confédération Mondiale des Activités Subaquatiques (CMAS) in Havana, Cuba. (See Chapter 8). We flew by Air Cuba from Madrid, but as the aircraft was an old turbo-prop Britannia, we were not quite sure if we would actually reach Havana without stopping for fuel at Gander, Nova Scotia. Cuba was still warm late in the year, so that ice-cold rum-and-Coke was an essential nutrient. Our hosts, the Cuban Academy of Sciences, arranged festive tours to a crocodile farm and various dance and music entertainments.

The conference was superbly organized by the Academy of Sciences. After a week of continuous scientific papers and committee meetings in Havana, we adjourned to Varadero for a series of geological dives on the reefs and rocks of that coast. The Cuban navy divers, who were excellently trained, did not turn a hair at being presented with nearly one hundred foreign scientists, speaking a dozen languages, all wanting to dive at once. They organized us into groups under the direction of Miguel Montanez from the Academy, not showing surprise even when told that the man in the wheelchair was going to dive too.

Figure 7.7. The author snorkelling down onto a coral reef near Varadero, Cuba, in 1970. The only way to descend is with very strong arm movements, since the lungs are fully inflated, and hence buoyant.

It was my first dive on a real coral reef (Fig. 7.7), where, although I studiously avoided touching anything, Natalka was determined to obtain samples of corals, which caused her some nasty poisoned burns as a result. The coral was magnificently rich in colour and branched texture, but, since that first experience, I have never developed a taste for just gawping at pretty corals. During the week in Havana, we had set up commissions to deal with different subjects – marine geology, biology, archaeology, and so on. At Varadero, we had the chance to discuss each subject separately so as to develop our own programmes. Since John Woods had disappeared in the direction of Tokyo (he had some problems with visas and eventually went the opposite way around the world to that which he had expected), I had to stay on and hold the fort.

* * *

Because I had an excellent employer, and a director, Sir George Deacon, who ensured that I kept my job and, because I received prompt and efficient cash payment of about £1,000 of insurance from the Guardian Royal Exchange, I was spared the complex and sometimes overwhelming problems that spinal injuries patients endure when they return to home life and try to find work. I was lucky.

Nevertheless, nobody should ever think that life is over because part of their body has been put out of action. I know cervicals – that is, people with broken necks – who would agree with me. A recent book by Henry Fraser describes how a young man who broke his neck at the age of 17 was able to build a busy life afterwards. Everybody I met since I came out of hospital has accepted that it is perfectly natural that I should want to dive, explore caves, fly in small aircraft, or go to Helgoland or the Aleutian Islands, and all, strangers and friends alike, have always helped. I sincerely believe that the opportunities for disabled people to lead an energetic and useful working life are greater now than ever before.

The word “independence” and phrases such as “preserving your independence” are used again and again in the context of disability. My activities described in this chapter give an illusion of independence which is, I hope, encouraging and optimistic, after my traumatic accident. But I must stress that it is an illusion for all sorts of reasons. At the heavy end of the scale, a paraplegic is always “nearer the edge” in terms of physical traumatic and health risks. With care and self-awareness, risk can be reduced to a level very close to that of a healthy person with no disability, but now and then mistakes or slip-ups will occur, and a rapid transfer to the professional care of the medical services becomes essential. That is, my ability to travel widely and present the illusion of independence is based on my knowledge that, if something goes wrong, the health services of Europe, the USA, and most other countries that I have visited provide a safety net and rescue service. I may have pushed my luck on occasion, or undertaken what looked like extreme risks, but that has usually been when I have the companionship of a team of strong divers, or members of my family, or both.

That brings me to the next stage of protection, support, and care. I have only my own experience to quote from, but my wife, Jay Kleinberg, and our children, Kirsten and Peter, provide a constant, albeit usually unobtrusive, layer of comfort, detailed support, and rapid reaction in emergencies. Superficially, I have not needed a “carer”, but, once again, I have been able to drive to work in Southampton, or go to a conference in London or Paris on my own, secure in the knowledge that I have a warm home to come back to and a strong family, and good neighbours, to provide support or minor medical repairs if I need them.

Finally, the most informal layer of support which is needed in the preservation of "independence" is the goodwill and kindness of strangers. I have found that in all countries, strangers in the street, minor officials, porters at hotel doorways, supermarket checkout staff, airport staff, cleaners in hotel bedrooms, and waiters in cafés and restaurants are, with a tiny number of exceptions, understanding, helpful, and supportive, without making themselves obtrusive. The obvious sense of care and helpful concern, whether I actually use the help offered, or not, is a continuous source of strength and reassurance for me.

No man is an island.

* * *

After settling down to work again during 1970, I had to think about my attitude to helping other disabled people. There was much publicity about a lobby called the Disablement Income Group, or DIG, provoked by a sense of public scandal at the high unemployment rate of disabled people. I drove to London to attend various meetings of sports committees and panels encouraging outdoor activities, including water sports, for disabled people. I wrote an essay entitled "The Incurable Optimist" for the Megan du Boisson Memorial Prize Competition organized by DIG, and won it. I tried to think how my experience and knowledge could be developed, either through committees, training programmes, or clubs, so that other disabled people could enjoy life as much as I did. I could see at once that simply advertising one's own achievements would not be welcome.

Norman Croucher, an established writer on sports for the disabled, at the time said:

> Disabled people don't like being told to emulate a so-called role model. Everybody is different, and the first reaction is to think: 'It's OK for him/her, but they don't have my problem.'

There is a lot of truth in this, and it would be insensitive and arrogant for any disabled person to take the line "If I can do it, so should you."

In all that follows I will never preach, seldom advise, and not even recommend, although I may complain. All I can do is describe what happened to me. If it helps someone, I will be happy. The only evidence I have is a statistical sample of one: my own experience. I am not an expert on disability or the myriad problems of the wide range of disabled people.

During the years 1970-1971, I became increasingly frustrated by the committees and their approach to disabled water sports. They talked about rowing, sailing, swimming, and canoeing, but never snorkelling or scuba diving. They always wanted to adapt boats in special disabled centres, rather than helping disabled people to travel widely where they could use any boat they found, or join any club anywhere in the world. Why should we be told to stay on safe tracks supervised by and approved by pedestrians? In their world, the disabled person was always on a leash, controlled by a specially-trained pedestrian. I wanted to be free, and I wanted to help other disabled people to be free. I decided that the bureaucrats I met were quite capable of running the committees the way that they wanted to: my attitude was disruptive. I could not help anyone that way. Eventually I stopped talking. Then I resigned. I needed to find another way to help disabled people who had not had the good luck that I had. I was physically fit, had a secure job, and good social support. It is difficult to imagine a more favourable set of circumstances in which to break one's back. Admittedly it is better not to break your back, but that is not the choice I had.

In later years, after I had acquired experience of snorkelling, scuba diving, and working on boats in many parts of the world, I tried to explain to British water sports charities helping disabled people that there were tricks and manoeuvres that would allow paraplegics to take control of their own safety and decision-making without the necessity for a professional carer or supervisor. I was rebuffed again, rudely, and abandoned that line of argument.

So how could I help other disabled people? My choice was very limited. Many disabled people are unemployed, so maintaining full-time employment and continuing to perform active research on the coast and at sea seemed a good start. This would act as a low-key encouragement to others – without any preaching – while demonstrating to my colleagues and other employers worldwide that paraplegics were employable and reliable. Over the last 52 years, I have been visibly sharing a professional work environment, and participating in conferences, workshops, expeditions, and committees, with thousands of people in more than 60 countries. My job was very demanding, so I had to devote myself to it entirely, choosing not to spend time on committees and campaigns for the disabled.

Consistent with the feeling that dedication to work is the priority, I find that the disabled paraplegic people I have admired in recent decades are the late David Tye, who ran a diving school near Oban in Scotland; Frank Williams, who ran a Formula One racing car company; Frank Gardner, a security correspondent for the BBC; Ade Adepitan, co-presenter on TV travel programmes; David Constantine, who builds wheelchairs for use in poor countries; Wolfgang Schäuble, a senior German politician; and John Hockenberry, who was a war correspondent for American magazines and is now a campaigner for disabled rights. The factor that appeals to me in all these cases is that the people just do (or did) a job of work, and made no fuss about their own disability. In terms of their projected image, they ignore their disability. Some of them consciously lobby for and support disabled causes, which I admire too.

The cause of maximizing equality and fulfilment for disabled people requires numerous kinds of lobbying, actions, and research. All are necessary, and some people are attracted to one kind of initiative, others to different targets. We need meticulous, persistent slogging in government committees to write new legislation; we need occasional demonstrations and protests and people who chain themselves to railings to highlight particular outrages; we need lobby groups who press for better access to buildings and transport; we need help for people suffering post-traumatic stress disorder; we need sports committees to develop access to sports for recreation and rehabilitation; we need the Paralympic Games; we need Prince Harry's Invictus Games; we need Eunice Kennedy Shriver's Special Olympics; we need legislation on employment and to counter discrimination; we need research on repairing the spinal cord; we need research on exoskeleton robotics; we need stunts and fundraising sensations such as rowing across the Atlantic or climbing mountains. Above all, there is the goal of increasing the proportion of disabled people in employment and providing special training for this.

For me, there was only one area where I could help. This is specifically at the margin, where a disabled person feels that they still need to be supervised or accompanied by a pedestrian professional carer but, with a little help and training, they could become completely free to travel and explore and make their own decisions, working full time, planning their own activities, and taking risks on the basis of their own judgement.

I felt that I could legitimately push the limits of paraplegic activities in the course of my work, surreptitiously using the attitudes developed in the Royal Marines (Chapter 2), demonstrating that harsh environments, rapid travel, living in places with no special adaptations, were all accessible to paraplegics as a routine. I was not interested in stunts or one-off dramatic achievements. I wanted to show that a paraplegic could say "Yes" if his or her boss telephoned at eight o'clock at night and said "Can you fly to Beijing on the mid-morning flight tomorrow? There's an important contract at stake." Or that a paraplegic could live for a week or even a month on a research ship, or an uninhabited island, without creating any special or expensive requirements. Just doing a job without any fuss.

If I could achieve these goals, I needed to develop numerous tricks and manoeuvres which might one day help other people. I could report back and say "If you want to do this, here are some tricks of the trade." Perhaps the committees and charities that had seemed so faint-hearted would adopt these ideas, which would spread them more rapidly than I could achieve alone.

Swimming, snorkelling, diving

Swimming is the most obvious physical activity which does not require mobile legs. The water supports your weight, while your hands can do the work. The problem for paraplegics, and I suspect other levels of paralysis, is that the body tends to adopt a "Lazy-Z" position (Fig. 7.8), with feet and backside floating on the surface, while the knees and head sink deep into the water. For a strong and determined swimmer this is easily overcome but it is hard work. For the weaker swimmer this may be the factor that deters them from trying to swim more than once or twice before giving up.

The solution for me was immediate. I always wear a mask and snorkel when swimming, whether in the pool or in the open sea. Maybe it is lazy to make life so easy for myself, but there are enough problems with paraplegia without accepting one that has such an obvious solution. With a mask and snorkel, the head can lie at a natural angle with the face submerged, breathing is easy, and you have the added pleasure of watching subsea life and the seabed.

For the first few years after my accident, I tried explaining this to physiotherapists and rehabilitation centres but finally gave up as it seemed a waste of time. They were never interested.

Snorkelling in the open sea and duck-diving down with strong arm strokes is an exhilarating experience (Figs 7.7-7.9), and I have enjoyed this form of exercise and adventure all over the world. I have also used it extensively in underwater survey work and research in shallow water (Chapters 6, 10).

Figure 7.8. 1987, Corfu, snorkel diving to check stones and possible artefacts. The project is described in Chapter 10. Note awkward position of legs and feet that causes extra drag when swimming.

Figure 7.9 After research in Crete in 1973 I joined George Legakis on a boat near Aegina. I have dropped off the cushion shown on the platform by the boat's rail, and am hitting the water while holding my mask and snorkel in place. It is quite safe for paraplegics to do this, but you have to be prepared to fall into an uncontrolled position in the water. When you resurface with mask and snorkel intact, you can get your balance.

Between 1971 and 1973 I conducted research using scuba diving and snorkelling in Turkey, Crete, and Cyprus (Chapter 6). In spring 1974, I was invited by Dr Elisha Linder, Haifa University, to run a course for disabled soldiers in Israel, as he had visited patients in a large hospital where one of them explained how he had once dreamt of learning scuba diving. The training course that I designed and supervised ran for one week, with the team carefully selected in advance, given medical checks, and some training in the use of mask and snorkel. With a supporting team of physiotherapists, diving instructors, and a qualified Israeli Navy diving doctor, we started their scuba training in the pool at Tivon, before migrating to a larger pool near Tel Aviv, culminating with all six disabled men – two double-leg amputees, four paraplegics – diving in the open sea at Akko (Figs 7.10, 7.11). I quickly established rapport and mutual respect with Dr Yehuda Melamed, of the Israel Navy, who supervised the safety of the disabled divers, and he quickly became – and remains – one of my best friends in the world. We published the results of the course in November 1974, describing techniques that have been developed and improved over the years by many sports diving organizations. The course in Israel was not the first time that disabled people had dived with scuba, but, so far as I know, it was the first time that the full medical and psychological factors during training were measured and recorded for benefit of others. Yehuda and I still receive requests for copies of the original research report, which suggests that very little study of this kind has been done since.

Diving with scuba using open-circuit compressed air (Chapter 8) is now an established sport for disabled people (Figs 7.12, 7.13), and most sports diving agencies provide courses. My experience is that the trained disabled diver should, if possible, be integrated into a normal diving club, where they are surrounded by their friends and colleagues who are strong and can help them over rough ground and rocks or into a high-sided boat. A group of wheelchair users together in a hazardous situation are at greater risk than a single disabled person in an able-bodied club. There are some disabilities which require such specialized care that diving is best done in a professionally supervised medical context, but this is not an area with which I am familiar. Diving with intensive supervision is also helpful as a psychological boost for seriously disabled people, as the experience gives them a sense of freedom and enjoyment.

Small boats

Cruising or rowing in small boats is another natural activity using arms and not much legs. There have been many successful adaptations of boats large and small for various classes of disabled people. This is admirable. My purpose here, however, is to show ways in which a paraplegic can use small boats without adaptation and in a completely spontaneous way at any holiday resort or anywhere else.

Small fishing boats: In 1970 and 1971, I needed to use small boats first in Cuba, and then in Turkey. In both cases, the boats were stable enough and strong enough so that when they were brought alongside the quay it was a simple matter for two men to lift the wheelchair with me in it over the rail and drop it down onto the deck. Since then I have used this method hundreds of times in dozens of countries, helped by fishermen and divers who just take a pragmatic view to solving problems (Fig. 7.14). Movement about the boat is restricted, but in a small boat this hardly matters.

Figure 7.10. Diving instructor and Israeli ex-soldier pupil. The trainee is paralysed from the chest down, and is learning to put on the aqualung that is lying on the floor of the pool. The trainee has already placed the weightbelt over his knee to stabilize his position. This training session was at a pool near Tel Aviv in 1974 (Photo Elisha Linder).

Figure 7.11. The disabled pupils, with some of the instructors and physiotherapists preparing for a sea dive at Akko, Israel, in 1974. I am in the middle to the right of the man standing up on crutches (Photo Elisha Linder).

Figure 7.12. Jay and I drove across Europe and flew to Israel in 1979. Courtesy of Howard Rosenstein who ran a diving school at Sharm el Sheikh, we dived on the reefs. I am diving and perhaps getting too close to abrasive or poisonous coral. Dr Yehuda Melamed is guiding my bare feet to make sure I do not scrape the skin on my legs or feet. I should have worn rubber bootees for this dive (Photo Howard Rosenstein).

Figure 7.13. Jay and I diving at Sharm el Sheikh in 1979, with escort divers. I am the one without fins, swimming with my arms (Photo Howard Rosenstein).

Figure 7.14. Farasan Islands, Saudi Arabia, 2004, en route from the main island to the island of Qummah. The small fishing boat has been adapted for passenger ferry work by fitting two enormous outboards and a nice carpet. The islands used to be the centre of the natural pearl fishing industry.

Inflatables: I first rode in a small inflatable in Crete in 1972. I transferred off the chair onto the ground and then slid onto the side tube of the inflatable. My friends folded up the wheelchair and laid it in the boat. Depending upon the size of the inflatable and the roughness of the sea, you can either sit on your wheelchair-cushion on the bottom or on the side tube of the inflatable, or stay in your wheelchair. In recent years I have taken cruises in wild places such as the Antarctic, Tierra del Fuego, and the Aleutian Islands, and found it comfortable and safe to remain in my chair in a large inflatable carrying a dozen passengers.

Small rowing boat or dinghy: This is the least intuitive form of water transport for someone with no waist muscles and no balance. The first time I tried it was with a 3 m (10 ft) fibreglass dinghy with a single thwart on a pond next to our house in England. I transferred down onto the ground and then wedged my cushion over the gunwale of the dinghy, half onto the thwart. While my young son held the boat steady, I slid across onto the cushion and dropped onto the thwart together with the cushion. I slotted the stubby oars into the rowlocks before making a tentative pull. As the dinghy moved away from the bank, I found that although I had no back support, and no waist muscles, I could use my body weight to lean backward, pulling on the oars until the stroke was almost complete, and then with a very rapid contraction of my arm muscles, pull myself upright again, then leaning forward before lifting the oars out of the water for the next stroke. Within minutes I was sculling about the pond, taking our children for trips. Since then I have rowed small boats in several different situations, where I have always found that there is a way of using the inertia of the oars in the water as a lever or fulcrum to maintain my balance and body position safely.

Hobie Cat: In 1985, I was attending a conference in Fiji when I was invited to sail in, or rather on, a Hobie Cat catamaran dinghy. Although I work a lot at sea, I was not familiar with this recreational craft. The only sitting area is a strong plastic trampoline strung between the catamaran hulls. It was quite safe and comfortable to sit on the net on my wheelchair-cushion, but I admit that I had to hang on for grim death as my host sailed gleefully at great speed on a choppy sea out to the edge of the coral reef and back. My teeth were rattling and my arms were tired by the time we regained solid ground. But great fun.

Surfboard with paddles: On a family holiday in Corfu in 1984 I found that one of the beach cafés nearby rented out little craft that were effectively surfboards with a small recessed depression for a seat. You sat on the board and used very light kayak-type double paddles to propel yourself. The proprietor helped me onto the board, where I wobbled for a moment unsteadily, before dipping the paddle into the sea to use it as a stabilizer. Within a few minutes, I found that I could use the paddles to maintain both my body balance on the board and to propel it forward using my body weight, first to give the force and then using the paddle to pull myself upright again. After a few experiments zipping across the bay and back again, I fitted a life-jacket on our three-year-old daughter, Kirsten, who could swim, and we went for wonderful voyages together, bouncing around the bay.

Kayak: Getting into a kayak is tricky, since it requires cunning and persistence to get your legs under the deck, before dropping into the seat on a soft cushion. I first tried this on a holiday with Jay and the children. In this situation, the backrest – or, lacking that, the cockpit rim – provides good support for the lower back. A double-bladed paddle is easy to manipulate and doesn't seem to require any special technique for a paraplegic to use. But on reflection, this was dangerous, as one's legs are trapped. A canoe for paraplegics should be completely open so that, in the event of a capsize, you can escape free of the hull. The Paralympic Games use open canoes, probably for this reason.

Pedalo/pedal boat: This is a common craft on lakes and on the open sea in calm holiday resorts. Some are propelled by a hand crank, others by foot pedals. Obviously the hand-crank type is best for paraplegics, but even with the foot type, and sharing the work with a pedestrian, you can have good fun, as the foot crank can be turned by hand with a bit of effort. In Greece, I used a pedalo together with Jay and three-year-old Kirsten. A family delight.

Travel

Foreign countries: International travel requires the disabled person to adapt quickly to different rules and customs. There are many good guide books for disabled travellers, with advice on access to trains, aircraft, hotels, taxis, etc., in different countries and cities. The situation has evolved steadily over the 50 years that I have been using a wheelchair, and I cannot summarize the vast volume of information available. For people who know that they will need specially adapted facilities, or help on the journey, there are many clubs and societies, not to mention websites, that can provide the contacts. The kindness of strangers is the dominant feature that strikes me after many journeys abroad. People in the street are almost always willing to help you up the kerb, or down a flight of steps, or to push up a long hill. But make sure that you only ask for help, or accept an offer for help, from someone who is fit to provide it. In addition, I have experienced taxi drivers and newspaper sellers who try to refuse payment, and people much poorer than me who cross the street to try to give me money.

Countless hotel doormen, train guards, taxi drivers, aircraft cabin crew, and functionaries everywhere have, through their empathy, facial expression, and body language, conveyed their support and goodwill with scarcely a word. It is impossible to explain how much strength this gives me. It is like a battery recharge again and again.

Hotels and restaurants: Most modern hotels have adapted rooms and bathrooms, but this can be a dubious advantage. I am sorry to report that some hotels convert only a room at the back overlooking the car park, with no sea view, and since the main adaptations are the width of the bathroom door and an enlarged space for the bath/

shower and lavatory, the room itself is often smaller than the standard hotel bedroom. This can be tiresome, even if it is not a serious problem. Since bathroom doors in larger hotels are often wide enough anyway, I have, on several occasions, asked to be transferred to a non-adapted room.

Restaurants in developed countries are usually required by law to be accessible to wheelchairs, and in poor countries people use their common sense and goodwill to make wheelchairs welcome. A tiresome change in recent decades has been the adoption by restaurants of tables with a single central leg, with a massive iron base to create stability. Unless the table is exceptionally large, the central leg and its base obstruct the wheelchair's foot pedals. This prevents you from getting close to the food unless you sit sideways, which is very awkward and antisocial. Such tables have become almost universal in the past decade, and have made eating out much more difficult. After decades of lobbying to get doors wider and removing steps so that we can get into restaurants, now we cannot sit at the tables. Four legs, Good; One leg, Bad.

Large boats and ships: Large cruise ships usually have one or two cabins adapted for disabled passengers. The lifts are either all big enough for wheelchairs, or there will be one lift adapted. Large ferries, like the Dover-Calais ferries, also are wheelchair accessible. Smaller cruise ships, like those that make so-called "adventure cruises," usually do not have wheelchair adaptations. Nevertheless, I have always found the staff on board ready to help me transfer into inflatables, to land on glaciers, visit outlying islands, and so on. I have been accompanied by Jay on such trips, and sometimes by the whole family. This means that I have personal help over ramps and storm sills. On Noble Caledonia cruises, the sailors were happy to manhandle my chair over the stern ramp into the inflatables. My approach to such cruises has been to explain my situation in advance, to assure the travel company that most problems can be solved pragmatically as you go along, but also to assure them that if some part of the activity is truly inaccessible for me, I will not complain. When I book on such a cruise where there is manifestly no adaptation for wheelchairs, I know what I am letting myself in for: I know when to back off and not ask for unreasonable provisions. Maintaining good friendly and working relations with the people you depend upon is the first rule. If a senior excursion guide on an adventure cruise needs a day or two to think over the risks of taking a wheelchair on a particular trip, give them that time.

Other large ships that I have worked on, those with no adaptation at all, include military vessels and civilian research ships. In these circumstances, I often depended completely on the assistance of big strong divers and sailors who naturally work in these conditions.

Submersibles and submarines: Access to underwater vehicles is not going to be a priority for many disabled people but, provided that there is a hoist to lift you on board, there is no serious problem thereafter. There are several tourist options now with submersibles having large acrylic sphere capsules, and I hope that disabled tourists can dive in them.

Airlines: Most airlines, most of the time, are very sensible, helpful, and adaptable. They don't insist on a one-size-fits-all draconian management routine for disabled travellers. They provide narrow "aisle-chairs" and paramedic assistants to help with access to your seat. Nevertheless, some staff can be unexpectedly obstructive, and I have to report that the most outrageous cases of discrimination against me as a disabled person have been from airlines. On occasion I have been left in an empty aircraft with both cabin crew and cockpit crew leaving me alone after arrival at my destination. In the extreme cases, I twice had to crawl out of the aircraft, dragging myself along the floor without my wheelchair, while the cleaners came and conducted their work without helping me. In contrast, when there was a radio message alleging a bomb on board, the pilot returned to the airport where we had taken off, and my neighbouring passengers carried me down the steps into a waiting bus, where we were surrounded by fire engines and ambulances. It was very efficient.

Light Aircraft: Small airliners and smaller private and charter aircraft are not adapted for wheelchairs, and often do not have narrow aisle-chairs to facilitate access. But I have found that airlines are very helpful. The small aircraft tend to be used in remote locations or on hedge-hopping short links, away from the hub airports, and the staff are used to solving problems quickly and without fuss. It is usually a simple matter of being manhandled into the seat, while the wheelchair may have to be disassembled to stow in the baggage compartment. For really small aircraft seating two to four people including the pilot, the main issue is access to the seat, since the cabin door is usually over the wing (Fig. 7.15). In New York in the 1970s, I twice enjoyed tourist flights around the city on helicopters. The access was quite easy by lifting myself from my chair directly through the cabin door.

Camping: I hate the word camping. It suggests boy scouts, bell tents, military tented campsites, music festivals, or – worst of all – commercial campsites providing showers, car parks, and a restaurant, just off a large road. In addition, there is the tired but oft-repeated sketch of a pair of incompetent campers trying to erect their tent in a strong wind, with jaded references to Baden Powell.

Figure 7.15. About 1971, with Dimitri Rebikoff's twin-engine aircraft, during refuelling in the Bahamas. Note that the cabin access is over the wing, which is strong enough for a person to walk on close to the fuselage. It is a simple matter to place a cushion on the wing, lift onto it, and from there into the cabin. The wheelchair then has the wheels and pedals removed for stowage.

I prefer to describe the way I survive in wild places as just living outdoors. It is probably more difficult nowadays, but until the 1980s, it was still possible to find a quiet place a few miles up into the hills, or on a beach, where you could light a fire, cook a meal, and sleep on the ground, either with or without a tent, almost anywhere in Europe. In the Mediterranean area, I prefer to sleep without a tent. A fit paraplegic can sleep safely on a foam mattress or an air mattress. Sometimes I slept on improvised layers of cushions and folded blankets, although I would not risk that now.

* * *

In writing this chapter I have been forced to think about my attitude to disability as never before. I have read magazines for disabled people for the first time, but do not feel at home with them, which is disconcerting. Many writers describe how joining a club for sports or other activities shared with other disabled people makes them feel accepted and relaxed. I explicitly recall one disabled writer saying that he can only really feel comfortable and happy when surrounded by people in the same situation who naturally understand one's problems, which able-bodied people can never do. Many societies and charities exist to provide these supports, and it is reassuring to see the sincerity, strength of communication, and advice that so many people find helpful. Funding for these organizations is desperately thin, and I hope that they can be strengthened.

At the age of 84, my body is showing signs of serious wear and tear. Over the years my impetuous nature has clashed with my sense of calculation and risk assessment, so that I have had accidents and collisions. I am now less independent in daily life and need a companion, usually Jay, every time that I travel long distances. At a domestic level, Jay has to help me change wound dressings and plasters, which, quite apart from the care and work involved, creates a sense of stress and worry for her. I am also more dependent upon the British National Health Service than when I was younger, and the quality of treatment has been truly remarkable, combined with a tolerance, or even encouragement, for my active way of life. In the winter of 2016-2017, I was rescued from really serious medical problems by the NHS, requiring three months in hospital, backed up by wonderful care from Jay and my family. My shoulders are now getting weaker, with strained tendons, so that I find it increasingly difficult to make transfers from my chair into a car or onto my bed.

In 1996, I could have retired from government employment to indulge in a hobby such as painting or joined a chess club. Instead I continued to work, first as a university employee, then in voluntary or short-term contracts, with a visiting research fellowship until 2018. In 2013, Jay and I went around the world via Moscow, Petropavlovsk, the Aleutian Islands, Seward, Alaska, Seattle, and home again. I have sometimes suggested to medical advisers or consultants that I should slow down, but they have always implied that I should continue working and travelling if it made me happy. It did and it does. But now I really do have to go slowly and be careful. Life after the age of 84 does require more care and medical attention, for which I am very grateful.

My routine now requires constant attention to potential pressure points, pressure sores, and slow-healing wounds due to the poor blood circulation in my feet. This creates extra work for Jay and uncertainty as to whether we will be able to attend social occasions, visit the theatre, or meet friends. Sometimes the sensible thing to do is to take

Figure 7.16. The man in the purple shirt is the mayor of Aghios Georghios, Lakonia, southern Greece. When we were doing research at Pavlopetri (Chapter 6) he always arranged for the village to host a party, and he always asked to dance with me. The young woman in the striped T-shirt is my daughter Kirsten (Photo Dimitris Sakellariou).

a break and rest the skin from any more stress. It is an unnatural restraint after a long and active life, but it is inevitable.

Looking back at my research career, I think that much of my most useful and productive work has been done since 1996, my 60th year, so that continuing to be active has been very gratifying. Employment still strikes me as one of the most important problems for paraplegics and other wheelchair users, since it provides all the benefits of income and independence, combined with getting out and about and meeting people. In recent years I have seen more frequently that companies like supermarkets and theatres employ wheelchair users at the checkout counters or front desks. Since these jobs create the interface between the company and the public, it is a wonderful example of the progress that has been made. I hope that other wheelchair users are employed in offices or workshops behind the scenes. There are also many more wheelchair users visible in British streets, just shopping, going to work, or taking the family and the dog for a walk.

There is a children's story that a spider was watching a centipede shimmering by with his legs rippling and surging in perfect wave-like symmetry. The spider says: "Tell me Mr Centipede, I only have eight legs, and I sometimes have trouble deciding which one to move next. You have hundreds of legs. How do you decide which one to move next?" The centipede promptly falls over with his legs all in a tangle. Writing this chapter has not been quite that uncomfortable for me, but there is no doubt in my mind that thinking more about my disability makes it less easy to get through daily routines which should be automatic. I find it much easier to ignore my disability if I can (Fig. 7.16). Like the centipede, thinking does not help me.

I make no recommendations to other people and have few items of advice as outlined above. My lifestyle suited me, and it was fun. Life has been good to me. I have tried to help others by my activities at work but worry that I have failed. Maybe someone will find my example useful. I hope it can help others. I only have a statistical sample of one. Every man likes to peel his own turnip, as said President Gomułka.

8

Scientific Diving: Technology and Safety

It was midwinter, 1993-1994: exact date unspecified. We sat opposite each other in two antagonistic negotiating teams of three men each. We were forbidden to take notes or minutes. We sat on either side of a small wooden table in the centre of an otherwise empty and very large meeting hall in the head office of the NERC in Swindon. The confrontation had been set up with a level of secrecy and a ban on recording that was reminiscent of a spy thriller. Perhaps the room had been searched for bugs.

A senior government representative, perhaps Grade 3 (Under Secretary), of the Department of Work and Pensions, opened the discussion for the other side, on behalf of the Health and Safety Executive (HSE) Diving Safety Inspectorate, with the following frank words:

"We are wearing sackcloth and ashes," said the leader of the opposing team. "We want to know how you have maintained safety for scientific divers without any serious accidents and no fatalities for thirty years. We can learn from you. The people who have been making your life hell, you will never see them again."

"Thank you," I said with emphasis. I momentarily had a vision of disgraced British Civil Servants banished to a power station in Outer Mongolia.

There were three spokesmen for the Diving Safety Inspectorate on his side of the table, and three from the NERC on our side. Our team was Dr Martin White, diving officer of the British Antarctic Survey, Richard Pope, safety officer of NERC, and me, diving officer for the Institute of Oceanographic Sciences, also representing the multi-agency UK Scientific Diving Supervisory Committee. The other team was led by the senior official from the Department for Work and Pensions, to which the HSE reported, flanked by two diving safety inspectors from the Department of Energy and the HSE itself.

Such moments of bureaucratic honesty and pure victory of common sense are not frequent in managerial life. The confession of failure by the "other side" was a joy to hear. It is difficult to convey to the normal public how a long-running vicious bureaucratic conflict arises, and how much trouble and danger it can cause before it is resolved. The usual end is a messy compromise, but in our case, the end was a fresh start with a new set of diving safety laws that have endured to this day. But how did the safest community of divers in the world become locked into a 20-year conflict with the British HSE, and why did that struggle spread through most British Commonwealth

In: Flemming, N.C., 2021, *Apollonia on my Mind. The memoir of a paraplegic ocean scientist*, 307-344. Leiden: Sidestone Press.

countries, and then through the European Diving Technology Committee, with acrid confrontational meetings, year after year, in Luxembourg and Brussels, Cape Town and Auckland, London and Edinburgh, and many other cities?

At that winter meeting in Swindon, after recovering from the shock of the total concession from the HSE, we set about elaborating a new regime for the management of diving safety in Britain, in which government statutory regulations provided the broad framework, within which each professional group and industry using diving had a say in its own safety management, based on experience. This led quickly – in a few months – to the first draft for a set of Approved Codes of Practice (ACOPS) for Scientific and Archaeological Diving, leading to negotiations that resulted finally in the Diving Operations at Work Regulations 1997. These were accompanied by the Guidance Notes and ACOPS which have survived, with routine revision, to this day, demonstrating both safety and the value of consultation with divers.

The problem started in 1974 and became a bureaucratic and technical battle of attrition. It would be easy to dismiss it as a typical Whitehall turf skirmish had it not been for the fact that at stake were the lives and safety of hundreds of diving scientists and students in Britain, as well as the careers of many dozens of senior university and institutional staff, diving officers, and diving supervisors. Ten commercial divers were killed in accidents on oil rigs in the North Sea in 1974, showing that strong safety regulations were certainly needed on the rigs. Many more lives were lost in other sectors of the diving industry during this 20-year conflict, but there were no serious accidents in British scientific diving and no fatalities. But each time the government agencies tried to make diving safer in the North Sea, on the rigs and pipelines of the offshore hydrocarbons industry, they made rules that, if obeyed, made diving more dangerous for people using other types of diving equipment. Each set of new regulations, amendments and advisory notices promoted by HSE during this period failed to recognize the type of diving conducted by scientists, and – if obeyed – would have made scientific diving with scuba more dangerous, not safer. We had to break the rules in order to be safe.

Many other countries followed the texts of British government diving regulations, especially throughout the Commonwealth. The technical misunderstanding damaged the scientific diving communities in dozens of countries. Around the world thousands of diving scientists were affected by the legislative conflict. One fatal accident to an employed scientific diver in the UK during these 20 years would, in addition to killing a good person, have brought the full force of the law down onto a community that was flagrantly in continuous dispute with the HSE, and would probably have resulted in court prosecution, with a diving officer – possibly me – going to prison. Scientific diving in Britain, and possibly in many other countries, would have been wiped out for a generation. If we had had to appear in court it would have been difficult to prove that we were using the best practical experience, the best-trained people, the best theoretical knowledge, and were occupying the high moral ground although we were in continuous dispute with the government safety agency, the HSE.

How did this situation come about? Why did mostly well-intentioned people on all sides allow the bureaucratic chaos to get worse and worse for two decades? Hundreds of marine scientists gave up diving during these years, causing several marine research institutes to close down their diving teams.

* * *

When I joined the BSAC in February 1957, I started to combine the joy of sports diving with scientific observation, since both activities used fins for propulsion while breathing compressed air from a regulator or valve that provided gas when you inhaled. In the civilian world, scuba equipment was not controversial, as it was often shown on TV as the norm for underwater work. Harry Gould, who was on the BSAC Council as well as being an active member of the London branch, suggested that I should start a BSAC branch at Cambridge University, and he arranged for Lillywhites, the famous sports store on Piccadilly Circus, to loan us diving cylinders. Harry, who was and is a suave and immensely successful chartered accountant and experienced diver, was determined to make the Cambridge branch a success.

The CUUEG used twin-hose Mistral regulators, and we hired large industrial cylinders of compressed air from British Oxygen Ltd, from which we decanted the gas into the diving sets. Webbing harnesses for the cylinders, weight belts, and the lead weights themselves were often homemade. Diving suits at this date were still exclusively drysuits, beneath which we wore warm woolly clothes, and the upper and lower half of the suit were joined together in a waterproof joint by means of a "cummerbund". Hans Hass, the celebrated maker of brilliant underwater films, agreed to be our club patron, and Marcel Contal, the chief diver of Club Mediterranée, to be our honorary vice president. After our first expedition to Apollonia (Chapter 1), other projects rapidly followed, with Bill Hemmings and John Gamble organizing biology research, myself exploring the geology and archaeology of submerged shorelines, and Keith Muckelroy later pioneering a new approach to shipwreck archaeology. Keith subsequently wrote one of the definitive books on shipwreck archaeology, emphasizing the changes that occur to a wreck after it arrives on the seafloor.

In the early days, we wanted keen scientists and archaeologists to join the club, but we tended to recruit people we regarded as "natural divers", people who seemed completely at home and relaxed underwater from the very first time they submerged their heads while breathing from an aqualung valve. The others dropped out of the course. This biased our membership to divers for whom accidents were less likely to happen, even in extreme circumstances. In contrast, sports diving training courses nowadays try to get everybody to complete the course, consistent with safety, and minimize dropouts.

In the following years, the diving club at Imperial College London, which was also founded in 1957, joined the BSAC with an active scientific objective, and new BSAC branches with research projects soon formed at Oxford and other universities. All over Europe, scientists and archaeologists were discovering the value of scuba in making the underwater world observable cheaply and safely. Diving exploration, combined with close-up photography of lurid beasts and beautiful fish, provided the story-line for countless TV programmes. Jacques-Yves Cousteau, Phillipe Taillez, and Jacques Dumas appeared in films and books. Anders Franzen had found the colossal wreck of the *Vasa* in 1956, still perfectly preserved in the freshwater of the Baltic. Italian divers found Roman and Greek wrecks by the dozen. Bruno Vailati persuaded many film watchers that sharks are really quite nice, once you get to know them.

James Dugan's 1956 book *Man Explores the Sea* described the beginnings of research in many countries where scientists had made forays underwater in clumsy helmets as early as 1900; but self-contained diving, with a portable gas supply of either nitrogen-oxygen in a rebreather, or air in open-circuit scuba, with the diver swimming with fins, was a revolution in accessibility. Underwater science became a routine at no excessive cost or complexity. Dugan's book and American dive magazines reported the research being done by Bob Dill, Willard Bascom, and Conrad Limbaugh, who introduced scuba diving as a research tool for the Scripps Institution of Oceanography in California. This evolution of techniques in research institutions confirmed that the science and the diving were bound together. Research work underwater should not be just an occasional excursion by a scientist who may only dive once or twice in his or her professional lifetime.

In 1960, as I was starting research for my PhD, the newly founded CMAS held its first World Congress in Barcelona. This was, and is, the affiliated body of national sports diving federations, which has been dominated largely by the French and Italians since its start. It does excellent work in ensuring compatibility and consistency of diving training standards for sports divers worldwide. Somebody in the BSAC upper echelons thought I should be a member of the team representing Britain in Barcelona, where I should talk about science to the international gathering of scuba divers. My mother was sceptical, saying, "Couldn't they find anyone more important than you?"

In Barcelona, where, as a hungry student, I revelled in the magnificent seafood at the grand dinners, I mounted a poster display that showed what different groups in the UK were doing, and I gave an illustrated talk about Apollonia. In the hotel bar, everybody chatted noisily. I was thrilled to meet Anders Franzen, and over beers, we speculated cheerfully about the future possibilities of research under water. I discussed underwater caves with Scripps Institution's Conrad Limbaugh and explained my ideas about ice age changes of sea level. We agreed to correspond. Unfortunately, Limbaugh drowned while exploring caves in southern France a few weeks later.

The expeditions conducted by Imperial College, Cambridge University, and soon other universities, established once and for all the concept of "scientific diving". We were not pretending to be commercial divers or contractors who could hire our rugged services out to any customer, but we were scientists, psychologists, geologists, and archaeologists who conducted our research in the normal professional manner, routinely using diving as a research tool, just as another scholar might use a microscope, an animal trap, or a cross-country desert-hardened vehicle. In 1962 the UK hosted the 2nd World Congress of CMAS, which was opened by Lord Louis Mountbatten, with Jacques-Yves Cousteau chairing the first session. The entire conference was devoted to exploration, discovery, and research. George Bass presented his work on a Byzantine shipwreck; Ed Link talked about mapping the submerged harbour of Caesarea in Israel and searching for shipwrecks; Hans Hass described a new theory of coral atoll formation; and Jacques Piccard described his dive to 10,911 m (35,797 ft) in the bathyscaphe *Trieste* in 1960. I gave a paper describing the mapping of underwater caves off Gibraltar and Marseille. Scientific work by divers in numerous disciplines had become a job and a feasible career. The conference finished appropriately with a sumptuous dinner at the hall of the London Livery Guild, the Worshipful Company of Fishmongers.

In the early 1960s the scattered British groups conducting archaeology under water, many of them entirely amateur, needed to consolidate and share expertise with agreed standards. The Institute of Archaeology, UCL, appointed Joan du Plat Taylor to promote underwater archaeology in Britain, and by 1962 she had designed training courses, followed in 1965 by the formal constitution of the Committee for Nautical Archaeology, of which I was a member.

Although the records of CUUEG for its early decades are not complete, I have found reports for the activities in 11 of the first 20 years – that is, up to 1978. During those years, in addition to numerous diving training projects in quarries, lakes, and on the British coast, CUUEG members carried out 26 highly original scientific research projects abroad, plus a further five in the UK. CUUEG conducted projects in France, Ireland, Greece, Italy, Malta, Libya, Yugoslavia, the Virgin Islands, Turkey, Tunisia, Algeria, Jordan, and Sudan. This geographical range shows the group's level of initiative and professional skills, involving research in psychology, underwater vision, geology, archaeology, many aspects of marine biology, coral bleaching, fish behaviour, and engineering. We can assume that the missing years would show similar levels of exploration. During the 1970s, dozens of CUUEG alumni worked as divers on the sensational survey and excavation of Henry VIII's warship *Mary Rose*, which was finally salvaged and raised to the surface in 1982. Other universities, especially Imperial College and Oxford, were very active, while Bristol, Exeter, and several Scottish universities developed strong diving research themes.

* * *

Dr Arthur Stride at the NIO had raised funding to pay for teams of divers from CUUEG during the early 1960s, during which we worked from a base in Polperro, a minuscule gem of a fishing port on the south Cornish coast. (The research on seabed ribbons of sand is described in Chapter 7.) The features we needed to observe and measure accurately were usually at 30-37 m (100-120 ft) depth, and sometimes 60 m (200 ft). After two seasons of work, it was obvious that nitrogen narcosis was seriously reducing the efficiency of otherwise trained minds while trying to work at depth. We achieved considerable success, aided by an electrically propelled diver-tug that helped us cover long distances on the seabed, and we published our results, but it was obvious that compressed air was not the right breathing mixture below 30-37 m (100-120 ft). This realization resulted in the heliox experiments that I conducted in Malta (Chapter 2).

During the spring of 1965, I attended conferences in Boston and Toronto, where I met Alan Krasberg and Jerry O'Neill, who had developed a demonstration model of an electronically controlled heliox rebreather diving system. It was a future technology that was bound to succeed. They worked for Westinghouse. Their device used a sensor that measured the partial pressure of oxygen in the breathing mixture and added oxygen or helium to adjust the mixture if it was not correct for the depth at which the diver was working. This was a brilliant advance, but the "bread-board" nature of their prototype set suggested that there was much more work needed before it would be reliable in commercial application. The mechanical heliox set that I developed with the Royal Navy was much simpler and less adaptable in principle, but it was very reliable and helped to make the point that heliox in self-contained breathing sets was very desirable (Figs 2.9, 8.1).

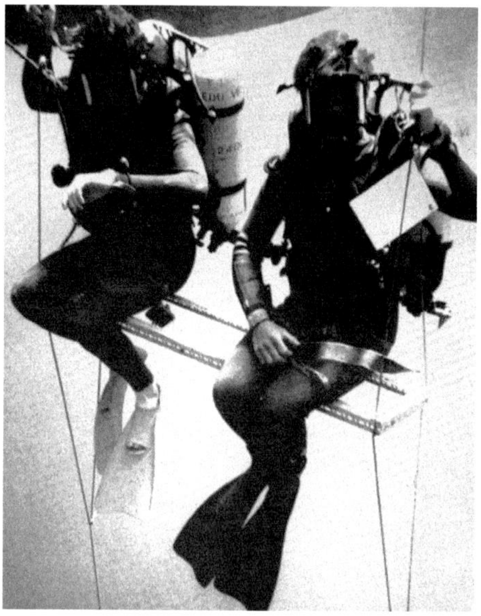

Figure 8.1. Two divers on oxy-helium conducting a stop after diving to 60 m (200 ft). Photo taken in Malta in 1965.

In Malta that summer there were 13 teams of research divers from British universities, including Imperial College, Oxford, and Cambridge, and participation by scientists from the Royal Navy. Much of the accommodation for the diving teams was shared in a spacious apartment with communal cooking, so there was plenty of time to discuss the future of scientific diving. John Woods was enthusiastic that we should set up a permanent base for research diving in Malta, which could be used year after year. The proposals were bandied about by Bill Hemmings, John Gamble, myself, Helen Ross, Mike Davis, John Lythgoe, Ed Drew, and others. We agreed to hold a conference afterwards at which the results of the various research projects would be formally presented as scientific papers, and John Woods arranged for us to hold the meeting at Imperial College in October 1965.

By the late 1960s, many of the divers who had started their research careers with vacation projects as undergraduates had moved on to complete their PhDs and thence into full-time professional research. A core group of several dozen divers supported the idea of a permanent association dedicated to scientific research using diving. After the 1966 diving season in Malta, a group of us met in London on 16th December to sign the Memorandum and Articles of Association of the new Underwater Association for Malta 1966 Ltd, a company limited by guarantee with no share capital. We appointed Bill Hemmings the first chairman and John Lythgoe as the honorary secretary. I agreed to organize the next symposium at the Royal Geographical Society in March 1967. John Woods continued to ensure a working base in Malta.

This is not the place for a history of the Underwater Association (as it was generally known), but it is astonishing to see the calibre of scientists who joined, and of those who attended from abroad to present their work at the annual symposia. Alan Krasberg gave a revolutionary paper in 1967 on the first-ever use of saturation diving in commercial engineering work; John Woods reported on his radically original research on internal waves and thermocline stratification in the sea using fluorescein dye tracers; Jean Vacelet came from France to describe his biological research work in underwater caves;

Alan Baddeley, later a Fellow of the Royal Society, presented his results on the psychology of divers breathing heliox; a special project team reported on the pollution caused in 1967 by the crashed supertanker *Torrey Canyon*. At subsequent conferences scientists described developments in electronically controlled rebreathers and research on the physiology of diving, and speakers came from Germany, Italy, South Africa, Norway, and Russia to report on discoveries in marine biology, seabed geology, and living in underwater habitats. Within ten years the membership of the Underwater Association (UA) grew to about 300 divers. When Bill Hemmings retired as chair in 1974, I succeeded him, and John Lythgoe took over from me a few years later.

The formation of the UA introduced an essential new level of communication and professional collegial spirit. A council of elected officers, who met several times each year, managed the Association, including the approval of research projects that could be covered by a special insurance policy. An editor produced a regular newsletter, and for over 20 years the proceedings of the annual conferences were published in books entitled *Progress in Underwater Science*.

Before the advent of competitive citations and computerized citation indices, it was not so important to publish in a well-known journal with a "high impact". As underwater science evolved we were all conscious that previous work had been published in a wide array of journals scattered through the disciplines, often in foreign countries, and was therefore very difficult to retrieve. To consult a paper in a foreign country, if there was no copy in the UK, you had to post a letter to the author or to a foreign library or publisher, hoping that they would send you a copy. By publishing a concentrated mass of research all using diving, and all easily retrievable between one set of covers each year, we demonstrated diving as a respected research method.

This was fully justified at the time, but eventually, as the logic of commercial academic publication citation indices and the attraction of high impact journals became irresistible, the book series dedicated to scientific diving and underwater science had to be abandoned. The UA and its annual *Proceedings in Underwater Science* lasted until 1992, when the UA merged with the Society for Underwater Technology, a much larger organization. Nevertheless, the need for professional management of the work of scientific divers and their safety continued to be a serious matter.

* * *

In the 1960s and '70s, industrial and commercial diving evolved rapidly, driven by the urgent requirements of the offshore oil and gas industries (Chapter 5). The growth of deep diving, supported by heavy submersible diving chambers, saturation diving, and deck storage and decompression chambers, was spectacular. Heavy industrial work, welding, cutting steel, rig repairs, inspection for damage and corrosion, pipeline inspection, and installing machinery were performed at depths of hundreds of metres. Divers had to live in pressurized surface chambers for weeks at a stretch, descending in submersible bells each day to the engineering work on the seafloor. I had, and have, enormous respect for the men who conducted these amazing dives and those who designed and operated the heavy engineering equipment that made it possible, but unfortunately, the tough professionals who did the work developed an exaggerated dislike and contempt for lightweight scuba diving. They called it "Mickey Mouse diving". I can understand their feeling that scuba diving to depths of 15-30 m (50-100 ft) is almost trivial compared with diving to 200 m (660 ft) in saturation for ten days, but the contempt went much further than that and was linked to an unshakeable belief that scuba diving was always dangerous. There are plenty of written statements of this unfortunate opinion. Scuba equipment is not suitable for work on oil rigs, but that does not justify the conclusion that it was always inherently dangerous and should be banned for all people at work everywhere. The correct response should have been to ban scuba on the offshore rigs or around heavy machinery and other heavy industrial operations in open water. But that was not done.

I had studied the different techniques for the COSG (Chapter 5) and had calculated the rate at which the frontier of diving would advance, by extrapolation from the depth and endurance of different commercial and military experiments achieved up to 1966. In 1967, I gave a paper on the rapid advances of diving and marine technology at a conference at Harwell. When I was taking questions after the talk, Dr Taylor, director of the Royal Navy Physiological Laboratory, stood up and said, "I thank Dr Flemming for that very interesting paper, but I note that he predicts that within a few years we will be conducting experimental dives to a depth of 1500 ft [457 m]. I can assure him that we in the Royal Navy have no intention of attempting dives to that depth, and I do not know of any other responsible research laboratory that is doing so either." Somewhat chastened, I took the next question as quickly as possible.

In 1970, I received a telephone call from Val Hempleman, chief diving physiologist at the RNPL, who asked, "Nic, you remember that table you showed at the Harwell meeting in 1967? Can you tell me what sort of laboratory chamber dive we should have done now, in 1970?"

I found the paper in a file quickly. "Let's see. OK, Val. You should just have done 1600 ft [488 m] in a laboratory compression chamber for two hours bottom time, or 1500 ft [457 m] for 10 hours bottom time. What's up? Why do you need to know what I predicted?"

"We have just had two divers compressed in the chamber down to a depth of 1500 ft and they have been at that depth for 10 hours in the chamber. They have just started decompression. Would you like to come down to RNPL tomorrow to see them?"

I hastily agreed, and next day I saw John Bevan and Peter Sharphouse inside the pot (compression chamber) through the porthole. They looked happy and healthy, even though they knew they would be confined to the chamber for more than a week as the gas pressure was slowly reduced and the percentage of oxygen gradually increased. Although this record stood for a few years, experiments on different gas mixtures and increased depths continued in several laboratories around the world until the early 1980s, when a new record of 2132 ft (650 m) was set for a chamber dive at Duke University in 1980. The deepest-ever working sea dive in 1988 was about 1700 ft (518 m), and the world record simulated deep dive in 1992 in a laboratory compression chamber was 2300 ft (701 m). The complexity, risk, and cost of diving to such depths is so great that almost all work deeper than a few hundred feet (over 100 m) is now done by robotic machines whenever possible.

During the 1970s, I made sporadic attempts to persuade research groups that semi-closed-circuit nitrox or heliox rebreathers would be more efficient than air, and safer, and I presented a paper on that thesis at the third CMAS World Congress in London in

1973. Bill Hemmings imported the "Electrolung" from the USA, a development of this system, for fish observations, because it releases very few bubbles, and does not disturb the fish. Electronically controlled rebreathers are now a very useful addition to the diving repertoire, but it is fair to say that it took several decades of trials by different companies before the complex sets became reliable enough, and easily maintainable, for everyday use. Several manufacturers now produce these sets and they are becoming popular in the sports market.

* * *

The headlong development of diving on oil installations in the North Sea resulted in a number of serious accidents, some of which were never fully explained, or involved foreign employers who flew the divers onto the rigs and flew the injured parties off the rigs. Official records show that between 1975 and 1983, 25 divers died in the UK continental shelf waters working on oil rigs, excluding inshore territorial seas, and excluding other deaths in the North Sea in the jurisdictions of other countries. Nine of those 25 divers were using scuba gear in unsuitable conditions. The UK regulating and licensing authorities became very alarmed and tried to update the old-fashioned and inappropriate Factories Act 1960.

Officials aiming to manage and improve the safety of diving had to cope with the industrial types of diving gear, which was primarily not scuba. Commercial divers working at modest depths used so-called surface demand (Figs 8.2, 8.3.), a type of gear in which breathing gas and voice communications reach the diver through a complex umbilical hose. The diver wears a heavy hood and mask or helmet that allows him or her to speak. Helmet-mounted lights and a video camera show the supervisors on the surface what the diver is doing. For very deep or prolonged dives the divers used a diving bell and surface chamber for living in while saturated with breathing gas (Fig. 8.4).

Figure 8.2. Diver preparing to dive using surface demand gas supply, with voice communications through an umbilical cable. Note lighting and recording video cameras mounted on the top of the helmet (Photo Daniel Pascoe).

Figure 8.3. Divers entering a free-flooding transport cage or wet-bell that will be lowered to the seafloor. This provides the diver with a working base, and the drag on the umbilical is only asserted on the length from the bell to the diver (Photo Daniel Pascoe).

Figure 8.4. Picture taken inside a pressurized diving bell at depth. The diver is supplied with heliox breathing gas and communications via an umbilical from the bell, and he is descending through the open hatch into the seawater at a depth of 140 m. On completion of the dive, the diver comes back into the bell, the hatch is closed, and the diver is transferred under pressure to a compression chamber on deck. This avoids decompression time until the task is completed, possibly after several days. There are three divers in the chamber to provide support (Photo Kester Keighley).

These complex types of surface-supply diving require equipment costing hundreds or even thousands of times as much as scuba gear.

Government officials presented papers on diving accidents and new methods of regulation at the recently founded Society for Underwater Technology, of which I was a council member and secretary (Chapter 9). The Department of Energy, which was responsible for offshore oil and gas licences, appointed retired Navy Lieutenant Commander Jackie Warner as inspector of diving safety. New regulations were urgently needed, and notwithstanding what follows, they greatly increased the safety of diving offshore on oil and gas rigs and pipelines in the subsequent decades.

* * *

In 1969 the engineers at the NIO ran trial cruises, launching the huge acoustic sidescan vehicle Mark 1 GLORIA from the Royal Research Ship *Discovery* (Fig. 8.5) (see also Chapter 7). GLORIA, which could survey a band of the seafloor 22 km (14 mi) wide on either side of the ship at a single pass, was towed 50 m (164 ft) below the surface and was a mainstay of seabed mapping for the next decade. The engineers had initially underestimated the difficulty of launching the 10 metre-long (33 ft) towed torpedo-like vehicle, which required the attachment of electrical cables and stabilizing floats that trailed behind the main "fish". Luckily, for the first sea cruise with GLORIA, there were several keen sports divers and snorkellers amongst the scientific and engineering team on board, who improvised ways of completing the launch and recovery, with snorkellers in the water supported by an inflatable dinghy. In discussions afterwards, particularly

Figure 8.5. Snorkel swimmers working from an inflatable to connect the cables and floats to the Mark 1 GLORIA. Later, the device was re-designed with a special launching ramp that avoided the need for divers (Archives, National Oceanographic Library, National Oceanography Centre Southampton).

with Stewart Willis, Ray Peters, and Roger Clements, we decided that we needed to have a small but well-trained team of swimmer-divers with available scuba gear on board the ship whenever GLORIA was being used for research. Instead of being the only scientist at the NIO who used diving in my work, I suddenly found myself responsible for a small institutional diving team, for which I had to report on its safety to the director. We quickly produced a set of house rules for safe diving and diver training. The launch and recovery procedures were improved, and over the course of a few years, the engineers designed later versions of GLORIA that could be operated without divers. Other research projects soon required diving support (Chapter 2).

* * *

John Woods, who had become a member of the BSAC Council, was aware of the collaboration between national diving federations through the CMAS. In 1970 the Cuban diving federation proposed the formation of a Scientific Committee within CMAS, and John invited me to stand for BSAC Council and support, with him, the Cuban initiative. The Cubans invited interested diving federations to nominate members who could present their scientific diving discoveries at a conference in Havana in September 1970, where the meeting would be hosted by the Academy of Sciences.

Enchanted by the chrome-bedecked, 1950s American motor cars, as were all visitors to Cuba, the participants at the diving science conference gathered in the lecture hall of a 1950s hotel, where we settled down to work, refreshed by iced rum and Coca-Cola. Countries represented included West Germany, France, Poland, Portugal, Spain, Italy, UK, Israel, Cuba, Norway, Holland, Chile, Japan, and several more from Central and South America. The conference started with formal speeches and presentations by dignitaries, including one speech in Hebrew by a young Israeli scientist, Nurit Gunderman, who was so beautiful that the audience, mostly male, gazed at her in rapt adoration. They burst into applause when she finished, although her speech had not been translated. Unfortunately, the talk concerned a serious tribute to a famous diver who studied coral reefs and had recently died, but this was only explained later. John and I collaborated with Miguel Montanez, the conference chairman, to make a logical programme from the wide range of papers submitted, dividing them into groups dedicated to marine biology, geology, archaeology, speleology, technology, and pollution control. By the end of the conference, these groups had settled into little professional clubs that we dubbed commissions. In the final administrative sessions, we agreed a constitution for the Scientific Committee of CMAS, with Miguel Montanez as president, further officers being Antonio Stefanon and Klaus Mueller, with John Woods as editor of conference publications, and me as conference organizing secretary.

We adjourned from Havana to the beach resort of Varadero for the last few days, where we enjoyed diving on spectacular coral reefs. The commissions held separate meetings at Varadero, each attended by seven to ten scientists. Alex Flinder, then chairman of the BSAC and a keen amateur archaeologist, pointed out that it was many years since CMAS had held a world congress, and proposed that the BSAC would host such a congress in 1973. When this proposal was accepted, we also decided to have a diving science conference as part of the congress in London, which I would arrange.

* * *

David Roberts, a very energetic, brisk, and forward-thinking colleague of mine at the NIO, worked on geophysical exploration, and with him I published several joint papers over the years. In 1971 David organized a one-day seminar on scientific diving and its applications, with published results. Researchers presenting their work included Helen Ross, Bill Hemmings, John Woods, Nigel Kelland, and Val Hempleman, while a range of representatives from the military, research, and commercial diving was in the audience. My contribution to this effort was to compare the way in which commercial airlines trained pilots, with the way in which the Royal Navy and commercial diving companies – and by implication, research institutes – recruited divers. In both flying and diving, about 20-40% of potential recruits who are accepted into training courses drop out or fail the course, incurring large expense and wasted staff time for employers. It would be a great improvement if a simple interview or short test could reveal the kinds of people who will complete their training courses, and then, above all, enjoy and continue their work. I was not the first person to consider this problem, but I thought that comparing diving and flying might reveal some key factors that others had missed.

Strangely, beyond some fairly obvious points about physical fitness and eyesight, reasonable mathematical knowledge, quick reactions, and a generally determined character, there is very little else that gives a strong positive indicator for an employable diver. In both flying and diving, a preference for adventure sports, such as climbing, canoeing, skiing, or sailing, rather than football or cricket, is a positive indicator, but not a necessary one. I am sure that some excellent divers and pilots are good at cricket. Helen Ross had already conducted psychological tests on divers, finding that their characteristics were undistinguishably normal.

Remembering my training in the University Air Squadron (Chapter 2), I compared that with what I could find about training commercial pilots. I found a distinction between pilots and divers: a trainee pilot undergoing a long course of lectures, successive exams, and flight skills evaluation will usually fail, if he or she is going to fail, because the instructor detects a weakness in exam marks or flight skills. The trainee will often feel hard done by or want to try again. They can be resentful, convinced they are doing well enough. In contrast, divers on a course of several weeks (or months in the case of SBS) tend to drop out voluntarily because they just don't like it. For a diver's potential employer, one of the most irritating circumstances is when the trainee drives himself/herself desperately to complete the course, but then never dives again, or at least not at work.

The most useful diver is obviously a person who enjoys it, is highly motivated by the spirit of scientific enquiry and discovery, is always ready to help with the chores and checks that are needed to maintain equipment, is prepared to dive in unpleasant conditions of cold and mud (provided they are safe), and can accept stoically lost time and delays when diving is necessarily curtailed through bad weather. It is difficult to spot this kind of persistent doggedness, combined with a useful intellect, in a short time. The people who continue diving year after year in variable and often unpleasant conditions seem to exhibit characteristics that are negatively correlated one with another in the general population. It is this paradoxical combination that makes them endure, rather than some unique excellence that differentiates them from the population as a whole. It would not be ethical to test the candidate by exposure to extreme discomfort or stress for a short time, and in any case that would not reveal the type of long-term endurance that we are looking for.

From this it follows that good scientific divers evolve and develop over months or years, and are seldom created by a short professional course unless the diver already has several years' sports diving experience. There are always exceptions, but people who have taken up sports diving in their late teens or early twenties, and then later realize that it can be useful in their work or studies often make the eventual transition to professional research diver in an almost seamless way. At some stage in that evolution, it may be necessary, depending upon national regulations, for the diver to undergo a formal test or course when they attain the status of an employee at work, and obviously an employed diver must obey employment law generally, but there is no need to create an artificial technical distinction between the sports scuba diver and the working scuba diver so far as the diving equipment and its safe use are concerned. The scientific diver will of course have many additional working skills and observational and recording techniques under water.

The 1971 seminar on scientific diving revealed that divers in different institutes and universities had developed specialized and ingenious techniques for their own research interests, but were using a range of standards and procedures which, in some cases, were clearly pushing their luck in order to get the work done. We needed a code of practice for scientific diving that would advise caution against diving procedures that introduced too great a risk, while maintaining sufficient flexibility to cope with the unusual circumstances that research requires.

By 1972 the membership of the Underwater Association was over 100 divers, involving dozens of UK marine laboratories and universities. The Association, as well as providing support and a forum for research, was soon implicated in the risks taken by the divers who were its members. The Association arranged an insurance cover for diving scientists, provided that the project leader gave a competent report in advance of what they proposed to do. But scientific divers as a community still had no status or published explanation as to how we managed safe diving. Either our employers, or an outside agency or critic, might question our standards of training and operations, especially if there were an accident.

We held a meeting at the diving centre at Fort Bovisand, near Plymouth, attended by 20 or more of the senior members of the Association, at which I chaired the discussion and revision of successive drafts until we had thrashed out a provisional "Code of Practice for Scientific Diving". The Underwater Association published this, and most research groups in the UK rapidly adopted it. Insurance companies specializing in the risks of diving provided favourable rates for divers who abided by the Code.

We produced a revised second edition in 1974, with Doug Miles as my co-editor, and the meeting on that occasion was rather melodramatic. There were regional power cuts of electricity at the time – during the so-called three-day week under the Heath government – so that in the shadowy vaults of the old Bovisand Castle gun rooms we had to illuminate the meeting with spotlights powered by car batteries. We looked like Bolshevik plotters planning a revolution, with our faces thrown into violent relief, casting huge black shadows on the walls. The NERC published this version of the Code, after which the group of institutional diving officers within NERC gradually expanded to include other agencies, becoming the new Scientific Diving Supervisory Committee

(SDSC). I chaired this for its first few years and was vice chair into the 1990s. It still exists and does excellent work.

* * *

Apart from the issues of personnel selection and training, there are further reasons why a working community of scientists should use training methods developed initially for sport diving. This justification includes both the equipment design and the technique itself. A diver who is working has responsibilities for safety which are more than those of weekenders, in the same way that a commercial traveller delivering sales products has more responsibility and insurance than a holiday-maker driving a car of similar size. But the car and the driving test are the same. Scuba diving gear sold to thousands, indeed millions, of sports divers around the world, is extremely reliable and well designed: there would be hundreds of damaging law cases against the manufacturers if it were not. Similarly, the methods of scuba diving training, and the various drills and safety procedures worked out by diving clubs and diving schools all over the world have resulted in manuals and guides that, if properly used, produce high standards of safety. The fatality rate for sports diving in the UK is of the order of one or two deaths per 10,000 divers at risk per year, which is comparable with other adventure sports such as riding, sailing, or skiing.

Scientific divers must work according to the current rules of employment and employer's responsibilities, but there is no need to change or convert the well-engineered diving gear used in the sports environment. Furthermore, diving equipment and training and safety techniques evolve slowly and continuously over the decades, so that the process of testing new ideas, dive computers, surface oxygen therapy, safety marker floats, mixed gas decompression tables, and so on, is trialled by hundreds of instructors, diving schools, and clubs before they are finally adopted. It would be difficult and very expensive for the small community of scientific divers to develop different equipment or different procedures. For most scientific work in depths less than 40 m (120 ft), scuba provides the ideal tool. Whatever allegations may be made that scuba is inherently dangerous, the facts prove the opposite, and the fatality rate for scientific diving using scuba in the UK has been zero for 60 years.

A distinguishing feature of the sports diving community in the UK is that all diving incidents and accidents are fully investigated before they are reported to the BSAC diving officers' committee, and published at the annual diving officers' conference. From these reports, describing in detail both the few fatalities and the dozens of minor incidents and near misses, divers everywhere can learn how to improve training drills and the conduct of dives in difficult conditions. Unfortunately, in the UK, the HSE takes the view that accidents and fatalities to working divers in all circumstances are commercially confidential. As a result, we and they cannot learn from those accidents.

The rapid and friendly communication between scientific divers in many countries provided a valuable stabilizing effect on the negotiations that followed later with the HSE. We could study and compare practices in many countries, and we used the CMAS itself as the co-ordinating body to compare equivalent standards of training.

The 1973 CMAS Scientific Conference on Diving Science in London attracted an audience of nearly 300 divers from 16 countries (many divers came from the USA, Cuba,

South Africa, and Japan) to hear 47 papers classified into ten different subject categories. We published and distributed the papers at the conference in a separate volume. The six technical commissions held their own meetings, each attended by eight to ten scientists. Since CMAS primarily represents sports divers, and since the BSAC is the national body for sports diving in the UK, it was obvious that the style and practice of diving, as discussed in the Scientific Committee, was based on scuba air diving, as taught by sports diving schools. Nobody would have thought that a different diving technique was applicable or relevant. Notwithstanding my appeal for the value of nitrox rebreathers, and Bill Hemmings' use of the Electrolung, scuba air was naturally the everyday method of diving for marine scientists in all academic disciplines.

The London 1973 CMAS Congress ended with a luxurious dinner for hundreds of guests in the Great Room of the Grosvenor House Hotel in Park Lane, attended by the astronaut Scott Carpenter and his wife. In 1965, for SEALAB II, he had spent 28 days living on the ocean floor off the coast of California (Chapter 5). I invited his wife to dance with me, which she did, and I recall the occasion with much enjoyment. As we settled down afterwards for what we thought would be a stage show or cabaret, we could hear the distant booming of drums, and then the tramp of heavy boots and the blaring of brass instruments, bugles and horns, trumpets and tubas. There were probably flutes and piccolos in there somewhere. To everyone's spine-tingling astonishment, the band of the Brigade of Guards, in full military rig, with scarlet tunics and towering bearskins helmets and twirling drum majors' sticks, came thundering into the hall like an irresistible storm of music. The very walls shook from the constant roar of sound. The floor trembled from the thudding boots. The chandeliers quivered. The audience sat spellbound while the band completed a convoluted routine of marching and counter-marching, before finally turning into the exit corridor and disappearing into the darkness.

The NERC version of the Code of Practice for Scientific Diving was hardly off the presses when the government started to issue a spate of new regulations which emerged from several agencies as they tried to respond to the accidents to commercial divers in the North Sea. First, in 1974 a Statutory Instrument (SI) entitled "Offshore Installations Diving Regulations" appeared (SI 1229). Since scientific divers very seldom visited offshore oil rigs, we noted this but did not react. The regulations greatly improved safety for divers on the rigs but did not affect us. I informed my colleagues in the Underwater Association, but nothing more. In 1975 the "Merchant Shipping Diving Operations Regulations" (SI 116) appeared, which caused much more trouble. The Regulations applied to any diving conducted by a working person from any British-flagged ship. Diving from the shore was not affected, but if divers, even sports diving instructors receiving pay for their work, were diving from a British vessel, strict regulations applied, requiring a decompression chamber, a doctor, and heavy commercial diving equipment. This resulted in a flurry of letters to Jackie Warner and complaints from the British Sub Aqua Club, and the directors of research institutes expressed their concern. The following year the government published "Submarine Pipelines Diving Operations Regulations" (SI 923) which, luckily, had very little effect on our work.

This sequence of regulations showed that the government, through the Department of Energy which was responsible for offshore hydrocarbons, was justifiably desperate to control the accident rate on North Sea rigs, and that some of the diving companies in those early days were moving their operations from one support base to another to escape the rules. On behalf of the Underwater Association, I corresponded with the various government agencies, from which I learned that to create an over-arching safety regime applying to all structures and places within UK jurisdiction, there would be a new set of diving regulations based on the Health and Safety at Work Act 1974. This would apply to all working people everywhere in the UK, and on the UK continental shelf, and on UK ships. The new rules would be written by the HSE.

The diving officers of the Underwater Association and NERC laboratories saw serious problems looming. The government was justifiably determined to stop the accidents in the North Sea which were getting repeated publicity, while the professional deep divers were unanimous that scuba at work was dangerous. Hints from the HSE showed that they were going to do everything possible to ban scuba or add compulsory expensive and cumbersome features that would make it useless. The first draft of the proposed Diving Operations at Work Regulations was circulated for comment in 1978. It was worse than we feared. Diving techniques were all lumped together, with the same safety procedures and back-up in all circumstances, using the heavy equipment of the offshore industry. If we accepted the new proposals our costs would be increased a thousand-fold and our efficiency reduced, with no increase in safety. Also, the draft assumed that diving was generally conducted in a remote and inaccessible location, such as on an offshore rig, with projects lasting weeks or months, so that a doctor was needed on site.

In contrast, scientific diving is often fairly shallow, less than 10-20 m (33-66 ft), with dive profiles combining depth and duration well within the safe limits requiring no decompression. Dive plans often require sporadic dives, or visits of one or two days, to many different coastal and nearshore sites where diving can be conducted from a small vessel. Stewart Willis and I had negotiated the purchase of a four-person two-compartment transportable decompression chamber to provide safety for the Rockall dives in the early 1970s (Chapter 2), and diving officers at any NERC laboratory could request the loan of the chamber for particularly remote or deep-diving operations, but the chamber could only be in one place at a time.

Far more people used scuba at work than the government had realized. Police, firemen, coast guards, the Army engineers, filmmakers, the BBC, and fish farmers all used scuba in situations that, if conducted sensibly, were perfectly safe. There were no recorded accidents to cast any doubts on this, though it soon emerged that there was a "cowboy" fringe of inshore divers who cut corners and did create fatalities. We quickly contacted the most responsible diving groups.

Our response to the 1978 consultative draft had to be professional, comprehensive, and negative. We allocated each group of clauses in the draft to a different diving officer to criticize and tear to shreds as logically and firmly as possible, without departing from demonstrable facts.

The diving officers critiqued every recommended aspect of buoyancy, tended lifelines, navigation, equipment, on-site chamber requirements, proximity to a chamber, underwater voice communications, medical examinations and fitness standards, etc., and showed them to include elements that were either factually wrong, impractical, or contributed to a reduction in safety for most working divers using scuba, including scientific divers.

The HSE civil servants had written the draft regulations entirely from the perspective of deep-diving bells or chambers and saturation diving on an offshore rig, or so-called surface-supply diving where the diver is connected to the surface by a strong hose that carries the breathing gas and an electrical cable for voice communications. We forwarded copies of our final critical document, which was over 100 pages long, to the ministers of the departments involved, as well as to the HSE itself, the Department of Energy, and various libraries. A long silence followed.

During this pause in events, I received a duplicated letter from the HSE, copied to about ten other people, saying that the HSE was very anxious to have the proposed draft regulations for diving authorized and adopted as a formal Statutory Instrument by Parliament as quickly as possible, because the enhanced training requirements would force trainees to go to the government-sponsored diving school at Fort William, which was in financial difficulties.

I read this several times to make sure that I had not misunderstood it. A government department was proposing to rush through as quickly as possible a set of draft regulations that, if wrong, could result in more deaths, in order to generate income for a diving school that had been set up prematurely without much thought and with no business plan.

I telephoned HSE to make sure that I had not misunderstood the letter. The embarrassed official at the other end said that, since I had been included on the mailing list by mistake, I should forget that I had received it.

Commonwealth countries began to copy the UK regulations. At the CMAS Conference in Brisbane in September 1977, I used the opportunity to make presentations about the way that scientists could combat the tendency in official circles to try and ban scuba for divers at work. This was repeated in Auckland, New Zealand, at a meeting on marine technology opened by the Prime Minister, "Piggy" Muldoon. Some of these trips were associated with the planning of the Cootamundra Shoal Project (Chapter 10). After my several years' service on the Council of the BSAC, the members elected me chairman at the Annual General Meeting in Plymouth in 1978. The club had a membership of about 25,000 divers at that time. The direct problems of scientific diving were not discussed in the BSAC meetings, but I told the national diving officer how much scientists valued the professional testing of new equipment and training methods that the club conducted.

Through various CMAS meetings, I had met several very experienced and effective diving scientists in the USA, although they were not particularly active in CMAS itself. Glen Egstrom, Mike Lang, Art Ulrich (co-editor of 1985 CMAS *Scientific Proceedings* held at the Scripps Institution of Oceanography in California), Patricia Masters, and others described to me the negotiations that were underway in the USA between the American Academy of Underwater Science (AAUS) and the equivalent to the HSE, the Occupational Safety and Health Administration, and the objections of the professional divers' unions. Between 1979 and 1982, complete exemption from OSHA Regulations was obtained for diving scientists in the USA provided that they were employed by and diving with an accredited academic institution with a code of practice for diving. The exemption was justified on their very low accident rate, lower than in industry, and the strong ethical and legal structure of academic institutions and the AAUS.

In 1980, I was invited to attend a diving workshop for sports divers and scientific divers in Nice, in the south of France. It was out of season but the more beautiful for that. In Nice, I met the doctors and diving officers who supported the vast population of local divers and tourist divers, as well as the staff of many research laboratories who were active along the south coast of France, some of whom inevitably experienced accidents. The doctors had discovered that when a diver shows signs of decompression sickness or pulmonary barotrauma, or simply is known to have surfaced too quickly, then the administration of pure oxygen to breathe on the surface greatly reduces the risk of serious symptoms later. I brought back the relevant publications for the BSAC medical committee and, after later data was obtained from the USA, the provision of surface oxygen in emergencies became standard first-aid practice in sports diving in the UK and many other countries. It was not approved by the HSE unless a doctor was present, as they said oxygen was a medicine. Since doctors were very seldom on site for scientific diving, this ruling made working scuba diving less safe than sports scuba diving.

* * *

While Lt Cdr Jackie Warner at the Department of Energy Diving Inspectorate was advising and copying documents to Commonwealth countries that might be modernizing their diving regulations, European agencies were naturally concerned with deaths in the other sectors of the North Sea. One result of this was the formation of an expert group known as the European Diving Technology Committee (EDTC), dominated by the offshore diving companies that worked on the oil rigs. In October 1980, the Commission of the European Communities Directorate for Health and Safety sponsored a meeting of the EDTC in Luxembourg at the Jean Monet building.

This two-day conference included papers by myself, working commercial divers, medics, Reg Vallintine, director of the BSAC, and many others. Vallintine's paper presented the safety record of sports diving, both in the UK and internationally, showing that the sports diving community had a range of technical committees and training organizations that could evolve with new technology or new medical information. Safety in fieldwork was upmost in our minds following the tragic death in the previous month of Keith Muckelroy while investigating a crannog in Loch Tay, although he was not using scuba. My talk showed that scientific diving using scuba at work in Britain had had no fatalities for the previous 20 years, and that internationally, known fatal accidents to scientific divers were very few: roughly between five and ten in more than 20 years, from a population of many thousands of scuba divers.

Then the HSE 1981 Diving Regulations were published with a focus on the diving equipment and support systems required offshore, but including the provision for exemptions for scientific diving that allowed us to use scuba, provided there was a high standard of training. Archaeologists negotiated a further exemption allowing them to work on underwater sites with a lower sports diving qualification, but that exemption was never used by the government laboratories and universities in the scientific community. The regulations were provisionally acceptable as a short-term compromise but unacceptable as a permanent definition of the status of scuba diving at work. The exclusion of scuba diving techniques from main clauses of the regulations implied that scuba was not a principal and equal technique that could be chosen by a working diver when the task and the conditions made it most suitable. It made scuba diving, and diving scientists, appear to be exploiting special exemptions to use less safe diving methods – exemptions that could be withdrawn at any time with minimal bureaucratic formality. Furthermore, this marginal status was a gift to those lobby groups trying to have scuba banned for anyone at work, and they used it with determination. There were clauses that permitted divers with years of experience to continue in employment and a new training standard and course for scuba at work that was designated HSE Part-IV.

The 1980s settled into a grinding war of attrition. An objective of the HSE Diving Inspectorate, and of the EDTC, seemed to be to define the requirements for dive conduct, training, and on-site equipment in such a way that scuba could not comply with the legal requirements without extreme added complexity and cost. For some unexplained reason, they did not take the simple step of banning it for use on offshore rigs and pipelines. Thus the regulations, and various directives that followed them, required cable-connected voice communication between the diver and the surface, the use of a surface lifeline or cable connection to a tended line, a fully kitted-up standby diver (with which we agreed strongly), the avoidance of the diver controlling his or her own buoyancy, and, in most circumstances, a relatively large diving team, with on-site decompression chamber and diving doctor.

There were many confrontational meetings that would have been farcical if lives had not been at stake. John Gamble, a marine biologist, and I had meetings with the Manpower Services Commission (MSC), which was responsible for defining the training syllabus for proposed two-week and four-week training courses for "at work" scuba divers. We explained that divers with several years of sports diving experience who wanted to transfer to employed status were the ideal recruits for scientists. Such people knew that they could endure diving in all conditions, even in winter. Courses should be designed to convert them to the requirements and discipline of legal conditions for a person at work, or to start them on the way to acquiring experience over a year or two. MSC took the opposite view. The idea of "amateurs" drifting or filtering into "employed status" without some distinct barrier or significant transition during which old habits were discarded, to be replaced with new training and new techniques, dismayed them. A rite of passage was needed. Every possible distinction and change in technology should be introduced to make the employed scuba diver different from the amateur. The HSE itself, and the industry at large, justifiably worried about amateurs doing work in semi-legal conditions. We agreed that people at work had to comply with all legal conditions that applied in terms of medicals, training, fitness, self-discipline, insurance, etc. But this did not require the invention of a perversely novel type of scuba diving or the rejection of sports diving experience.

"Ah," said the MSC spokesman, "all training is the same. Once you know how to structure a training course, it doesn't matter if it is piano-tuning or chicken-farming, the course structure and psychology are always the same. The professional working diver must use distinctive technology."

John Gamble and I gave up and left after a short meeting.

On another occasion – I cannot remember whether it was with only HSE representatives or included the MSC – Jackie Warner expounded forcefully his view that allowing the diver to control his or her own buoyancy was dangerous because they might surface rapidly under an obstacle. The representatives of the police divers and professional film cameramen looked embarrassed, but the representative of the Army Royal Engineers Diving School at Marchwood exploded with invective that is seldom heard in Civil Service meetings. At that date, the use of ABLJs and buoyancy "wings" attached to the diving cylinders had already been standard for more than 20 years and had saved many lives. We did not reach agreement.

A similar stand-off applied to the diver controlling his or her duration on the seabed by looking at a watch and a cylinder-contents pressure gauge. Jackie maintained that this was a dangerous practice, and that such decisions should only be taken by the surface dive supervisor or line tender sending rope signals down to the diver. You could not trust the person underwater. To provide the diver with an emergency back-up signal when it was time to ascend, Jackie recommended using cylinders with J-valves that introduced a resistance to breathing and cut off the gas supply when there was still a pressure of 50 atmospheres available. The diver then pulled a lever that opened the valve completely, providing more air for a controlled ascent. In sports diving this crude technique had been abandoned in the early 1960s, more than 20 years earlier, when gauges measuring the remaining pressure in the cylinders became standard fittings attached to breathing regulators.

In all cases, the UA and SDSC noted the views of the HSE and MSC, but we refused to adopt training standards, diving practice, or equipment that we considered dangerous. Some researchers needed to conduct dives that were so unusual that further special safety arrangements were needed, and hence a special exemption from HSE. It was very difficult for HSE to judge these requests, since they usually came from the world expert in a particular type of research, making it impossible to find a different expert to be referee. When we complained about the delays that followed, we were told that the delays did not matter, since while divers were not diving they were obviously safe. We could not agree with that as a policy.

Diving skills and techniques continued to evolve. Accidents in the Royal Navy submarine escape tower at Gosport had shown that training for submariners in free ascents, of the kind that I had enjoyed so much in Corfu (Chapter 4), had produced two fatalities, while free ascent training in the US had caused several deaths. A consensus of experts advised that we should not train scuba divers in free ascent technique. That raised the anomaly that, if an emergency escape routine had not been practised in advance, was it reasonable to recommend it in an emergency? The answer is obviously no. Free ascent requires a lot of confidence and self-control, which is possible only if you have practised it first. That is not the case in an emergency with no prior training.

Established divers who had been doing marine research for many years were allowed to continue after the enactment of the 1981 Regulations, provided they obtained a certificate confirming their previous qualifications and experience. This was the so-called "Grandfather Clause". As the years passed, more young scientists wished to join the working research community, and they had to convert their sports diving training with a short compulsory course, or undergo the full Part-IV commercial scuba course. These courses generated complaints from scientists who, in some cases, were shocked by the primitive treatment of physiological theory, sexual discrimination against women, the poor and unreliable mathematics of instructors, and the occasional excesses of training which was treated like a military survival exercise. Some scientists resigned from the courses in protest or gave up, with good reason.

Although the introduction of strict safety regulations at national level, and the work of the EDTC, were supposed to make standards transferrable and recognized between countries, at least in Europe, they had the opposite effect. Different countries adopted widely different standards of training, ranging from the view that the only body qualified to train divers was the national navy, through to complete acceptance of advanced sports diving training for people at work. Some countries required the use of nationally recognized decompression tables that were not used in other countries. The British government took the view that the Employment Medical Advisory Service (EMAS) could not trust diving medicals conducted by doctors in other countries, which meant that a visiting scientist to the UK had to undergo an expensive medical examination in Britain conducted by a recognized British diving doctor. Since it was logical that this medical check had to be made several weeks or months before the planned project, this effectively prevented international collaboration.

The various research committees of the European Commission Directorate General XII, "Research", were in despair, as previously easy-going international collaboration between research diving teams simply came to a stop. We held several meetings in Brussels to try to solve these problems, but the idea of mutual recognition of national standards was difficult to sell politically, and the scientists did not have the clout, or the time, to do it. The old days of groups of scientists assembling informally before diving together in common research projects (Chapter 4) had gone forever. In 1980 the CMAS Scientific Committee started issuing certificates of competence to highly qualified divers working in research institutes and universities, but this was a stop-gap that did not carry governmental authority. I am happy to say that these problems were eventually solved by a new European Scientific Diving Committee in the late 1990s.

Figure 8.6. Diver returning to an ice hole under the Antarctic ice (Photo Martin White, courtesy British Antarctic Survey).

Through my connections with CMAS and the Intergovernmental Oceanographic Commission (Chapter 9), I was encouraged by Dr Selim Morcos in 1983 to produce an international Code of Practice for Scientific Diving that could be supported by UNESCO. Starting from the NERC code, and the works of the AAUS and scientific diving organizations in several other countries, Michael Max and I generated a first draft that stacked up to over 300 pages. This was done by correspondence with contributors all over the world, using ordinary-postage surface mail. The draft included topics such as diving in the Dead Sea, diving under ice in the Antarctic (Fig. 8.6), diving on towed fishing nets, working with radioactive chemicals, diving in mountain lakes at very high altitudes, diving from a research ship in the open ocean, diving in flooded chambers in deep underground mines, diving with sea mammals, and so on. It also provided a condensed guide to the laws applying to working scientific divers in different countries. The agreement on the final text was reached at a meeting with diving officers and chief divers from major institutions around the world. Gregg Stanton at Florida State University Tallahassee hosted the meeting at Wakulla Springs in November 1986, and UNESCO covered the travel costs.

The editorial review at Wakulla Springs included 13 diving science experts from different agencies in the USA, Canada, the UK, Ireland, and Norway, although the initial contributions came from a much wider range of countries. Of particular value were the contributions from Philip Sharkey (University of Rhode Island), Hans Roverud (University of Bergen, Norway), and David Dinsmore (NOAA National Undersea Research Program). Gregg Stanton and I duplicated the papers before the other diving officers arrived, and piled them around the meeting table with a thick stack for each person. I was worried that these august, no-nonsense characters would be aghast at the pile of secretarial waffle, quickly losing interest or showing their impatience. Their response was the reverse of what I feared. As we ploughed through page after page, people kept exclaiming at the good ideas shown by the contributors, with the manifest evidence that so many recommendations were based on long experience with practical guidance on the best methods. The editors added many refinements during our two-day meeting, after which, in very good spirits, we adjourned to dive in Wakulla Springs, in the grounds of the hotel.

This huge karstic sinkhole, which is 90 m across, is renowned for the clarity of its water, combined with the danger of numerous tunnels and caves where prehistoric animal bones tempt the unwary diver to explore too far. As I plunged in and gazed across the pool, I was astonished to see the far wall of the sinkhole as clearly as if there was no water in between. Looking downwards, I felt for the first time in my life that I needed a parachute to prevent myself falling to the bottom. Seawater was never this clear. You could see every twig, leaf, or snail-shell 30 m (100 ft) below. I descended slowly, enjoying the incredible view, but had no difficulty in avoiding the caves and side tunnels. The death rate for cave divers in Florida is unbelievable. It was a fabulous dive for pure pleasure, with no need to take pointless risks.

UNESCO published the Code of Practice for Scientific Diving in 1988 and quickly translated it into Chinese. The English language edition was widely used in Europe, the USA, and elsewhere.

A remarkable aspect of this decade of continuous conflict was that the heads of the UK government agencies and universities employing divers gave us full support. In

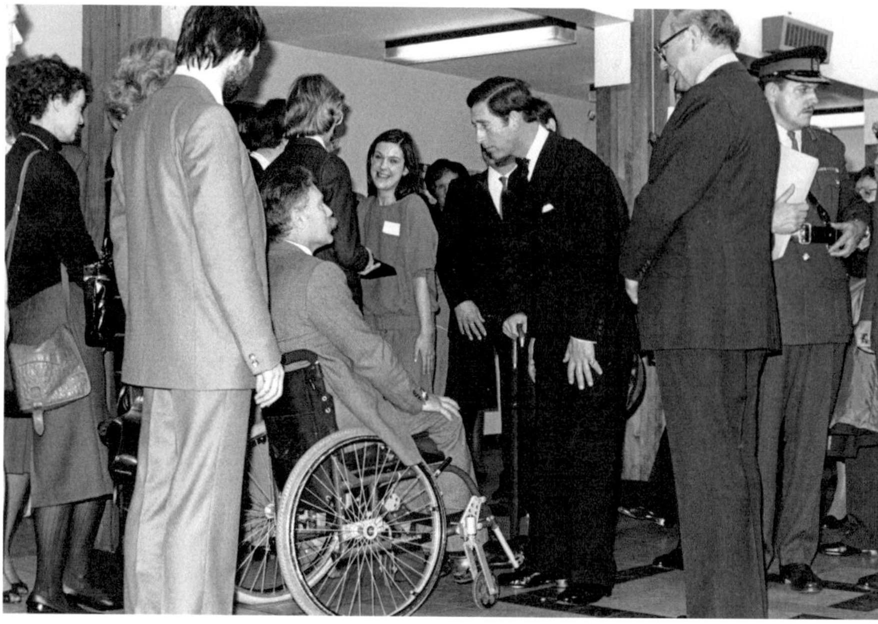

Figure 8.7. At the start of an Operation Raleigh cruise HRH Prince Charles attended to wish the participants good sailing. Colonel Blashford-Snell is on the right in military uniform.

the current era of risk-aversion, which was already prevalent in the 1980s and '90s, this required a great deal of confidence and courage for all parties. The individual diving officers, the heads of agencies, and their safety officers, the Committee of Vice Chancellors and Principals (of universities) (CVCP) all considered that we had the right attitude and the moral and ethical justification to refuse to obey government regulations which we thought were dangerous. I reported from time to time to the successive chairmen of NERC, including Sir Hermann Bondi, and found that they were unwavering in their support. At a more modest level, I found myself acting as adviser on the use of scuba in diving research conducted as part of Operation Raleigh, founded in 1984 by Colonel John Blashford-Snell. Operation Raleigh organized adventure-research projects in dozens of countries, providing young people with valuable experience (Fig. 8.7).

The moral dilemma that we faced during these years, between complying with the law or maintaining maximum safety, was so extraordinary, and so sustained, that I wish to place on record the names of at least some of the diving officers in the UK who maintained our campaign, and managed diving in their institutions year after year with no accidents. The team of institutional diving officers during the confrontations described above included Jim Atkinson and Phil Lonsdale from the Millport Marine Station, Cumbrae, Martin Dean (Archaeological Support Unit), Richard Hartnol (Port Erin Marine Laboratory), Chris Frid (University of Newcastle), Valerie Flook (Unimed Scientific), Alan Gale and Martin Sayer from Dunstaffnage, Mark Holt (Sea Mammal Research Unit), Paul Somerfield (Plymouth, after 1993), Bobby Forbes (Edinburgh and Orkney), Doug Miles (Institute of Hydrology), John Gamble (NERC Coastal Marine Research Institute, Plymouth), and Martin White (British Antarctic Survey), the diving officer from the Freshwater Biology Association whose name I have not recorded,

Keith Hiscock, Orielton Field Centre, and the safety officer from the Vice Chancellors and Principals Committee. I apologize for the above list being so male dominated, but there were a fair number of women divers in the various institutions represented, with occasionally a woman serving as diving officer. Unfortunately, I do not have all the names for every year in my files. When HSE made an inappropriate set of recommendations about women diving during pregnancy, I asked the women divers to form their own working group to write their own set of rules, which they did.

* * *

In late 1993 my worst fears came true when the story of the horrific deaths of young divers gradually leaked out into the scientific community. The situation was made worse because the deaths had occurred in 1992, with the details having been concealed, and I had to extract information over many months against the apparent unwillingness on the part of the authorities and HSE. I learned first that four divers had been killed recently in HSE-approved schools, and then that two of these had been scuba divers. I did not realize at first that these two deceased divers had been students. To begin with, I did not react strongly to the rumours, but this changed as I slowly discovered the facts of the case.

What follows is supported by the records that I kept at the time, dozens of letters exchanged between research institute diving officers, my daybooks of meetings, minutes of meetings, and printed contemporary documents. While the events are seared into my memory, that alone would not be sufficient to justify putting this story into print. It is an unfortunate story.

The SDSC was chaired at this time by Dr Jim Atkinson of the Millport Marine Research Laboratory, the University Biological Station, Isle of Cumbrae, Scotland. I was vice chairman, still an active member of SDSC, attending every meeting, and serving as the regular negotiator with the HSE, often accompanied by other diving officers and chief divers. The SDSC provided a range of support for institutes and universities, reviewed and edited the Code of Practice, arranged loans of equipment, and provided advice on projects and training. There was plenty to do in addition to the long-running conflict with HSE. All the major government civil laboratories employing diving scientists (NERC, MAFF, DAFS, SMBA, FBA, HI, SMRU, etc., etc.) and most universities and their safety officers either attended meetings or were on our mailing list for newsletters, reports, and minutes of meetings. There was one notable exception, Plymouth Polytechnic, which became Plymouth University during 1992. They received occasional papers but did not respond.

The Plymouth student divers, mostly engineers and biologists, conducted their training at the commercial diving school located in the Napoleonic-era Fort Bovisand, with its convenient small harbour attached. The course was the HSE-approved Part-IV scuba diver training course intended for people at work. Since Fort Bovisand was also the home of many large industrial diving chambers and training systems, it is possible that the Polytechnic/University would have found it embarrassing to be associated with SDSC, an organization that was in such obvious conflict with the HSE. Be that as it may, Plymouth was the only academic or research establishment in the UK, so far as I am aware, that rigorously adhered to the training standards and equipment mandated by the HSE Diving Inspectorate. The rest of us rejected many HSE requirements and used the NERC Code of Practice instead.

At the SDSC meeting in Edinburgh on 22nd October, 1993, the diving officers discussed the cumulative damage caused by frequent increases in restrictions by added regulations, the failure to accept surface oxygen therapy and other improvements, threats to withdraw the Scientific Diver Exemptions before the proposed new regulations came into force in a few years' time, the mandatory requirement for through-water ultrasonic communication, etc. Jim Atkinson, the chairman, proposed that it was time to confront the HSE with some sort of legal complaint, stressing that they were actually reducing the safety for working scuba divers, who were now required by law to adopt practices that were less safe than those used by sports divers. Jim's opinion was supported by the safety officer representing the CVCP.

After the meeting, when I returned to the Institute of Oceanographic Sciences (IOS) in Surrey, I considered how to mount such a complaint. It occurred to me that the HSE Diving Inspectorate was failing in its duties under the Health and Safety at Work Act 1974. Would it be possible to confront them and request some kind of judicial review on these grounds? I consulted the 1974 Act, of which the opening phrase is: "An Act to make further provisions for securing the health, safety and welfare of persons at work". That was clear enough. HSE was acting in a way that had the opposite effect, at least for a substantial number of divers at work.

During the next few days, I finally learned through the BSAC that the two divers killed diving on scuba in 1992 had been Plymouth students under training at the commercial school at Fort Bovisand. I was angry and horrifically shocked by this news. I tried to obtain information from HSE but was told that the enquiries into the accidents in 1992 had been commercially confidential. I was shocked again, since the BSAC publishes all accident and incident reports in detail, with professional assessment of the causes, and recommendations to improve future training and conduct of dives. So much for the view that commercial diving is professional and amateurs are careless "Mickey Mouse" divers. I wrote at once to the SDSC and complained about the secrecy practices of the HSE. Other diving officers supported my views.

I telephoned the Plymouth University diving officer, who was quite open in describing what had happened, as far as he knew, which was not very much. I also obtained other verbal descriptions from various involved parties during the following weeks and months. Putting these together, I learned that two divers on a course had had fatal accidents on different occasions, two months apart, though with some similarities. The divers were on rope lifelines. One was found on the seabed with his mouthpiece out and drowned. The other surfaced much too quickly and experienced fatal pulmonary barotrauma. I do not know the names of the deceased divers, and the coroner's report is confidential to the families. I apologize for the distress this analysis may cause to anybody related to the deceased.

In both cases, in accordance with HSE Part-IV standards (and contrary to the NERC Code of Practice), the divers were using cylinders fitted with J-valves that cut off the air supply while there is still a substantial volume of air in the cylinder. The diver is then supposed to pull a reserve lever on the cylinder behind his/her back, which releases the remainder of the breathing gas to make it possible to breathe again. After that, the diver should ascend slowly to the surface. During the following year, I pressed HSE to conduct a proper enquiry into what had happened and, after a long delay, there was a partial enquiry into the human factors of all components of the HSE Scuba Part-IV training, to which I gave some evidence.

My suspicion regarding these accidents is that, by the 1990s, the highly efficient improved diving regulators in use provided easy-flowing inhalation with very little resistance to breathing right up to the last drop of air before the cut-off came into effect. Back in the 1950s and '60s, when breathing regulators had only one or two moving parts, and could be mended with a large screwdriver, the breathing resistance on inhalation increased steadily as the cylinder pressure dropped towards that of the cut-off valve. Thus the diver had several minutes of warning, and many successive breaths, before reaching the point when inhalation became impossible. With modern 1990 regulators, I suspect, this warning time was much less, possibly only one or two breaths and that may have triggered panic in the student divers. Since no sports diving group had used the obsolescent J-valves for several decades by the 1990s, and no university or research group associated with the SDSC had used them for a similar period of time, one could only surmise that the accidents had probably been caused, or at least exacerbated, by the outdated recommendations of the HSE. In support of this assessment, I note that the divers on the *Mary Rose* wreck in the early 1980s found that cylinders donated to the project by the French Spirotechnique company, fitted with J-valves, cut off the breathing gas so abruptly that it was difficult to reach the reserve lever before being completely without air. They fixed the J-valves permanently open, and relied, as usual, on their contents pressure gauges.

Legally, nobody directly involved in the Plymouth fatalities had done anything wrong. The college had sent fit and competent students to an approved HSE diving school. The school had conducted training as required by HSE; the accidents were investigated by HSE; the school was not blamed, nor was the university. It was exactly as the SDSC had anticipated and feared for several years: a situation where all boxes had been ticked, but unfortunately, there were two dead bodies. I prefer fewer tick boxes and no dead bodies.

In early November 1993, I decided that a formal confrontation with the HSE was essential. After informing Richard Pope at NERC, I wrote to a Queen's Counsel whom I knew and put the situation to him. I did not mention the fatalities at Fort Bovisand, about which I still did not have many details, but outlined to the QC the distressing reduction of safety caused by the statutory regulations. Was it possible to prosecute the HSE, or request judicial review, on the grounds that they were negligent under the 1974 Health and Safety at Work Act? It seemed risky to sue them under the legislation that they themselves were supposed to enforce. My friend replied that there were legal firms that were expert in this kind of judicial review, giving me several contact addresses. He assured me that we seemed to have a good case.

A week later I received a telephone call from an official in the HSE with whom we had frequently had frustrating confrontations, during which, over the span of two hours, he explained, protested, excused, and conceded all sorts of points of dispute and conflicts. He promised that the Scientific Diver Exemptions would not be withdrawn prematurely, and that many other issues of repeated conflict could, he assured me, be resolved amicably. I was puzzled by this change of tune, copied a record of the discussion to the SDSC diving officers, and sent a copy to the HSE for confirmation. The

SDSC members insisted that the concessions were not sufficient, were purely informal, and that we had to keep up the pressure.

A few days later I received an extraordinary telephone call from Richard Pope, the safety officer at NERC.

"Nic, may I talk to you in absolute confidence?"

"Yes of course. What on earth is it about?

"I have received a very strange call from HSE. From the highest level in the Department of Work and Pensions. They oversee the HSE. They want a secret meeting."

"What on earth? Secret? Why?"

"They want a meeting that is completely off the record. No minutes, no agenda, no announcements, no report. Nothing. Nobody is to have any paper or notebook with them. No recorders. No secretarial support."

"Well, I suppose you can arrange that. I'll attend if I'm invited. But what's caused this request?

"I think it is their response to your consultation with "m' learned friend."

"You mean counsel? QC? The barrister?"

"Yes. I think that is what has happened."

"But Richard, he did not do anything at all. He just replied and gave me some contact addresses."

"Maybe. But I am pretty sure that the news has got through to the HSE somehow, and it has really put the frighteners on them."

* * *

Such was the precursor to the meeting which I summarized at the beginning of this chapter. The officials in charge of regulations for diving safety in the UK conceded totally that our methods of maintaining diving safety amongst diving scientists had been proved successful, and that their approach of trying to force all diving and divers into the same technical "box" had failed. They proposed a system of regulations that would accept the principle of ACOPS. The statutory regulations, which are government laws and are usually only altered at intervals of decades, would be very simple and all-embracing. The new regulations would require each professional community of divers to work with HSE experts and lawyers to agree to a code of practice embodying the experience and judgement of the senior divers within that community. The ACOPS could be reviewed annually, and updated or altered in response to new diving equipment design, new medical discoveries, or new experience in diving operations.

During 1994 and '95, SDSC diving officers held several very friendly meetings with HSE officials and their legal experts, during which the ACOPS evolved rapidly to a format acceptable to all parties, adapting ideas from the existing NERC and UNESCO Codes of Practice. The new Regulations became law in 1997.

With the confidence that the management of diving safety was settling into a reasonable pattern at home in the UK, Phil Lonsdale and I confronted yet another meeting of EDTC in Luxembourg in April 1994 with weary resignation. It seemed that the effort of trying to force all diving into one set of prescriptive rules would never go away. Bureaucrats, the trade unions, the professional diving schools, and the heavy diving offshore community all wanted it. Amongst many other side-effects, such regulations would eliminate all the small players, and force up prices and profits.

Phil Lonsdale and I were the only scientists there from any country, so far as I know, all other delegates being either from the offshore industry, trade unions, doctors expert in diving medicine, or government officials. There were four parallel sessions, dedicated respectively to medicine, training, operations, and technology. Phil agreed to cover the training session while I attended operations, since these were the two sessions where we would have to work hardest to protect science.

The 150 delegates met first in a daunting, cavernous conference hall where the glass windows of the overhanging interpreters' booths seemed to be in danger of toppling onto the audience. We all wore headphones since there was simultaneous interpretation for every European language. The meeting did not start well. A minister from Luxembourg delivered a short but rousing speech, written for him by his advisers, which included the exhortation to eliminate dangerous diving techniques such as scuba diving. The audience laughed their appreciation. Phil and I hunkered down for the battle.

When the four parallel sessions started we split up into different rooms, each seating about 30 to 50 people. The operations session was chaired by a senior diving officer for the COMEX company, one of the most successful and respected oil-field diving companies in the world. The room was crowded. In the gossipy milling about over coffee and biscuits before the session was called to order, I distributed copies of the 1979 third edition of the NERC Code of Practice for Scientific Diving to every delegate. I also had a few copies of the UNESCO International Code of Practice that I had edited with Michael Max. The meeting started with a *tour-de-table* for people to introduce themselves. I was relieved to note that there were several representatives of companies working in shallow water, docks, canals, and lakes.

The chairman divided the work so that we started first talking about necessary policy for on-site safety management during diving operations, including auditing safety performance and revising performance and procedures. Speakers from different countries immediately revealed that they already had a wide range of different cultures and philosophical views, and although the discussion was interesting, it was not conclusive. The Norwegian spokesman admitted that the diving regulations in his country amounted to 500 pages. Some delegates pleaded for a flexible approach, with simple safety plans for shallow diving operations of short duration, with the Dutch suggesting that 65% of working diving is at depths less than 10 m (33 ft). The maximum depth that divers are permitted to dive breathing air varied from 40 m to 60 m (131 ft to 197 ft) in different countries, but the meeting converged on 50 m (164 ft) as being the safe limit. We agreed that risk analysis is a necessary step.

The minimum size of diving teams caused much controversy, with a range from three to six people, depending upon the type of equipment being used. The team of engineers and experts needed to run a saturation diving project in deep water is obviously much bigger, and the discussion was impressively technical. Much discussion related only to offshore heavy diving on rigs, but several of the participants brought out the wide range of gear being used by small commercial companies, alluding to the activities of many hundreds, perhaps thousands, of divers working in coastal waters and inland. They stressed the use of scuba by the police, fire service, and other rescue services. I found it easy in this context to describe, without being at all provocative, the essential work done by diving scientists across Europe, and the fact that they used scuba. I drew attention to the Code of Practice, and several participants referred to it with approval.

By the end of the first day, I thought that we had made quite good progress. Having compared notes, Phil and I felt that we were making our mark well.

Next morning we filtered into the meeting rooms to grab one more coffee before taking our seats. I accosted the Dutch delegates, who were adamant that there were hundreds if not thousands of small diving companies scattered across Europe working in shallow water. We had a lot in common and decided tacitly to make sure scuba was given a good reception. At this stage the meeting was discussing mostly on-site preparation for the dive, diving team composition, supervision from the surface, preferred decompression tables, practising emergency drills, and environmental factors limiting dive safety. The experts agreed that the real incidence of sub-clinical symptoms from decompression sickness was much higher than reported, due to people being afraid that they would be barred from subsequent dives, and hence lose money. We recommended that the actual dive profile of each dive should be recorded electronically. Within a few years, dive computers worn by scuba divers performed this automatically.

Again, comparing notes with Phil that evening, we seemed to be doing fine.

I was quite unprepared for what happened on the third day. The chairman opened the discussion in the operations meeting.

"Welcome everybody. This morning is our most important session regarding operations at sea. There is only one question we need to resolve and that is the banning of scuba for divers at work. Any comments?"

I tried to look as if I was not shocked and swivelled my eyes surreptitiously to see how other people were reacting. The Dutch shallow water divers who had been so helpful on the previous days looked inscrutable.

"No comments?" said the chairman, "Let us proceed."

"Mr Chairman," I intervened hurriedly. "Yesterday we agreed that several thousand employees of small diving contractors throughout Europe use scuba safely in shallow water, often in inland conditions. Throughout the past two days we have established the good safety record of working divers using scuba other than on the rigs. There are very few fatalities. Also, over one thousand European scientists use scuba in conditions controlled by major laboratories and universities with a very low accident rate. Surely we can discuss scuba as one of the numerous types of safe diving that working organizations can choose to use, given proper training and management?"

The chairman bristled and glowered.

"I am a diving officer of COMEX, one of the most successful professional diving companies in the world. In the past five years, there have been four fatalities for divers using scuba on the rigs where we dive. That is not acceptable, and scuba must be banned for working divers."

"Mr Chairman," I responded, "I am directly responsible for 300 diving scientists in UK laboratories and universities, and indirectly responsible for the standards used by over one thousand divers in Europe and elsewhere through the CMAS Scientific Diver system. The UNESCO Code which I edited is used in dozens of countries outside Europe. In the UK we use almost entirely scuba in our research, and in the past 30 years we have not had one fatality. Not one. Who is professional? You or me?"

The chairman was expressionless this time.

"We will discuss the banning of scuba diving at work."

"Mr Chairman, I am reporting you to the president of this conference as an unfit person who is not competent to chair this session. I am leaving the meeting now and will go straight to the president's office. I will see you in the coffee break."

I bundled my papers into my briefcase and left the meeting room. Nobody else moved. It was the only time I have ever stormed out of a meeting in protest. I have no idea what happened in the conference room after I left.

The president's office was a large panelled and glazed room opening off the main foyer of the conference centre. I knocked on the door and went in. He was probably a junior minister in the Luxembourg government. His desk was suitably vast, with no documents visible upon it. After I had explained my concerns as simply and calmly as possible, the president gave his judgement.

"I understand your worries. In the coffee break please come here with the session chairman, and I will do my best to resolve your concerns."

Back in the foyer, seated by a huge wall of glass overlooking distant gardens in spring bloom, I sipped a cup of coffee, and read the newspapers until the sessions ended. People came streaming out of the meeting rooms noisy, chatting, laughing, busy. The chairman of our session saw me, and came towards me with a broad smile, both arms outstretched as if to hug me.

"Nic, Nic, let's forget it. There is nothing to worry about. Nobody is going to ban scuba at work. Come back into the meeting after coffee and we will just get on with business."

He held out his hand to shake. Not quite sure, I looked at him very firmly but did shake his hand. The issue was passed. I did not ask what had happened in my absence, but the rest of the meeting passed with no dramas or conflicts. EDTC never tried to bundle all types of diving into one prescriptive set of regulations after that, nor did they try to ban scuba at work.

* * *

Later in 1994, the HSE published the Offshore Technology Report (OTH 94 447) explaining the conclusions reached in the expert enquiry about the inadequacies of diver training during HSE-approved Part-IV scuba training courses. At the time, I skimmed through the report, noted some conclusions, copied a few pages for my records, and then forgot it. The report was too late to save the lives of the young students who had been killed in 1992, and the process of writing the ACOPS was well underway after the HSE had had a radical change of heart the previous winter. The report seemed rather wordy, anodyne, and past history. Too late.

I recently downloaded the whole report from the web: it is devastating. It is 56 pages long, written in a calm, deliberately bland style, combining a soup of business jargon, psychological circumlocutions, and Civil Service euphemisms. But certain phrases and judgements leap off the page with an abruptness that almost mimics the vulgar expletives of the Army diving instructor during our meeting years earlier with Jackie Warner and HSE when discussing diver buoyancy control (see above). The OTH report, wise and thoughtful, focuses on the human reaction to a range of stresses and emergencies, and the extent to which the diver during training felt that he or she was in control of the situation. Quotations follow.

On page 25:

> Need for training in self-rescue. [This is not a phrase I have ever used, but its meaning is obvious.] There is no direct reference to any form of self-rescue or survival procedure in the HSE Part-IV Standards for Assessing Diver Competence document.

To put it bluntly, the course required the diver to be excessively dependent, and to think that they were dependent, on the surface control. They were deemed to be, in my phrase, "a puppet on a string".

Again (page 26) air-sharing and the advisable use of octopus rigs are also referred to:

> Training in self-rescue could include the teaching of air-sharing (which is not referred to within the HSE Standards for Assessing Diver Competence Part-IV) and training in free ascent. ... training in air-sharing will provide the trainee with a valuable self-rescue skill.

Regarding J-valve cut-off devices (page 29):

> There are a number of obvious reasons why a diver may experience inhalation resistance to his air supply, (and so) this is not an unambiguous or benign form of signal.

Exactly.

> From a behavioural point of view, interfering with the main air supply is an extreme and primitive warning device .... The air supply is not fully returned or restarted until the lever is pulled, in this sense the reserve signal "fails dangerous" rather than "fails safe". For such a system to "fail safe" the air supply would have to be designed to automatically restart.

Then, in this immortal phrase (page 31):

> The message (sent by the J-valve cut-off) has an implicit and very clear meaning for the human. The message, which arrives with appropriate emotional labels, is that the diver will shortly suffocate unless he gets more air, a larger increase in message priority than is operationally required.

Quite. Though I would not have put it like that myself.

Discussing the HSE prejudice against the use of cylinder air-contents pressure gauges (page 30):

> ... divers should be trained to continually maintain an awareness of their cylinder pressure status. This can be achieved by the use of a pressure gauge attached to all scuba sets. Arguments concerning the availability of this system to operate in nil

visibility conditions are fallacious "red herrings" – it is irrational to reject a device that is effective 99% of the time, and to replace it with no device at all – after all, the diver will not be misled by the gauge, he will know when he cannot see it.

There is much more in this vein. The document, taken as a whole, is a vindication of our criticism of the HSE Part-IV as it was defined in the early 1990s, while confirming the safety of scuba diving given proper training with modern equipment. But all too late. The factors that failed in the HSE Part-IV course definition were exactly the issues that the UA and SDSC had criticized time after time in confrontations with the HSE. We had repeatedly refused to adopt training methods and practices that we knew were dangerous.

* * *

The existence of numerous relatively small professional groups using scuba, all in slightly different ways, was inconvenient for those bureaucrats and safety desk-drivers who wished to have a quick tidy solution to the industrial death toll in the North Sea by simple legislation. Several of the fatalities on the rigs in the British sector were caused by professional divers using scuba in various dangerous situations when it was not appropriate, and, as I pointed out earlier, the obvious response would have been for the companies themselves, or the HSE, to ban scuba on the rigs. The diving companies and rig operators were free to do so. For reasons that I do not understand, nobody thought to do that, and the prejudice against scuba that was so prevalent in the heavy diving industry was transferred to officialdom as "scuba diving is inherently dangerous".

In multi-agency meetings and discussions with the HSE, Department of Energy, MSC, and the EMAS, the scientific divers found that our natural allies were the police, fire service, Army Engineer divers, coast guards, and BBC and media filmmakers. These organizations were not "diving contractors", hiring out their services, but professional bodies that used diving from time to time, with their own staff diving often in rather idiosyncratic ways to solve problems, using their in-house knowledge and skills. Furthermore, we were not bidding against each other for contracts. In all cases, there would be a very strong internal system of mutual respect, professionalism, and pride in getting the job done. For the HSE writing regulations about "Diving Contractors", these groups were just a nuisance, though how the HSE thought that they could defy the police, I cannot understand. There were, of course, some groups of divers using scuba such as scallop fishermen and fish farmers who had a higher accident rate in those days, which again tends to illustrate that corporate culture is a major factor, not just the type of gear being used.

During my time in the military, I learned much about the management of risk and the prevention of fatalities, while preserving initiative, whether in the SBS with oxygen diving or the RAFVR flying light aircraft. I learned how a corporate structure can first train, and then delegate and trust an individual to take responsibility for his/her own life and possibly the lives of others. The person who delegates trust also must have good reason or evidence to show that the trust is justified. When a crisis strikes, however, the diver or pilot is the only person who can take instant action and save the situation within seconds. (This is what OTH 94 447 calls "self-rescue skills".) In sports diving, the delegation of responsibility and trust is also used, but in most cases the responsibility is entrusted to a buddy pair with a diver leader, rather than to a single person. In the case of the SBS engaged in clandestine operations, there would be no back-up, or it would be very remote from the battlefront, and that is not acceptable in civilian circumstances. However, the Special Forces military approach of training and trust is quite easily adapted to include surface cover boats, standby divers, surface medical support, radio or mobile phone contact to a basecamp, a recompression chamber in remote locations, and so on.

The key point in almost all aspects of scuba diving is that the diver on the seabed is provided with instruments, data, depth gauge, cylinder-contents gauge, decompression computer, a means of buoyancy control, etc., and is expected to make intelligent decisions during the dive. The dive should be planned in advance, and the diver should stick to the plan so far as possible, but that still leaves a lot of room for observation, thinking, and responsive decision-making. It is this freedom to make decisions that seems to have caused so much condemnation from the bell divers and legislators. They seemed to reject the idea of the diver thinking under water, which, after all, is the main purpose of scientific diving.

It is commonplace in recent press discussion of "'elf 'n' safety gone mad" that modern society, due to the popularity of no-win-no-fee legal action, has become much more risk-averse than it was a few decades ago. This is certainly true, but the rapid improvement in diver instrumentation, including the reliability and performance of scuba diving gear, has ensured, by and large, that the actual risk of accidents has not increased in the past 50 years: in fact, it has probably declined. The public who wish to engage in adventure sports accepts the remaining risk. But where employment is concerned, the attitude requires stricter accountability.

The jargon phrase "tick-box culture" did not exist in the 1960s or '70s. From the point of view of a professional person taking a moral and ethical position about protecting the life and safety of employees, colleagues, or subordinates, the purpose of a safety culture is to avoid accidents. That is the beginning, middle, and end of the argument. If there are no accidents or deaths over a decadal period or more, with hundreds or thousands of people at risk, you must be doing something right. In contrast, the tick-box culture is designed – usually, by people who are not in a position to make professionally-informed decisions – to ensure that if and when an accident or death occurs, it is possible to show that nobody is to blame. That is a different objective.

In the run-up to the 1981 Diving Operations at Work Regulations, and particularly during the decade 1980-1990, the proliferation of advisory notices, forms, recording books, log records, compressor maintenance forms, etc., which were all compulsory, was designed to show that everybody was doing their job, as defined in law. Sometimes it was difficult to carry enough bits of paper in waterproof files when working in an inflatable boat in a breeze and a choppy sea. Because it is difficult to define all possible dangerous situations in advance, the law and regulations must be fairly predictable, and therefore the steps that have to be checked and ticked have to be constrained into a standard pattern, which does not always fit the real world. This invites contempt for the system, with people ticking the boxes because they are there, not because they have thought about real risk. An HSE supervisor or inspector can then visit and, without having any real experience of diving, check all the logbooks, saying that the system is perfect! I have seen it happen. Are the best techniques of diving really identical from

Shetland to the English Channel? I doubt it. How much more would the inspector have learned by sitting down with the institution's diving officer for an hour or two over a good supply of coffee and really finding out what the local problems are, and what steps they take to counter currents, tides, fog, overfalls, rip currents, local ferry boats, or the leopard seals under the ice?

I actually heard someone at one of these meetings saying that a "lifeline enables you to recover the body if the diver was killed". Is that its main purpose?

During the whole 20-year conflict, the openness of the scientific community to the exchange of information, training methods, codes of practice, and investigation of accidents at CMAS scientific symposia meant that we always had a common purpose. We exchanged newsletters and conference reports with several other countries. I and other diving officers could go to meetings anywhere in the world and state, reasonably accurately, the numbers of divers at risk, their requirements, the laws in different countries, and the accident rates. This was greatly helped by the UNESCO Code of Practice. The BSAC annual diving officers' conference report on accidents is exemplary. In contrast, the world of professional commercial diving contractors is a murky zone of dark secrets. No accident can be examined publicly in detail, and even when the students were killed during training at an HSE-approved training school in 1992, I was unable to obtain any official explanation. The enquiry, such as it was, was deemed to be commercially confidential. I despise this attitude.

In 1994 I started to try and understand why scuba was so safe in spite of the allegations that it should not be. Why does it look dangerous to divers used to being supplied with breathing gas from the surface, and to bureaucrats who try to understand diving without doing it? The obvious factors that cause alarm or scepticism are the limited duration of the gas supply, the lack of permanent connection to the surface, the lack of voice communication between divers and with the surface, and the divers' sheer mobility. Since accidents are so rare, some additional factor or factors must be creating a supportive and stable safety network, even if its existence is not obvious.

I recently discovered in one of my early 1994 notebooks a tabulation of the number of ways that a scuba diver, with a buddy and a surface boat with standby diver, could respond to numerous everyday situations and exceptional emergencies. For many situations, there are usually three options to choose from and sometimes five or six. That made me think that the resilient safety of scuba diving is created by a multiple redundancy network. Instead of safety being determined by a chain of strong rigid links, none of which must ever break, scuba safety is maintained by a flexible network, rather like a weight hanging from a suspended tennis net. If something goes slightly wrong, you first try this response, and if it doesn't work you try another. For the well-trained diver, one or two steps like this usually return the dive to complete order and tranquillity, or the buddy pair can surface under control at a safe rate. If you cut a strand of the tennis net, it remains strong; if you cut two or three, they must be close together to seriously weaken the whole net. To cause it to break or tear you may have to cut five or six strands in a single row.

At first I wanted to commission a complete mathematical model of scuba safety, to represent this positive control network in terms of probability of elements failing, and the probability of a simple corrective decision stopping the spread of the problem. That turned out to be too complicated. Engineers often do failure-risk analysis, studying the chances of things going wrong, but to turn this on its head and study the safety of a complex network going right was not easy. I discussed the problem with our new friends at HSE and they suggested that they could fund a study of accidents in scuba, to see how frequently different parts of the equipment, or decisions by divers, fail and cause accidents or fatalities. Since the number of accidents in scientific diving is too small to be statistically significant, we agreed to study all sports-diver scuba accidents reported in the UK and USA. Nigel Bellamy at the Paras Ltd company accepted the contract to make the study. The assumption was that the scientific diving community would be safer than the sports divers because of better training, albeit on the same equipment.

The BSAC study of accidents and fatalities early in the history of sports diving showed that accidents were seldom caused by a single dramatic factor like a heart attack, an exploding cylinder, or being struck by lightning. Usually two or more causes contribute to a serious accident. In order to quantify this problem, I asked Nigel to classify accidents by the number of causative factors that he could identify for each event. We agreed the maximum number of separate causes that might occur would probably be five or less, and designed the study accordingly. The statistical sample combining UK and USA sports diving data was over 1000 accidents, including 268 fatalities. The data extended from 1990 to 1994 for the UK, and from 1992 to 1994 for the USA, with an average fatality rate of 2 per 10,000 divers at risk per year.

Nigel's final report, in December 1996, contained countless tables and graphs showing the frequency of different kinds of accident and their causes. These included: too-rapid ascent, mask leaking, repetitive diving, electro-mechanical failure, inadequate monitoring of air supply, previous medical condition, entrapment in a cave, obesity, failure to share air supply, failures on the support boat, failure of communications, entanglement, alcoholic hangover, and many more. The table showing the number of independent causes typically involved in a single accident was the one that surprised me. Most of the fatal accidents had five or more contributory causes.

I had expected the multiple-causes table to show a frequency maximum at perhaps two or three causes. The fact that a fatal accident typically involved a combination of five or more factors – such as failures of equipment, training, and decision-making, and unfavourable changes in environmental conditions – was consistent with my redundancy model of safety. This tended to show that, contrary to the expectations of the critics, in order to provoke a fatal accident using scuba gear, you have to be either very unlucky, very badly trained or, to put it bluntly and a bit unkindly, rather careless. With good training, the system is strongly self-correcting, thus avoiding accidents, even if it means terminating the dive so as to surface in an orderly way.

* * *

My American colleagues, if they have read this chapter, will be bug-eyed with incredulity that so many well-intentioned people failed so completely to resolve a relatively simple problem for 20 years. Scuba diving on offshore oil rigs and other platforms of the hydrocarbon industry in the North Sea should have been banned in 1974. The solution of the Advisory Codes of Practice, when it was adopted in 1997, has many similarities to OSHA's recognition almost 20 years earlier of scientific diving as a professional, self-disciplining community. Each professional sub-section of working diving in the UK is now trusted to help design its own strict and workable safety regime. The accident rate in commercial offshore diving is now very low and has been declining for years. Sometimes it takes a long time for a good idea to cross the Atlantic. Scientific diving in the UK, Europe, the USA, and globally now prospers and flourishes, although ripples of the old anti-scuba conflict recur from time to time. In Britain, the SDSC, now chaired by Martin Sayer, maintains a steady oversight of the standards and discipline of scientific diving, while in Europe, the European Scientific Diving Committee provides a continuous structure of collaboration and review of safety. If you look on the websites of the HSE and the different organizations described in this chapter, you will see tables of equivalences between standards in many countries, including equivalence between working and CMAS amateur qualifications, and the ACOPS for several professions, including scientists. All the troubles seem to have blown away and perhaps civilisation reigns. I hope so. The present teams of scientific divers deserve much credit for this, since good order does not evolve or persist by chance.

9

From Law of the Sea to Ocean Forecasting

The first time I heard a professional climate scientist speak with real gut-wrenching fear in his voice was at a working meeting in the year 2000. It was a technical workshop attended by about 100 experts at Den Haag, in the Netherlands. The speaker gestured to his graphs on the screen with the laser pointer.

"This is terrifying," he said. Then he let his hands hang down limply by his side with his shoulders slumped in the manner of a defeated boxer whose opponent has been declared the winner.

"The forecast world temperature by the end of the century is a rise of three to four degrees. And it is worse than that. Because there are uneven rates of increase, with pauses and even reversals, the politicians and the public will keep expressing doubts and relaxing. They won't have a consistent policy. It is terrifying."

It was not the end of his talk. The emotional fear had gone from his voice by the end and our discussion then adopted the usual evasive terms of "concern", "negative impacts", and "mitigation policies". These were the phrases that we had been using since the mid-1980s, though with increasing conviction that human activities were causing dangerous climate change.

During the 1980s and '90s, I worked regularly as a participant in large teams of oceanographers, meteorologists, diplomats, and computer modellers to build up a global network of instruments that could measure the properties of the sea every day and send the results to feed huge numerical models requiring some of the largest computers in the world. We worked on the network of instruments and not on the modelling. It was the task of the Global Ocean Observing System to provide reliable calibrated numbers, and to provide them fast: other experts then did the calculations and forecasts. Hence my shock when I heard that tone of fear. Familiarity with bureaucratic weasel phrases can deaden one to the realities. By 2021 the impact of recent floods, hurricanes, forest fires, oceanic warming, and glacier melting has now, at last, caused public alarm.

It all started for me in 1968.

In: Flemming, N.C., 2021, *Apollonia on my Mind. The memoir of a paraplegic ocean scientist*, 345-382. Leiden: Sidestone Press.

* * *

"Don't work too hard," murmured Sir George Deacon as he ambled into my office in the late afternoon on a Thursday in late March 1968. It was his standard greeting, which could dangerously disarm a recipient who might be unaware that Sir George was a brilliant scientist and a Fellow of the Royal Society. I assured Sir George, my boss, the director of the NIO, that I was not working too hard, which seemed to please him. In those days scientists often worked at the Institute until midnight, and the Institute main doors were never locked.

"How would you like to go to New York on Sunday?"

"I like New York. What is it about?"

"Read this, and tell me what you think of it tomorrow."

He dropped a thick wodge of typed official paper onto my desk. The top margin was stamped in red "Restricted". A mild level of security. I agreed to read it, as he ambled out of my office and I packed up my files to drive home.

This exchange was the start of an intellectual adventure that lasted 39 years, zig-zagging through 45 countries from China to Chile, Sweden to Singapore. I plunged into a world of committees, workshops, UN agencies, European directorates, space agencies, satellites, and huge international oceanographic experiments. It started with the Law of the Sea but ended with a Global Ocean Observing System. During this adventure, I was sometimes the chair or director of some component organization or committee, but just as often a minor spear-carrier in the back ranks of boffins and advisers. We were creating new global scientific organizations and making new discoveries that are now essential parts of the international effort to save the planet from the crisis of climate change. It was exciting, serious, and sometimes funny.

The document given to me by Sir George described the forthcoming first meeting of the United Nations Ad Hoc Committee on the Peaceful Uses of the Seabed Beyond the Limits of National Jurisdiction, which was due to start the following week. In it the British Foreign and Commonwealth Office sought to establish a workable UK policy. It looked as if the FCO had been taken by surprise by the emergence of a possible new regime of the Law of the Sea, with potentially sweeping implications for shipping, coastal defences, fishing, pollution control, access to oil and gas beneath and beyond the continental shelf, offshore mining, the definition of an area of the seafloor to be controlled by the UN, and possibly a new regime for the control of coastal seas. The previous major agreement was the Treaty on the Continental Shelf in 1958. But that had left some controversial issues unresolved, including even the definition of the outer limit of the state's rights on the continental shelf. The proposed new British policies were sketchy, hesitant, and unconvincing. I decided to telephone discreetly my old contacts in the COSG (Chapter 5) for ideas the next morning, before reporting to Sir George. The UN drafts and the UK comments on them were wrong about the probable rates of improvement in marine technology, being grossly over-optimistic about the possible mining of manganese nodules from oceans 6 km (4 mi) deep (Chapter 5). It would need a lot of work to flesh out the policies and achieve a realistic estimate of new technology and offshore exploration.

Quick telephone conversations the next morning convinced me that I could bring useful experience to bear on the drafting of UK policy. I told Sir George that I was happy to fly to the meeting at the UN Secretariat. I would be an adviser to the UK Mission to the UN, participating in the Economic and Technical Working Group of the new committee, with an office desk at the UK Mission to the UN in New York.

Henry Darwin met me at Kennedy Airport on Sunday wearing tennis shorts and shoes as he had come directly from his afternoon recreation, perhaps playing squash. The weather was a crisp sunny 10 degrees C (50° Fahrenheit), so I don't think he had been playing tennis. Henry was legal counsellor to the UK Mission to the United Nations, with previous experience in nuclear test ban treaties at a very high security level.

"Welcome, welcome to New York!" Henry beamed. "We've just finished an Outer Space Committee. You're the scientific expert on the seabed, right? I don't know anything about the seabed. We'll have to work fast. I'll drive you to your hotel, and then at six there's a reception at the Mission. We'll walk over there from your hotel. I'll come to the lobby at 5:45."

My first impression of Henry was a serious mistake. He was a fraction over six foot (1.83 m) tall, a bit gangly and loose-limbed, with a very round boyish face and an infectious guileless smile. His casual attitude and his appearance in tennis shorts misled me completely. I wondered how we were going to conduct effective negotiations if the top man had made so little preparation.

Having put on my best suit, I walked with Henry the few blocks to the UK Mission to the UN on Second Avenue. The occasion was designed to economize on the FCO reserves of sherry, combining a farewell party for the specialists attending the Conference on Disarmament of Outer Space with a welcome party for the experts on the Peaceful Uses of the Seabed. Since the Seabed Committee had two working groups, one on economics, technical issues and science and one on law, while the space conference had multiple sub-committees, there were various complex introductions. The ambassador's wife, noticing me surveying the scene and clutching my drink thoughtfully, ushered me briskly towards a dapper figure who was similarly at a loss, saying cheerfully:

"Dr Green from Outer Space, I want you to meet Dr Flemming from the Seafloor."

She smiled hostessishly and skipped away to rescue other stray guests.

Dr Green and I stared at each other wondering if she realized how funny her introduction had been, and then we both burst into laughter. We had a gossip about our respective disciplines and interests. Henry caught up with me.

"You'll be working mostly with Steven from our diplomatic staff as your minder in the Technical Working Group, and I'll watch over the Legal one. Steven will brief you on the procedures and rules for submitting written statements to the meeting and security of documents in the Registry. Documents for verbal presentation will always be drafted the day before and checked by Steven or me. All documents and your typewriter must be handed in to the Registry whenever you leave the office. Come back to the Mission at 9 pm, and we'll work on strategy until midnight with the Mission staff. OK?"

The strategy session was an eye-opener, a rapid introduction to the world of diplomacy and committee tactics. We were estimating the positions to be adopted by other major countries while assessing the position adopted by the UN Secretariat, who were supposed to be neutral, but of course were not. We worked out the probable time during the two-week meeting when different issues would come to the boil, when the UK preferences would be most likely to be acceptable, and which countries of the 35 members were most likely to support our views on which items. Besides the agenda, there were stacks of official documents already submitted to the Committee by various UN bodies and briefings prepared by the Secretariat. I was given another thick file to read and digest. Henry had an excellent grasp of all the major issues, both scientific and legal. I should never have been so daft as to think that the great-grandson of Charles Darwin would not be on top of his job. Henry explained that the South American Pacific coast countries, had already claimed 200-mile fishing limits unilaterally, having recently arrested US fishing vessels within that area, provoking predictable outrage. Ecuador would be chairing the Legal Working Group, and their purpose was to stall any progress for as long as possible, while encouraging all coastal states to consider their self-interest in claiming as wide a limit for national jurisdiction as possible. In this way, the South American countries, known in UN jargon as "the Latins", hoped that their 200-mile claim would gradually become the norm, rather than an outrageous exception.

In flat contradiction to this policy, the UN Secretariat and the Maltese, led by their Ambassador Arvid Pardo, hoped to persuade coastal states to accept narrow coastal sea and seafloor jurisdiction, thus maximizing the seafloor beyond national limits, which would then be governed directly by the United Nations, with any profit from licensing helping to fund the UN itself. Every country that had taken the trouble to analyse their own position would have found different angles to support different interpretations, advantages, and disadvantages, between these two extremes. We were not supposed to discuss military matters, but every country with a navy would also be working out the military implications of different regimes that might be agreed. Where would it be permissible to place listening devices or mines on the seafloor? What was legitimate defence against encroaching submarines?

This first two-week meeting in New York needed to set up the two working groups with carefully defined objectives, mutually agreed terms of reference and separate lists of topics to discuss. All 35 member states could appoint delegates to both working groups. During this first meeting, all the delegates and advisory experts would meet together, so that the working groups only started debating separately in the later meetings in the summer and autumn. We had to submit our final report to the General Assembly of the UN at its session in October 1968.

Henry said that I should take the UK delegate's chair in the conference room, sitting at the microphone, while Steven would sit behind me as my FCO policy minder, also watching over legal matters when they arose if Henry was not present. We agreed to keep our powder dry for a day or two until the mood of the meeting was clearer, and I should report back regularly to Henry at the end of the day to warn him if I thought we should prepare an official UK statement on any topic.

Soon after midnight, we felt secure and comfortable that all the members of the team, both Mission staff and advisory experts, knew their duties. Over a final cup of strong coffee, I began to relax.

Henry tilted his chair back onto its back legs, looked thoughtfully into the far distance, and said: "That looks OK for the next two weeks. Now we need to look five years ahead when the committee has ground itself into a stagnating stalemate of unresolved conflicts, with all actions frustrated. We must plan what will our strategy be then? What do we really want?"

He took a swig of coffee and grinned. UK policy at the UN was in good hands.

* * *

The next morning the Committee met for its opening plenary session when important bureaucrats and ambassadors made long speeches. The conference chambers in the UN Secretariat building are large airy rooms with a semi-circular array of curved rows of about 100 pale wooden desks, each fitted with microphones and earphones. They face towards the focal point of the Chair and the bureaucratic staff on a raised tier of stages fitted with booths, lecterns, and work desks for the rapporteurs. The interpreters gaze down from glass-walled rooms high above the main space so that they can see who is talking. After delivering rousing speeches, the VIPs departed, leaving us technocrats, lawyers, and national mission staffers to do the work.

The Sri Lankan chairman, Ambassador Hamilton Amerasinghe, opened the real business with the provisional objectives for the two working groups, and an explanation of procedures, circulation of report documents, and so on. The meeting still had to formally confirm the creation of the two working groups. Representatives of UN agencies devoted to fisheries, transport, safety at sea, and pollution made sturdy defences of their existing roles and spheres of interest. Confronted with a new zone of influence, every agency needed to protect its own turf, preferably showing that it naturally should be the dominant force on the new turf. Amerasinghe then threw the meeting open for comments or statements by national delegations.

Anodyne and soothing diplomatic speeches followed, usually from the smaller countries amongst the 35 nations present. The big countries held their fire or spoke very briefly indeed. Then the delegate from Ecuador, who was also chair of the soon-to-exist Legal Working Group, raised his hand to speak. The chairman acknowledged him.

"The Distinguished Delegate for Ecuador has the floor."

I will summarize briefly the gist of the very long speech. "Mr Chairman," he began, "my delegation considers that there is no need for the Economic and Technical Working Group since the boundary between national limits of jurisdiction on the seafloor and the international area has not been defined. It is illogical to discuss the technical or scientific nature of an area that has no agreed boundaries. The Legal Working Group will consider the arguments for the definition of these boundaries, and when that has been agreed, the technical issues can be assessed. In the circumstances, Mr Chairman, we propose that the Economic and Technical Working Group should be suspended or abolished until the Legal Working Group has reached its decision on the limits of national jurisdiction on the seafloor."

Amerasinghe listened with a calm face. The Secretariat should have got wind of what was going to happen, but maybe nobody had warned him. If the Ecuador proposal was accepted, half the members of each team of experts and diplomats from the 35 countries represented in the room could pack their bags and go home that afternoon. Science and technical expertise were not needed.

Furtive whispering broke out on all sides. Eventually several delegates did ask for clarification as to what the Ecuador delegate really intended, but it was only too obvious what he intended. I discussed this unexpected blow frantically with Steven, my minder, and with Vince McKelvey, the USA expert from the US Geological Survey on my right. The USA had a large delegation, but like the UK they had put their boffin in the front seat. I scribbled a paragraph on my notepad and showed it to them. Would this solve the problem? Steven nodded to go ahead and speak. Try it. The Chairman responded to my raised hand.

"The Distinguished Delegate for the United Kingdom has the floor."

"Mr Chairman, Distinguished Delegates, we appreciate the point made by the Distinguished Delegate from Ecuador, but there is a logical solution to the problem he has raised. This Committee is manifestly concerned with the deepest parts of the ocean floor, vast areas several kilometres deep, as well possibly as parts of the continental shelf or slope. The Economic and Technical Working Group can consider issues in the deep ocean and shallower waters, including the economic resources, or their cost of recovery, while stating very clearly that none if its observations imply any assumptions about jurisdictional limits and national boundaries. We can state that principle in the first paragraphs of the Working Group report. At all stages of the examination of economic, technical, and scientific data, the Working Group can state that there are no assumptions about legal boundaries, and we present our observations purely as a service to the Committee as a whole, so that we can all be aware of the nature of the seafloor that we are discussing. Thank you Mr Chairman."

Several delegations asked to speak after that intervention, and the Chairman carefully managed the lively discussion that followed. It would have been unheard of to put the contentious Ecuadorean proposal to a formal vote but it gradually became obvious that the majority of delegations supported a compromise along the lines I had suggested, and the Ecuador proposal was rejected. Nobody had to pack their bags. The meeting got underway.

The Final Report of the Committee to the 1968 UN General Assembly includes phrases of the kind that I suggested at several places in the text, and the Latins included a specific reservation of their own stressing the same point again at the end.

* * *

I will not try to describe or even summarize the Machiavellian manoeuvres that followed throughout that first meeting in New York, nor the following two meetings that I attended in New York in June and July, and then in Rio de Janeiro in late August 1968. Legal experts have written many technical books about the origins of the United Nations Convention on the Law of the Sea, which was finally agreed in 1982, while the official documents of our 1968 meetings can be found on the web. My participation in Rio was complicated by the fact that I was doing diving fieldwork in Greece that summer, and so had to fly from Athens to Lisbon to Rio, returning via New York to Athens. In addition to the formal presentation of written delegation statements followed by technical and often very positive discussion, several events occurred in the margins of these meetings which are worth recording, some amusing and some not amusing at all.

The Economic and Technical Working Group got into its stride at the New York meeting in June. Roger Denorme, the Belgian chairman, set a cracking pace, making no attempt to hide his wish to get business completed urgently so that the agreed boundaries of the coastal states' jurisdiction would be as narrow as possible. This was in direct opposition to the intent of the Ecuadorean chairman of the Legal Working Group, which was meeting separately in another conference room. As the meetings progressed from day to day, it became clear that the speakers for most delegations assumed that exploitation of the seafloor under thousands of metres of ocean water, by commercial enterprises or government agencies, would be developed far more quickly than I thought was realistic. I reported this to Henry Darwin at one of our evening meetings.

"The discussion in our technical group is being warped by the assumption of very rapid progress in dredging for manganese nodules in thousands of metres of ocean depth, but that is not going to happen for decades. It will happen one day, but only in response to reduction of costs, radical new technology, and the right world prices. At present we have steady progress for oil production into deeper water, but it is much less rapid than people think. The maximum seawater depth of production of offshore oil today is less than 100 m (330 ft), and oil has an assured market. There is no shortage of manganese or nickel. Perhaps delegates are reluctant to say this because it is contrary to so much commercial publicity. Not to mention enthusiastic lobbying. What are the advantages or disadvantages of our injecting some truth into the debate?"

"You mean," said Henry, "that you don't want the UK to be accused of stealing the kids' football!"

We talked it through. Recognizing the slow pace of advancing technology in the economic production of oil or manganese-nickel ores from deep water would remove the motive for extreme haste which was so much in the interests of the UN itself, and of the enthusiastic followers of Arvid Pardo, the Maltese ambassador who had initiated the meeting in the first place. It would also tend to play into the hands of the Latins who wanted a maximum width of coastal state jurisdiction or exclusive economic access. But what was the advantage to anyone, the UK and all the other states, of continuing months of debate on seriously false premises verging on delusion? Law written in haste might simply be bad law. The truth would become apparent eventually, perhaps soon after an unworkable set of international laws had been approved. The realization that the emperor had no clothes would be more disruptive the longer the truth was delayed.

Fifty years later, licences have been granted for survey and production of manganese nodules in the International Area of the Seabed, but no commercial production has taken place. The technology is still too expensive. The risk of pollution is also extreme. The maximum depth of production of oil from floating offshore platforms today is in 2700 m (8800 ft) of water, so the rate of increase of depth capability for oil has been an average of about 50 m (160 ft) per year. Impressive, but not a melodrama. And still not as deep as most manganese nodules, which are at 4000-6000 m (13,000-20,000 ft) depth.

Under Henry's guidance, I wrote a careful paper based on my analyses of technological possibilities in the ocean during my work for COSG (Chapter 5), updated for the ensuing two years of exploration and offshore investment. I read the paper formally to the Committee the next day. The first reaction was, as Henry had predicted, that the UK for inexplicable reasons wished to undermine the justification for the very existence of the Committee and its worthy purpose. But the discussion evolved, maturing quite quickly as engineers, mining experts, and oceanographers from other countries began to produce facts and figures that fitted with my cautious projections. It was almost as if the technically experienced experts had, up to that point, been afraid to dispute or question the enthusiasm of the lobbyists who anticipated a dramatic exploitation in deep water so soon that it required immediate political action. The recognition, however sadly, of the slow advance of drilling and dredging into deep ocean water also removed the illusion of a quick and limitless source of funds for the UN itself.

After this awkward moment, the meeting progressed constructively from the UK point of view, but there was an amusing incident just before the end. The meeting was due to finish on Tuesday, 9th July, so delegates wanted to make sure that editing and adoption of the final report were conducted speedily. This would ensure a clear holiday on Thursday, 4th July, preferably with the Friday also free, creating a four-day break. It was not unusual for the adoption of a final report to be so contentious that meetings worked late into the night and through weekends. They wanted to avoid that. On the afternoon of Wednesday, 3rd of July, we were still slogging paragraph by paragraph through our 20-page draft report, making minute alterations and adjustments to the text to reflect the sensitivity of every delegation.

On this occasion, the Distinguished Delegate for the United Republic of Tanzania had decided to attend for the first time, where he took the hitherto vacant seat between the UK and USA. He was a New York lawyer who had, it was rumoured, never been to Tanzania. As we worked carefully and respectfully, trying to accommodate objections and queries, he repeatedly requested the insertion of the following phrase: "For the benefit of all mankind, especially taking into account the needs of the developing countries." This phrase had been written early in our deliberations and was used several times already in the draft report. The Tanzanian delegate raised his hand again and again requesting that the phrase be inserted once more.

Amerasinghe protested.

"With due respect to the Distinguished Delegate, this phrase has been used many times in the text, and inserting it yet again will render it trivial and meaningless."

"Mr Chairman, I am sure that every delegation in this room supports this statement, and I have yet to hear anyone speak against it."

"You request that we insert this phrase again?"

"Yes, Mr Chairman."

The Chairman gestured helplessly and looked around the conference chamber for signs of agreement or rejection from the ranks of delegates. Silence.

"Mr Chairman," said the Delegate for Tanzania, "I propose that we put my request to the vote."

There were gasps of horror from several UN old-hands who knew what this meant. Late nights and no holidays for anyone over the Fourth of July.

The Chairman consulted fervently with the UN Secretariat staff. Then he took the microphone again.

"The Secretariat has requested that I inform you of the implications of our taking a vote on this proposal. Relevant documents would have to be translated into all the UN Official languages, and the procedure of taking the vote itself would have to be conducted with interpreter staff present and providing their services, together with neutral observers from the UN Secretariat staff. To prepare the necessary arrangements will go well into the night, and will almost certainly require most of tomorrow morning as well. The cost to the UN will be very considerable. I do request that you reconsider this proposal."

The Delegate for Tanzania stood up and left the meeting hall, to return a few minutes later with a large buff-coloured book, a couple of inches thick, which he placed prominently on his desk. Everybody had been whispering, muttering, and consulting their diaries during his absence. Mobile phones did not exist, so it was difficult to inform families.

Without turning my head, I endeavoured to read the title on the back of the large book. It was the "Procedures and Regulations for Conducting Votes in the Agencies and Subsidiary Bodies of the United Nations". The sight of this book provoked an even greater sense of alarm amongst the Secretariat staff on the podium.

Steven tapped me on the shoulder. "Hang on a minute. I'll be back in a second."

He hurried out of the room while the chaotic consultations and whispering continued. Steven returned clutching three enormous brick-orange volumes which he placed on the UK desk. The stack of orange volumes was about six inches high. I furtively read the inscription on the spine of the top one: "Charter, Constitution and Regulations of the United Nations Volume 1."

The Chairman called the meeting to order, explained again the dilemma and costs of putting the proposal to a vote, and turned questioningly to look at the Delegate for Tanzania.

"Mr Chairman, I withdraw my request for a vote."

"Thank you, the Distinguished Delegate for Tanzania. Let us proceed with paragraph 134."

The meeting finished at five o'clock, as scheduled. Everybody could enjoy their holiday.

As we left the meeting room I turned to Steven.

"What happened there? What on earth did you do?"

"Superior bureaucratic psychology, old boy! Just bureaucratic psychology."

Perhaps we really do need diplomats.

* * *

The splendidly florid Copacabana Hotel on the seafront in Rio de Janeiro was the venue for the third meeting of the Seabed Committee. The diplomatic delegates and advisory experts now knew each other quite well, so that the mood of the meeting should have been friendly and relaxed, with the added joy of swimming and a great view. The staked-out political and tactical positions of the committee members were known, so that interventions or proposals were tailored to be acceptable, so far as possible, without provoking extreme reactions. We would only risk direct confrontations when we knew we had strong supporters. The meeting lasted from the 19th to the 30th of August, and the UK delegation gathered over the prior weekend to be briefed by our head of delegation, Christopher Audland. Although there were still some controversial points to be sorted out, with continued need for alert cultivation of allies, the final report for the UN General Assembly was taking good shape. Tactics and timing were everything. The first two days went smoothly (Fig. 9.1).

Without warning, Russian heavy transport aircraft landed in Prague airport in Czechoslovakia at 11 pm local time on 20th August, disgorging light tanks and Special Forces troops, while the Warsaw Pact armies drove their tanks across the Czechoslovakian border and converged on the capital. By midday on 21st August, Prague was under military occupation by the USSR and its allies, and they had killed some of the citizens who took to the streets. Protesters set several tanks on fire.

Our minor committee, way down in the UN hierarchical pecking-order, was one of the few UN bodies in session at the time. Within hours, the news of the Russian invasion reached Rio. Throughout the night of 20th August, the heads of delegations consulted hectically with

Figure 9.1. Rio de Janeiro August 1968, at the 3rd meeting of the UN Committee on the Peaceful Uses of the Seabed Beyond National Jurisdiction. Left to right: Christopher Audland, Hamilton Amerasinghe, the author second from the right in the foreground, and unnamed delegates.

their embassies in Rio, who were exchanging frenzied coded cables with their capitals back home. By the time we gathered for breakfast in the dining room on the 21st, a mood of frozen diplomatic protocol and reserve had set in. Through the sunlit windows, the view of the most beautiful beach in the world was no longer beautiful. Greetings were cautious, and casual friendly remarks between delegations were non-existent. Everything depended on who said what in the opening session of our meeting, and how the many Third World countries chose to respond. Would they take sides in the Cold War, or try to stay neutral?

Christopher Audland was tight-lipped at breakfast, scribbling notes for his statement to the Committee. We felt wretched at what had happened, bringing a sudden end to the so-called Prague Spring, which, combined with the slogan "Communism with a human face", had been promoted by Alexander Dubcek in Czechoslovakia. For a few months there had seemed hope for a better world. Viewed from the USSR, it was a step too far. Communism itself was under threat.

When our meeting started, the chairman begged members to keep their political comments as moderate and restrained as possible. He requested us to be brief and to remember the real purpose of our Committee. There is no record in the minutes of the speeches or debate that followed. The USSR claimed that the intervention had been requested by the good Communists in Prague. They were only trying to help a neighbour. The USA and the west European delegates discreetly but firmly criticized what they saw as an unprovoked military invasion. Czechoslovakia said nothing, but Poland and Romania made neutral comments that, by implication, let the USSR off the hook. Third World countries kept quiet.

The most astonishing statement came from Professor Alexander Yankov, the delegate for Bulgaria. Yankov was professor of law at the University of Sofia. He was a short, slightly chubby, very energetic man, always ready with a kind word, or a helpful phrase to make the drafting of our report go smoothly. On this occasion he spoke slowly, very clearly, without emotion, dissecting the Russian case for the invasion from a strictly legal position. Without actually criticizing anyone, he gave a merciless analysis based on international law. He drew quietly to his conclusion, leaving the USSR's pretensions effectively valueless, then thanked the chairman for allowing him to speak.

Everybody was thunderstruck. Cold War politics at this level, even in our little committee, was not a game. East European countries of the Soviet Bloc were expected and indeed compelled to support the USSR with no quibbles. It was probable that Yankov's speech would result in his being recalled to Sofia, if not to Moscow. His family might be punished, or banished to a desolate town east of the Urals. He would almost certainly lose his professorship.

Nothing happened.

Yankov cut a lonely figure for the rest of the meeting, since he could not be seen to fraternize with the "Western Friendlies", which would compound his crime, while he was simultaneously ostracized by the Soviet Bloc.

Over the following years, I got to know Yankov as a good friend at many Law of the Sea meetings, and we always greeted each other with a bear hug and kisses on both cheeks. But I never asked him how he had dared to make that speech and why he escaped retribution. I did discover that he was always ready to do or say unconventional things, and his sense of fun caught up with me on several occasions. Some years later, a month after he was appointed Bulgarian Ambassador to the Court of St James, I queued in the embassy reception hall, having been invited to celebrate the Bulgarian National Day. A uniformed master of ceremony formally announced each guest. As the reception line moved forward I approached the new ambassador. He extended his hand, smiled, bent forward, and said "I don't think we had better kiss here." I last met him in Hamburg in about 2000 where he had been appointed to chair the Internationaler Seegerichtshof (International Tribunal for the Law of the Sea). A wonderful man.

Back to Rio.

The two working groups produced well-argued reports agreed to and adopted unanimously by their members, but the Committee was unable to reconcile some of the inconsistencies and contradictions between them, or to agree that the UN itself should manage or benefit directly from the seabed area beyond national limits. The final report explained the discrepancies and the broad agreements on general principles. The main points that we presented to the UN General Assembly later in 1968 were:

1. The seabed should be used for peaceful purposes, and military use should be limited.
2. There is an area, yet undefined, of the seabed beyond national jurisdiction which should not be subject to the sovereignty of any single nation.
3. A precise boundary between national and international areas of the seabed should be established.
4. Exploitation and exploration of the seabed should be in accord with international law and specifically the Charter of the United Nations.
5. Exploitation of the seabed should be for the general benefit of mankind.
6. There should be proper safeguards against pollution of the sea.

On one of the last days in Rio, Christopher Audland and I were jogging along the beach in the early morning sunshine before breakfast, running on the firm wet sand where the waves lapped at our feet. The sand squidged up between our toes. A few young people were already playing volleyball. Both the men and the women in Rio were so beautiful that it changes one's conception of what it means to be human. I felt more than usually gauche and northern. We were not running seriously to keep fit, just relishing the cool air, the silence of the city before the traffic built up, and the joy of getting away from the claustrophobic intensity of the final days of work.

"As a scientist," said Christopher, "don't you find it absurd that every country represented here has selected the most intelligent and highly qualified experts they can find, paid for them to come here for two weeks, and then we all neutralize and counter-balance each other so precisely with incompatible logical demands that nothing useful can be achieved?"

"Not all," I said. "From a scientific point of view that is exactly what I would predict."

* * *

I returned to Greece via New York, stopping briefly at Atlantic City to visit Barbara Guggenheim's family at their summer holiday home. I presented Barbara with a yellow-gold kite shaped like an eagle which I had bought in Rio, where hundreds of kites like this were flying over the town. We flew it on the beach, with a flip-flop attached to the tail to make it balance in the strong onshore wind blowing off the North Atlantic. A warm tropical touch of Rio in bracing Atlantic City. We had been sitting on the beach chatting for a while when a large gun-draped cop came galumphing across the sand.

"Is that your kite, sir?" From my sitting position, his boots seemed very big.

"Yes, Officer. I am holding the string, but it belongs to this young lady."

"Get it down, sir, it's dangerous!"

"It's been up there half an hour, Officer, and nobody has complained. The wind is very strong. It is very stable."

"Why is there a shoe hanging on it? It's over the boardwalk. It could fall on someone's head. It's dangerous."

"A flip-flop would hardly hurt anyone."

"Get it down now, sir! Do you hear me? Now!"

"All right, all right."

We dragged down the dangerous kite and went back to her parents' house for an evening meal. I caught the overnight flight from Kennedy Airport to Athens and from there drove to the Mani Peninsula on the south coast of the Peloponnese to continue diving.

* * *

After two weeks diving on underwater ruins on the coast of the southern Peloponnese, I drove north through Yugoslavia, picking up a Czech hitchhiker called Tomasz at the Greek border. He had been enjoying the freedom of Dubček's "Prague Spring" to tour the antiquities of Greece. Now he did not know if he had an independent country to go home to. We discussed his problems, and I said I would take him as far as he wished.

Tomasz had an expert eye for mushrooms in the forest where we spent the nights off the main road. One night, as I drove slowly down a forest track searching for a good place to stop, the headlights suddenly reflected blindingly from the polished aluminium fuselage of a fighter jet aircraft. There was a red star painted on the tail fin. Tito's air force had dispersed its fighters to remote airstrips in case of war with Russia.

Soldiers with machine guns rested on the ground in the shadows. They did not reach for their guns.

I dimmed the headlights, turned slowly to show no guilt, and drove slowly to another glade in the forest. Tomasz rose at dawn, and we cooked gorgeous mushrooms for breakfast. I dropped him at the Austrian border. I hope he made a good decision about going home, or not.

* * *

The next session of the Seabed Committee was in February 1969, once more in New York. The meeting took place against the constant background publicity caused by the Santa Barbara oil leak off the coast of California, where almost 100,000 barrels of oil gushed from a blow-out on the seafloor near a Union Oil platform in the Dos Cuadros field. (In 2021, it is still the fourth-largest oil spill ever in American waters.) The repeated attempts to drill down and intersect the submerged well to block off the leak were described in garish detail in the daily papers and TV programmes. After several failures, the well was blocked off. Many thousands of seabirds were destroyed by the pollution, while oil washed up on miles of popular bathing beaches. The delegates in New York reacted with contrition to the stream of bad news, which revealed the technical limitations of controlling a well in less than 60 m (200 ft) water depth, nowhere near the thousands of metres that they had previously discussed so glibly. The failure for many days and even weeks to combat the pollution made its mark. This real-world event had a very sobering effect on our meeting, and led through contorted bureaucratic pathways to influence the US Natural Environmental Policy Act in 1970, and even the establishment of the UN Environment Programme in 1972.

I worked as an attached expert for the UK delegation to the Seabed Committee for two more meetings in 1970, but the transition to a much more political committee that would plan a new regime for the Law of the Sea was already in sight. At that level, the discussion was 90% political and legal, so that a new generation of high-powered legal experts started to take over, and I merged back into the world of support services and research. I had no further direct input to the nascent Law of the Sea drafting or preparations, apart from one last meeting in Port of Spain, Trinidad. The United Nations Convention on the Law of the Sea was agreed in international law in 1982, came into force when signed by sufficient number of states in 1994, and was signed by the UK and ratified in the British Parliament in 1997.

* * *

The February 1969 seabed meeting in New York was followed almost immediately by the first Oceanology International conference and trade exhibition, which the Society for Underwater Technology organized in Brighton, at the Metropole Hotel, in March.

Before I left New York I just had time to give a talk on underwater cities in the Colony Club, arranged by Margaret Poor, while my accommodation was provided by Alfred Poor at the Knickerbocker Club. This was an aspect of Manhattan life that I had not previously experienced. The trip was also memorable, in a way that was unforeseeable at the time, as, during a brief excursion to Washington, DC, to give a talk at the National Geographic Society, I met Jay Kleinberg in the mêlée at the snowbound railway station. She was an attractive young woman conducting historical research. We shared the last remaining copy of the Sunday *New York Times*, beginning a friendship that lasted for years. Ten years later we were married (Chapters 3, 10). Returning to the UK, I found in Brighton that the Chairman of the OI Conference, Professor Gordon Tucker, from Birmingham University, had been taken ill with influenza, and I would have to chair the whole Conference and exhibition.

Oceanology International 1969 was the first of a series of technical trade exhibitions and ocean research conferences that has been running biennially to this day. It started in Brighton, but since 2000 it has been held in the ExCel Centre in east London, which is a much bigger but bleakly soulless venue. The genesis of this hugely successful series goes back to the foundation of the Society for Underwater Technology (SUT), which a small number of my colleagues and I established in 1967. The members appointed me as Honorary Secretary of SUT at its foundation, and I held that post for about ten years.

The purpose of SUT was to bring together the skills, technologies, and knowledge of scientists and engineers who work in the sea, and who provide the technology and services that everybody else relies upon. The big companies that make their profits from oil, gas, shipping, trade, gravel extraction, provision of electricity across sea basins, fisheries, shipbuilding, fish farming, port construction, waste disposal, and so on, need hundreds of smaller companies and technical consultancies to provide accurate seabed maps, samples of sediments to test their bearing strength, cores drilled into the seabed to study the rocks, seismic surveys to search for oil-bearing strata, diving companies to weld pipes together, improved acoustic devices to track fish shoals, and dozens of other technical services and predictions. What is the highest wave that will strike this oil platform in the next 50 years? Will the sand move away under this pipeline and leave it suspended in mid-water? Will this anti-fouling paint protect a ship's hull without poisoning the surrounding seawater? These smaller companies in turn need high-precision instruments to do their work, combined with the latest scientific oceanographic data. Government regulatory agencies need exactly the same data and services to fulfil their duties.

SUT started as a small, almost niche, organization, but has grown consistently over the decades so that it now has a membership of nearly 2000 experts belonging to ten branches in six countries, including the USA, Brazil, and China. There are 170 corporate members. It has its own offices in London and goes from strength to strength. I retired as the honorary secretary in about 1980 and its subsequent prospering has been entirely due to my excellent successors and the office staff.

Back in 1969, SUT teamed up with a professional exhibition organizer to launch the first OI, which was due to run for five weekdays. We expected two or three thousand people to walk through the exhibition halls, with a few hundred registrants for the conference. In the event, all hell broke out.

Many thousands of people came to the exhibition, in spite of the stormy weather with flurries of snow. Local colleges and schools brought large groups of students who were not there to conduct business with the exhibitors, although I am sure the experience did wonders for their education. They were welcomed. Exhibitors found themselves doing excellent business with droves of potential commercial partners from all over Europe and further afield, but began to run out of brochures and documents that had been taken by students or the general public. Journalists came from far and wide. Everybody was frantically trying to order more documents, but the snow was blocking roads, and deliveries were late. The registration numbers for the conference were twice that expected so that I had to book larger meeting rooms, and frantic orders were placed with the printers for hundreds more copies of the bulky three-volume printed conference reports. A special lorry had to be rented. Again the snow-blocked roads delayed deliveries and tempers became frayed. The vast stack of books arrived just in time on the second day.

In spite of the appalling weather, the thousands of participants found a rich variety of food, drink, and diversion in the well-known entertainment districts of Brighton. Every seafront café and bar was crammed with our visitors poring over catalogues and technical documents as they talked and made deals. The advantages of not being in London were apparent. Many of our foreign guests fell in love with Brighton, thus ensuring the success of the follow-on conferences. The mayor invited the conferees to a grand reception in the Brighton Pavilion, which further delighted everybody. The foreign visitors were astonished by the outrageous architecture and chinoiserie decorations. Even engineers have souls.

The Russian participants, about 20 people, were living in a huge oceanographic research ship anchored for the week just outside the three-mile limit of the UK territorial waters. The vast black silhouette set in a silver-grey sea was intermittently visible from the Brighton Parade through drifting clouds of snow. The patriotic British manufacturers of a large hovercraft, a ship that floated above the sea on a cushion of air, provided the shuttle transport for the Russians every day. By the third day, the wind was so strong that the breaking surf prevented the hovercraft from taking the Russians back to their ship. They stood huddled together on the beach, close to the pounding surf, like a herd of penguins waiting for the Antarctic winter. With relentless suspicion, they refused to believe the explanation that a bus had been arranged to take them to Shoreham harbour, where a larger boat would take them out to their ship. It must be a capitalist plot. The snow drifted, the Russians were obdurate, the sky darkened, the organizers pleaded. Eventually the cold must have gotten through their thick coats and earmuffs and they agreed to board the bus.

The conference barrelled along, gathering momentum with a rising sense of exciting science, pioneering success, and good business deals (Fig. 9.2). There was standing room only for many of the presentations, but the speakers were happy. At the final banquet dinner, we were entertained by a Scottish jazz trumpeter of extraordinary musical agility, after which our guests from so many countries dispersed happily. By the time that we had cleared up the conference debris next day, I was so exhausted that I only just managed to drive home.

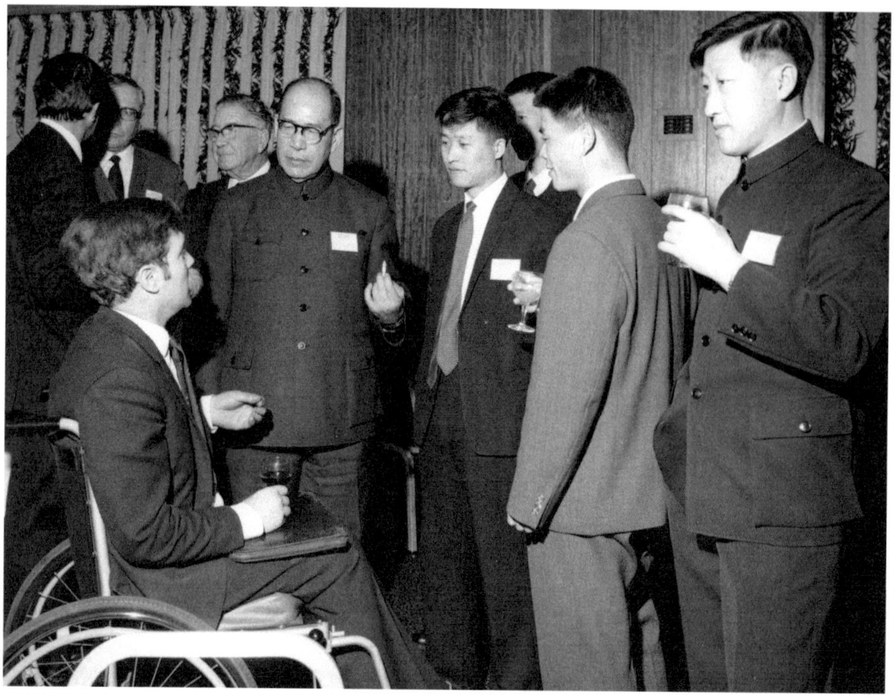

Figure 9.2. The author talking with the Chinese participants at the Oceanology International Conference in Brighton in 1972.

* * *

During the 1970s, most of my research effort was devoted to the study of sea level change in the past 3000 years (Chapter 6). But I continued to spend much time promoting the aims of SUT and working with the marine industries. The US Vice President's Commission on Marine Science, Engineering and Resources, the Stratton Commission, had reported in January 1969, publishing a large book entitled *Our Nation and the Sea: A Plan for National Action*, which included 126 recommendations, the most significant of which was the establishment of the NOAA. A scientist from the USA had presented a summary of the Stratton Report at the Oceanology International Conference in Brighton. Other countries were beginning to undertake similar assessments of their marine industries and services. I attended meetings of the Marine Technology Society in the USA, using the trips also to visit Ed Wenk, John Craven, and other luminaries of the marine technology world there. I copied some of the policies of the Marine Technology Society when planning meetings for SUT. Offshore oil and gas developed rapidly in the North Sea, with larger reserves discovered further north in deeper water towards the Shetland Islands. New technology enabled platforms and drilling to work in depths greater than 100-200 m (330-660 ft). Although the deepest drilling and production was usually in the USA, or sometimes off Brazil, the technology required in the stormy North Sea, with its strong tidal currents, forced continuous innovation by British companies. SUT was constantly in the forefront of discussions and plans, establishing its first branch in Aberdeen, which was, by now, the focus of the offshore industries in the North Sea.

By 1979, the British economy was in a mess; we had survived the four-day week; Mrs Thatcher was elected prime minister; and expenditure on science was threatened with large financial cuts, with the result that several scientists at the NIO were made redundant. The director of the NIO, by then Henry Charnock, asked me to collaborate in developing an oceanographic data service to assist industry, to be called the Marine Information and Advisory Service (MIAS). This would combine staff from the Surrey laboratory with those at the Bidston laboratory near Liverpool, supported by a large (by the standards of the day) computer database on a Honeywell 66 computer, centralizing all the numerical oceanographic data that we could get hold of from government departments and industry. I was sad to abandon my fieldwork on submerged archaeological sites, but every good seam of research runs out of productive ore at some point, and it was a good time to stop. It was obvious that my background knowledge of commercial and industrial marine operations, combined with my work on the Law of the Sea, would be ideally suited to the new project.

In 1980, I was appointed director of MIAS, entering the vast and bewildering world of large-scale computer data management at the national and international level. I held this post until 1987, by which time the data centre became formally the British Oceanographic Data Centre, which remains its title to this day.

This work led to my participation in the International Oceanographic Data Exchange Committee (IODE) of the Intergovernmental Oceanographic Commission (IOC) of UNESCO, where I became chairman at the Moscow meeting in 1986. Throughout the 1980s and early '90s, I was surrounded by computer experts, both men and women, who can only be described as brilliant and dedicated geeks and nerds, deeply committed to the science of oceanography and reliable data archiving, at which task they were much cleverer than me. I enjoyed every minute of it. I was able to understand just enough of the actual bits-and-bytes work that they were doing so that I could discuss their problems with them, agreeing on their work schedules, and then act as an interface with the world of funding agencies, users, customers, and oil rig operators who did not want to know anything about bits-and-bytes. Just tell me what the highest wave is going to be in the next 50 years, please! If the predicted maximum wave height is estimated too low, then the oil rig will be badly damaged in a large storm; if it is too high, the rig construction will cost millions of pounds more than necessary.

In 1984, in the midst of the miners' strike, a team of three Chinese oceanographers visited the UK to see how the MIAS data centre was organized, and to study our methods for distributing data to users. They planned to expand ocean data use in China. Professor Ho Wen Feng and his colleagues were amongst the first visitors from China to be allowed to meet immediately with their British hosts, rather than going first to the Chinese embassy in London. Jay and I invited them to dinner at our house on the first evening, which was a Sunday. On the hour-long car journey from Gatwick to their hotel in Haslemere, I explained that the miners' strike was not the beginning of a Communist revolution, and that they should read the newspapers carefully every morning.

To my embarrassment there was a local power cut that evening, so that our house, which was at the end of a narrow, wooded lane, was in darkness for a while, lit only by candles, until somebody trimmed back the tree that caused the problem. It happened every few months. Again, I assured Professor Ho that this was not the collapse of capitalism. Soon the lights came on again. After drinks, we went into the dining room

Figure 9.3. *In 1984 a team of three Chinese oceanographic data experts, led by Professor Ho Wen Feng on the left, visited the NIO, to work with the author on methods of running a national ocean data centre. The figurehead above the door is that from HMS Challenger that conducted the first global survey of the world oceans from 1872 to 1876.*

Figure 9.4. World Data Centre meeting at Tianjin, Peoples' Republic of China, 1991. Standing on the author's right is Iouri Oliounine from IOC, and on the author's left is Professor Ho Wen Feng, Director of the Chinese Ocean Data Centre. Dr Vyacheslav Smirnoff, the Director of the Russian Ocean Data Centre is directly behind the author. No Americans attended this meeting, as they were not allowed to fly during the Gulf War.

where Jay had prepared a festive table, complete with two silver candelabra. Professor Ho looked at them with a twinkle in his eye.

"These candles are for decoration, no?" he asked.

"Yes, they are decoration," I said.

"But you might need them!" he said, with just the right ironic emphasis.

I had to admire his sense of humour, and I had the pleasure of working with him over the next few years (Figs 9.3; 9.4).

* * *

The pious hopes and noble intentions of the United Nations Convention on the Law of the Sea and the deliberations of the United Nations Environment Programme can only produce benefits if actions are planned with accurate and up-to-date measurements describing the whole world ocean every day. In the 1980s, no system existed to provide such data.

The word "data" has an innocuous, neutral, technical façade that the public have got used to in recent decades. This is especially so in the realm of the big technology companies that collect information about millions of people. It has become a shorthand for something that we know is important, but the details can be left to someone else to figure out. In the world of oceanography, data means numerical values from instruments measuring the temperature of seawater from the surface to the ocean depths, salinity of seawater, surface wind speed and direction, current velocity and direction at all depths, surface wave height and direction, ice cover, dissolved oxygen, suspended sediment composition, and numerous other characteristics which scientists can use to forecast what will happen next. These data must be measured at millions of locations daily, and the numbers transmitted to computer models as quickly as possible.

For many years the popular TV programmes about life in the oceans have included the phrase "We know more about the far side of the moon than we do about the floor of the ocean (or simply the whole ocean)". This phrase originated in the 1960s after the first Russian satellite photographed the side of the moon that is never facing the earth. The first Russian pictures, rather fuzzy, were in 1959, but the many improved orbital satellites going around the moon soon built up an accurate photographic image of the whole surface.

Provided that the criticism of our knowledge of the sea refers only to the topography of the seabed, the statement is probably correct. We can see the whole moon with optical detail, and hence map its surface to an accuracy and resolution of a few inches or centimetres. By comparison, the vast surface of the seafloor cannot be photographed with natural light, and even acoustic bathymetric measurements can usually only be run on ships' tracks several kilometres apart. Sometimes the separation of tracks is many tens of kilometres. Consequently, there are large patches of the ocean floor which have not been measured accurately.

On the other hand, if we include knowledge of the geology and sediments of the seafloor, then marine geologists have obtained hundreds of thousands of samples, probably millions, of cores and grabs and dredges, while only 382 kg (842 lb) of moon rock have ever been recovered by the six USA moon landings between 1969 and 1972.

If the allegation that we know so little about the sea is extended to the water itself, as it often is, then the statement becomes seriously misleading, plain wrong. Since the moon has no atmosphere or sea, there is nothing to compare directly with the miles-thick layers of water overlying the planet Earth. But if you think that we still know nothing about the inside of the sea, please look at this European website for the Mercator Ocean international forecasting bulletins: http://bulletin.mercator-ocean.fr/

This information bulletin can tell you the condition of the seawater today at the surface and at depths of 30 m, 100 m, 300 m, 1000 m, and 3000 m, in terms of salinity, temperature, current speed, and density, with extra details for the coastal and semi-enclosed ocean basins like the Mediterranean or the Caribbean. Not only does the bulletin display maps and animations of the conditions inside the ocean now, but you can request forecasts just like the meteorological description of the weather. The same website, and related sites, can provide forecasts of ice cover and thickness, wave height, wave direction, sound velocity, plankton density, sediment suspended in the seawater, and so on. Similar data are available from agencies such as NOAA and the UK Met Office.

In short, almost anything that any person or any business needs to know about the ocean surface and the interior of the ocean in 3D can now be obtained in reasonable detail free of charge on the web. Oil companies, fishing fleets, surfers, yacht skippers, flood warning agencies, can all use these data. Of course, there is always a need for more detail, more accuracy, and extra parameters like dissolved CO_2 or phosphate, light intensity, pollutants or contaminants. International investment and collaboration is adding progressively to the range of data provided. Companies and government agencies can also obtain specially detailed services for a fee. These websites are concerned only with the ocean's physical and chemical properties, but others provide comparable data on its biology, its countless species of plants, animals, bacteria, and viruses. Knowledge can never be described as complete, but the existing databases make a good start.

How did this huge system of information-gathering, many thousands of instruments permanently in the ocean, many satellites measuring the sea from space, and centralized computer models, come into existence? The fact that it does exist is almost unknown to the general public, never being mentioned in popular TV programmes about the sea. Perhaps the desire to preserve terrifying mysteries in distant places is coming into play again, like Peter Simple and his dismay at the taming of Rockall (Chapter 2). The recent films by David Attenborough and political conferences and protests have at last provoked genuine determination to act on marine pollution and global warming in the ocean, but there is still a disconnect between that anxiety and the measurements that can describe what is really happening and hence enable the right response.

* * *

If we look back 100 years or more, research in the ocean has been the purview of navies and a few institutes and national research centres, as described in Chapter 5. During the 1970s and '80s, the growth of the offshore oil industries and industrialized fishing created a greater demand for reliable forecasts of undersea conditions, but this remained a specialized and limited market, usually at a local level. These information services did not make national news, or get into major policy magazines. From the public point of view, oceanography was – incongruously – a political and economic backwater.

Much has changed in the past 20-30 years. The anthropogenic factors causing changes in the climate, biodiversity, habitats, and pollution of the global ocean are now the subject of articles in every newspaper and serious weekly magazine. Questions and concerns that were speculative in the extreme in the 1960s (Chapter 5) had become serious scientific research areas by the 1980s, emerging as major political and economic issues in the 1990s and later.

As these driving forces increased in political importance, so the demand for reliable information on which to make decisions also increased. Politicians and environmental agencies needed to understand better every aspect of natural process and human intervention in the deep ocean, with information instantly available like a weather forecast on land. With accurate descriptions and reliable numbers to feed into computer models, they can make decisions about long-term impacts, with estimates of damage or benefits. Frequently repeated model runs and forecasts of the state of the ocean would also enable policy-makers to check whether changes in legislation or conduct of marine industries are having the desired effect.

Collectively the task involves many factors. How big is the marine economy of the world, and what is the value of each of the environmental benefits, or the cost of losing them? What is the chance that present trends, natural or economic and industrial, will cause damage of a certain magnitude to industry or the environment, and at what cost? To what extent do local forecasts depend on having a picture of the whole world ocean? Anyone watching TV weather forecasts is familiar with the Atlantic-scale maps, and occasional global maps, that help explain the weather reaching Europe in a few days' time, but to what extent does ocean forecasting depend on global data? Do we know enough about the ocean and its chemistry and wildlife to understand what is happening now, or to make predictions for a few hours, days, or years in advance? If not, what measurements are needed, and how often? If the science shows that we need more accurate measurements on a global scale, what would the cost be, and how would we process the information and convey it to the people who use it? Would the benefit justify the cost? Benefit of course is not restricted in meaning to pure cash: there are moral and environmental benefits that extend into the future with immeasurable or unquantifiable beneficial effects. Some risks should not be taken.

Bureaucratic and administrative difficulties added to the scientific and political ones. Organizations large and small, UN agencies and advisory panels, had existed for many years collecting ocean measurements carefully and slowly, checking their accuracy over several months, publishing scientific papers, and then archiving the data where it could be used by scientists who requested it years later. Now they had to respond to the demand for many more observations to be made every month or even every day, with the data processed and verified within hours and delivered to customers who were not scientific experts and might find it difficult to understand. To say that this was a culture shock would be an understatement. We had to assess, redesign, and test every aspect of the chain from instrument design to sampling density, rapid – so-called real-time – data transmission, automatic quality control, computer numerical modelling, and design of maps and data products for users. We needed innovation at every step.

John Woods (Chapters 4, 8) became NERC Director of Marine and Atmospheric Science in 1986. He was one of the few people who could see the potential for the change in oceanographic sciences to establish an operational real-time service. Such a service would, given its economic benefit, justify greatly increased investment in the fundamental science, creating truly global ocean models. The benefits would feed both ways between the basic science and the application for human benefits forming a virtuous circle. Other scientists who shared this ambitious view included the late Walter Munk (who, in October 2017, celebrated his 100th birthday at the Scripps Institute of Oceanography), Noriyuki Nasu from the University of Tokyo Ocean Research Institute (who had researched deep ocean geological cores), Jim Baker, then Director of the US NOAA, Neville Smith at the Bureau of Meteorology in Melbourne, Pierre Papon at the Institut Français de Recherche pour l'Exploitation de la Mer (IFREMER) in France, and Mel Briscoe in the USA. Amongst the general community of marine scientists, there were both enthusiasts and sceptics, with the latter in the majority.

* * *

As I write this, the European Space Agency (ESA) satellite Sentinel-3B is being prepared for launch at the Russian Plesetsk cosmodrome 800 km (500 mi) north of Moscow. The satellite's fuel tanks have already been filled with hydrazine. The satellite is part of the European Commission Copernicus/Environment programme, and its sensors will measure hundreds of thousands of observations each day from the surface of the ocean, including the topography and altimetry of the sea surface, its temperature, the surface wind strength and direction, wave height, and so on. Sentinel-3B is the most recent of many dozens of ocean-observing satellites launched in the past 40 years, and it joins other orbiting ocean-observing satellites launched by ESA and other national space agencies. The typical lifespan of a satellite of this kind is about five years, so that many space missions have come and gone since the first specialized satellite marine mission, SEASAT, in 1978.

In June 1978, NASA Jet Propulsion Laboratory launched the earth-orbiting satellite named SEASAT. It carried a suite of novel downward-looking instruments that would test the feasibility of measuring many properties of the ocean from space. Potentially it could provide complete global maps in enough detail to join ocean data and atmospheric data together into a single global forecasting model. SEASAT carried instruments that measured the speed and direction of winds at the sea surface, the temperature of the water at the sea surface, the height of waves, the amount of water vapour in the atmosphere, the occurrence of floating sea ice, and ocean topography, which includes undulations in the sea surface caused by variations in gravity, the strength of currents, and atmospheric pressure. It was an astonishing achievement which demonstrated technologies that scientists and engineers have improved continuously in many ocean-observing satellites in subsequent decades.

Downward-looking satellites, so-called earth-observing (or ocean-observing) satellites, can only make observations in a fairly narrow path or swath directly under their orbit. This means that they have to build up a picture of the state of the globe over the course of many orbits, and hence only return to the same spot after a long delay. The interaction of height of orbit, swath width of observation on the sea surface, repeat period before returning to the same track, and inclination of the orbit to the equator usually means that we need several identical satellites to provide sufficient cover and continuity. The ocean changes rapidly, with fluctuating eddies and gyres of the order of 100 km (60 mi) across. Over the decades after 1978, there were continuous efforts to obtain enough international funding to launch identical satellites and program their orbits to be complementary and produce consistent data.

To the dismay of oceanographers all over the world, the SEASAT satellite broke down completely in October 1978, after operating for only 106 days. The measurements obtained by scientists had been excellent, but a massive short circuit brought the experiment to a sudden stop. Almost immediately rumours circulated that the instruments had unexpectedly revealed military secrets, such as the tracks of submarines, and that the satellite had been deliberately switched off. This conspiracy theory is still on the web. It is difficult to believe since many similar sets of instruments have flown on satellites in the following decades, often with greatly improved accuracy. If SEASAT was revealing secrets, why did the same inconvenient revelations not appear with the later satellite missions? In spite of its short life, SEASAT paved the way successfully for all that followed. Scientists and space engineers have ensured a continuous increase in the range of marine characteristics that could be measured from space, so that modern space missions have measured the sea surface temperature, global gravity field, fluctuations of sea level, occurrence and movement of oil slicks, currents, sea ice, ocean colour, phytoplankton density, and even the salinity of the surface seawater.

From the mid-1980s onwards, as head of MIAS, UK delegate to the Intergovernmental Oceanographic Commission, chairman of IODE (Fig. 9.5), and chairman of the World Data Centres Ocean Data Panel, I was peripherally involved in the planning of the instrument suites and data formats for the ERS satellites launched in 1991 and 1995. At meetings of the Global Ocean Observing System (GOOS) planning committee and other climate and ocean groups, we discussed the programmes for Topex-Poseidon (1992), SeaWiFS

Figure 9.5. Delegates of the International Oceanographic Data Exchange Committee of the Intergovernmental Oceanographic Commission meeting in the UN Secretariat building in New York, 1984. Dieter Kohnke, Chairman of the Committee at that date, is standing in the centre. I was elected Chairman of IODE in 1986, at the next meeting which was in Moscow.

(1997 to 2010, to measure ocean chlorophyll productivity), Jason-1 (2001, followed by Jason-2 in 2008 and Jason-3 in 2016), Quickscat (1997), Envisat (2002), and Cryosat (whose launch failed, 2005). There were also meetings of the international bodies that coordinated all satellite launches, at which the oceanographic community pleaded, sometimes successfully, to ensure that its preferred sensors were included on future launches. I was a very small cog in these big wheels, but it provided a fascinating insight into how a vital element of global oceanography has grown to its present established status.

An essential step in the evolution of operational oceanography from space was to transfer at least some missions from a science research budget to an operational budget, where agencies accepted repeat missions of similar instruments to provide economic or environmental management benefits. EUMETSAT, the European agency in Darmstadt, Germany, that coordinates remote sensing for the European national meteorological offices, hosted a conference that I organized in 2000, at which experts from many industries – offshore oil, shipping, safety at sea, fishing, pollution management, oil slick monitoring, and so on – presented the collective case for including oceanographic operational missions within the EUMETSAT terms of reference. EUMETSAT now routinely supports the analysis and distribution of data from operational marine observing satellites such as the Jason-2 and -3, Sentinel-3, and later Sentinel series.

* * *

The culture shift from oceanography as a scientific research discipline to an operational suite of agencies delivering information, maps, and data products every hour of the day required a huge change in the structure and objectives of dozens of agencies, both national and international. Politicians and research councils were reluctant to found totally new bodies, so we had to find ways of building on existing programmes and agencies, while grafting on the minimum of new structures to handle the real-time operational information flow. There was a constant jockeying for position, with some scientists worried that they would lose their budgets, or that the release of data quickly to the public would undermine their own future ability to publish original papers. After one tense meeting, the president of a well-known university in the USA exclaimed in anger: “Some of these oceanographers think they’re a fucking priesthood.”

During the 1980s, the concern about global warming because of the increase in CO_2 in the atmosphere provoked fears generally about climate change, and it was obvious that changes in the ocean itself had to be included in climate predictions. Much of the CO_2 emitted by combustion dissolves in the ocean, and the incoming solar heat which is absorbed in the ocean is transported around the globe by the currents and rotating eddies. Cold water from the ocean depths can be transported long distances on the seafloor and surface hundreds of years later to influence the atmosphere. In short, the ocean has underwater “weather” inside it, with warm and cold rotating eddies, fronts, and stratification of layers of different densities, analogous with the weather in the atmosphere. How could the integration of atmosphere and ocean be achieved in computer models? Very tentatively the idea of an operational GOOS that could provide the missing technology, staff, instruments, data, and investment, began to take shape.

My boss at NERC, John Woods, was one of the leading professionals with the skill to envisage and help to design GOOS.

Public anxiety about global warming was fanned by environmentalists and equally down-played by the oil companies and the coal industry. Both sides tended to overstate their case. Governments found it difficult to ignore or assess the conflicting claims and interests of the many factions. The first meeting of the Intergovernmental Panel on Climate Change (IPCC) was in 1988, from which the first Assessment Report appeared in 1990. During the 1990s, politicians and industry treated the IPCC with much scepticism, and the scientists involved became cautious about what they claimed was the contribution of human sources of CO_2 to global warming. Somewhat surprisingly, the 1990 report did say confidently that human sources of CO_2 were adding to natural processes. Quoting from the executive summary:

> We are certain of the following: there is a natural greenhouse effect...; emissions resulting from human activities are substantially increasing the atmospheric concentrations of the greenhouse gases: CO_2, methane, CFCs and nitrous oxide. These increases will enhance the greenhouse effect, resulting on average in an additional warming of the Earth's surface. The main greenhouse gas, water vapour, will increase in response to global warming and further enhance it.

As the conflict with industry intensified, subsequent IPCC reports became more tentative, and paid more attention to the causes of uncertainty. There was a reason for this. The fossil fuel energy companies adopted a position of "case not proven". They published analyses that drew attention to possible errors, finding a few scientists to defend their position, while they spread doubt and uncertainty in any way they could. The confusion made it difficult for governments to take a strong position and decide on new regulations. The tactic of the oil and coal industries was similar to that used by the tobacco industry in the 1960s and '70s to conceal the harmful health impact of tobacco. Nevertheless, as the decade progressed, the evidence for human influence on climate became incontrovertible, and IPCC reports after 2000 have strengthened the case further.

In the UK, the House of Lords Select Committee on Science and Technology recommended in July 1986 that the government should establish a Co-ordinating Committee on Marine Science and Technology (CCMST). John Woods, as director of marine and atmospheric science at NERC, arranged for Sir John Mason, the recently retired director of the Met Office, to be the chairman, and asked me to serve as secretary to the committee. The members included Sir Anthony Laughton, Director of the NIO/IOS, professors of marine sciences, marine biology, and engineering from several universities, and representatives of government departments such as the Department of Agriculture and Fisheries for Scotland (DAFS), Department of Energy (DEn), Ministry of Agriculture Fisheries and Food (MAFF), Foreign and Commonwealth Office, the Research Councils, NERC, Scientific and Engineering Research Council (SERC), Ministry of Defence (MoD) and the Cabinet Office. We first met in January 1988. Subsequently we met every month, publishing our main report in 1990, with several more specialized, narrowly focused reports in the next two years.

For part of the CCMST study, we estimated that all marine industries and services together contributed £16 billion per year to the British economy, equivalent to 5% of the

GDP. The national annual expenditure on marine research and technical development was £400 million, only 2.5% of the value added to GDP. The costs of research were almost equally shared between government and industry. About 1750 professional scientists and technologists were employed in universities and government laboratories, but they were scattered between too many small sites, with insufficient capital installations, and sometimes with outdated equipment.

To estimate the economic importance and priorities of operational oceanography in the UK, I conducted a survey of 50 companies and agencies which were potential users of UK operational ocean data or GOOS data. They provided rankings of their preferred information selected from over 100 ocean data types, with rankings of requirements in terms of spatial and temporal resolution, accuracy, forecast periods, etc. These ranked tables helped to establish the national priorities in developing operational oceanography, and to design global systems which would be important to the UK marine industries.

CCMST strongly supported the decision taken by NERC and the University Grants Commission to concentrate several existing laboratories, satellite remote sensing capacity, computer modelling, and research ship support at one site on the coast. This ambitious move to Southampton, creating a new National Oceanography Centre, would give Britain an oceanographic focus with a strength comparable to that already existing in Germany, France, and the USA. John Woods masterminded this important concentration of research skills. As a research strategy, CCMST stressed that the UK should participate to the maximum in measuring the whole global ocean and making computer predictive models, which would be a benefit to industry, services, saving life at sea, and protection of the environment. This policy would also ensure that Britain was engaged in the research needed to understand climate change.

The CCMST later morphed into the Inter-Agency Committee for Marine Science and Technology, for which I continued to be the secretary until 1992, and then it changed again into the Marine Science Co-ordination Committee, which continues to this day.

* * *

In 1988 Jim Baker was appointed by IOC to explore the possibilities of GOOS, and I had meetings with Jim in Washington, London, and Paris to discuss plans. At a critical meeting of the General Assembly of the Intergovernmental Oceanographic Commission in Paris in the late ’80s, the wonderful common sense of Alexander Yankov came into my life once again. As a member of the British delegation to IOC, the meeting appointed me during the two-week General Assembly to chair a drafting group on the need for a GOOS. I had to submit the decision of the drafting group as a resolution to the General Assembly and, if approved, it would be in the final report. Yankov chaired IOC at this date and, although the creation of GOOS was hotly contested by some delegates, he assured with diplomatic finesse that the assembly approved my report.

Notwithstanding the continuous diplomatic shuffling, realpolitik, and compromises, the international meetings sometimes had intriguing or amusing events in the margins.

In 1988, I attended a meeting of world data centres and experts in ocean data management in Moscow during which Jay and I were accommodated in a hotel for the Academy of Sciences and their guests. During the meeting, the participants travelled by bus to Obninsk, (Fig. 9.6), a closed research city, where we were able to descend into

Figure 9.6. Delegates at an Ocean World Data Centres meeting in Obninsk, USSR, in 1988. Jay Kleinberg is standing on my right wearing dark glasses.

an ancient salt mine that had been converted into a geophysical listening laboratory to evaluate earthquakes and underground atomic bomb tests. Glasnost was taking hold and, as one Russian lecturer explained, with a smile, "If I had given you this lecture last week and showed you these maps then, I would have been sent to prison."

One evening, Iouri Oliounine, the secretary of IODE, invited Jay and me to join him with several others from the oceanographic community for dinner in a fashionable new Moscow restaurant. The attraction was not the wine list, or a rare vodka, or a new chef, but Coca-Cola, which, as we discovered when we arrived, was served in huge glass jugs standing on each table, gradually warming up and losing its fizz. Not to worry, it was Coke, and the wine and vodka and food were excellent.

A small band played lively tunes in the corner beside the dance floor, which was surrounded by the crowded tables of diners. Jay danced with me and there was enough room for me to spin around on my wheelchair. An exotic party of guests occupied the long table two away from ours. The men had dramatic black moustaches and craggy eyebrows above prominent bony straight noses. They wore wide-sleeved jackets which might have concealed knives or guns tucked into their belts. The women looked as if they were well able to manage the men. They had long flowing black hair, scarlet painted lips, broad cheekbones, and were of muscular and athletic build. Tonight they were celebrating, but there was no doubt that when they had to do a day's work in the farm fields they could compete with any man. Perhaps they came from Chechnya, Georgia, or even Azerbaijan. Their table was covered in empty vodka and wine bottles.

As the dance band stopped, and while Jay and I returned to our table, one of these exotic beauties leapt to her feet, took a red rose from a vase of flowers, placed the rose between her teeth, and came sashaying towards me. Jay sat down and abandoned me.

The beauty was swaying her hips in sensuous thrusts as she raised her hands above her head and clapped. Her brightly coloured skirts swayed and billowed as she invited me to dance. The band struck up again. I balanced on my back wheels, dancing happily, but I secretly hoped that my dancing partner had agreed her plans with her ferocious male companions. There was no hail of bullets or flashing knives.

Other dancers came onto the floor, but gave us plenty of clearance as I spun and pirouetted on my chair, while my partner swayed and skipped with agility. We established a communicating mutual rhythm. As the music stopped, she removed the rose from her teeth and kissed me on both cheeks before prancing back to her companions, who clapped wildly. Perhaps she had done it for a bet. I refreshed myself with warm Coke.

* * *

John Woods, who was in touch with the World Meteorological Organization (WMO) and had monitored the plans for the Global Climate Observing System (GCOS), was concerned that operational oceanography could just disappear into the meteorological monolith if there was not a clearly separate GOOS. Traditionally, national meteorological offices, coordinated by the WMO, collected weather information by radio from ships, and combined this in less than 24 hours to provide marine meteorological forecasts of wind strength, visually estimated wave height, and risks of ice or fog. If the processing and calculation of the forecast took any longer, it would be useless. Since this service depended on the standardized routes taken by ships, and a thinly scattered system of weather ships and drifting buoys, the accuracy of the global service was, by modern standards, primitive. A parallel programme, managed by IOC, consisted of a device known as an expendable bathy-thermograph (XBT), which was thrown overboard from a few ships to measure the change of temperature with depth in the upper ocean. This again depended mostly on routine commercial ships' tracks. Huge areas of the sea were never measured. But with its goal of creating the GCOS, the WMO had the advantage that it was grafting its new objective onto a worldwide operational forecasting system that already paid for itself. There was an established community of users for shipping forecasts and general weather forecasts. To our disadvantage, as we tried to invent GOOS, there was no such pre-existing population of users familiar with real-time ocean data. We had to discover their needs, and then engage our potential customers by showing them the value of the new types of information.

John and I attended the Second World Climate Conference in Geneva in November 1990. The conference was in the palatial glass and marble tower block that is the head office of the World Meteorological Organization. We did our best to sound out support for a specialized GOOS, and to make the point that there were major aspects of the ocean below the surface which meteorologists were in no position to measure and study scientifically.

During the conference, a speaker from Woods Hole Oceanographic Institution gave a dramatic presentation of the first measurements telemetered from a new kind of oceanographic neutral buoyancy float. The technology was an advance on the previous float design invented many years earlier by British oceanographer John Swallow. The new float was a metal cylinder less compressible than seawater, so that it could be programmed to sink slowly to a depth of 1000 m or 2000 m (3281 ft or 6562 ft), measuring the water salinity and temperature as it sank, stop at the programmed depth, drift with the deep current, and then ascend again, measuring the water as it went, and then dump its recorded data and position by satellite to a laboratory on land. The cycle of sinking, drifting, and rising back to the surface was repeated hundreds of times until the batteries expired, possibly lasting for several years. Manufacturers in various countries adapted and changed this revolutionary device over the years. It now provides the technology for the Argo Programme, which, in 2018/19, uses 3800 such profiling floats that continuously measure the changing internal structure of every ocean basin in the world.

Bert Bolin reported on the IPCC meeting from 1988, and the evidence he showed for human alteration of climate gave a strong motivation to agree on a detailed monitoring programme. But the Second World Climate Conference struggled to reach agreement and issued a Ministerial Declaration that disappointed everyone. The environmentalists and those scientists most worried about climate change objected to its tentative level of commitment. Governments were worried about costs and about restrictions on their industries. The fossil fuel industries were terrified of potential criticism and the possible future transition to renewable energy sources. However, the GCOS was created formally in 1992, and this led to the UN Framework on Climate Change and the Kyoto Protocol in 1997. The Kyoto Protocol was adopted in Japan on 11th December, 1997. When the US Vice President Al Gore was interviewed at a press conference on arrival in the USA after signing the Kyoto Protocol, and being challenged that it would never be ratified by Congress, he replied, "The American public is not stupid."

Back in the UK, John asked me to prepare a report that would set out the arguments in favour of GOOS, showing that a global analysis of marine data, measured continuously and repeatedly in real-time, was essential to provide scientific understanding for ocean models, and would support the generation of local forecasts to help industry, governments, and other agencies. Extending the techniques that I had used for analysing the UK strategy for marine science and technology, I searched for global information of the same kind. John recruited an international team of globally respected oceanographers to review my draft, and Walter Munk hosted the meeting at the Scripps Institution of Oceanography, La Jolla, California, in August 1992. In parallel with this activity, IOC formally decided in 1991 to create GOOS in collaboration with other UN agencies, which it announced at the UN Conference on the Environment and Development in Rio de Janeiro in June 1992.

Our group of 12 scientists, each with different marine disciplines, met in the Institute of Planetary Physics at Scripps, which is a fabulously modernistic timber and glass building cantilevered out over the cliffs at La Jolla, providing ever-changing moody views of the Pacific Ocean. We dissected, revised, and improved my original draft during the course of three days, before circulating it to more than 100 oceanographers around the world requesting their comments. The Intergovernmental Oceanographic Commission published our final report early in 1993 as "The Case for GOOS". We assessed the main uses of the ocean and coastal seas grouped under the headings of the Geophysics of the Ocean Floor, the Coastal Seas, Ocean-Atmosphere Climate, Science and Technology, Living Resources (like fish, shellfish, and seaweed), Health of the Ocean (concerning pollution and loss of habitats), and finally, Shipping and Defence. We estimated that it would take 15 years to create a working GOOS, with much of the time devoted to global experiments which were already in preparation, while computer modellers would produce new algorithms and operational services as trials or prototypes.

The economic case for investing many millions of dollars in GOOS derived from the calculation that the wealthy G7 countries were already spending about $10 billion per year on all aspects of marine research, of which they allocated about one-tenth to marine environmental research, and biological, physical, and chemical oceanography. Thus a baseline of measurements for GOOS, costing about $1 billion per year, already existed, but with no plan to manage the measurements in an operational real-time way. The G7 countries at that date operated about 200 civilian research ships in all the world oceans. Marine data already available to the public domain was costing $70 million per year, as a good start.

We estimated that the annual global value of marine industries and services was conservatively in the range of $600-800 billion. To this one should add the unquantifiable value of environmental and ethical benefits. The existing operational meteorological weather forecast system cost about $2 billion per year. If the expenditure on operational oceanography increased to the same level as meteorology by 2010-2015, doubling the existing expenditure on marine environmental measurements, then the resulting increase in efficiency within marine industries in the order of 1% would pay for the investment many times over.

Forecasts and predictions for the coastal seas, where most economic activity occurs, together with the positive and negative environmental processes, showed serious errors caused by our ignorance of the processes happening in the open ocean. Only global ocean models permitted reliable forecasts, as with the atmosphere and the established daily weather broadcasts. But the multiple eddies in the ocean, only 100 km (60 mi) in diameter, compared with the typical meteorological features of 1000 km (600 mi), meant that the ocean models had to be much more accurate and fine-gridded, taking more computer power.

The scientists at the La Jolla meeting knew only too well that the design of GOOS in detail still depended on the results of several large global experiments that were either already under way or would start in the near future. The Tropical Ocean Global Atmosphere experiment (1985-1995) focused around an array of multiple lines of deepwater instrumented buoys moored across the equator in the Pacific Ocean, from central America to Papua New Guinea. This vast array of more than 70 buoys moored in water 6 km (4 mi) deep has been updated and is still working as the TAO-TRITON array, providing data to the Tropical Pacific Observing System. The World Ocean Circulation Experiment (WOCE, 1990-1998) consisted of a global pattern of multi-year research ship cruises, requiring measurements of the seawater from the surface to the ocean bottom in every sea basin. The Joint Global Ocean Flux Study (1991-2000) was an international study of the way that carbon and carbon dioxide are transported through the atmosphere and ocean. The Global Energy and Water Cycle Exchange project (1995-2000 and beyond) was designed to understand the flow of water and solar energy on the globe through the atmosphere and ocean surface. The scientists working on it have continued the project to the present day. The Global Ocean Ecosystem Dynamics project (1991-2010) sought to understand the processes in the world ocean ecosystems and their response to climate change, with the ability to forecast ecosystem response. And the Climate and Ocean Variability and Predictability and Change project (1995-2018), whose name more or less defines itself, is still in existence.

To put it mildly, the design of where GOOS instruments should be placed, what properties of the ocean should be measured, what satellite instruments were needed, and how to create the best computer models, was going to build on good scientific knowledge. GOOS could not be rushed into existence on the basis of ignorance and enthusiasm.

The cost of this colossal series of experiments was borne by the international co-ordinating bodies of the United Nations agencies, and the International Council of Scientific Unions, plus, of course, the commitment of national governments and agencies to provide research ships, satellite communication networks, computers, and laboratory staff. We had to show that the benefits would be worth this cost.

* * *

IOC gradually built up the collaboration between countries and existing international programmes to make the foundation for GOOS, while maritime agencies in European countries collaborated to make a strong regional component, known, inevitably, as EuroGOOS. European collaboration was essential because we had the necessary skills already scattered between the countries, but not a central focus that could compete with the USA or Japan. The British Met Office, MetéoFrance, and IFREMER provided the main mass of computing power, and Pierre Papon, director of IFREMER, was very keen to see EuroGOOS launched as soon as possible. We set up EuroGOOS as a technocratic cross-border association between agencies such as departments of fisheries, environmental management, coastal defences, offshore energy, and marine science, rather than through a government-to-government treaty. This made EuroGOOS more flexible, with multiple expertise from each country, and able to take quick decisions when needed. Pierre Papon chaired the first meeting of European agency representatives in Rome in December 1994, acting as interim chairman. The meeting elected John Woods as the first

Figure 9.7. The first meeting of the European Global Ocean Observing System at Den Haag, Netherlands, in 1996. John Woods, Chairman of EuroGOOS, is the tallest person in the front row.

chairman of EuroGOOS, and appointed me as the full-time director of the EuroGOOS office. Michel Gauthier, from IFREMER, became my scientific assistant.

EuroGOOS started with 12 member agencies, but within two years we built up to 22 national agencies from 14 countries. In 1996, we held the first EuroGOOS conference at Den Haag, Netherlands, attended by over 300 people (Fig. 9.7). Today EuroGOOS has 41 member agencies from 19 European countries, and is registered in Brussels under Belgian law as a legally constituted non-profit foundation with no commercial goals.

The purpose of EuroGOOS was to promote and launch very detailed networks of recording instruments and forecasting services in the local seas such as the Baltic, North Sea, Aegean, and so on. These needed the input of extra data provided by the bigger models in the Atlantic and globally in which the European agencies were major participants. You only have to see the everyday weather forecast to appreciate the importance of the Atlantic dimension, and, because of the measurements being made by US agencies, we maintained close collaboration with colleagues in the USA.

In 1998, I still wanted to obtain stronger evidence from the wider European market that would confirm the complete range of types of forecasts that users needed. In practice, we expected small intermediary companies to spring up to interpret and provide easily grasped visualizations for the end user. This did indeed happen. We devised a survey questionnaire, translated into Spanish, Greek, and Italian, and distributed it to 155 marine organizations in Denmark, the Netherlands, Spain, Greece, and Italy, as well as the UK. The recipients were invited to identify their own requirements from a list of 136 measured sea properties, and were asked to describe their own activities from a list of 116 ways in which people work at sea or on the coast.

The responses were amazingly professional and careful, showing that answering questionnaires is not a boring chore if the questions are really useful. Recipients will take a lot of trouble if you can show people that you understand their problems. But I was astonished to discover that every one of the 136 variables was wanted by someone, except for information about manganese nodules. Furthermore, they wanted the data now, accurately, and with forecasts. There was indeed a strong demand for operational data from a well-informed user community. Their requests included the obvious physical variables such as sea surface temperature, salinity, currents, and wave height, but phytoplankton, chlorophyll, nutrients, and oxygen concentrations all appeared as requirements in the top 40 ranking. More surprising were the requests for optical light intensity, anthropogenic noise underwater, dissolved oxygen, and deep ocean carbon storage. GOOS was going to be busy satisfying these discerning customers.

* * *

The agencies at a global level continued to jostle and compete for advantage in the fight for funds, and there was considerable objection to the rapid growth of EuroGOOS, and to some of the aims of GOOS itself. The skirmishing between the big bureaucratic agencies, scientific budgets against operational budgets, economic goals against pure science research goals, government independence against UN agency supervision or oversight, continued. The turf wars threatened to delay GOOS indefinitely. Several countries set up and funded their own "global" programmes with different acronyms as they fought for the bureaucratic high ground.

In 1996, the GOOS Joint Scientific and Technical Steering Group, of which I was a member, decided that a major book-sized report was needed, bringing together the multiple projects and programmes already in existence that could contribute to GOOS. It would include the ideas and plans that had been produced by several different agencies, the valuable national and industrial programmes, and establish once and for all the logical reason for spending large sums of money on integrating GOOS. The steering group appointed John Woods to chair the drafting review group; I was asked to serve as editor; while Peter Ryder, a retired senior scientist from the UK Met Office, conducted literature research and wrote the first draft.

Ryder gathered an astonishing mass of information on different marine industries, their dependence on marine subsea forecasts, and the new technology that could provide the data they needed. The book was profusely illustrated. As it neared publication, I was copyediting thick stacks of eight-page proofs in full colour, spread all over our dining-room table. The volume was published as the *GOOS Prospectus 1998*, and set out in technical detail the way in which so many components could gradually be coordinated, using modern electronic communications, to make an integrated system. The logical case started with the numerous UN conventions which demanded more prompt and efficient information on the sea to make decisions in real-time, showing how the various institutions and projects – some of which had been in existence for many decades – could transfer data to GOOS without losing their unique skills or strengths that depended on specialized academic membership.

By now, more countries had made estimates of the proportion of their GDP attributable to marine industries and services: the USA estimated 2.6% or $109 billion per year; Australia 3.9% or Aus $9 billion, and the UK 4.8% or £27.8 billion. The conclusion is that the total benefit produced by all marine industries and services is typically 3-5% of GDP in all developed countries, and in some cases, much more. Some 20% of oil and gas was produced from offshore, worth, in 1990, $135 billion. Already we could see that offshore renewables would eventually develop, but we did not anticipate correctly how rapidly many countries would build offshore windfarms. The Dong offshore field of turbines was built in Danish coastal seas in 1991, but the technology accelerated rapidly only after 2010. Overall, Ryder estimated that the data and forecasts provided by GOOS could improve the efficiency of marine economic activities by 1%, with the proviso that this figure could vary regionally by a factor of two, that is, within the range 0.5–2.0%. To this should be added the more difficult valuations of ecological and aesthetic protection of the sea and wildlife.

Ryder listed all the ocean-observing satellite missions, including those already launched and those planned up to 2010. The total number of satellites, both approved for funding and proposed, was an astonishing 82 launches.

New technologies proposed included long-range autonomous underwater vehicles, subsea instrumented gliders, profiling neutral floats, moored buoys with instruments fixed at many depths on mooring cables, and drifting buoys on the surface. In all parts of the world we could show that industries and governments needed local observing instruments and coastal models to maximize the benefit of the data acquired in the open ocean and the information from the models that made use of them. That is, the processes detected in the open ocean would be used to control the outer boundaries of coastal models, making them more realistic and giving them longer forecasting limits.

Twenty years later, the *GOOS Prospectus 1998* is still a useful description of GOOS and its benefits between one set of covers. On the web you can find the GOOS homepage, an up-to-date outline of the different expert panels, and especially the list of Essential Ocean Variables that are measured and analysed continuously worldwide.

* * *

The completion of the *GOOS Prospectus* in 1998 showed a way to resolve the endlessly simmering conflict between the short-term, real-time "quick and dirty" information services for industry and coastal management, on the one hand, and the multi-decade "pure science" objectives of climate forecasting on the other. (I am parodying extreme attitudes here. The two depend on each other, but many people still needed persuading.) The clear truth was that many of the instruments and innovations needed to achieve the two goals were similar, although the short-term system needed extra instruments and models in the coastal seas, while the big ocean climate models required extra instruments in the remotest parts of the globe. Also, climate research and modelling required greater control of the quality and accuracy of the data, with more frequent instrument calibration, and screening of vast data flows to eliminate errors and identify gaps. Because climate researchers did not have to produce results within 24 hours, they naturally had time to do these checks. But the immediate benefits to be obtained from the commercially-useful data should be seen by governments as offsetting the expenditure needed for long-term climate forecasting. I wrote a paper explaining how this could work, which I presented at the Ocean Observations Conference in St Raphael, France, in late 1999. Jim Baker was intrigued by my argument, and invited me to join a drafting team headed by a senior NOAA economist, to write a paper justifying to the US Congress their inter-agency financial contributions to GOOS, known as the Integrated Sustained Ocean Observing System.

The OceanObs meeting at St Raphael concluded with a boisterous party for the scientists and technologists at a nearby monastery, which the monks had abandoned long ago and was run as a commercial vineyard and winery. After a tour of the wine cellars, we dined luxuriously in a vast stone-vaulted hall. As we sipped coffee at the end of the meal, the unmistakable music of Offenbach's *Galop Infernal* burst from loudspeakers, and six "cancan girls" appeared on a stage dancing with the traditional high kicks and flouncing skirts. At the finale the women ran amongst the dining tables, kissing many of the guests before fleeing through the exit doors.

My neighbour at dinner, an elderly Japanese professor turned to me with a serious expression on his face.

"Is it true that this is a religious building?"

"Well, it was. But the monks left a hundred years ago. It has been a commercial winery now for many decades."

"But," said the professor, "it does seem odd to hold a party like this in such a building."

"You have to realize that this is the extreme south of France. We are in the Mediterranean world," I said, trying to defend the organizers of the conference. "Maybe in the north of France they would not have invited the dancers like that."

The professor was not convinced. Nor was I.

* * *

The process of banging heads together and connecting the activities of different agencies and panels and sub-groups that had, for many years, protected their intellectual privacy, continued remorselessly. At a purely bureaucratic level, this collaboration and convergence of ideas is perhaps best symbolized by the formation in June 2001 of the Joint Commission on Oceanography and Marine Meteorology (JCOMM), reporting to IOC and WMO, both of which are UN agencies, though of differing scale. Of course acronym mergers like this solve nothing unless the experts in all the associated institutes, agencies, and laboratories really want it to work. In my estimation, they did and still do. We held the first JCOMM meeting for two weeks around Midsummer's Day at Akureyri in northern Iceland. The tulips and daffodils were just coming into bloom. The JCOMM appointed co-presidents, one of whom – the co-president for oceanography – was Savithri Narayanan, who directed the Canadian Oceanographic Data Centre.

On Midsummer Eve, a small group of us took a boat out to sea, surrounded by whales and dolphins and millions of seabirds. We watched the sun never set on the Arctic Circle. The sea everywhere was boiling with fish jumping up or swirling just below the surface. The torrential screaming of whirling birds diving like javelins to catch the fish, combined with the repeated leaps of whales and dolphins, made the boat-load of tiny humans seem irrelevant, mere flotsam. We drifted in the half-light on a pullulating primaeval ocean that did not belong to us. The sense of our fragile insignificance was frightening. My scalp tingled with a slow fear.

I retired from my post at EuroGOOS at the end of 2001, but retained membership on some GOOS committees and working groups, as well as becoming involved in the ocean component of the EU Global Monitoring for Environment and Security programme, and serving on panels evaluating applications for European Commission funding of marine research. I edited some of the EuroGOOS conference publications and books produced by IOC on marine observations and forecasts. These are the sorts of things that retired academics do. The most challenging task for me was to apply cost-benefit analysis in more detail to the myriad marine observing technologies and their outputs for industry and government services. This study was called "Sustained Efficient Production of Required Information Services", or SEPRISE. EuroGOOS commissioned the book and submitted it to the EU Copernicus programme in the Directorate of Research, DG XII, in 2006. Very conveniently, Nicholas Stern's comprehensive analysis of the risks and potential damage from climate change was published during 2006 while I was working on the SEPRISE assessment, so that I was able to include the results of some of his calculations.

* * *

My experience of decades spent engaged partly in bureaucracy leads me to an optimistic saying: "One good piece of paper can stop ten bad pieces." I use the word bureaucracy in its rare non-pejorative sense.

The world of expertise needs to have some decisions and agreed goals written down. Specialists have only themselves to blame if they avoid all committee work, letting the decisions be taken by ill-informed pen-pushers and keyboard-tappers. It is a necessary pact with the devil. A scientist should be doing science. But a bit of time spent on committees making bold plans, defining goals, raising funds, and setting standards of safety or training may prevent the selection of irrelevant goals, or the evolution of rules that make science impossible. It is a murky compromise, but one that I have found worth accepting at times. A well-written, technically accurate document that is recognized as sensible by scientists or engineers can avoid dozens of mistakes, revisions, explanatory supplements, appeals, and complex exemptions that arise from a badly drafted one. Everybody's life becomes easier, and they can get on with their work.

Even back in the 1990s we were already aware of the paradox that climate research scientists and policy-makers were clocking up an awful lot of air miles, generating huge personal carbon footprints, thus shamefully contributing yet further to climate change. At the time it seemed unavoidable. Twenty years later this paradox has not changed. A study of the attitudes of climate scientists to travel shows that they use the same mileage in proportion to their commitments as other scientists and the general public. Once you know really well the people you have to work with, it is easy to continue the work by e-mail, Skype, Zoom, and supported by Dropbox and other ways of sharing large quantities of data. But meetings are needed from time to time, and when negotiations become fraught, the only way to reach a consensus is around a table. When every word and comma matters, when a single phrase can obliterate or support somebody's hidden agenda, you have to be able to see the facial expressions and body language of every person involved, not just the one who is speaking. Electronic video conferences still seem clunky and do not produce complete understanding between the participants.

This chapter has treated, more or less in chronological order, an evolving series of national and international organizations, conferences, committees, UN agencies, and European bodies which developed over the 50 years from 1968 onwards. In the next phase, the United Nations began, in September 2018, to plan an extension of the Convention on the Law of the Sea to control the environmental impacts of deep-sea mining, bottom trawling, and other destructive activities, while creating large areas for complete conservation of the ocean. They will need vast quantities of reliable scientific data if they are going to make the right decisions. To satisfy this increasing demand for observational measurements throughout the ocean from top to bottom, and from space, the United Nations has launched the Decade of Ocean Science for Sustainable Development (2021-2030).

Information on the disastrous consequences of climate change is now available to the public from many conservation and Green lobby groups. The crises of global warming, increase of desert areas, loss of cultivatable soil, rise of sea level, coastal erosion, extinction of hundreds of species of plants and animals, loss of biodiversity, droughts, intensification of floods, loss of glaciers, melting ice caps, and other threats are now explained and debated in national parliaments and reported through all kinds of media and social networks. The urgency of finding solutions to these problems, countering the threats, and reversing the degradation has not been exaggerated. The prospect is terrifying.

A panicky response may please some politicians with grand gestures. But success will only come from a huge improvement in the technology of carbon reduction with alternative sources of energy, and the provision of staggering volumes of environmental data so that we can understand speedily what is happening in detail all over the globe. It is essential to discover rapidly what technologies, schemes, and laws produce benefits, and which are failing or making matters worse. We are racing against the clock, and without measurement and data on this scale we will get the decisions wrong and possibly waste decades, resulting in disaster.

In most cases the organizations and conferences in which I participated have produced official reports that are publicly available, and can be consulted by readers seriously interested. In some cases such reports have been followed later by books, commentaries, analyses, or counter-proposals. The intellectual challenges of these meetings and the problem of condensing the ideas and arguments into accessible reports kept me continuously engaged. I enjoyed the experience, which was at times exhilarating, and led to my friendship with many scientists much cleverer than me.

One meeting of the GOOS Steering Group at Vina del Mar, just north of Valparaiso in Chile, in mid-March 2001, had a special frisson. We had just gathered for coffee and biscuits in the committee meeting room on the second day when the building started to shake slowly and wobble. It was not a rapid vibration, but a curious shaking and trembling with several thrusts and lurches each second, which went on for a minute or more. I moved swiftly under a strong roof beam by the door, on the assumption that the middle of the floors would collapse first in an earthquake. Some others did the same, while most stood frozen and puzzled with their coffee cups poised in mid-air. When the earthquake stopped, the conversation resumed, and we sat down for our meeting. An hour later, the delegate from the Department of Geophysics of the University of Chile came round the table distributing framed certificates showing the position of the magnitude-5.3 earthquake that had occurred a few kilometres offshore from the city at two minutes past nine o'clock on 15th March. The focus was at a depth of 46 km (29 mi) below the seabed. Plate tectonics was doing its thing. That evening I went along the esplanade at the seafront to see if there was any damage, but the rows of high-rise apartment buildings with glass balconies were intact. At ground level, shop display windows were undamaged. A man with a harp sat in the street playing local melodies. They have good building regulations in Chile.

* * *

I retired from research and administration connected with GOOS nearly 20 years ago. I was in the middle of a maelstrom of change in the 1980s and '90s, travelling to 45 countries, as we made the transition from treating oceanography as an esoteric science, to seeing it as part of a global system of measurements that provided model-based computer forecasts on every timescale from days to centuries. That ability is now more important than ever. The threats of climate change are more sharply defined, and the need to understand where there are limits or tipping points that could be irreversible is almost terrifying. It is possible, but still unknown as to probability, that a gigantic melting of glaciers and ice shelves, could transmute planet Earth into a new and hostile climate from which it would be impossible to escape. So much of the knowledge of the sea, and so much of the operational aspects of it, are still strangely concealed from the public. The media and popular science writers do not seem to be able to find a way to present these great sweeping ideas in a simple and graphic manner. Perhaps they are still hypnotized by the idea of the unknowable mysteries of the sea. I hope, notwithstanding the rather nerdish spirit of parts of this chapter, that I have provoked interests which the reader will follow with curiosity.

10

Stone Age Lands Beneath the Waves

I snorkelled slowly round the island of Aghios Petros (Fig. 10.1), one of the Northern Sporadhes islands in the Aegean. First along the northern side of the island, round the rocky tip, then back slowly along the southern side towards the main shore of Kyra Panagia. The sea was flat calm, the water gin clear. The sun beat down on my back, cooled by the slosh of water as I worked my arms in slow sweeps. The circuit was less than 400 m (0.3 mi) and took about 20 minutes. I was searching the seafloor around the island for signs of a Neolithic prehistoric settlement. I was not sure what these would look like. The remains, if they existed, would be 6000-7000 years old, but would not include any squared

Figure 10.1. The island of Aghios Petros in a small bay in the island of Kyra Panagia, with, in the background far right, the island of Alonissos. I swam round this island several times in 1981 searching for submerged Neolithic remains on the seafloor. Note that vegetation grows very close to sea level on the sheltered end of the island, and is much higher above the sea at the far end where it is exposed to bigger waves (Photo Jay Kleinberg).

In: Flemming, N.C., 2021, *Apollonia on my Mind. The memoir of a paraplegic ocean scientist*, 383-426. Leiden: Sidestone Press.

stone blocks or major signs of buildings. Mud bricks don't last in seawater. Nobody knew exactly what we should be looking for. I stared downwards and hoped.

The slow circuit showed me that each bit of the seafloor looked just about what I would expect, given its exposure to the open sea, or shelter, supply of sediments, and proximity to the shore. There was an icy-cold freshwater spring spouting up from the seabed in a mirage of transparency close to the rocky beach. I could rule out large areas of bare rock or pure white sand. But what about the rest? Much of the bottom was a thin covering of green algae on bedrock.

Let's go round again and think very hard.

Where the coast of Aghios Petros was closest to Kyra Panagia, leaving a watery gap of only 100 m (328 ft), there was a rocky submerged isthmus with scattered pebbles and stones. This area would have been a dry ridge when the sea level was lower 7000 years ago, making Aghios Petros into a small peninsula jutting out into the bay, with an ample supply of freshwater nearby. A good sign to start with. Then slowly searching, scanning, peering through the blue along the steep side of the island where the seabed was a bit darker than I expected. Then round the exposed southern headland of bare white rock swept by larger waves. The edge of the land vegetation started much higher above the sea here, a sure sign of storm waves in winter. Round the headland and back along the southern shore again. Scattered patches of weed on grey-white rocks and pebbles. Nothing really odd.

I turned once more and started swimming very slowly along the northern shore. I stared downwards, searching and searching for clues. There *was* something wrong where the seabed was dark with gravel, dark mud, cobbles, and clumps of weeds. Some of the weeds must have been growing in a fairly thick layer of mud or sand. The seafloor here was sloping gently from about 3 m (10 ft) down to 10 m (33 ft). How did that coarse dark material get there? Perhaps it was a fossil remnant of soil? Why did it extend 30 m (100 ft) out into the sheltered bay in the lee of the island, extending for many metres along the side of the rocky shore?

I lifted my head out of the water and saw that I was just opposite the early Neolithic settlement on Aghios Petros that was being excavated by Nikos Efstratiou. This suggested that the original area of the settlement had extended several tens of metres further out under the sea, which was what Efstratiou, to his credit, had surmised, and why he had invited me to join his team excavating the site in 1981.

The cover boat kept close above me, and the rower helped me to scramble over the gunwale. It was a simple wooden rowing dinghy with rope loops on thole pins for rowlocks, and we rowed the 1 km (0.6 mi) back to the archaeologists' basecamp. Here there were tables and chairs arrayed in the shade of a few trees. There was a securely closed tent for storing food – proof against marauding goats and donkeys – our cooking stove, and our scattered sleeping tents. A sort of archaeologists' village. The island of Kyra Panagia was otherwise uninhabited, apart from one farming family who visited in summer to look after the goats for the monks of the Holy Monastery of Megisti Lavra on Mount Athos. The entire island belongs to that monastery. (The derelict monastery on Kyra Panagia was restored in 1992, and there is now a monk in office.)

Efstratiou had assembled a conventional shore-based team of prehistorians, experts in ceramics, dating techniques, pollen analysis, and palaeozoology, which he had supplemented at the last minute with six snorkel divers, including myself, my wife Jay,

Cathy Giangrande from University College London Institute of Archaeology, and three others. Jay was three months pregnant with our first child and so was being careful about the extent of her diving. I asked Nikos if we could excavate a small sample from the anomalous area of the seabed.

"We need to check the ethics of sampling with the representative of the Museum of Volos who is acting for the Department of Antiquities," he said. "We have a permit to excavate, and the whole purpose of inviting the snorkel team is to explore the seabed, but we must proceed carefully."

A hurried debate in Greek followed with the Inspector from Volos, who reached the conclusion that we should take one or two samples within a small area of the bottom, to be examined by washing and sieving. After that a further decision would be made.

"The permit includes the whole area of Aghios Petros island itself and the surrounding seafloor," explained Nikos. "As you know, the richness of the deposits and imported trade objects and ceramics on this small island makes no sense at all unless the occupied area was much bigger, and that is why I have asked you and the other divers to join us. The permit allows for that. But we must start by taking a sample just 5 cm (2 in) thick off the seafloor, from the area you have identified. Then we will see what happens."

Rigid, square frames called quadrats, 1 m or 0.5 m (39 in or 20 in) on a side, are standard kit on archaeological surveys, so I instructed one of the divers, a strong swimmer, to take a 1 m quadrat down onto the dark area of the seabed with a sack and a trowel, to scrape a thin layer off the bottom into the sack. I hovered above the diver as he did this work. He had to dive down to 8 m (25 ft) depth repeatedly, holding his breath while he scooped gravel and mud into the sack, then surfacing quickly, taking another breath, before diving down again. He worked hard, bobbing up and down many times, kicking furiously with his fins (Fig. 10.2). When the sample was complete, we dragged

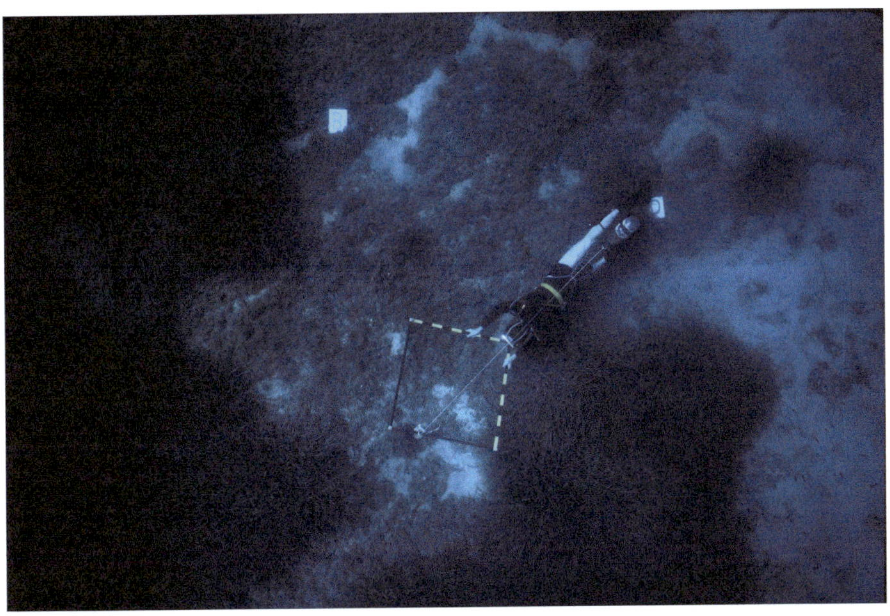

Figure 10.2. Snorkeller on the seabed laying out numbered squares for survey and sampling of Neolithic deposits. This area produced the Neolithic remains shown in Figures 10.3 and 10.4 (Photo Jay Kleinberg).

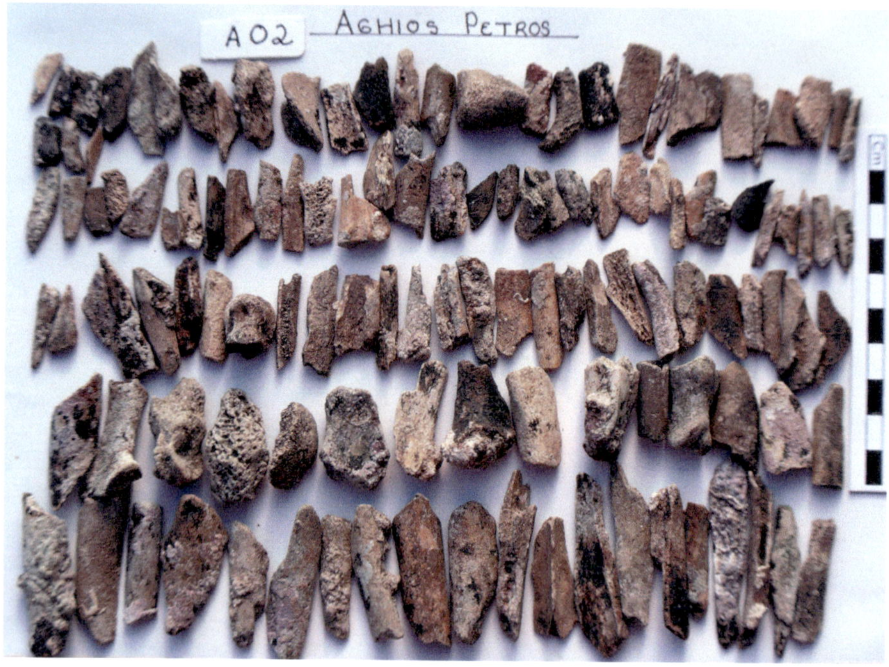

Figure 10.3. Broken bones of sheep and goats which divers found on the seafloor in the area of the submerged Neolithic settlement, off the island of Aghios Petros (Photo Jay Kleinberg, permission of Nikos Efstratiou).

Figure 10.4. Stone tools of obsidian, flint, and quartz found in a 1 m quadrat surface sample recovered from the submerged Neolithic settlement at Aghios Petros, 1981 (Photo Jay Kleinberg, permission of Nikos Efstratiou).

the heavy sack back into the boat. The main party who had been watching from the shore walked back over the hot rocky slopes to the camp. There were trees high in the hills, but close to the shore there were very few, so shade was a luxury. Somebody hauled me up into the small rowing boat that was our only research ship, and we rowed back.

After a cold drink and changing to some warm clothes I waited while the experts sieved the sack of mud, gravel, and debris from the seabed. The results were truly astounding (Figs 10.3, 10.4). The hundreds of fragments of bones, worked flints, flakes of obsidian, pottery, and a small clay figurine in our first sample were success beyond a fanciful dream. The sack of mud had looked so normal, with nothing odd until the washing and sieving revealed the hidden treasures in the matrix. The experts said the bones were a mixture of goats and sheep, which had been killed and cooked. The pottery fragments were Stone Age Neolithic, about 7000 years old, and also Bronze Age. The mud itself was very fine and could have been relict soil. Nikos' theory was confirmed.

* * *

The two weeks of snorkelling and exploration at Aghios Petros had arisen unexpectedly in the summer of 1981. I was visiting the diving equipment exhibition of the BSAC at the Crystal Palace when I was hailed by a brisk American woman with long blonde hair who insisted that she had some important matters to discuss with me. I was puzzled, but we had lunch together. My new acquaintance, Cathy Giangrande, explained that she was a member of the diving archaeology group at the Institute of Archaeology of University College London, where one of the Greek project leaders had found that his excavation was creating a curious intellectual problem.

Efstratiou's problem arose because the small area of ruins on Aghios Petros contained obvious trade goods and sophisticated pottery both from the mainland of Greece, and from Anatolia, modern Turkey. A settlement of one or two families, even with competent boat handling and seamanship, could not routinely have conducted such long trading voyages. Nikos had therefore speculated that, at the time of occupation, the settlement could have been much larger, extending over the seafloor around the island when the sea level was lower. Cathy Giangrande wanted to know if I could help organize a team to investigate this question.

There was no permit to use scuba diving tanks, and in any case, the transport to Kyra Panagia was so unpredictable and unreliable that a scuba expedition would have been difficult to plan at short notice. Cathy, who was clearly not one to waste time, could recruit a team of a few snorkellers, with some survey gear, and a theodolite, but they lacked the special skill to plan the project logically and provide Efstratiou with a clear answer.

Jay Kleinberg and I had married in 1979, and we set up house in England in 1980 (Chapter 3). After discussion with Jay, we agreed to join the team to work for Efstratiou. Jay is an experienced diver and fin-swimmer, but pregnancy would obviously require her to be extra careful with her dives. Cathy and three snorkellers from CUUEG departed first, to survey the outline of the coast around Aghios Petros, and to take soundings of the water depth to map the shape of the seafloor before Jay and I joined the party.

The fishing boat that Jay and I boarded at Alonissos chugged its way across the strait to Kyra Panagia, before it headed further up the island chain to Psathoura, carrying some tourists. They were keen to see the island reputed to be the home of the seductive sirens who sang such sweet songs that sailors happily crashed their ships rather than pass by. We could see the island of Aghios Petros from the beach where we landed. The other archaeologists helped us carry our gear and tent up to a nice sheltered spot. I had already read that the ancient settlement on the tiny island was one of the earliest then known on the Aegean islands, with a permanent settlement more than 7000 years old. Ceramics showed cultural contacts with Thessaly, Anatolia, and the Balkans, and a semi-autonomous development by 6500 years ago. When the settlement was founded, the sea level would have been 10-20 m (33-66 ft) lower than at present.

Figure 10.5. Cathy Giangrande with the notebook recording the fixes taken by the theodolite, while the snorkeller in the water marks the corner of the grid of nylon cords on the seafloor at Aghios Petros (Photo Jay Kleinberg, permission of Nikos Efstratiou).

The advance team, using the theodolite, had already surveyed an accurate plan of the coastline adjacent to Aghios Petros, including that of the island itself, as well as conducting soundings to measure and plot the water depth all around the island. They were pleased with that progress, but when we arrived, they were still frustrated by not having identified any ancient remains in the water.

After I had helped to identify the area of rich Neolithic remains on the seafloor, we moved rapidly to the next stage of the plan. Cathy, assisted by several of the snorkellers, assembled a grid from white nylon cord, brought for this purpose, extending to a total size of 50 x 35m (164 x 115 ft), with cords across at intervals of 10 m (33 ft), dividing the whole array into 10 m (33 ft) squares, with one row of 5 x 10 m (16 x 33 ft). This fragile array was cautiously pulled out into the sea from the shore of Aghios Petros, with the boat and a snorkeller pulling it gently seaward, while a team on the island fed out the cords and prevented tangles (Fig. 10.5). When the whole grid was floating on the surface, we attached stones to the corners, sank it over the Neolithic area, and then pegged down the intersections of the grid. I say "we", but of course 90% of this work was done by Jay, Cathy, and the snorkellers Tim Alexander, Nick Lowe, and John Wood.

In spite of the outrageously primitive methods used, we now had a reliable grid of squares over the archaeological deposits, enabling snorkellers, during the following week, to photograph and measure every 10 m (33 ft) square. They then probed the bottom to detect sediment thickness to bedrock and continued to take surface samples

Figure 10.6. Aghios Petros, the day after donkeys and goats broke into the food tent at night and ate all our stores, except for a sack of rice and a sack of spaghetti. The whole wall of the tent was torn away. I look worried (Photo Jay Kleinberg).

from quadrats. Because the yield from the 1 m (39 in) quadrats was so rich, the later samples were only 0.5 x 0.5 m (20 x 20 in). Every quadrat produced bones and ceramics. I snorkelled above the site to sketch each grid square and record the distribution of sediments, gravel, and weed, to check against the photo-mosaic later. This proved to be a useful double-check given the roughness of our methods. Seven thousand years ago, the Neolithic village or cluster of houses stood on the crest and slopes of a peninsula jutting out into the bay, bordering a sheltered creek, completely protected from storms, and provided with a freshwater spring. The Neolithic sailors were experts and picked the ideal spot for their settlement.

Then one night, donkeys and goats broke into the food tent (Fig. 10.6). This was especially unfortunate because the divers worked hard and were very hungry. The donkeys and goats trashed all the fresh food, smashed many tins and boxes, and left only a sack of spaghetti and a sack of rice undamaged. The supply boat was not due to visit for several days, but we had no radio contact with Alonissos. Although one of the snorkellers was a good spearfisher, our diet became very boring.

When the farm family up in the hills heard of our plight, they slaughtered a goat and invited us to dinner. My friends helped me onto a donkey, where I sat astride, and somebody else carried my wheelchair. Our procession climbed the steep, rocky track to the farmhouse, which was, in truth, a series of open shelters and half-walled-in rooms in the forest. It was ideal for summer living, but not a good place to be in winter. The family piled great dishes of food on the table, feta cheese, olives, tzatziki, humous, potatoes, aubergine, salad with raw onions, and goat meat stew with beans, followed by fresh fruit and coffee. We ate by the light of paraffin lamps. Copper jugs of wine washed down the food. Those of us who were happy to eat goat had a very good dinner indeed, but nobody was left hungry. Afterwards, the old grandfather played exquisite and mournful music, always tragic, on an instrument like a large zither. It had numerous taut strings laid out on a flat board so that the player produced the notes by striking it with hammers. In the dim light, Grandfather struck the strings with passion. The zither accompanied his nostalgic tales, sung in a strong voice that held us motionless. The music rose and vanished into the forest of the night. As the songs finally drew to a close it seemed that we had been holding our breath, as if hypnotized, for a long time. The journey downhill in the darkness must have been meandering, but the flasks of good wine we had drunk enabled us to avoid disaster.

Our findings under water at Aghios Petros proved a point that I was beginning to formulate after years of searching for prehistoric remains in the sea since my discoveries at Pavlopetri (Chapter 6). Both settlements proved that terrestrial prehistoric archaeological deposits could be preserved intact and found under water in soft, unconsolidated layers on the continental shelf, in formerly dry areas inundated at the end of the last ice age. The great majority of prehistorians believed that this was not the case. The determining factor in protecting the evidence from destruction was the surrounding topography and the physics of waves and currents. The rich deposits under the sea at Aghios Petros were in the lee of a rocky island in the corner of a sheltered bay in a larger island that was part of an archipelago blocking large waves from every direction. I had already seen signs of such preservation during a visit to Israel in 1975, while stories were beginning to come in from other parts of the world, especially from the Baltic.

The year after our work at Aghios Petros, Cathy Giangrande and I were involved in another project together, but its inception was before we had met. In the 1970s, Keith Muckelroy, a young archaeologist recently graduated from Cambridge, and an ex-member of CUUEG, studied two Bronze Age wrecks on the British south coast, one at Langdon Bay, Dover, the other at Moor Sands, Salcombe. Muckelroy and his fellow divers recovered hundreds of bronze objects, including axe heads. These discoveries inspired me with the conviction that ancient seafaring and ancient shorelines could both be studied in European waters. Muckelroy visited my office at the NIO several times during the late 1970s, and he repeatedly sought advice on where there were active programmes of underwater archaeological research and asked for introductions and references. He had an excellent First Class degree but still found it surprisingly difficult to gain long-term salaried employment. I helped as much as I could. He dived on and studied Dutch East India Company wrecks, Spanish Armada wrecks, and a submerged town in Cyprus. He was finally given employment at the National Maritime Museum in 1980, but tragically died in a snorkelling accident in Loch Tay in September that year. Keith Muckelroy's mother wanted the royalties from his books to be used to help promote marine archaeology. Starting in 1982, Cathy and I worked with her to establish a small charitable trust in his name, which was incorporated in 1985. As trustees, we made grants to support diving archaeology projects by amateurs and undergraduates and awarded an annual prize for achievements of exceptional quality. The trust itself was wound up when the money had been disbursed within a few years, but the Keith Muckelroy Memorial Award was subsequently administered by the Council for British Archaeology, and biennially since 2009 by the Nautical Archaeology Society. Muckelroy's death was a terrible loss to maritime archaeology, and I like to think that, had he lived, he would have contributed to the study of prehistoric shorelines, as well as to shipwrecks.

* * *

The old saying that "Revenge is a dish best eaten cold" has no agreed attribution or origin, nor do I wish to revenge myself on anyone. But a nice paraphrase would be to suggest that "Success is a dish best eaten cold". Sensational social media websites and popular newspapers confirm that too much success too early in life is often destructive. In contrast, it may be best to let success mature with time. But for how long? How much patience do you need? Fifty years?

In 1962, I predicted that undersea caves would reveal prehistoric occupation of the submerged lands around the continents known as the continental shelf (Chapter 4). Our ancestors must have lived on these vast exposed coastal lands when the water from the sea was piled up in the glaciers and ice caps on the mountainous regions of the continents, especially North America, Scandinavia, and the Alps, causing the sea level to drop by more than 120 m (360 ft).

I thought at the time that within five or ten years, divers would be excavating underwater caves revealing thick strata of prehistoric tools, debris, bones, and charcoal. This turned out to be false, but other submerged prehistoric settlements began to emerge here and there around the world, confirming my underlying assumption. The evidence was skimpy and scattered, but I clung to it. The evidence showed that people had lived all over the then-exposed continental shelf, creating settlements or occupation sites or discarding chunks of stone tools, which could survive the rising of sea level after the ice ages. Now, 55 years after my first over-optimistic guess, it is reasonable to claim victory. Marine archaeologists have identified thousands of prehistoric sites on the seafloor around Europe and off the coasts of the USA, South Africa, Taiwan, and many other parts of the world.

But I was wrong about the caves. The evidence survives best on the open seafloor, in areas of slight undulating relief or dotted with islands, not in caves. Believing that the shallow seas – out to a depth of about 120-130 m (393-427 ft) – had once been dry land occupied by our ancestors virtually ensured that I would find some of the relict traces myself. It also enabled me to encourage and promote many other researchers who felt that their discoveries were being ignored, rejected, or trivialized. Sometimes, when there were gaps of years between new sites being discovered, I felt that I was clutching at straws, but finally, around 2000, the evidence was overwhelming. I retired from my professional bureaucratic appointments in 2001 (Chapter 9), and then the fun really began.

* * *

When I explored undersea caves for evidence of sea level and global climate changes over the past million years (Chapter 4), I was always on the lookout for signs of prehistoric occupation, but I realized after several years that these were unlikely. There are thousands of caves to explore under the sea, but storm waves crashing into a cave mouth during inundation would have destroyed human traces near the entrance. If anything ever survives it is under heavy roof-falls of rock or far back inside tunnels or around corners. When I explored underwater cities, including some uplifted, around the Mediterranean coast (Chapter 6), I did not consider possible prehistoric ice age settlement earlier than 5000 years ago.

The melting of the ice caps produced a steadily rising global sea level which stopped, more or less, around 5000 years ago, and I am interested now in the human, and pre-human, settlements and debris that were inundated by that global rise and are now offshore. Since there were several ice ages during the past million years, submerged prehistoric sites might exist from much earlier than the coldest peak of the last glaciation 20,000 years ago.

By the mid-1970s, I had accepted that there was no quick solution to the search for ice age seabed archaeology. Pavlopetri was almost 5000 years old, and that showed how the local topography of rocks could protect a settlement for much more than 3000 years. But could primitive settlements and scatters of flints survive for 10,000 years or more under the sea?

In 1972, George Bass published his comprehensive book *A History of Seafaring Based on Underwater Archaeology*. As the title implies, it is mostly about ancient ships and early boats as revealed by the study of shipwrecks found under water. From 1960 onwards, Bass and his colleagues greatly enhanced the study of ancient shipping by applying the standards of land archaeology to shipwrecks, revealing unexpected details of cargoes, ship construction, and crew accommodation. In the Introduction, Bass wrote:

> Before there were either farmers or shepherds there were sailors; before man settled in villages and began to make pottery, he had learned to construct some sort of seagoing craft.

This magnificent generalization, which surprised many archaeologists who had studied the agricultural revolution of about 10,000 years ago, had a quite special implication for me. The oldest boats found anywhere are log boats usually on land, and dating from 8000 to 10,000 years ago, when the sea level was still 30 m (100 ft) below present. The oldest boats found as shipwrecks are less than 5000 years old. But the trade in goods like obsidian from volcanic islands, and the first date of occupation of Australia, all confirm that people were using boats tens of thousands of years ago, and the shores they departed from and the shores where they landed are all now deep in the ocean.

My love of landscape and sense of place forced me to visualize again and again those drowned coasts, the cliffs and islands, the borderlands of marshes and creeks where early tribes kept their boats and made their first settlements. I saw in my mind's eye the sailors embarking from a sheltered creek that is now 10 m or 20 m (33 ft or 66 ft) deep in the sea, navigating to a lost shore that was good for hunting or fishing, and then returning with their prey to their settlement. The materials they used were sparse, with a lot of wood and bone, and the primitive boats would rot very quickly unless covered with thick mud. The boats themselves might never be found, but the harbours they used, the hearths where they cooked, the flint factories where they knapped tools, the huts they lived in, the flint axes, the fishhooks, the burials of the dead – any of these might survive and reveal human activities on the continental shelf throughout the low sea level periods during the last million years. Since the area of the global continental shelf is similar to that of Africa, the land that has been drowned and forgotten is enormous. Compared with the discovery of a shipwreck, I felt that the human occupation of the continental shelf and the archaeological mapping of a migration pathway between continents was a bigger game. In spite of slow progress this belief endured.

I remained convinced that the rising sea would not destroy all signs of prehistoric human occupation. Most sites would be destroyed, yes, but not all. Various natural influences interacted with one another. The physics of coastal processes and the way that waves changed the seafloor in shallow water showed that erosion was not the universal result. Beaches could be stable or accretional as waves moved sand shoreward. I knew that historically a few items of pottery and an ancient harpoon had been trawled-up from the North Sea, and my observations convinced me that sediments were piled over a prehistoric remain as often as eroded from under it. The destruction was never 100%. Headlands eroded and bays filled in with mud and sand, burying archaeological remains for thousands of years. An archipelago of irregularly shaped islands would necessarily contain some bays and patches of seafloor that had been totally protected during the inundation by the rising sea, whatever the direction of storms and waves. Changes of currents, tides, or shifts in sediment supply could then lead to prehistoric settlements being exposed anywhere on the seafloor shallower than 130 m (420 ft).

Archaeologists were ready to admit that the continental shelf could have been occupied by humans, and probably had been in some places, during ice age low sea levels. But, having made that concession, nobody seriously suggested that it made sense to go out searching for human remains or artefacts that might have survived undisturbed under water. The technology in the 1950s and 1960s was inadequate to map the seabed accurately, let alone to detect small buried prehistoric tools or hearths. Finds that came up by chance in fishing nets were insufficient to build any model of who lived where at what time. Land archaeology had already advanced to the stage where hundreds of sites could be analysed from a single prehistoric period so as to build complex models of culture, family structure, diet, migration, evolution, response to environmental change, and so on. A few random finds from the seafloor could be dismissed as antiquarian curiosities, not modern archaeology. Just quaint souvenirs.

As Mark Twain said of the weather, "Everybody talks about it, but nobody does anything about it." To draw a parallel with the archaeology of submerged occupation sites, suppose that in 1900, archaeologists had said, "Africa was probably occupied by some prehistoric people, but it is a difficult place to work, very hot, with lots of diseases. We can just assume that the same sort of prehistoric people lived there, in some places, as we find for the equivalent ice age periods in Europe and the Middle East. There is no need to find deposits or to conduct excavations in Africa, which would be very expensive." We know now what a terrible decision that would have been. Africa contains prehistoric remains that are unique, utterly different from other continents, probably holding the secret to the origins of *homo sapiens*.

It is unkind to draw a comparison between the continental shelf and Africa for the significance of its prehistory because we know now how special Africa is, but the area and the timescale make for a genuine parallel. The area of the global continental shelf is 26 million square kilometres (10 million sq mi), which is 5% of the global surface, and 17% of the existing global dry land area. Around Europe the shelf adds 40% to the existing land area. Archaeologists realized soon after 1900 that prehistoric artefacts trawled by chance from the seabed showed that it had been occupied during glacial periods, but there was no intellectual incentive to do anything about it. My conviction was that a steady effort at studying known finds, and a constant alertness for new ones, would gradually demonstrate that well-preserved and even stratified prehistoric sites from all ages could survive on the seafloor. If every chance find offshore was dismissed as out of context and thrown away, a well-documented record of who lived where could never accumulate. The scepticism was self-reinforcing. Every chance source and fluke should be followed up and recorded professionally. It would just take a bit of time.

It was against this sceptical background that I persisted throughout the 1960s and '70s to think that, since archaeological occupation settlements and their deposits should survive undamaged in a few locations, then it was worth looking for them. Because I was not an archaeologist and was not paid out of archaeological funding, nobody could tell me that my ideas were foolish or misguided.

Pavlopetri itself provided the first intellectual clue, if unappreciated at the time, since it was a complete town built without squared blocks or cement, but it survived in a few metres of water, with some ruins right in the sandy beach. Modern waves and winter storms had not destroyed it. All generalizations about the destructiveness of a sea level rising over fragile remains were nullified by the protecting rocky reef, ridge, and small islands. Additionally, large parts of the underwater town were covered by banks of sand for much of the time. Just as at Aghios Petros, it was local topography and local wave physics within a few tens of metres around the neighbourhood that determined survival.

In 1975, I was visiting professor at the University of Haifa, both lecturing to students, and conducting coastal fieldwork (Chapter 6). Professor Elisha Linder had invited me to lecture for the two winter terms. My purpose was to examine the coast from the border with Lebanon to the border with Egypt, since Gaza was still administered by Israel at that date. Although most of the coastal ruins were not prehistoric in date, some very clearly were, so that the survival of such ancient evidence on the shore, and in shallow water, was manifest. Ernst Wreschner took me to see the ruins at Newe Yam, where Neolithic house walls of uncut stones extended out from the sandy beach into the sea in shallow water. The low gradient of the beach, combined with offshore ridges and islands, meant that waves blown by storms over half the length of the Mediterranean as far as Sicily could not destroy the fragile walls. Avner Raban at the University of Haifa was also exploring submerged prehistoric traces at this time on the Israeli coast, but he did not publish his observations until a few years later.

In the mid to late 1970s, the news spread about two karstic salt springs in Florida, Little Salt Springs and Warm Mineral Springs. In both cases, the water level was linked to the sea by underground channels, and during the ice ages, the level in the springs or sink-holes had dropped many tens of metres. Starting in 1958, a succession of scholarly divers, amongst them William Royal, Eugenie Clark, Carl Clausen, Wilburn Cockrell, and Larry Murphy, found prehistoric materials including human skeletons dating from 10-11,000 years old. Meanwhile, Jim Moriarty, Pat Masters, and others published the discovery and raising of grinding stones from depths of as much as 20 m (66 ft) off the coast of San Diego County, California. In both cases, I visited the sites a few years later to meet the original researchers.

Larry Murphy and Sonny Cockrell also published a thought-provoking paper in which they described the discovery of a palaeo-American burial under the beach sand at an exposed location on the Atlantic coast of Florida, near Cape Kennedy. I listened to their talk at an underwater archaeology conference in San Antonio in 1975. This reinforced my conviction that an understanding of the physics of water movement, waves, and tides, would help to determine which ancient deposits survive and which don't. The beaches of eastern Florida are exposed to horrific winter storms from thousands of miles across the open Atlantic, not to mention hurricanes, and yet the beach sand with a low gradient had not been disturbed to a depth of more than a few metres for centuries and millennia. This is further evidence that a rising sea level does not destroy everything if the local landform conditions, sediments, and gradients are right.

My confidence was growing, though it was by no means secure. The trail was leading someplace interesting but very slowly. In 1979, I worked with Pat Masters at Scripps Institution of Oceanography, La Jolla, California, to plan a conference on submerged prehistoric archaeology throughout the globe. We invited speakers from everywhere that submerged traces had been reported, however tenuous. Pat raised the funds to host the conference at Scripps in October 1981, including a dinner in the aquarium where we were watched throughout the meal by circling sharks with big teeth.

The speakers provided an excellent variety of papers describing evidence from Alaska to Australia, from Greece to Sweden. The range of discoveries presented was exciting, but a sceptic could still have said that the material was so thinly scattered, coming from different dates and cultures, that it did not provide the basis for modern archaeology, nor would it justify much expenditure of research funds. Academic Press published the collected papers, which were well reviewed by the usual journals. After that the contents became a sort of "sleeper" *agent provocateur*, providing a surprising outlook to one young archaeologist after another, but without changing the views of the top professors, with one honourable exception (to be named later).

I reported our findings at Aghios Petros to the Scripps meeting, but also presented an analytic review paper in which I tried to find the common factors in all the known prehistoric sites that had survived under water: why had they survived? The common factor throughout was the local topography, such as a lagoon, barrier island, sand-spit, re-entrant bay protected by a narrow entrance, estuary, archipelago, or rock outcrop, that protected the archaeological material at very close quarters from attack by waves and currents. In spite of the fact that the sea has been at its present level to within half a metre (20 in) or so for about 5000 years, prehistoric remains submerged in a few metres of water had, in the listed cases, all survived without destruction. But you had to map and plot the local terrain with extreme accuracy to see why the human artefacts were still there to be found.

A bonus to this deduction was that, if prehistoric remains of wood and stone could survive centuries or millennia when they are, at least in theory, exposed to the greatest risk of destruction by waves, then, as the sea level continues to rise and the water gets deeper, the settlements and artefacts will probably be safe, since wave action is much weaker at the bottom in deep water. This mechanism of survival of prehistoric settlements in shallow water, protected by the local topography, killed off the proposition that all prehistoric sites would necessarily be destroyed if they were exposed at the same level as the surf zone for a prolonged period. That greatly increased the probability of finding sites at any depth in the sea anywhere on the continental shelf.

At the final session of the conference, after field excursions along the Californian shore, the delegates were invited to give their ideas and views. This session contained a perfect highlight. Geoff Bailey, who had reported on his work in Queensland, Australia, and who is now a professor at the University of York, proposed that the subject we had been discussing was primarily important because he was convinced that people in the Palaeolithic more than 10,000 years ago had extensively exploited marine resources – shellfish, fish, marine mammals, and the fertile lagoons and floodplains on the coast – and that this productivity had contributed to the development of prosperity, surplus wealth, and complex cultural settlements, before the invention of farming. Bailey said that archaeologists had not yet found such early settlements because they are now all under the sea. I was thunderstruck. This was a legitimate archaeological reason for unifying continental shelf research beyond mere souvenir hunting and curiosity.

* * *

A dramatic target for seabed prehistoric research is the submerged land between modern continents, such as the Bering Strait between Siberia and Alaska, or the Arafura Sea between Australia and Indonesia. People lived on these lands 10,000 years ago and earlier. During the late 1970s, I was introduced to Arthur Weller, later Sir Arthur, who had made his fortune in the insurance industry in Australia. His previous philanthropic work included support for the rescue of the early steamship *Great Britain* from her wrecksite in the Falkland Islands in 1970, and subsequently the construction of a sail training schooner, and a replica of Captain Cook's ship *Endeavour* in the 1980s. In 1980, Sir Arthur set up a charitable research foundation to support a diving expedition to study the seafloor north of Darwin, where I hoped to find evidence for an ice age crossing of peoples into Australia. Previous geological surveys showed that there were many submerged islands and archipelagos, creating protected channels or closed valleys on the seafloor between the present Australian coast and the Indonesian island of Timor.

We called the project the Cootamundra Shoal Expedition. Sir Arthur's vision and generosity enabled us to proceed without financial worries. I was very grateful for his enthusiasm. Sadly, he died in 2011.

I made visits to Sydney to see Arthur, and, after travelling with him to look at potential diving vessels, I continued on my own to Darwin in 1981 to make an aerial survey from a light aircraft. Sitting next to the pilot, I gazed down at the mangrove-covered northern shores of Bathurst Island, north of Darwin, imagining the submerged land stretching out towards Timor beneath the waves. There had always been a deep oceanic trench close to Timor, between Timor and Australia, so the first people to cross must have been competent navigators, but we had, and still have, no idea of the kind of craft that they used. In 1981, Darwin was still suffering from the aftermath of the great hurricane, Cyclone Tracy, but the city was rebuilding, and local support for the offshore expedition was unqualified.

The evidence in 1982 from archaeological discoveries on land showed that the first people arrived in Australia (and Papua New Guinea) about 50,000 years ago, while more recent dating technology suggests that an earlier date of 60,000 years ago was possible. For much of the period between 60,000 and 20,000 years ago, the time of the glacial maximum, the sea level was between 50 m (164 ft) and 130 m (427 ft) lower than at present. There are no corals building modern reefs in the Arafura Sea (nobody knows why), so that the ice age landscape from periods of low sea level should survive in its original shape and topography. Two hundred miles (320 km) out from Darwin there was a shallow submerged landmass more than 100 miles (160 km) across, of which the shallowest point was the small Cootamundra Shoal.

As I was chairman of the BSAC at that time, I tentatively asked HRH Prince Charles' office if the Prince, as president of the BSAC, would graciously agree to be a patron of the Cootamundra project. He agreed. Arthur chartered the specially-designed 21 m (69 ft) dive support boat MV *Reef Explorer*, to be based in Darwin from 23rd April to 31st May, 1982, at a time when the hurricanes were supposed to be least threatening. Sea crocodiles, sharks, stinging box jellyfish, and sea snakes were local hazards, but the alert diver could avoid the latter dangers with a little attention, and crocodiles very seldom strayed far from the shore. I recruited a team of divers and participating experts from the UK, the Australian National University Canberra, the Western Australia Museum, Fremantle, and the University of New South Wales, military divers from the Royal Australian Navy (RAN), and Mike Davis, an ex-CUUEG doctor from New Zealand. Our land accommodation was in the Larrakeyah Barracks, an Australian Defence Force base, while the *Reef Explorer* was docked at the RAN berth in Darwin harbour. Peter Scoones (sadly now deceased) joined us to shoot underwater film.

When the Cootamundra Shoal Expedition put to sea, our baby daughter Kirsten was five months old. Since she had already had snorkel diving experience at Aghios Petros before birth, Jay and I thought that she would be quite adaptable to life on board a diving vessel, which turned out to be the case (Fig. 10.7).

On 25th April, we loaded a 1-ton two-man recompression chamber onto the foredeck of *Reef Explorer* and welded it in place. We sailed on 1st May. The navigation and position-fixing problems on this project outstripped everything I had previously experienced. There was no GPS yet in operation, and the main work area was more than 200 miles (320 km) offshore from Darwin, with no land breaking the surface anywhere closer than that. Cootamundra Shoal itself rose to a shallow point at a depth of 15 m (49 ft) but did not break the surface. The US NASA Landsat satellites provided the best possible source of accurate fixes (accurate by 1982 standards), but a reliable fix depended on waiting for a succession of satellites to pass overhead, which took many hours, while the computer in the Motorola communication unit on board the ship improved the accuracy of the fix with each pass. That meant we would have to anchor overnight, or even for 24 hours, to get the most accurate positions.

Using the Landsat satellites, we fixed four positions, each with an error of about 10-15 m (33-49 ft), within 5 km (3 mi) of Cootamundra Shoal itself, placing moored buoys at these points. We could then use radar and horizontal sextant angle on the buoys to fix any number of local positions. It took 5-10 Landsat satellite passes, with *Reef Explorer* at anchor, to obtain an accurate position for each buoy. By anchoring the buoys at known points, we made an open-ocean navigation exercise similar to a coastal one.

Our onboard archaeologist was Rhys Jones from the Australian National University. Jones (Fig. 10.8) was a rugged archaeologist who seemed totally indestructible until he died of cancer in 2001. During the Cootamundra project, he happily dived into muddy seawater with zero visibility without having any previous open-water diving experience. For this dive he was accompanied by two experienced escort divers. He took a sample of mud from the bottom, close to Bathurst Island, when we were sheltering from a storm and was as excited as a schoolboy with his achievement. During our surveys offshore,

Figure 10.7. Kirsten in her bouncy carry-cot aged 5 months, in the saloon onboard MV Reef Explorer, 200 miles offshore in the Arafura Sea, north of Darwin, Australia. I am working on dive records, while, in the background, Mike Davis is inserting stitches in the toe of a diver, Carl Shaw, who cut it on the edge of an aluminium dive boat (Photo Jay Kleinberg, Cootamundra Shoal project).

around the Shoal, he watched carefully as the map of valleys incised into flat-topped banks emerged. When the semi-enclosed topography of narrow valleys and bays became clearer, Jones stomped up and down the deck exclaiming, “If you took the bloody water

Figure 10.8. Rhys Jones, a distinguished archaeologist who researched Aborigine sites in Australia, comparing gravel from sediment samples with Jay Kleinberg after a dive.

off this place, I would find you an Aborigine rock shelter in 20 minutes!" We were on the right trail.

The expedition consisted of three 10-day cruises at sea, with two days in between, and a week in advance and afterwards for stores and management. The schedule intended to exploit the best season for the weather, after the hurricane season and before the trade winds built to gale force. That was the theory. But storms came in unexpectedly. The divers working from inflatables and aluminium dinghies were tossed about by the white-crested waves, and it became difficult to hold the dinghy close to a mooring buoy to pick them up when they reached the surface. The weight of the decompression chamber on the foredeck forced the bow of the *Reef Explorer* to plough straight into the steep foaming waves, without riding over them. Finally, the storms forced our small ship to shelter in the lee of Bathurst Island for several days. We did not lose useful diving time, since there were interesting drowned valleys and submerged landscape features close by which no one had surveyed since the 1870s. We radioed the Aborigine Council on Bathurst Island to inform them of our work, requesting permission to dive in the Council-controlled area. They said that they knew that their ancestors had walked across the continental shelf, and they wished us well in our discoveries. For much of the sea area around Bathurst Island, the sky was hazy with smoke. Seasonal burning of the forest cover is part of the Aborigine way of surviving in this wild landscape, so that smoke often drifted for tens of miles offshore, even being included in the marine weather forecast.

For all dives on the Shoal, the divers entered the water from inflatable boats or from aluminium dinghies powered by outboard engines. Then they descended directly down buoyed shot-lines dropped on the dive sites. In good weather, the dive dinghy, with a standby diver, could hold station close to the buoy, watching the divers' bubbles, ready to pick them up quickly on ascent. The dinghies communicated with *Reef Explorer* by handheld radio sets. The underwater visibility over the banks around Cootamundra Shoal was 10-20 m (33-66 ft), which is rather less than one would expect in the open ocean. The divers saw sea snakes at times, but these were slow-moving or motionless on the seabed. We were not troubled by other venomous or dangerous creatures, but we never relaxed our vigilance.

The divers worked in pairs at depths of 30-60 m (100-200 ft), using a range of techniques to sample the rocks and sediments on the seafloor, including corers, powered rock drills, sledgehammers, and, when the seabed was very resistant, explosives (Figs 10.9-10.14). The same species of coral that build reefs were often present, growing singly like small cabbages on the seafloor, but they were always tens of metres apart and did not coalesce into a reef. We spent so much time discussing and arguing about this strange circumstance that eventually I had to ban the topic at the dinner table. Much of the seabed was covered in an encrustation of calcareous algae several centimetres thick. Divers broke this through with a heavy sledgehammer, or with explosives, and found earlier phases of coral growth underneath, dating from 7000-9000 years ago, in the depth range 20-28 m (66-92 ft). Jay, who maintained a better resistance to nitrogen narcosis than most of the divers, conducted some of the deep dives to 60 m (200 ft), while Kirsten stayed on deck, oblivious to the details of the day's activities.

Figure 10.9. Diver at 60 m depth near Cootamundra Shoal placing a quadrat and working on samples of the calcareous algal crust. No modern coral is visible, but fossil corals under the crust are of the reef-building species.

Figure 10.10. Underwater video camera with shipboard display being used to film divers breaking samples of calcareous algal crust off the seafloor at a depth of 45 m near Cootamundra Shoal.

Figure 10.11. If the rocky crust covering the seabed was too tough to break with hammers or a rock drill, we used explosives, provided by, and implemented by, staff of the Royal Australian Navy (Photo Jay Kleinberg).

On the calmer days, when work was fast and efficient, the map of the seabed evolved on the echo-sounder chart, while we navigated backward and forward in zigzags over the steepest terrain. This pattern of tracks revealed the nature of the cliffs and slopes on the sides of the valleys, which seemed to have a topography rather like fjords. They were deep right from the mouth to the closed end, sometimes even with a deeper basin at the closed end where there might have been a lake. This was consistent with freshwater solution of the limestone rock during the ice age low sea levels, possibly with underground drainage through caves and tunnels called karst solution. Divers installed current meters on the seafloor so that we could record the tidal current for 24 hours. A towed video camera lowered deeper than 60 m (200 ft) showed overhangs in some of the submerged cliffs but at depths below the safe diving limit. The dive dinghy was usually out on a buoy while we did the echo sounding, staying in contact by radio. Like Jones, but without his profound knowledge of the Aborigine occupation patterns, I could sense the landscape below the ship, as if the water could be taken off.

I was on the bridge for most of the daylight hours (Fig. 10.12), but Cootamundra Shoal was a lonely place. We did not see another ship during the whole month, apart from a brief visit by one of the RAN patrol boats from Darwin. The weather was less steady than we expected, often windy, with the *Reef Explorer* rolling and pitching badly. We were able to conduct echo sounding even when it was too rough to dive from the dinghies, but when the motion of the ship became too much to sight on the buoys, and the spray flew over the wheelhouse, then we ran for cover in the lee of Bathurst Island.

The 18 divers, some of whom joined the ship for one cruise only, achieved 230 person-dives, with 51 dives in the depth range of 40-60 m (130-200 ft). Forty-two person-dives required decompression, which the divers conducted by making stops in the sea during their ascent. The recompression chamber was not needed to cure any bends, but it was an essential back-up safety factor. If we had not had the chamber on board, and if a diver had developed a bend, it would have taken probably a week to transfer them to a recompression chamber on shore, as there was no chamber in Darwin. That was an unacceptable risk, and it would have been illegal to dive without the chamber on board.

As we mapped the drowned landscape of steep-sided valleys, we also identified some curious mini-reefs of dead coral that had started to grow on the edge of the main platforms at a depth of 36 m (118 ft) as the post-glacial sea level rose and the ice caps melted. They grew a few metres high and then they failed to develop fully into modern reefs that would reach up to the modern sea level. They just grew for a few years and died. We had no explanation for this. Why had the corals at Cootamundra failed to develop and keep up with the rising sea level in the same way as those at the Great Barrier Reef, 1000 miles (1600 km) away on the east coast of Queensland at the same latitude?

During one of our brief breaks onshore for refuelling, Jones took Jay and Kirsten into the Kakadu National Park, accompanied by Pat Baker (Fig. 10.13), during which trip they drove across the shallow deltas of the North and South Alligator Rivers. They found many Aborigine rock shelters with paintings inside. Australia is fundamentally an urban country, with the great majority of the population living in a few very large cities. But in Darwin in 1982, there were still great hunks of broad-shouldered sun-darkened men, wearing shorts, huge boots, and hats swinging with rows of corks, queuing in the bank to deposit their latest finds of opals or rubies or gold in the bank vaults.

Figure 10.12. Reef Explorer rolling heavily in a choppy sea, with the author on the bridge taking radar range and bearings during an echo-sounder traverse (Photo Jay Kleinberg).

Figure 10.13. The Northern Territory has its moments, and road signs like this are not uncommon along the coast road where you have to cross the North and South Alligator Rivers. Pat Baker displays the sign.

By the end of our third ten-day cruise, we had discovered and mapped submerged terraces and small fossil reefs that can be tentatively correlated with sea levels at depths of 25 m (82 ft), 34 m (111 ft), 48 m (157 ft), 59 m (194 ft), and 90 m (295 ft). The 48-59 m (157-194 ft) sea levels would leave large areas of the banks at an elevation of 10-25 m (33-82 ft) above the sea, while the lower sea level correlated with extensive semi-closed drainage basins and karstic topography. We never expected to discover human remains in one month of work in a sea area that had never been previously studied or mapped at the accuracy required. But we demonstrated that the geology, terrain, freshwater supply, climate, vegetation and availability of coastal seafood species were sufficient to support a Palaeolithic ice age human population during periods of low sea level. In this we succeeded. This does not prove that people did live on the Arafura-Sahul shelf, nor that they travelled there from Indonesia, but the conditions on the shelf were ideal for them if they did attempt to make that journey.

At global sea levels lower than 40 m (131 ft), about 80,000 to 15,000 years ago, the valleys were perfectly sheltered from oceanic winds and waves, creating miles of calm shores, overgrown with mangroves and mussel beds. The steep rocky slopes below 60 m (197 ft) had overhangs that were ideal rock shelters of the kind that are often found in the Northern Territory on land. We also showed that the techniques available, even in the 1980s, were adequate for this kind of research in the open ocean. Because the banks had a flat upper surface at 30-36 m (98-118 ft), most of the diving had to be deeper than this, which pushed our diving capabilities to the limit. In 1985, I tried to organize a project using a submersible from the Hawaii Underwater Research Laboratory, but, in spite of the project being strongly supported at HURL, the distance to Australia was just too much to make the project economically feasible on their research budget.

On return to the UK, I gave a lecture at the Royal Geographical Society, and Jay and I were invited to Kensington Palace to give a slide show to HRH Prince Charles. He particularly enjoyed the picture of Kirsten in a baby-bouncer on board the dive ship. Nobody has revisited the Cootamundra Shoal in the past 39 years, but interest has increased in the last decade in the prehistoric remains on the coast of northwest Australia, including the nearshore islands.

In 2020 Jonathan Benjamin and co-authors published the results of five years' work by a team of experts from several Australian institutions combining the skills of multi-spectral satellite remote sensing, aerial photography, lidar, photography from drones, acoustic bottom surveys, diving, seabed sampling, and underwater photography. After meticulous analysis of the combined data types, and successive refining of their search areas, they discovered scatters of more than 250 stone tools on the seafloor in 2.6 m and 14 m (8 ft and 46 ft) depth at Murujuga, in the Dampier Archipelago on the northwest Australian coast. This was a rare example of using purely deterministic logic with a predictive framework to narrow down the search area and find Stone Age tools on the seafloor. The finds are in shallow water, but the discovery proves that submerged prehistoric artefacts can be found in the tropics, and are the first underwater stone tools found in Australian seas.

This exploration will continue in a wider area of the northern continental shelf of Australia, around the submerged basin named the Bonaparte Depression, and possibly returning to the Cootamundra Shoal.

* * *

In the 1980s and ’90s, as I was increasingly drawn into the world of science policy and international treaties (Chapter 9), I continued to find some time for my fascination with prehistoric settlements under the sea. Jay and I took a holiday in Brittany, where we visited the obscure island of Er Lannic. This is a small, dry hillock of land adjacent to a submerged standing stone circle, like a small Stonehenge, in the bay of Morbihan (Fig. 10.14). Most of the upper circle is visible at low tide, but it is covered at high tide. There is another stone circle completely submerged next to it.

Divers discovered a deposit of Palaeolithic stone tools under the sea at Cap Lévi, near Fermanville, 10 km (6 mi) east of Cherbourg, during the laying of a small waste outfall pipe in 1970, and the local divers, led by Jean Allix, recognized their importance. They found the flint tools lying in the lee of the massive granite reef off Cap Lévi, at a depth of 20 m (66 ft), and brought some to the surface. Frédéric Scuvée, a regional archaeologist who understood their great age, encouraged local sports divers and off-duty navy divers to search for more. Gradually he assembled a collection of more than 2000 artefacts which were all at least 50,000 years old, possibly older. His co-worker, Jean Vérague, a geomorphologist, considered the site from the point of view of ocean and tidal currents, exposure to storm waves, changes of ice age sea level, possible earthquakes, and sediment movements. Meanwhile, Scuvée and his archaeological assistants had classified the dozens of different types of tools into the numerous categories used by French prehistorians. The flints and cores and small debris known as debitage collected by the divers even allowed the archaeologists to re-assemble a complete chunk of flint, showing how the original flint-knapper had sat there chipping away to make the tools needed. This showed that Neanderthal people had resided at the site, and lived and worked there. It was not a temporary shelter with a few abandoned flints. They

Figure 10.14. Er Lannic, Golfe de Morbihan, Brittany. This standing stone circle is mostly exposed at low tide, as in this picture, but almost completely submerged at high tide. There is another stone circle under water to the left of the picture (Photo Jay Kleinberg).

probably collected shellfish from the shore or made hunting excursions across the floor of what is now the English Channel to the great Channel River that drained northern Europe during the low sea levels.

Scuvée and Vérague published their final book-sized report in 1988, combining archaeological interpretation with geology, climate analysis, and sea level change. It was a decade ahead of its time in pure thoroughness and attention to detail. This astonishing document was, and is still, so innovative and pioneering that it deserves to be much more widely known. Since it was printed locally in Normandy on cheap paper, and only a few hundred copies (at a guess) were circulated officially by the Service Régional de l'Archéologie Basse-Normandie, it still hovers in the half-shadows of formal academic publications. As far as I know, it has never been digitized or stored electronically.

Basse-Normandie is not a well-funded region for archaeological expenditure. There are many Palaeolithic occupation deposits on land. Nobody could see the need to spend money studying one more that just happened to be under 20 m (66 ft) of water in a location where dangerous currents poured across the rocks around Cap Lévi at high tide, even though the seabed was comparatively sheltered down below. Further academic interest did lead to re-classification of many of the finds, while echo sounding and sub-bottom profile tracks produced a better understanding of the seabed after they were published in 2005. Nevertheless, the true significance of the site, its extreme age, its depth in the sea, and the completeness of the preservation compared with other known offshore prehistoric sites in Europe were not fully appreciated. It was five to ten times older than any other known underwater site at the time, and two or three times deeper. In the 1980s, it was the only known prehistoric site in the sea from before the last glacial maximum, about 20,000 years ago, and showed that Palaeolithic deposits of tools could survive post-glacial inundation, and possibly multiple phases of inundation without being destroyed.

This discovery vindicated my theories, assumptions, and prejudices, which is a nice feeling for a scientist. But most of my time from the mid-1980s to 2001 was devoted to bureaucracy (Chapter 9), and I was not able to follow up the original publication. I expected that the French authorities would conduct further underwater research, with comprehensive mapping, measurement of the boundaries of the site, spatial distribution of tools on the surface, and stratigraphic excavation, but this did not happen.

In 2007, in an attempt to find out what had been discovered at Cap Lévi in the past two decades, I searched for the addresses of Scuvée and Vérague. I drew a blank, since none of my French archaeological contacts could find any record of them. Their work had been so provincial that the mainstream academic institutions did not register their existence. There were no clues on the web. I challenged a young French student, Victor Fèvre, whom I had met recently at a conference, to track them down. He had impressed me with his quick thinking. Within a few days, Victor replied by e-mail. Frédéric Scuvée had died, but his widow said that Vérague survived at a known postal address in Normandy: he did not have e-mail. I wrote to Vérague, who, in a handwritten letter, agreed to meet me a few weeks later, and I invited Victor to join us. Vérague had stipulated that we should meet in the car park of an abandoned café at a point on the road between Cherbourg and Fermanville, where there were important geological features to inspect.

Jay and I drove our Honda people-carrier to Poole, crossing the next day to Cherbourg, where we checked into a small hotel by the port and met Victor. Cherbourg is sadly run down since its glory days as an embarkation port for the great ocean liners going to New York, when it was all chandeliers and marble-floored dining saloons and you dressed for dinner. Ten minutes before the appointed hour to meet Vérague, we arrived at the dusty abandoned car park by a derelict, boarded-up café, and waited. There were no other buildings within sight. There was no traffic. Shades of *North by Northwest*. We waited. But no light aircraft came buzzing out of the hills with guns blazing. The autumn sun was bright, and the choppy sea below us twinkled with diamond glints on the waves which were not quite breaking into white horses. At the precise scheduled hour, to the minute, a small Renault sped into sight, turning abruptly into the car park. Vérague was a short, intensely busy-minded man who, clutching rolls of maps, launched immediately into explanations and plans for what he wanted to show us. He spread maps on the bonnet of his car. The subsequent tour of the local coast, culminating at Cap Lévi, was valuable from a technical point of view, but details are not needed here. Vérague showed us that the coast was more or less stable from the point of view of tectonic earthquakes. There had been very little tectonic earth movement in the past 100,000 years.

We drove down a twisting farm road onto Cap Lévi itself, where there was a lighthouse to warn of the dangers. Wind-blown grass covered the headland, sloping down to tumbled blocks and ridges of dark-red crystalline rock. A submerged ridge of granite stretched seawards for a kilometre (0.6 mi) or more, creating a hazard of terrifying proportions. The tidal range in the area is 6.5 m (21 ft), so that large pinnacles and spiky peaks emerge at low tide. One peak, known as Biéroc, stays above the sea even at high water. The tidal currents are vicious, often over 5 knots, and at intermediate tidal levels the water roars over the half-submerged reef in turbulent foaming torrents. That is on a calm day. The effect of a storm wind off the Atlantic, with huge breakers rolling over the reef, is almost beyond imagination. The seabed Palaeolithic settlement was immediately at the foot of the slope to the east of the Biéroc, at a depth of 20 m (66 ft) below mean sea level, where it is protected from the maelstrom on the surface. It is also protected from the impact of waves from the open Atlantic coming in from the west. Once again, local topography determines survival or destruction.

We moved on to lunch in a charming restaurant in Fermanville, where I raised the question of future work under water. Vérague was convinced that the authorities had decided to ignore the extraordinarily rich site, and that his work in the 1980s was forgotten as worthless.

"Surely they must understand that it is a unique occurrence on a European, if not global scale?" I insisted. "The work you did was pioneering, but modern acoustic survey instruments and diving techniques could discover so much more. We need to know the full extent of the Palaeolithic deposits."

"They will ignore you," said Vérague. "They will nod and be polite and promise to think about it. But nothing will happen. It is best not to disturb them."

I discussed the problem with Victor, who thought that we might be able to contact the archaeological authorities in Caen. Vérague looked uncomfortable, so we changed the subject. We relaxed over brandy and coffee, before Vérague departed, expressing his sincere gratitude for our interest in his work.

Victor telephoned Caen to see if we could arrange an appointment. Dominique Cliquet at the Service Régional de l'Archéologie Basse-Normandie kindly agreed to meet us at 15:30, so we drove immediately to Caen. In the oak-panelled library at the Service d'Archéologie, we sat and tactfully skirted around the issues involved in proposing further diving at Cap Lévi. It was diplomatically impossible for a visiting British non-archaeologist even to suggest obliquely what the French archaeologists ought to do.

First praising the work already done by his department, I explained the huge interest of the submerged remains from a British and European perspective. In his reply, Cliquet stressed that Basse-Normandie was very poorly funded from Paris, so that the underwater deposit had to take its place alongside hundreds of Palaeolithic sites on land. Many were threatened by agriculture and modern developments. I suggested that IFREMER, the French National Oceanography Centre nearby at Brest, could provide an acoustic survey vessel that could map the topography of the site using multibeam acoustics. But the importance of this was not obvious to the French group. Several follow-up dives had been made in 2000 and 2002, when more flint tools had been recovered. Describing these tools accurately had been done. From cores, sediment analysis, and pollen, the main occupation could be dated to 80,000 to 100,000 years ago. In view of the costs, that was sufficient. There was much still to discover, but there were other priorities.

"Viewed from Britain," I said, "prehistoric settlements on the floor of La Manche are very significant because they show possible tribes either entering or leaving occupation of the British Isles, depending on climate change and ice ages, or the extent of ice cover to the north." I explained that all such sites were important to us, even if, from a French point of view, the Cap Lévi site was just one more of many. I suggested a joint Franco-British diving project, with the UK divers being provided by Garry Momber's Maritime Trust group at Southampton, and explaining that we would raise much of the money needed as well, some from European sources. French academics should have control of the objectives and of the publications, but we could provide some divers with archaeological training. IFREMER might help with a new acoustic map.

This proposal was accepted, and Momber took the lead. He assembled a team from the Hampshire and Wight Trust for Maritime Archaeology, the Association pour le Développement de la Recherche en Archéologie Maritime (ADREMAR) in France, and the Centre National de la Recherche Scientifique, backed up by local volunteer divers near Cherbourg. The divers worked in July 2010, making 27 dives, which had to be planned to avoid the worst tidal currents. The speed of the flood tide was potentially extremely dangerous. The visibility at the bottom was a murky 3-5 m (10-16 ft) even at still water. The divers fixed a network of five datum benchmarks, consisting of strong galvanized steel pitons 1 m (40 in) long banged into the seabed, and connected in a triangular network defining three measured triangles with sides of 30 m (98 ft). This geometrical grid enabled all subsequent observations and sampling to be plotted exactly. Previous trenches from 2002 could be fixed on the grid. The divers recorded the nature of the seabed and the distribution of lithics using photography and video. They noted that the surface artefacts originated from sub-surface layers, and were revealed by gradual erosion of the finer sediments that resulted in their exposure. Many of the lithics recorded looked as if they had been freshly uncovered with minimal signs of degradation or wear. The rate at which the site is changing is unknown but should be monitored to assess the deterioration. Notwithstanding the 60 m (197 ft) extent of the survey grid, the full length and width of the area of Palaeolithic remains is still not known accurately. Nor has the complex and rugged shape of the protecting cliff been mapped in detail. The ridge itself may have been occupied by prehistoric people, giving both shelter and a vantage point overlooking the Channel Valley.

The Fermanville-Biéroc site is very rich in archaeological material from an important period for understanding hominin dispersal and adaptation to a worsening climate at the end of the last interglacial, which was 125,000 years ago. The re-colonization of Britain after a period of approximately 100,000 years of abandonment is now thought to have occurred 35,000 years earlier than previously believed. New evidence points to Neanderthals crossing the Channel/Manche during a glacial low-stand 100,000 years ago, not 65,000 years ago, as previously thought. Understanding Neanderthal capabilities, preferences, and population movements during this period will help us understand their distribution in Europe, social organization, and ultimately their demise.

The Fermanville-Cap Lévi story had a happy ending, although there is still a great deal more to be discovered there, including the size of the community, the extent of their hunting and fishing, when the site was occupied and abandoned, and how the materials have survived so well, in spite of the slow erosion.

* * *

In 1995, the NIO moved to Southampton, into a beautiful, purpose-built laboratory complex overlooking the harbour, combining there with other research institutions to form the National Oceanography Centre (NOC). I discovered that we shared the building with the Hants and Wight Trust for Maritime Archaeology (now the Maritime Archaeology Trust). I became one of the trustees. One day in 1999, Garry Momber, then one of the Trust's most active divers, and now its director, came into my office looking coy and reserved, clutching something mysterious in his closed hand.

"We've been diving again at Bouldnor Cliff," he said, "on the north shore of the Isle of Wight, where there are thick deposits of peat and some drowned tree trunks. A lobster had dug a hole, a sort of burrow, and thrown the debris out. And in the pile we found this."

He opened his hand to show several small worked flints. His embarrassment evaporated and he was suddenly full of excitement and ideas. This was the beginning of the excavation of a submerged Mesolithic-Neolithic settlement at a depth of 11 m (36 ft), where the survey and excavation has continued every summer for more than 20 years. As a trustee, I went out in the dive support boat several times when the divers were working there and helped recover their finds when they surfaced. The remains at Bouldnor Cliff are not the centre of a village with domestic houses, but a busy waterfront for launch and recovery of canoes, landing fish, woodworking, maintenance, repairs, some hearths, and the dumping of tools and debris. The main village with domestic dwelling was probably further inland. The mapping, photography, and excavation at the Bouldnor Cliff gave me the experience of seeing an area of submerged human activity revealed as successive layers were excavated or eroded each year, and reading about it through regular reports. The growth of knowledge year by year was like a slow-motion frame-by-frame strip cartoon. Both lithics and organic samples could easily be brought back to NOC for conservation. Through various committees and funding programmes, I continuously lobbied for support for this project, which was poorly financed for the first few years.

The Bouldnor Cliff remains are at least 8000 years old, and there are timber planks forming a floor, and other worked timbers suggesting that that the villagers were possibly building logboats close by. Remains of food and grain suggest that a species of wheat was brought from the Middle East much earlier than expected. Bouldnor Cliff provides a rich deposit of occupation levels, timber posts, planks, hearths, string, food remains, and flint tools, demonstrating an excellent state of preservation in British waters. Unfortunately, in spite of the protection from open ocean waves, the tidal currents rip through the narrow channels of the Solent so that the human materials are continuously being eroded. Measurements by divers in 2020 showed that the submerged outcrop of clay and peat containing Mesolithic deposits had been eroded back by a distance of 7 m (23 ft) in 17 years, a rate of 0.4 m (16 in) per year. A photogrammetric survey recorded the exact shape of the submerged vertical face in 2020. The divers collected flint tools that had fallen from the outcrop and cut a carefully excavated trench from which they took both core samples and box samples for analysis in the laboratory. One truly beautiful discovery was a large flint adze, made by an expert flintknapper (Fig. 10.15).

The Fermanville and Bouldnor Cliff sites show that stratified, prehistoric occupation levels can survive many thousands of years in the rugged ocean conditions of the Channel, with strong tides and potential exposure to storm waves from the Atlantic, as long as the local configuration of the coastline provides adequate shelter. Nevertheless, we now know that both sites are eroding.

Figure 10.15. Diver at a depth of 11 m at Bouldnor Cliff, off the Isle of Wight, examining a freshly discovered flint tranchet axe or adze, which is 12 cm long, and dates from about 7000 years ago, location Cliff II (BC II). This tool was found in 2020 amongst hundreds of flint tools and fragments of flint washing out from the top of an old sand bar, beneath a deposit of peat. Just 400 m away there are cut timbers. The occupants of the settlement could have been making dugout boats, though the adze was used more for trimming and scraping wood, and not for gouging out (Photo Brandon Mason).

* * *

The diving science group at the University of Cape Town had been active members of CMAS for many years (Chapter 8), so I was delighted to receive from them an invitation to present a paper at their annual conference in 1995. By this date, the evidence that prehistoric, and even Palaeolithic, tools, artefacts, and deposits from occupied sites could survive multiple inundations from the ice ages was strong. I prepared a lecture explaining that the Agulhas Bank, extending 100 miles (160 km) south and east from Cape Town, was a natural place to find Palaeolithic deposits. There are many large caves on the shore very close to sea level on the coast of South Africa, rather like Gorham's Cave in Gibraltar (Chapter 4). Those that have received the most excavation by archaeologists include Blombos cave, Ysterfontein, Pinnacle Point, and Klaseis River caves. I intended to advise the local divers to keep in touch with fishermen, seek for chance finds in the sea, and to search the beaches in front of the known prehistoric caves on land.

It was my first visit to South Africa. We gathered in the bar at the university the night before the conference started, where I met colleagues from previous meetings in Europe, including Peter Zoutendyk and John Field. In the half-light of the bar, against a background of noisy chatter, a man that I did not know cautiously introduced himself, asking if we could have a discussion.

"My name is Bruno Werz," he said. "I run a small archaeological diving company and we have had licences to excavate several Dutch East India shipwrecks near here. Last month we found this deep in the sand beneath one of the wrecks."

Figure 10.16. Bruno Werz found these Palaeolithic Acheulean handaxes buried in the seabed sediments in Table Bay, South Africa. They are probably about half a million years old (Photo Bruno Werz).

He reached into his briefcase and produced a pear-shaped Acheulean handaxe bigger than his hand (Fig. 10.16). It was a classic example of the kind of tool used for many hundreds of thousands of years in the Palaeolithic. He passed it to me for inspection. I turned it over in the dim light and felt the edges, which were still sharp. I needed to revise my lecture and congratulate the local divers, rather than advise them what to do.

"This is beautiful," I said. "Have you published anything about it? How deep was it under the wreck? Let's find a table and sit down." We struggled to find a place but finally sat away from the crowd.

"Well, the local newspaper published a picture of it last month, but nobody at the university has shown any interest," said Bruno. "One of my divers, using an airlift, kept the nozzle digging longer than he should have done, and went below the level of the wreck, digging a pit down to the bedrock, a metre or more lower. This Palaeolithic tool was on the bedrock. But I am a shipwreck archaeologist and I honestly don't know anything about prehistory. I just recognize it for the kind of tool that you can find in the caves around here on shore."

"Bruno, if this artefact was *in situ*, embedded in an earth deposit, under the sand, then it is the oldest human artefact ever found under the sea by many thousands of years. This could be half a million years old. It proves that an object lost by prehistoric peoples on the continental shelf many glaciations ago can be buried in the soil, and then survive the sea level going backward and forward over it for hundreds of thousands of years. It is a complete revolution! It should be published in a big international science magazine like *Nature*."

Bruno looked nervous and took a gulp of beer.

"I couldn't write that. The university is not interested, and I wouldn't get any help. I don't see how to publish it, and I don't know the context in other countries. Oh, incidentally, one of the divers found another handaxe yesterday."

"Was it in a similar condition on the bedrock?"

"Yes."

"We must publish this, Bruno. It is globally significant. I can help you write the paper, but you must be the first author, of course. The tricky bit is working out what the tectonic earth movements have been in Table Bay over this very long time span. The tools could have been lost on land, and sunk under the sea by earthquakes. That would not prove that people lived on the continental shelf. I can work on that bit. Can you get more data on the exact deposit on the bedrock? Was there any soil? Are there any terrestrial fossils or microfauna in the soil? Could the tools have been carried offshore by local rivers or streams?"

Bruno looked more optimistic.

"I can talk to the divers again, and get more details from their logbooks. We will have a pretty good record of the contact with the bedrock, but there's no sample from the contact zone. Just visual inspection and notes."

I thanked Bruno for telling me about his amazing discovery, and for showing me the handaxes. The rest of the conference was interesting, but paled into forgetfulness in comparison with Bruno's information. We submitted a joint paper to *Nature* in 1998 showing that there had been no land subsidence in the area, that the tools had been found in a red oxidized soil in contact with the bedrock, and that they were sharp-edged, so could not have been rolled downstream by rivers. They had been discarded or abandoned where the divers found them.

Two of the *Nature* reviewers recommended publication with some clarification requested, but the third dismissed the discovery as pure chance, out of context, and asked how we could possibly envisage making any further similar finds. Since that was clearly impossible, said the reviewer, the paper should not be published. It was rejected.

Nature had published several highly speculative papers on the so-called southern prehistoric migration route from East Africa, along the Asian shore to Australia, before the last glaciation, 20,000 years ago. These papers had provided no factual evidence from the shore or below sea level. Werz and I had provided the scientific world with a tiny clue that might substantiate by similar discoveries elsewhere whether the migration route existed or was a speculative fiction. We had shown that the necessary definitive data could survive on the continental shelf. But one referee did not like it.

Swallowing our pride, we submitted the paper to the *South Africa Journal of Science*, where it was published in 2001.

* * *

Although my work with operational oceanography at the pan-European level was absorbing, interesting, and rewarding (Chapter 9), I retired at the age of 65, which was at the end of the year 2001. In January 2002, I started a new life, planning, amongst other things, to write a book about prehistoric settlements on the continental shelf. In March 2002, I visited the Oceanology International Exhibition which was, by that date, held at the ExCel Centre in east London. During my wanderings around the exhibit stalls, I met Quentin Huggett, a tall, bearded, bespectacled marine geophysicist, who had left the NIO some years earlier to set up a company, Geotek, which had prospered greatly. Quentin greeted me with his usual big grin.

"Nic, I've got a great proposition for you. Come on. Let's grab a beer at the café and talk it over."

We found a rickety table that would not balance, purchased sandwiches and beer, and settled down to business.

"Geotek has got a fascinating contract from the Department of Trade and Industry" (DTI), said Quentin. "There is a big new round of licensing for offshore petroleum exploration, and they need advice on environmental impacts. The companies tendering for a licence have to show that they will protect every conceivable aspect of the marine environment, on the seafloor, in the water column, at the surface, and in the air. It's big stuff."

"It sounds very complex," I said. "But how does it affect me? You said you had a proposal of some kind."

"Wait, wait. The DTI list of environmental factors is unbelievably long. Sea birds, sea mammals, toxic chemicals, acoustic noise, aesthetic impact. Everything."

"I can believe it."

"Geotek has got the contract for specifying all the issues where surveying, drilling, exploration, and pipe-laying could impact the environment of the seafloor, or disturb sediments that stayed in the water column."

"OK, I follow that."

"One of the impacts on the list is possible disturbance of the prehistoric archaeological remains within the sediments or on the surface of the seafloor."

"That is really unexpected," I said. "Where did they get that idea from? DTI is not renowned as a great centre of intellectual enquiry."

"I don't know where it originated," Quentin said. "But they are very serious. It may be the result of European directives. I think it is because the gravel dredging industry has already come up against this concern, and they have an agreement with the British Marine Aggregates Producing Association, to report all archaeological finds, including prehistoric tools. The DTI wants a set of guidance notes for oil companies that is equivalent to the BMAPA rules, complete with the identification of prehistoric hotspots that should be avoided if necessary."

"What is the area that they are licensing now?"

"It's the northern North Sea, but they want guidance for the whole North Sea, as they did not include this factor, prehistory, when they issued licences for the southern North Sea."

"You think I could help?"

"I know you could. We need a report by late May, and you need to attend a workshop session to discuss drafts at Newburgh, north of Aberdeen, in July."

"I could do it," I said. "But obviously I would have to consider the whole North Sea, not just the UK sector. Oceanography and prehistory don't stop at the median line. I might need to include data from the Dutch, Germans, and Danes. Perhaps Belgium, Norway, and France as well."

We agreed a contract fee and parted in good cheer.

My report did indeed include discoveries from the eastern sectors, and I was able to show widespread finds of ice age animal bones all over the southern North Sea, mostly landed in the Netherlands, where there was a big community of amateur palaeontological bone collectors. Using bottom beam trawls with chains that drag through the seabed, fishermen trawled-up mammoth teeth and tusks, and bones from bear, wolf, pig, deer, and so on, on nearly every trawl, bringing tons of ice age bones from land animals into the fishing docks every year. This was the latest development of the finds that were first reported in the 1970s.

I wrote in my report that in shallow coastal and intertidal areas, where ice age sites were most likely to be known already, possibly eligible to be protected by conventional antiquities legislation, individual sites should be strongly protected by existing laws. For offshore waters, I concluded that industries should report their plans in advance, and should be required to monitor the seabed for prehistoric sites, report any findings, and be prepared to accept protection and designation of important sites if they were discovered.

Pre-disturbance analysis should estimate the possibility of prehistoric sites existing in the area of the licence and, if the authorities deemed it necessary, the operator should carry an archaeologist on board to monitor progress. There should be no blame, and no penalty, if the drilling, dredging, or pipe-entrenching did damage an archaeological site that did not have previous protection. There were very few areas that should be banned to industry. Given the difficulty of finding prehistoric sites in deep water, tens of miles from the shore, and, since it was probable that the first contact or disturbance of the site would not destroy it, there were many advantages in allowing industry to proceed as normal. Academics would never have discovered the sites otherwise. I expected academic uproar from this recommendation, but the scholars and agencies accepted the common sense of the idea. No academic could charter large offshore vessels to dig in the centre of the North Sea, and we could not blame industrial companies for doing so, nor for their finding prehistoric remains that we would not otherwise have known about.

Geotek hosted the review conference at Newburgh in a hotel close to the marshes and mudflats of small tidal rivers. The sky was rich with birds and endlessly atmospheric with ever-changing light over the sea and dunes. As the hotel did not have rooms accessible for a wheelchair, I stayed in an adjacent lodging house on the very bank of a winding creek. DTI accepted my recommendations. They put the report on the web with the other environmental criteria for offshore concessions and licences.

The data that I had accumulated became the focus of a small conference or workshop. I convened an organizing team and we invited colleagues from both sides of the North Sea. English Heritage paid the overhead costs in London, and the Council for British Archaeology published the papers in a special volume. The workshop, held in May 2003, attracted 58 people from five countries. Prehistorians and diving archaeologists described their finds – far more than expected – while governmental experts from the different countries explained their regulations and strategies for managing the offshore prehistoric heritage.

After the conference, I drove across Holland to meet as many of the North Sea palaeontological bone collectors as possible. I met Dick Mol, who specializes in mammoth skeletons (Fig. 10.17); Jan Glimmerveen, who has an exquisite collection of archaeological finds from the North Sea; and Klaas Post in the village of Urk, who collects and sorts out the huge quantities of bones brought to him by fishermen. Klaas, who showed me around the warehouse stacked full of ancient animal bones, later helped us to arrange a magnificent display of the finds from the North Sea, which we mounted

Figure 10.17. This well-preserved skull of a mammoth was trawled from the North Sea in 2006. Note the intact molars, and one complete tusk (Photo Persburo Flakke – Wim van Vossen jr).

in Brussels for members of the European Parliament in 2010. This display included a replica of a Neanderthal skull fragment, and a mammoth tusk 3 m (10 ft) long, which needed two strong men to carry it.

The Parliamentarians were happily convinced that modern supertankers and oil rigs are cruising in the waves over the hills and valleys where our ancestors, tens of thousands of years ago, were hunting mammoths and wild boar.

Meanwhile, in November 2002, the NERC announced that it would provide funds totalling £2 million for projects conducting research into the Environmental Factors in the Chronology of Human Evolution and Dispersal (EFCHED). Geoff Bailey and I, in partnership with Abdullah Alsharekh from King Saud University, Riyadh, Saudi Arabia, proposed a diving programme to search for evidence of prehistoric crossings of modern humans from Morocco to Gibraltar and from East Africa to Yemen and Saudi Arabia. With Garry Momber, we dived at Gibraltar in 2005 (Chapter 4), after which, in 2006, we made a series of dives around the Farasan Islands on the Saudi-Yemen border in the southern Red Sea. Abdullah and I selected the Farasan Islands by reviewing the various groups of islands along the Saudi coast. Most of the Saudi Red Sea coast is relatively straight, with very minor indentations, waddis, and occasional small islands or reefs offshore. The Farasan Islands, close to the Yemen border, stand out as a widespread archipelago of dozens of large and small islands spanning an area 93 x 56 km (68 x 35 mi), which would provide both a likely area for prehistoric occupation and numerous protected channels and embayments between the islands, well sheltered from storms.

If there was an ice age southern route across the Red Sea, it is certain that the Hannish Islands, close to Bab el Mandab at the extreme south, and the shoals around them, provided the narrowest point. But this channel off the coast of Yemen has been plagued with piracy and war for decades. Since it was impossible to dive there, the Farasan Islands, 300 km (186 mi) to the north, provided the next best terrain, especially as the shelf was wider so that we would be looking for signs of a larger resident population, not small groups in transit. Our task was to demonstrate that ice age peoples lived on that wide shelf, or at least that it was an ideal landscape to live in.

The oil company Saudi Aramco generously provided a large oil-field support vessel, MV *Midyan*, for the work at Farasan, with participation of divers from the Saudi Navy, King Saud University, local civilian dive groups, and the Hants and Wight Trust (Fig. 10.18). The EFCHED project formally ended in 2006, but Geoff Bailey had, by this time, obtained grants for several further seasons of work using smaller local vessels. A final geophysical survey was conducted using the Hellenic Centre for Marine Research's (HCMR) research vessel *Aegaeo* in 2015.

The Farasan Islands consist of massive, slightly uplifted, fossil coral platforms from the last interglacial, 120,000 years ago, resting on unstable sedimentary strata which are perforated in many places by upwelling salt. This extraordinary situation, known as "salt karst tectonics", was caused by the complete evaporation of the saltwater in the Red Sea 30-24 million years ago, during the very early stages of Arabia splitting away from Africa. Beds of salt thousands of metres thick formed first, and then coastal deposits of sandstone and limestone formed on top of the salt. Under the weight of the overlying rocks, the salt began to extrude like toothpaste, fracturing the rocks above, bubbling up almost like volcanoes, and then dissolving away to leave huge circular depressions. The map of the Farasan archipelago shows repeated patterns of circular depressions and

Figure 10.18. Measuring a 5000-year-old shell midden on the edge of a fossil coral reef on the Farasan Islands, near the border of Saudi Arabia and Yemen, in the southern Red Sea. The Aramco support ship Midyan is supporting the divers on this project in 2006.

linear troughs where upwelling salt has dissolved. Circular pits, hundreds of metres deep, which would have been huge freshwater lakes during the ice ages, occur on the continental shelf or between the islands.

I visited the Farasan Islands twice on reconnaissance trips, experiencing temperatures in the high forties centigrade (~120° F). I fell in love with the slab-like rocky landscape, where the beaches of the minor islands coruscated with the debris of mother-of-pearl from the pearl fisheries of a hundred years ago. When the wind blew, the mother-of-pearl dust rose in opalescent clouds. The hospitality and kindness of the fishing families on Qummah island created an unforgettable memory of when little boys ran into the water to place upturned boxes so that my friends could lift me more easily from the boat onto the beach.

On the Farasan Islands, the research teams, led by Bailey, discovered over 2000 ancient shell middens, vast pyramidal dumps of empty shells discarded after their contents had been eaten by people 5000-6000 years ago. At that date, the sea level was already close to present, and the islands were separated from the mainland. Nevertheless, the rich food supplies of the coastal zone were similar at lower sea levels. There is a gazelle species that presumably reached the islands during lower sea level, when they were connected to the mainland, and gazelles are still present as a protected species. Other food sources include a rich inshore and intertidal marine environment with great variety of fish and marine molluscs, turtles, seals, and migratory birds. For several days during March or April each year, the longnose parrotfish school in large

Figure 10.19. Diver surveying cliff overhang at a depth of 60 m near the Farasan Islands, on the Saudi-Yemen border. The diver is breathing trimix from the big cylinders on his back, and carries nitrox for decompression in the smaller cylinders.

numbers as they move into the shallow water of the inner bays, where people harvest them in huge quantities by wading into the water with nets. All of these resources were present during the prehistoric period when the sea was lower, and prehistoric people would have found a comfortable living.

The diving team searched below the headlands and cliffs on the main Farasan island and some of the outlying islands, especially Qummah, where they found cliffs with undercuts at 10-20 m (33-66 ft) depth. In other locations they found terraces and cliffs down to 60 m (197 ft) depth. For the deeper dives they used a trimix breathing gas of oxygen, helium and nitrogen, breathed on open circuit, not in rebreathers (Chapter 8) (Fig. 10.19).

* * *

I first met Dimitris Sakellariou in December 2007, at a conference in Stabiae, south of Naples, where we discussed the technology available for studying underwater archaeology. In a conference break we visited the breathtakingly beautiful derelict Roman palaces high on a terrace overlooking the sea, where the riot of creepers and flowers sprouting from the buildings emphasized their lost and crumbling glory. Sakellariou is a geologist and engineer at the HCMR, near Athens. At the final conference banquet, he demonstrated his skills at Greek dancing, and I played an accompaniment to the music by making resonant booming notes on a conch shell with the end cut off. I don't know who provided the conch shell. You don't find them everywhere. Sakellariou told me about his surveys in the Aegean and invited me to his institute. In January 2008, I gave a talk at HCMR on the potential for finding seabed prehistoric sites all around Europe. At our meeting afterwards with the director, George Chronis, Sakellariou and I explained our thinking for further collaborative research. Chronis threw his arms wide apart, gestured dramatically, puffed on his cigarette and exclaimed, almost shouting:

"I can see it! I can see it! A huge European project from Norway to Greece, studying the prehistory of the whole continental shelf!"

We were encouraged.

Sakellariou and I chaired a session on submerged prehistory at the Third International Conference on Underwater Archaeology (IKUWA3) in London in July 2008. We established a European planning group which we called Deukalion, after the Greek legendary equivalent of Noah, who survived a great flood. The Deukalion Group met several times, gaining approval in late 2009 for a project funded by the European Committee for Science and Technology (COST) with a four-year grant to study and coordinate seabed prehistoric research.

The COST project was called Submerged Prehistoric Archaeology and Landscapes of the European Continental Shelf (SPLASHCOS), with Bailey as chairman and Sakellariou as vice chairman. Our tasks were to collect all the information we could on known subsea prehistoric sites in European waters; to document them with their date, culture, and depth; to promote training courses and transfer of technology; to hold workshops and conferences conducting analysis of sites and integration of discoveries; to encourage new, funded diving on prehistoric sites at the national and regional levels; and to conclude with a major conference and book publication.

While we were planning SPLASHCOS, with the most exquisite timing, a British gravel dredger excavating aggregate for concrete discharged its cargo at the SBV Flushing wharf near Antwerp, where an off-duty chef named Jan Meulmeester was searching for fossils. He observed Palaeolithic handaxes tumbling down the piles of the discharged gravel, and collected 28 of them between December 2007 and February 2008. By May 2008, the discovery was in the press, and in early 2009 an Anglo-Dutch joint archaeological workshop was able to examine the tools at Amersfoort, where we met and congratulated Meulmeester. He was delighted but rather surprised at the scale of response by so many academics.

This discovery, which came from the dredging concession area known as A240, 11 km (7 mi) to the east of Great Yarmouth, led to the excavation of the seabed by marine archaeologists from Wessex Archaeology, a non-profit trust. The cores that they extracted contained flint tools and debitage from pre-humans dating in the range 70,000-360,000 years old. This extraordinary deposit in the sediments of the North Sea, at a water depth of 21-35 m (69-115 ft), is the oldest Palaeolithic find offshore anywhere in the world, apart from the handaxes found in Table Bay. Buried in the sediments, the A240 relics have survived multiple sea level transgressions and regressions over many tens of thousands of years. Wind-blown sand, river mud, and seabed dunes, drifted over the abandoned tools and protected them from the storm waves of the North Sea thundering and crashing above them.

SPLASHCOS had its formal launch at a meeting in Brussels in November 2009. There was a real sense of excitement at bringing together professional archaeologists with young researchers and amateurs from 25 countries, with plans for practical training dives on different coasts. Over the next four years, SPLASHCOS pooled the skills of archaeologists, geophysicists, oceanographers, and palaeo-climatologists from over 100 research institutions, while circulating a mailing list of over 200 correspondents. While SPLASHCOS remained active, from 2009 to 2013, there were 23 research projects funded by a wide range of agencies, amounting to a total financial support of over €20M. I attended almost all the meetings devoted to the ancient terrestrial environment below sea level, from Amersfoort to Zadar, and led the editorial team that produced the final book on that subject.

Dr Ehud (Udi) Galili organized a technology transfer workshop at Atlit Yam to give students from other countries in SPLASHCOS the experience of diving on the well-preserved remains of the seabed Neolithic villages on the coast of Israel. I attended to give some background lectures. The course included supervised diving work, underwater survey and mapping, data recording, sorting and analysing finds on shore, underwater photography, lectures, visits to archaeological laboratories in Haifa, and a tour of the vaulted Palaeolithic caves in Mount Carmel. The experience brought back old memories but also illuminated the amazing progress that we had made in 36 years. In 1975, I could not have dreamt of a collaborative programme like SPLASHCOS, nor a workshop training session of the sophisticated kind run by Udi Galili. The young students were exhilarated to dive daily on the submerged villages, with their intact standing stones, burials, and freshwater wells (Fig. 10.20, 10.21)

SPLASHCOS concluded with a conference at Szczecin, on the Baltic coast of Poland, in September 2013, at which the various working groups presented their reports and discoveries. The scope was impressive, showing the advances that could be made by combining modern prospecting technology, accurate bathymetric maps, sub-bottom profiling, and diving by trained archaeologists. The database compiled by Hauke Jöns and Moritz Mennenga, with support of a team of data experts and Dik Schaap at MARIS SeaDataNet, showed an astonishing inventory of 2672 documented prehistoric sites under water or in the intertidal zone, with data from 19 countries. More sites have been added since the Szczecin meeting.

Most of the finds reported are single or multiple artefacts recovered by fishermen, dredging, or divers, without firm context, other than the location and depth. More than 400 sites have been classified as remains of settlements, with fireplaces, cultural layers, wooden structures, post-holes, floor-matting, and tools such as spears, fish-spears, fish-traps, canoes, and paddles. The preserved material is very different from that which would be found on land from the same Mesolithic and Neolithic periods, including many objects made of wood, bone, antlers, leather, and fibres for string and ropes. Only a few underwater sites have produced human remains, but these will help to understand human evolution and migration routes. Through my connections with the Brussels research committees of the European Commission, I was able to ensure that future maps of the seafloor produced in one of their programmes will contain reference to the distribution of prehistoric landscapes and archaeological sites. Geoff Bailey edited the final SPLASHCOS inventory of known seabed ice age archaeological sites and settlements that was published in April 2020 (Fig. 10.22). An amazing achievement.

The community momentum generated by SPLASHCOS helped to secure more funding for research in the North Sea for Belgian and British scholars. Tine Missiaen, Simon Fitch,

Figure 10.20. Diver on well at Atlit Yam, Israel. The top of a well designed to provide freshwater is now at a depth of 15 m under the sea. The well has been excavated by divers, and the sandbags have been placed round the edge to stabilize it (Photo Itamar Grinberg, excavator Ehud Galili).

Figure 10.21. Divers measuring the standing stones, Atlit Yam, Israel. The neat outer circle consists of sandbags to protect the site (Photo Itamar Grinberg, excavator Ehud Galili).

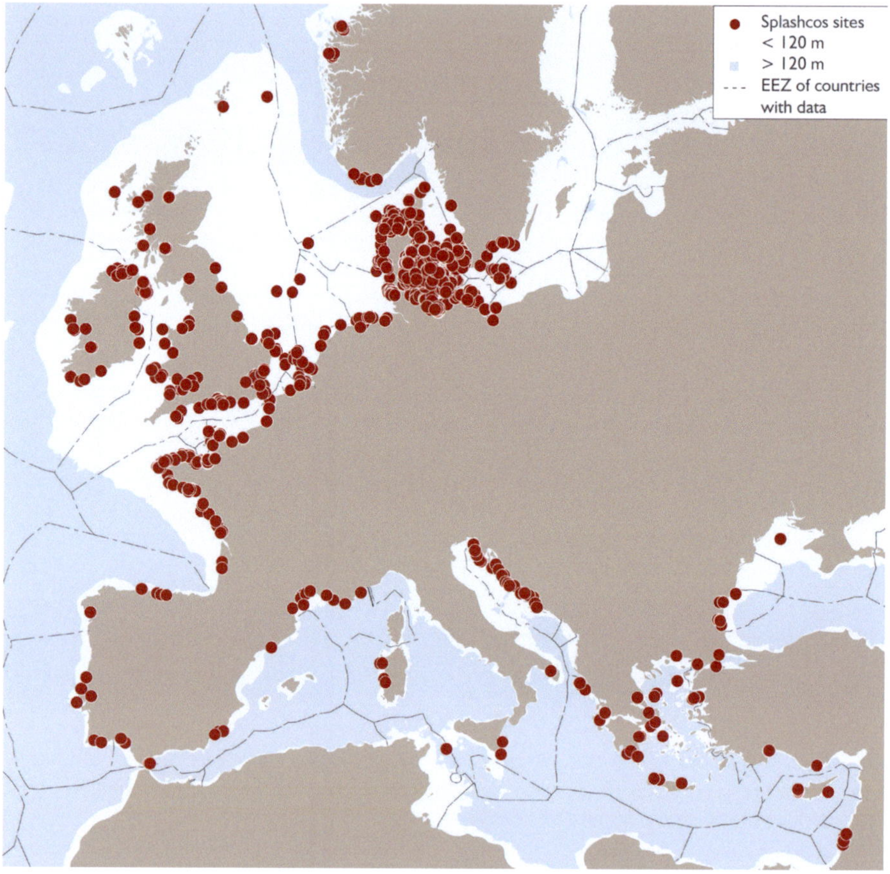

Figure 10.22. Map showing the distribution of submerged prehistoric sites in the coastal seas of Europe. The compilation of sites ranging from 5000 years old to nearly 1 million years old was made during the SPLASHCOS Action. The dashed grey lines show the limits of national marine boundaries, and the lighter shade of colour for the sea shows the width of the continental shelf to a depth of 200 m (Map by Moritz Mennenga).

and Vince Gaffney used a wide range of modern technologies to explore around Brown Bank, a shoal in the central North Sea, approximately halfway between Lowestoft and Haarlem. The bank is at a depth of 16-22 m (52-72 ft) and is approximately 25 km (15 mi) long and 3 km (2 mi) wide. In recent decades trawlers had repeatedly recovered Pleistocene mammal bones and occasional bone tools of Mesolithic date from the bank, but the artefacts seemed to have been eroded and transported there from a nearby stratified deposit.

A series of projects was conducted in 2018 and 2019 to identify the submerged river banks and lakes where the artefacts may have originated. Starting from regional maps of the submerged Mesolithic landscape constructed from industrial acoustic surveys by oil company contractors, they narrowed the areas of interest using very high-resolution sonar, and then sediment cores. Shallow penetrating vibro-cores provided important environmental data showing layers of ancient peat. Step-wise analysis of findings and the refined sampling of targets showed increasing likelihood of suitable human occupation conditions. The dredge and core samples on highly focused sample areas finally recovered Mesolithic worked flints, debitage, and chunks of wood. This novel achievement will lead to increased investment in ship-time and seabed survey and excavation. Not only does it tell us about the Mesolithic life in the marshes and wetlands 8000 years ago, but it is the first time that marine archaeologists have been able to work from geophysical and palaeo-environmental data in deep water, 20-30 m (66-98 ft), to recover anthropogenic material in primary deposits. The proof that this can be done will lead to a greatly expanded range of projects in the North Sea.

This project in the relatively deep turbulent waters of the central North Sea shows that the accumulated expertise from SPLASHCOS has enabled collaborative groups of researchers to tackle problems that would have been unthinkable ten years ago. Technology has improved, and, above all, the persistent step-by-step work by scholars with different skills has enabled archaeologists to survey and describe the submerged terrestrial and coastal environments with unprecedented detail. The potential for discovery of human and pre-human deposits *in situ* beneath many parts of the North Sea floor is now very exciting.

* * *

Can we now answer the questions that were still uncertain after the Scripps conference in 1981? Does seabed prehistoric archaeology coalesce into a logical and synthesizing discipline, with research objectives and established methodology? Or is it still an assemblage of random curiosities and trawled-up antiquarian goodies? Can we start to reconstruct whole communities that existed on the continental shelf during low sea levels, together with their technologies, seafaring skills, social structure, and change through time? Can we find sufficient settlements and scattered remains on the seafloor to define migration routes?

The answer is positive, if a little tentative. We – many hundreds of divers and archaeologists working together over many years – have proved that it is a real science, producing significant intellectual results. Academics now have hundreds of sites and published papers to work on, both in Europe and other parts of the world, but humility is appropriate since the majority of sites are still found by chance. Only a few sites, some of them described above, have been found by archaeologists directing searches without previous chance discoveries. Nevertheless, the ability of the academic community to respond quickly and efficiently to chance finds has been effective again and again, with excellent results. The chance discoveries can be fitted into the previously mapped ancient landscape and linked to other known prehistoric remains. It may not be flattering, but it works.

Viewed globally there is still a deficit of prehistoric remains offshore between the Tropic of Capricorn and the Tropic of Cancer. This deficit is probably due, at least partly, to a very low effort at searching for prehistoric objects offshore, but it means that we have, at present, no means of checking the validity of the so-called southern migration route from Africa to Australia along the Asian coast at low sea level. The discovery in 2020 of submerged stone tools by Jonathan Benjamin's team off the Dampier Archipelago is the first demonstration that we can find human artefacts in the tropical seas, but the paucity of remains may also reflect a very different exposed environment during low sea levels, followed by the growth of coral reefs, mangrove forests, and large, muddy deltas.

Because of improvements in acoustic technology and in data storage and data search tools, we can map the seafloor much more accurately than a few decades ago, and store vast quantities of data on computers, so that everybody can consult the data, and zoom in on the local details that they need. At least it is becoming a possibility. Because prehistoric peoples hunted and foraged over a wide area, and sometimes migrated seasonally, we need electronically searchable seafloor maps at scales of 100 km (60 mi) range with accuracy locally down to 1 m (39 in) or so. That is demanding.

Groups of submerged sites in the southwest Baltic and off Israel's coast are sufficiently numerous and varied to provide data on responses to changes in the environment, food availability, climate, and sea level. Discoveries in the North Sea, the English Channel, and the Gulf of Mexico are approaching the same level of detail. This means that there is still a great need for new discoveries and new ways of finding sites. Since we know that, even in the best circumstances, some key sites are slowly eroding away, this is urgent. Groups of deposits and settlements now offshore provide evidence for population distribution and resource exploitation that can be integrated into the same type of analysis applied to the onshore sites, treating them as a continuum. For example, it is logical that one should study the site distribution from southern England, the southern North Sea, and the Netherlands as a single extended culture at times of low sea level, exploiting a continuous landscape. In the academic literature, I am seeing an increasing number of papers discussing the seabed prehistoric finds off the coast of China, and a growth of interest in Indonesia.

During my career I have dived in, or supervised diving projects in all the zones where there were, or may have been, significant prehistoric human sea crossings at low sea levels. That is the Strait of Gibraltar, Southern Red Sea, and Timor-Australia. I have not been involved in diving projects for the Bering Sea crossing, but to try to understand the environment better, Jay and I sailed in an Expedition Cruise ship, MV *Caledonian Sky*, from Petropavlovsk, Kamchatka, to Seward, Alaska, cruising through the Aleutian Islands, in 2013. It was sobering to experience the Arctic environment of volcanoes, glaciers, and the wildlife of bears, eagles, and whales, not to mention millions of seabirds. We explored glaciers and caves using inflatable small boats. The people who lived in Beringia and hunted mammoths and seals at the maximum glaciation 20,000 years ago were a tough lot. Regional teams have now been conducting underwater research in these areas for a decade or more, but it will take years to accumulate sufficient data to understand the ancient climates, the vegetation, and who may have lived there, or exploited the coastal route from Siberia to North America.

Research on human and pre-human hominin migration using genetic evidence from DNA illuminates many aspects of how and when we spread around the globe. But DNA evidence usually only shows that people with known biological traits existed at two different points at different dates, not how they made the journey, nor what route they took, nor the adaptability needed to transit different environments. New evidence from the seafloor may help to refine the models of migration already postulated on the basis of DNA and palaeo-genetics. DNA study of the modern population in Europe and the Middle East has been sufficiently detailed to show that one route of the agricultural revolution was carried westwards along the north Mediterranean coast, suggesting that fishermen can be farmers after all. Human genetic data also suggests that the process that we think of as migration was often a slow and confusing mixture of movements backward and forward, with frequent dead-ends and occasional failures. Averaged over thousands of kilometres and thousands of years, the net rate of "migration" was only a few hundred metres per year.

* * *

When did people first display the organization and collaboration to make long-term communal settlements and social structures? The archaeological assumption for several decades, based on plenty of dated evidence on land, has been that organized concentrations of population worthy of being called permanent settlements evolved after, or contemporaneously with, the development of agriculture about 10,000 years ago. It appeared probable that the need to store seasonal produce from cultivated land in a secure way required a hierarchical or communal effort, resulting in permanent storage buildings and habitations, linked by language and rules. Inverting this argument, it seemed that concentrations of population and ceremonial public buildings would not occur in a hunting-foraging culture. This is now known to be incorrect, with the astonishing example of a megalithic temple at Gobekli Tepe, near Urfa in southwest Turkey, where a sophisticated society that lacked agriculture nevertheless constructed megaliths at the date of 11,000-12,000 BP. This is a single example, but it shows that ice age peoples could collaborate and engage in integrated non-subsistence social projects in the Palaeolithic, before agriculture.

Building boats, even a simple dugout log canoe or log raft, suggests an assertion of purpose that is consistent with social links. At a guess, there may also have been coastal communities that captured seasonal migrations of tuna or marine mammals, leading to communal structures. Collaboration was also necessary to build large numbers of rafts or boats for sea crossings. A migration to Australia or Papua New Guinea would have required a minimum of 100-200 people within a short interval of time to avoid population collapse from disease, accidents, and predators.

The low-lying coastal plains, rich in both terrestrial and marine food resources, provided a pathway for hominin diffusion into northern Europe about one million years ago. The discovery of fossil worked flints in the cliffs of East Anglia, and fossilized footprints in a submerged beach nearby, show that a family with children was living on a marshy estuary. The geological beds that contain prehistoric tools and animal fossils continue under the water into the open North Sea. Offshore archipelagos or groups of coastal islands were important for early experiments in seafaring and maritime trade, and as stepping stones in the coastwise expansion of early populations around the Mediterranean and into northwestern Europe. The earliest occupation of Crete and Cyprus, which could prove Palaeolithic sea voyages, is now subject to intense research.

There is still no instrumental technique that gives an indication when prehistoric remains are present on or under the seafloor. No device detects shaped flints distinctly from normal cobbles and stone; no device indicates that a timber has been cut and is not just a tree trunk. Scientists are testing devices now that may solve these problems. The next decade will decide whether seabed prehistory takes a permanent place in the world of archaeology or whether it fades away again into a niche curiosity.

There is still so much to discover.

11

Requiem for Apollonia

After 1959, as the years drifted or scrambled by, I never expected to see Apollonia again. The thought never occurred to me.

The young Colonel Muammar Gaddafi, so absurdly good looking, used his support in the army to overthrow the government of Libya in 1969, expelling the king, and the remaining British troops. His subsequent smuggling of heavy weapons to the IRA in Northern Ireland, the shooting of policewoman Yvonne Fletcher in 1984 outside the Libyan Embassy in London, and then the Libyan sabotage of the Boeing 747 that crashed on Lockerbie in 1988, confirmed that Libya was a hostile state. As he became older, and in spite of claiming to live frugally in a Bedouin tent, Gaddafi's appearance became increasingly raddled and dissolute, accompanied by weirdly eccentric behaviour. Libya was out of bounds for decades.

In 2001, I received in the post one of those travel brochures that shows the Sydney Opera House, the Taj Mahal, and the Rose Temple at Petra in Jordan apparently visible simultaneously without leaving the deck of your cruise ship. I was about to throw it away when I saw a more serious brochure behind it from Cox & Kings Travel, offering tours of Libya. Prime Minister Tony Blair was cultivating Muammar Gaddafi at the time, so that the tourist attractions of the great classical cities of Libya became a fresh product for the more enterprising travel agencies.

Cox & Kings restricted the archaeological tours on offer to the obvious cities of Tripoli, Sabratha, Leptis Magna, Cyrene, and home again. The internal journey from Tripoli to eastern Libya was by direct air flight. Other trips included escorted southern desert drives into the interior Sand Sea.

There was a footnote saying: "Customized tours requesting to visit other sites can be arranged at prices to be agreed."

My mind went into a rapid buzz. Jay had accompanied me often as a full team member on expeditions in very rough countries, sometimes with the added risk of finding that the local people carried guns. Our children were in their late teens and would be useful on a rugged trip in a hot country. Why not try to negotiate a family holiday driving along the whole coast of Libya, starting at Sabratha in the west near the Tunisian border, looking at coastal ruins all the way, and ending up at Apollonia? There was no harm in asking since the worst outcome would be a simple "no".

I wrote to Cox & Kings, proposing the family holiday trip along the coast, making a party of four tourists, explaining also that, having mapped Apollonia many decades

In: Flemming, N.C., 2021, *Apollonia on my Mind. The memoir of a paraplegic ocean scientist*, 427-470. Leiden: Sidestone Press.

ago, I would like to view coastal ruins along the shore, snorkelling or swimming over the underwater buildings. I promised to report all observations, finds, drawings, and photographs to the Libyan Antiquities authorities. I also agreed to accept continuous supervision from an expert appointed by the Libyans.

Cox & Kings Travel said they would have to consult their associated agency, Dar Sahara Tourist Services in Milan, which resulted in silence for a few months. After a polite reminder, they asked me to provide a complete itinerary, suggesting ruins that we wanted to swim over, number of days along the track, towns where we could stay overnight, and number of days staying at Apollonia itself, or nearby in Cyrene, or El Beida, or Shahat. Using the Barrington Atlas of the Greek and Roman World, publications of the British Society for Libyan Studies, and all the sources I could find, I sent a proposed itinerary, illustrated with maps and charts. I purchased guide books and modern maps from Stanfords in Long Acre near Covent Garden.

Several procrastinating letters and reminders followed over many months. Then in spring 2003, we struck gold. The Dar Sahara Agency in Milan had arranged with Sukra Travel in Tripoli and Bright Star Tours in El Beida, for two Toyota Land Cruisers to be driven by desert-trained expert drivers, allowing us to drive on beaches anywhere along the coast, accompanied by a senior archaeologist, and representatives of the Government. The tour would be most conveniently enjoyed in the first weeks in early September, starting precisely on 27th August. The itinerary was carefully spelled out, including the towns where we would spend the nights in hotels.

This was promising, but the accompanying letter from Cox & Kings carried an ominous warning. Reading between the lines of a letter which had been checked carefully by company lawyers, I deduced that they, Cox & Kings, could provide no guarantee as to our freedom to explore or photograph anything. We might be supervised by KGB-style police guards who controlled our every movement. We might have a wonderful family holiday, or we might feel like prisoners on parole. Cox & Kings could take no responsibility for what actually happened.

I consulted the family, who all agreed that it was worth the risk. As the price quoted by Cox & Kings was acceptable, I paid the whole sum up front.

In my usual way for planning coastal trips looking for underwater ruins (Chapter 6), I prepared files on each major ancient city, plus many small settlements, clipped maps together into A4 booklets and arranged everything so that we could enjoy the fun of swimming and snorkelling in utterly wild places, while finding any ruins that might be under water.

Suitably armed with Libyan visas, sun-cream, big floppy hats, rugged carry-on bags, masks, fins, and all the rest of our luggage for a trip of two weeks, we boarded the flight to Malta on 27th August. From there, after a break of a few hours, we flew with Air Malta in an almost empty aircraft to Tripoli, landing at sunset. Porters carried me in my wheelchair down the gangway steps, where I felt the daytime heat rising from the hot tarmac, although the sun was already on the horizon.

The Toyota Land Cruisers were parked next to a small group of men, from which a middle-aged man of kindly appearance came towards me to introduce himself.

"Good evening, Dr Flemming. My name is Moktar Hawel. I am the representative of Bright Star Travel, and I am in charge of your visit to Libya. You are welcome. I will accompany you throughout the trip. I think it will be interesting for all of us."

"Thank you, Moktar, you have arranged a wonderful tour for us. We are looking forward to it. Can I introduce my wife, Professor Jay Kleinberg, and our children, Kirsten and Peter." Everybody shook hands. Moktar introduced us to Mr Abdul Hamid Abussaid, who was to be our archaeological supervisor. He was the retired chief archaeologist from the Ministry of Antiquities, a man of culture, wisdom, and philosophy as we discovered, whose help over the following weeks was essential to our success. He was an elderly, short, chunky man radiating a sense of calm. I guessed that he had suffered much and seen much suffering in his lifetime. Finally, we were introduced to the Man from the Ministry. We never learned his name, but he held the sheaf of security clearance papers that enabled us to pass through police and military checkpoints with minimum trouble. He had a pinched, miserable face and narrowed eyes but turned out to be friendly and helpful throughout the trip.

"You should meet also our two drivers," said Moktar, "who are from Sukra Travel. They are trained in desert driving, Ali, and Abdul." We shook hands. The drivers were both large, muscular men who exuded the authority derived from their command of the vehicles, rather like the captains of ships. During the first two days, while we were visiting the traditional tourist ruins of Sabratha and Leptis Magna, I detected that driving on tarmac roads was very much beneath their dignity.

The airport floodlights had come on by the time we loaded the gear into the Toyotas, with my wheelchair on the roof rack. We drove to the tower-block hotel by the port in central Tripoli. Jay and I were in one vehicle with Mr Hamid, the archaeologist, so that Kirsten and Peter were with Moktar and the Man from the Ministry in the other one.

Having checked in smoothly at the hotel, all negotiated by Moktar, we parted in the lobby. "I'll pick you up again at nine o'clock with a six-seater to go to a restaurant nearby," he said. Dinner the first night was somewhat stilted and formal. We sat together at a large table, but nobody quite identified the right verbal conventions or mutual respects. The meal itself combined European and North African dishes, with spicy Libyan soup, macaroni or couscous, main dishes of mutton or chicken, a range of vegetables, fruit, and sweet Arab cakes. Alcohol is forbidden in Libya, so we drank flavoured soda or non-alcoholic beer, which was quite good. Later we found that many small restaurants on the coast served fish. After returning to the hotel, we agreed to start with a trip to Sabratha, 68 km (42 mi) to the west, at nine o'clock the next morning.

* * *

In the morning sunshine, we drove west through many small towns and villages along the main coast road, bordered by typical Mediterranean arrays of concrete houses and shops of one or two storeys. Ancient Sabratha was a Phoenician trading station linked to Carthage, trading in the products from Africa to the south. By the 4th century BC, it was specializing in trans-Saharan trade. Carthage was destroyed by the Roman Scipio Aemilianus in 146 BC, after which, although the culture of the city was strongly influenced by the Greeks, Sabratha came under Roman influence, eventually being rebuilt and developed by the Emperor Marcus Aurelius and his son Commodus. Most of the large monuments that make the city so attractive today are from the Roman period, although there was serious destruction by earthquakes in the 4th century AD. The Roman theatre that was excavated and reconstructed by Italian archaeologists in the 1930s stands to a height of three superimposed columns (forming, properly, the skene and proskenion), and dominates the city dramatically. The theatre could seat 5000 spectators and is the largest in Roman Africa. After 365 AD, Sabratha became Byzantine Christian, with a surrounding city wall from that period, before occupation by conquering Vandals. Justinian briefly re-conquered Sabratha in the 6th century.

We arrived at Sabratha in mid-morning. The ruins cover more than 21 hectares (60 acres) with well-preserved theatres, temples, the forum, baths, mosaic floors, a grid of streets, and many of the public features of a large Roman city. Our party of Mr Hamid, Moktar, one of the drivers, and the four family carried swimming-gear and cameras in bags as we zig-zagged through the astonishing ruins and reconstructions. Most of the paths were quite navigable for my chair, but the driver was always ready to help over steps and fallen blocks of stone. At the beach, I could see that the Temple of Serapis, Justinian's Basilica, and the Temple of Isis were close to the shore, with the ashlar foundations of the Temple of Isis being partly undercut by erosion (Fig. 11.1). Several streets and buildings, including baths, are cut off abruptly as if there has been a bit of land loss. An offshore reef of fossilized dunes was protecting the city from major storm attack, and the modern curators had tried to protect the beach in some places with concrete, but slight erosion was still taking place. Although there were some fallen blocks of masonry on the beach, and in the shallow water, the absence of any blocks beyond 2-3 m (6-9 ft) from the shore is evidence of no submergence.

It is a bit of a mystery as to why there are not more obvious harbour constructions, quays, docks, or warehouses at Sabratha. In 1966 a team of CUUEG divers led by Bob Yorke studied the underwater site and the offshore reef. They noted that drains and

Figure 11.1. At Sabratha, near the Basilica of Justinian and the Temple of Serapis. The ashlar walls of buildings break off at the waterline. The corners of the buildings have been eroded and collapsed into the sea. But the foundations of buildings do not continue into the water, and the masonry has simply fallen into the beach. The columns of the Temple of Isis are just visible in the background (Photo Jay Kleinberg).

gutters carrying water off the Temple of Isis seemed to be at the right height in relation to the present sea level. Aerial photographs show that a row of drowned fossilized sand dunes extends 8.5 km (5 mi) to the west, gradually diverging seawards from the modern shore. Much of this ridge is submerged, but it rises above the sea sporadically to create a chain of islands. A submerged rubble mole, just 30 cm (1 ft) below the sea surface, partially closes the gap between the islands opposite the west margin of the city. The ridge converges close to the present shore at Sabratha itself. Yorke's team found that the island immediately opposite the city had been covered in Roman concrete for a length of 180 m (590 ft), which had improved the shelter for shipping in its lee. Scattered blocks adjacent to the concrete suggested that there had been buildings on top of it. Bob's divers also found a triangular area of scattered columns and masonry massifs on the seafloor out to a distance of 100 m (328 ft), leaving a gap of only 40 m (130 ft) between the buildings and the concrete-covered island.

The overall effect showed that the citizens had built a few structures in a sheltered area between the island and the shore that could probably support the loading and unloading of ships. But it seems rather inadequate when the ability of harbour builders was so much more advanced already in the Roman era. Why nothing better? Why no docks or jetties for warehouses?

I turned to Mr Hamid.

"Is it acceptable for Jay, Kirsten, and Peter, to swim over the submerged area between the shore and the reef?"

"Yes, it is permitted."

The others kitted-up and I watched them snorkelling around the bay where Yorke had reported masonry blocks on the seabed. The tracks of the family's bright-coloured snorkels criss-crossed over the ruins, with occasional splashes as someone duck-dived to the bottom to inspect something or take a photograph. When they returned to dry land, they described seeing blocks, but not distinct walls, just a massif of close-fitted masonry. The photographs later confirmed this, so that it looked like a quay foundation, not a road or house. The upper courses had probably been pillaged when the harbour and the city declined.

There is no evidence for sea level change, but there are no accurate indicators, so that the error is potentially plus or minus 40 cm (16 in) in this case.

After the swim, we drove into the modern town for lunch. During the discussions both in the ancient city and over lunch, I sensed the continuous alertness of the Libyans as to how much help I needed, or did not need, but also a general sensitivity to our response to what we were seeing. Did I need help over a step? They always judged it correctly. Did we find the ruins impressive? Did we like the food? This concern was reassuring, but I also sensed behind it the possibility that they had suffered frustrating or humiliating experiences when their previous clients had sometimes treated them in a condescending or dismissive way. I made sure that nothing I said could be interpreted in this way, and no member of the family made that error. After lunch we visited the museum to see the breathtaking mosaic from Justinian's Basilica, which covered a huge floor area with myriads of motifs in hundreds of brightly coloured panels. The Roman antiquities of Libya are comparable to Rome itself.

Leptis Magna, 130 km (80 mi) to the east of Tripoli, was our main visit next day. This is one of the most perfectly preserved Roman cities anywhere in the Mediterranean.

Figure 11.2. At Leptis Magna the entrance to the city is celebrated with this magnificent arch, built to honour the Emperor Septimus Severus (Photo Jay Kleinberg).

Like Sabratha, Leptis was originally founded by Carthage, then greatly developed by Septimus Severus, in whose honour there is a magnificent arch as you enter the city (Fig. 11.2). The most expensive coloured stone for columns and construction was imported from all over the Empire. Unfortunately, by the later 3rd century AD the city was frequently subject to Berber raiding parties, but it was reinstated as a provincial capital by Emperor Diocletian. It was captured by Vandals in 439 AD, and, although brought back into the Byzantine Empire in 533, it was finally overwhelmed by the Muslim expansion along the North African coast in 647, after which it was abandoned.

The harbour at Leptis was and is magnificent in its grandiose ambition, perhaps over-ambitious, comprising curved docks and quays embracing a basin 350 x 300 m (1150 x 980 ft), with integrated mooring stones and bollards, backing onto roads and storage areas. One can imagine the sight and noise and smell of such a basin crowded with moored vessels carrying every conceivable cargo, including herbs, spices, and pickled fish. Thanks to freshwater springs in the region, the export of grain and oil to Rome was immensely profitable. In the harbour, the lowest mooring stones are 1.2 m (4 ft) above present sea level, while others at the back of the quays are higher up. A lighthouse tower marked the entrance to the basin on the western side. The harbour was constructed by digging out and widening the mouth of a small river now known as the Wadi Lebda. As so often in North Africa, the sand brought down by the wadi, perhaps assisted by wind-blown sand and migrating dunes, gradually filled in the basin, which had no natural flushing to the open sea. Today the harbour appears to be on dry land. The British, in the 19th century, removed 22 granite columns from Leptis, 15 marble columns, 10 capitals, entablatures, and sculptures which ended up as a romantic feature of re-invented "ruins" in the park at Virginia Waters, 30 km (19 mi) to the west of

Figure 11.3. Mr Hamid, our archaeological guide, explaining the entrance to the Frigidarium of the Hadrianic Baths at Leptis Magna. Kirsten, Peter, and I are listening carefully (Photo Jay Kleinberg).

London. But this pales into insignificance compared with the 600 columns from Leptis "donated" earlier to Louis XIV for his palaces at Versailles and in Paris. Notwithstanding the removals by Europeans in the 18th and 19th centuries, an astonishing number of magnificent buildings has survived.

Mr Hamid gave us an informative tour of the city (Fig. 11.3), concluding at the hippodrome and amphitheatre overlooking the sea. From there we climbed down to the shore and scrambled along the rocky coast back towards the harbour mouth. The crumbling remains of the city where it reached the shore had collapsed in places, scattering blocks and columns on the rocky wave-washed terrace, but there were no foundations or cuttings under water (Fig. 11.4). Since none of the buildings I could find gave a really accurate relation to sea level, the overall evidence from Leptis shows no relative change of sea level to within an error of plus or minus 50 cm (20 in).

During our tour of Leptis Magna, we had lunch at the nearby visitors' centre, where there was a small museum. One of the exhibits was a Palaeolithic handaxe, which made me conscious yet again of the paucity of prehistoric evidence from this coast. As the tour came to an end, Mr Hamid explained that there was one more surprise.

"You have to visit the villa of Sellin," he said. "I know it is not on the list you asked for, but it is impossible to pass it by."

After driving back westwards a few kilometres, the Toyotas bumped down a dirt road towards the sea, halting at the edge of a low cliff. Tiny waves, mere ripples, whispered on the beach. The sun was low in the sky, but there would still be another two hours of light. The domed baths of the single-storeyed Late Roman villa, surrounded by smaller rooms, walls, and gardens, were perched close to the sea as if the occupants wished to enjoy every mood of the seasons, storms, calms, bright mornings, and peaceful sunsets.

Figure 11.4. The rocky shoreline at Leptis Magna, below the cliff by the amphitheatre. Buildings very close to the sea have collapsed due to wave attack, but the foundations do not continue across the rock terrace, and there are signs of rock cuttings or quarrying on the terrace itself, but not under water.

With its domestic architecture, friendly family layout, exquisite mosaics and frescoes, built on the brink of a low cliff with the sea washing the beach below, the villa of Sellin is a perfect real-world example of that bucolic image that Diolé had imagined of the rich retired Roman citizen, sipping wine, at ease with the world. The garden itself is perfect beyond belief, with beds for flowers and herbs separated by small stone walls. Far more than at Pompeii with its many grand houses, I sensed the comfort and pleasure of ordinary domestic family life at the time of Roman prosperity. One has to remember that slaves did all the hard work around the house and on the estate lands surrounding it, but I hope that the owner of Villa Sellin was kind to his slaves.

The villa was only discovered under the sand and excavated in the 1970s. It had 20 rooms, a library with bookshelves, dining rooms, walled gardens, baths, and summer rooms that faced the gardens. The separate components were linked by a portico decorated with a rich array of mosaics depicting mythical events, intermingled with scenes from nature and decorative patterns. There are copious tritons, nymphs, plants, and birds, and wall paintings of gladiators. One scene shows a horse race in the circus at Leptis Magna.

* * *

As the sunset, we turned east again, to spend the night at Zliten, a small port town. Once we had driven east of Leptis, we were moving into terrain not normally explored by European visitors other than professional archaeologists and oil engineers. It had always been a difficult and dangerous coast, where there were no major Greek or Roman ports, but just minor settlements and fortresses or defensive posts along the shore. The ancient coastal towns had been concentrated in the Three Cities around Tripoli (Greek: *tri* + *poli*, πόλη), and the eastern five cities around Cyrene, known collectively as the Pentapolis. In between was 700 km (566 mi) of wild coast with scarce freshwater springs, no good natural port or shelter, and very little vegetation. In many places, the dunes of the desert simply carried on down to the beach. There might be a few ridges of fossilized dunes, either just on shore or in the shallow water, then merciless blue sea to every horizon. The Gulf of Sirte was a devastating place to be shipwrecked.

While we had been at Leptis, over lunch in the restaurant linked to the big tourist site, Moktar had turned to me to ask some questions, rather nervously.

"The food here is quite good, I hope you agree, but as we drive east, spending more time on the beaches, and without big hotels and restaurants, would your family be prepared to eat food that we cook for you in the open? On the beach? It might be possible to find small cafés and eating houses, but I think this would be best. Would it be all right for you? We will buy fresh food in the markets each day."

Moktar was so apprehensive as he asked this question that I recalled my thoughts from our conversations at Sabratha. I sensed that he had perhaps experienced previous tour parties where some people had behaved arrogantly or demanded a standard of comfort, service, and European-style food, which obviously was impossible in the poorer parts of the country. On the other hand, maybe Moktar was just being very considerate and polite, as was his nature.

"That sounds perfect," I said. "We are all used to living outdoors, so fresh food cooked on an open fire will be delightful."

"The drivers are very good cooks. But they also make good salads, and we will buy fruit which we will wash very carefully. We should be able to buy fish sometimes. Can you eat goat meat?"

"I'm OK with goat," I said, "but my wife certainly cannot eat it. It would be OK sometimes, provided there is an alternative. Lamb would be better if you can buy it."

"That is very good," said Moktar. "I think you will find that Ali and Abdul prepare excellent meals, and we can make camp-stops on the beach as if we were in the desert. They will like that."

He was relieved that this negotiation had gone smoothly. When I told the family they were pleased with the new plans.

Zliten was a modest town with crowded, run-down streets near the waterfront, but avenues of palm trees in the main approach roads. It is now, after the civil war, one of the central assembly towns for migrants seeking to cross the sea to Italy. Already in 2003 there were small gangs of Africans sitting tired and exhausted in the dusty streets by the port, hoping that someone would provide a boat for their escape. The hotel had no lift and no bedrooms on the ground floor. Peter and Kirsten, ably assisted by Ali and Abdul, whisked me up the stairs after dinner, and as easily rolled me down for a dawn start next day. The friendliness of the hotel staff, the cheerful service in the restaurant, and the support of our guides were perfect.

The small archaeological remains that we visited on the long coast from Zliten to Benghazi do not merit description in detail but, to me, the names are magic: amongst others are Thubactis, Euphranta, Charax, Ad Palmam, Zacazama, Philaenorum, Boreum, and Euhesperides, that is Benghazi itself. For the next two days, we drove steadily further east and south, where the desert became hotter and the few towns smaller and more desolate. The tarmac coast road was controlled at intervals of a hundred kilometres or so by fortified police or military checkpoints, with chains and other obstacles that could be raised rapidly across the road to stop vehicles trying to pass. At each checkpoint, the armed guards eyed us suspiciously until the Man from the Ministry went into their offices with his great bundles of duplicated forms. After a few minutes he would return looking very pleased with himself. After one such check he said, “Someone up there really likes you. I have never seen papers approved so quickly.”

We had lunch on the beach (Fig. 11.5), in preparation for which even the Man from the Ministry busily took part. For most of these stops, the Toyotas were arranged with a screen stretched between them to provide shelter from the midday sun. The drivers now lowered the air pressure in their tyres and drove happily over soft dunes onto the beach sand, confident that their skills were really needed. We spent one night at Sirte, in a very smart modern building, which I think was intended as a conference centre. But the hotel restaurant was not functional yet, so that we ate the evening meal of fish at a seaside restaurant in the town.

The desert dunes became larger and more mobile as we drove south, with huge ridges of active sand blending into fossil ridges, either on the shore, or offshore, creating chains of small islands. Although the evidence was never totally reliable, erosion terraces, beach-rock, fossil dunes, natural lagoons, and archaeological remains all provided signs that there had been no change of sea level in historic times to within a general error of the order of plus or minus 50 cm (20 in). Mr Hamid quietly mentioned Muslim buildings and ancient sites from several hundred years ago, which we visited with real interest. He did not emphasize the fact that I had not included these on my list of places to visit,

Figure 11.5. Cooking on a remote beach in the Gulf of Sirte. The Man from the Ministry is preparing vegetables, while Kirsten, who has just been swimming, and Mr Hamid, watch with approval. The debris of plastic bags was a constant factor everywhere in Libya (Photo Jay Kleinberg).

but the point was well taken. They were not on the shore, but were well worth a visit, to get the sense of the development of this inhospitable region by the Arabs.

As we turned directly east at last near Ras al Ala, Mr Hamid drew our attention to an abandoned Italian prison camp by the beach. We had to drive a few kilometres from the main road towards the sea, but there, at the very southernmost and hottest point on the coast, devoid of any shelter or vegetation, were the unmistakable vestiges of a typical prison camp, with wooden guard stations and gun emplacements on rickety towers, miles of barbed wire collapsing into the sand, and sun-baked timbers and cracked planks projecting from the dunes. To imagine living in such a place, with prison treatment, bad food, and no medicine, struck terror into one's body. Mr Hamid explained quietly the various resistance bands and guerrillas that had fought against the Italians in the 1930s and '40s, and how they had been interned in this camp for many years. There was not a hint of bitterness or condemnation in his calm voice. How could he be so forgiving? The place haunted me all the more because of Mr Hamid's calm. I did not ask whether any of his relatives had been there, or even himself. He showed no rancour. His philosophical resignation and the acceptance of unpleasant human nature seemed an example of boundless tolerance. British, Germans, and Italians had stormed backward and forward for decades across his country, killing and bombing, but he seemed to show no resentment. I am sure he had strong views, but he did not think he needed to flaunt them.

For tens of miles on either side of the abandoned prison camp, the beaches stretched as pure white sand from horizon to horizon. Occasionally as we drove on the beach, we saw the rotting hull of a wrecked boat, and, where the fossil dunes provided a hard foundation, there might be a fisherman's hut, a few small boats drawn up on the sand, or some nets drying in the sun. The desolation was pervasive. The phrase "unspoilt beaches" does not apply to situations like this. The symmetry of white and blue was aesthetically attractive, but the sense of menace and risk was overwhelming.

* * *

We crossed oil terminal pipelines at Al Brayqa (El Brega), turning north towards the ruins of Boreum on a bold headland, and then to Benghazi for the night. Boreum was interesting because the main fort and major buildings were high on the hill overlooking the sea, but from that summit we could look down on many walls and foundations on the shore, though not extending under water. The hotel at Benghazi was the epitome of expensive appearance for its own sake, with countless mirrors, glittering fake-gold-plated frames and gewgaws, and chandeliers everywhere. It had a lift and comfortable bedrooms. In the foyer, a large video screen was displaying Gaddafi making a long speech, but the few guests did not pay it much attention. We were led to understand that nobody took such speeches seriously. Moktar explained that we had recently passed the anniversary of the 1969 revolution, and I recalled that some buildings in the bigger cities had been hung with complete wall coverings in retina-bashing green, accompanied by the slogan, "Another 34 years".

The next morning, Jay went for an early walk in the cool dawn air but was molested by a male passer-by. She was wearing a flowing dress, long sleeves, and a head-scarf, but was obviously European, which seemed to justify the attack. She screamed and beat the man off. He fled from her strong reaction. Over breakfast at the hotel, Jay told Moktar, who reported the case to the police, but we heard nothing more. We were advised that in Cyrenaica, women had to be more careful and preferably chaperoned by a man, in contrast with Tripoli, which was much more European in attitudes and dress.

The Toyotas drove rapidly into the hills of the Gebel Akhdar, where the bridges that I remembered as rusting World War II military concoctions had been replaced with gleaming white modern suspension spans. We arrived in the late afternoon at Shahat, the modern town close to Cyrene, where we checked into our hotel. Moktar lived in Shahat, so he spent the next few nights at home with his family. Mr Hamid, I think, had a house there, or stayed with friends, and the drivers found accommodation elsewhere. The hotel had no lift, but the ground-floor saloon was busy with men playing backgammon and dominos. Whenever I appeared in the foyer, half a dozen well-built locals would leap to their feet and offer to drag me upstairs. This was very welcome and was executed with great humour and good spirits. Peter learned to play their games of dominos, but I am not sure anyone really agreed on the rules. The food was simple, but filling, with every day a choice of two soups: lentil soup and Libyan soup. The waiter was very punctilious at explaining the content of the soups, but after a day or two began to see the joke, and allowed us to choose from what we already knew. The hotel staff were kind and helpful at all times.

In the morning, fresh with sunrise, we boarded the Toyotas for the return to Apollonia. We curled past the hundreds of rock-cut tombs on the street that slopes down from Shahat and negotiated the hairpin bends as we dropped from one terrace to the next. At the final turn, the beloved view, so well remembered, opened before me (Fig. 11.6). Forty-four years disappeared into a time warp, and I felt as if I had never been away. The chain of islands far below, the village of Marsa Susa, the little fishing port, and, slanting away from the shore, the islands that enfolded the underwater city: how much would have changed? We descended the last hill and motored towards Susa, where the modest skyline was now dominated by the reddish-ochre monstrosity of a hotel of six storeys (Fig. 11.7). I cannot imagine that it is ever fully booked, or maybe nobody stays there at all. The village had expanded eastwards and inland so that the ruins on land and the Hellenic city wall were almost surrounded by modern houses.

Driving through the village, we arrived at the massive gate tower of Apollonia still poised as if it would fall into the sea. But it had not fallen. To the right-hand side we saw young Europeans sitting on a flight of steps with notebooks and drawing pads. The steps led up to a small building. We disembarked, to find that there was an ongoing excavation supervised by André Laronde, professor of Greek History and Director of the Institut d'épigraphie grecque at the Sorbonne, and this building was their dig office. Students pulled me up the hill to the building, where Laronde greeted me effusively. My poorly remembered French seemed to blossom again in response to this unexpected linguistic emergency, so that we were able to hold a brief, but fluent, scholarly discussion. Laronde professed great respect for my early work, but his own excavation was exclusively on the acropolis hill, not in the water. We exchanged business cards very formally, while we sipped our tea.

Our party then drove along the track at the edge of the cliff to the centre of the archaeological site, opposite the islands. The drivers rigged an awning between the Toyotas, spread rugs on the ground, and prepared coffee.

We were home.

Figure 11.6. Looking down from the terrace part way down the slope of the Gebel Akhdar over the coastal plain to the modern village of Marsa Susa, and, close to the right-hand margin, the two islands that mark the outer limits of the underwater city (Photo Jay Kleinberg).

Figure 11.7. The shoreline at Apollonia looking westwards to the columns of the Christian basilicas, dominated by the modern six-storey Hotel Al Manara. This tragic construction is typical of the neglect of the antiquities at Apollonia by the official agencies (Photo Jay Kleinberg).

We relaxed. The north wind breezed out of a blue sky, bringing the breakers piling over the Grotto Reef, churning over the Block Fort towers, and throwing spray high in the air around the islands. But the islands provided partial shelter so that the waves were less rough on the beach. After absorbing the landscape, trying not to notice the ghastly hotel, and relishing the pure sense of the great city under the waves, I suggested that we just get in the water for a token swim over the ruins closest to the shore, hoping for a calmer day later. Peter and Ali lifted me into shallow water, where I donned my mask and snorkel and started swimming towards the Nine Quays. Jay and Kirsten donned their fins to swim beside me. The low parallel lines of ashlar blocks, greeny-black through the cloudy water, were as strong and solid as forty years ago. There seemed to be more sand around the ruins than I remembered. Once again the sense of the past almost crushed me with its intensity. Two thousand years ago, these huge parallel structures were teeming with people, but what were they doing? I cruised slowly over other low walls, but the choppy water and poor visibility meant that I would have to postpone a serious swim. I gestured to Kirsten that we should head for the beach.

Returning to the shore, I warmed up in the sun while I discussed the ruins with Mr Hamid. The drivers, assisted by the Man from the Ministry, cooked a fine lunch of fish with vegetables and salad. Peter was sitting in the shade reading Jules Verne's *The First Men in the Moon*; Kirsten and Jay discussed cameras and photographs. We were still using Kodak reversal film in the Calypso Phot underwater camera. After lunch, Jay wandered along the shore towards the acropolis, climbing around it to look at the theatre on the far side. In 1959, there had been the short stump of a young palm tree growing in the middle of it. Jay reported that the palm tree was now full height and looked very healthy.

I felt at home, but the experience reawakened in me the dissatisfaction that had consumed me at the end of 1959: there was so much to be done. The whole city needed surveying and mapping to modern standards. The strange dockside buildings needed to be re-assessed in the light of what archaeologists had learned about ancient harbours in the past four decades. Acoustic mapping was needed, and a planned set of excavations in different parts of the city, with close-up study of details of construction. The sequence of buildings one on top of another, reflecting a thousand years of constant change, needed proper investigation and dating if possible. I didn't voice these concerns to Mr Hamid, but we did discuss the principle of doing more work underwater with Libyan archaeologists and students.

We spent some time on each of the next three days at Apollonia, while extending our exploration along the coast to the west, and enjoying guided tours of Cyrene, with explanations of each building by Mr Hamid. Where Sabratha and Leptis Magna are impressive in their grandeur, symmetry, and civic boastfulness spread out for all to see at a glance, Cyrene blends into the mountainside, hugs the landscape, and in its ruined state, is partially concealed by trees and patches of forest. The trees, which make the city so attractive, were planted by the Italians between the wars. The remains of temples, palaces, baths, theatres, markets, the agora, and sports stadium extend over a vast acreage, but one has to find each building by climbing and clambering over the uneven ground. The Temple of Zeus is fixed in my mind because the rows of huge standing columns are reflected in the towering pine trees growing close by on all sides, enclosing a grassy glade with the temple in the centre. On the undulating southern slope

Figure 11.8. Sculpture of a warship prow at Cyrene, showing upper and lower bronze rams for attacking enemy ships. Both rams made of stone have a trident 3-bladed structure (Photo Jay Kleinberg).

of the hill, parts of the city have been excavated, where the well-preserved ruins and reconstructed buildings overlook the steep descent of the Gebel Akhdar towards the sea.

There is a unique sculpture at Cyrene comprising a replica in stone of a warship bow showing two stone replicas of the cast bronze rams mounted so as to attack enemy warships, breaking their oars and staving in the planking (Fig. 11.8). This image helped my colleagues and I later when trying to understand the function of a bronze ram that had been found in the sea near Tobruk. Some archaeologists have used this stone model to understand how bronze rams were mounted and used, but since the assemblage of stones had been reconstructed by archaeologists, the argument is somewhat circular. The arrangement of the upper and lower rams exhibited at Cyrene, with a fair amount of cement padding, is what archaeologists expect, not direct evidence from the past.

* * *

Mr Hamid had an extraordinary knowledge of the small sites along the coast to the west that fascinated me so much. He was the perfect guide. There had been no road along the coast in ancient times, so each small settlement or depot had to communicate with its neighbours by sea when there was a heavy load to carry.

The ruins that we visited, from Apollonia westward, were the island that Martin discovered on the last day in 1959, Brak Nota, Sidi Amer, Phycus, an unnamed site, then Ausigda, and another unnamed island to the extreme west beyond El Haniyah. Martin's Island was just 3.7 km (2.3 mi) west of the ancient city, one of six just offshore, and you can see on Google Earth the rectangular tanks incised into the solid rock. We did not swim out to it on this trip, but the island was probably joined to the mainland when the sea level was relatively lower in Roman times.

Brak Nota is the modern name for a curious combination of an ancient site on the shore close to two huge, collapsed karstic dolina, which are now filled with freshwater. At low sea level, the groundwater dissolved out two vast underground caves, the roofs of which have fallen in to create lakes with a steep rocky edge. The closest of these lakes to the shore is only 200 m (650 ft) back from the beach, but it is so inconspicuous, concealed by bushes and tall reeds, that we had passed along the tracks several times without my seeing it. The ancient settlement was on a rocky ridge 5-10 m (16-33 ft) high projecting, as so often, almost parallel to the shore, and protecting a bay on its southeastern flank. Given the dominant winds from the northwest, this was a sensible location to choose for a small harbour with no additional breakwater. The crest of the ridge was covered in a maze of ancient olive presses, silos, tanks, and cisterns, suggesting a huge amount of storage for agricultural products. In the lee of the ridge, close to the sandy beach, there was a complete row of ashlar blocks just awash in the shallow water (Fig. 11.9). It was breezy on the day we visited, so the waves were breaking into a rolling surf 100 m (330 ft) offshore on the sloping rock. The run-up of the surf then sloshed up the slope to the edge of the ancient storage tanks. The sea must have been relatively lower when the storage spaces were being used. I sketched rapidly. As usual, we cooked lunch on a driftwood fire, while the snorkellers, Jay, Kirsten and Peter, searched the seafloor to check for any ruins that might be totally submerged.

Sidi Aamer, just over 9 km (5.5 mi) further west, consists of a Muslim shrine on one side of a gulley by the sea, and a Roman fortress tower, several courses high, on the other side. Mr Hamid said that there had been a fortified Roman village nearby inland. The gulley did not provide much shelter but could have been used to pull a boat alongside in those rare days of calm weather.

Figure 11.9. Brak Nota. At the top of the slope to the left are many rock-cut storage tanks, circular silos, and an olive press. They are splashed by the breaking waves. At the waterline, exposed to every wave, there is this line of well-fitted ashlar blocks. The settlement has submerged by about 2 m (Photo Jay Kleinberg).

* * *

Our visit to Phycus with Mr Hamid provided a very different perspective from my previous visit in 1959 (Chapter 1). My study since then of hundreds of ancient coastal ruins (Chapter 6), combined with the published reports by other archaeologists working on other sites, meant that I now knew which critical factors needed to be measured most accurately: quarry cuttings, road levels, storage tanks, salt pans, and so on. Phycus receives several mentions in the ancient literature as an active port in the 1st-5th centuries AD, although the original foundation is older, since pottery from the 4th century BC has been found on the site. The natural form of the bay provides shelter from the dominant northwest winds, but there are no remains of docks or mooring quays, and no artificial breakwater. It is difficult to imagine large grain-carrying ships being loaded here, even if the grain had been ferried out on lighters.

The orientation of my 1959 map of Phycus was slightly wrong, twisted too far anticlockwise. Correcting for this showed that the cuttings, submerged roads, and buildings are open to the water only to the northeast, with a massive, rocky barrier 6-10 m (20-33 ft) high protecting the city to the north and northwest where the dominant waves come from. On the day we were there in 2003, there was a very light wind, no white horses, and the waves were just the low swell left over from the day before. Kirsten and Peter measured the depth of the water at nine locations on the submerged roads, giving an average depth of 92 cm (36 in) at 12:30 pm. We put a marked stick and tape in a position where we could measure the water level every hour, to see if there was a measurable tide in the area. The mean sea level dropped 30 cm (12 in) during the next five hours. This seemed excessive as pure tide, which I would have expected to be only about 20 cm (8 in). A change of wind direction, barometric pressure, or wave action might have been acting in concert with regular tidal action.

We made a camp by the two Toyotas perched at steep angles on the high sand dunes that covered most of the ruined city. Kirsten and Peter climbed all over the rock cuttings, exploring the hidden chambers and rock-cut rooms. Jay climbed down the cliffs to take photographs. We ate lunch under a shade supported by two sticks of driftwood and explored the shoreline around the city afterwards. There were tanks cut in the rock on the waterline, some ashlar blocks, and many occurrences of "chocolate bar" criss-cross incisions from the extraction of quarried building blocks. Within the drowned road system, the rock-cut walls had a deeply incised solution notch at present sea level, with coralline algal growth and *trottoir* encrustations extending to about 1 m (3 ft) above the notch (Fig. 11.10). This suggests that the site has been geologically stable for a long time, several centuries, although it was submerged by about 1.5 m (5 ft), with an uncertainty of plus or minus 0.5 m (20 in) since the Classical period. During our exploration, Jay photographed some beautiful flowers growing in the sand, while Peter somehow tamed a most unpleasant-looking insect. We drove home after a good day's work.

On another day we approached Ausigda, the settlement to the west of Phycus, by driving inland and travelling through the hills from Cyrene, before descending to the coast along the great Wadi Kahuff. Looking down into the valley from the road, we could see forests of huge pine trees growing at the bottom, which confirmed an ample supply of groundwater even at the end of summer. The mouth of the wadi spreads out into a

Figure 11.10. Phycus, west of Apollonia. Peter has been measuring the water depth in the cut channels that form roadways, now submerged. The stratification in the cut cliff shows the internal bedding structure of large sand dunes that have been turned to stone over tens of thousands of years (Photo Jay Kleinberg).

Figure 11.11. Campsite at Ausigda. The Toyotas are parked with a large sunscreen taut between them, and rugs and cushions on the ground in the shade. The drivers are cooking lunch while I work on my notes and sketch plans from the ruins. The Byzantine buildings are at the top of the sandy slope (Photo Jay Kleinberg).

broad delta of sediments, crossed by many braided channels, flanked by the extensive coastal plain on both sides.

The highest part of the headland at Ausigda, 15 m (50 ft) above sea level, was covered in ruins half-buried in the sand. Mr Hamid said that it had been a small Byzantine village. A church dating from about 500 AD stood on a secondary hillock, set back from the sea. Mr Hamid explained that the wide coastal plain from here to Apollonia was dotted with Roman settlements, with the buildings tending to be on small hills.

Jay and Kirsten ran down to the beach, sheltered as always from the northwest wind, and explored the seabed by snorkelling as far out as a small island. They climbed onto the island but found no cut blocks and no cuttings in the rock. I snorkelled in the shallow water, finding natural rock ledges covered in a thin dusting of sand.

We quickly discovered the body of a dead camel rocking gently in the surf a few hundred metres from our proposed campsite. But there was no smell, and nobody was going to use the seawater for anything but swimming. We kept as far away as possible and ignored it. The beach was quite flat, so the drivers arranged a classic back-to-back campsite with rugs and cushions spread on the ground (Fig. 11.11). I warmed up in the sun, writing notes and sketches, while the drivers and the Man from the Ministry cooked a lunch of lamb chops, peppers, tomatoes, and salad, followed by fresh fruit and ice-cold drinks. The ruins of the Byzantine village were very distinct where they protruded from the sand, but there were no cut blocks, either *in situ* or fallen, lower than a height of about 3 m (10 ft) above the sea. There was no evidence for subsidence or a rise of sea level, but the margin of error was such that one could not exclude a change either way

Figure 11.12. The end of the road. As the sun was almost setting, we reached this far west but had to turn back. I wanted to try and visit these islands, but it was too late. Google Earth shows a large rectangular storage tank cut in one of the islands. Very tantalizing (Photo Jay Kleinberg).

of about 1 m (3 ft). As I jotted down this rather vague conclusion, two gorgeous blue-spangled kingfishers flew low over the beach before settling on a rock by the headland.

We drove further west from Ausigda in the late afternoon, following a twisting coast road over red dust and gravel, through sun-scorched husks of villages with concrete houses, bravely decorated with plants in tubs and newly planted trees. The population of the coastal plain was increasing, with fresh-ploughed farms strewn with shacks and sheds in the open land, and villages beginning to sprawl with new buildings.

Eventually even the Toyotas had to stop amongst criss-cross goat paths, ridges of rock, and soft banks of sand and soil. There were several Roman dams across the wadis near the end of the track, with ruins visible on the far side of the valley. The islands offshore looked as if they had been connected to the mainland when the sea level was lower, and there was a straight line in the water that could have been a causeway. It was very tantalizing, but there was no time to swim or dive. We could probably have gone further, but it was late in the day, and the prudent drivers decided that we had to turn back. There were scattered small islands still further west, just black silhouettes against the calm white sheet of sea that reflected the glare of the setting sun (Fig. 11.12). I longed to go further to explore the islands, but I knew it was unwise. We returned to Shahat through the high hills of the Gebel. Checking on Google Earth years later, I found tanks cut into one of the islands that I had seen from the end of the track. One tank was a perfect large rectangle, just awash, so that in calm water it could be seen clearly, but in rough weather it would have been swamped by the waves. There must have been subsidence of a metre (3 ft) or more.

* * *

The number of closely spaced, small ancient settlements along the coast west of Apollonia became more and more astonishing as Mr Hamid took us to one obscure site after another, but every one was rich with buildings close to the water, foundations in the sea, and a host of storage tanks, silos, plaster-lined cisterns, and inexplicable cuttings and rectangular hollows, sometimes connected by cut channels. The volume of these storage spaces was out of proportion to the size of the settlements, indicating that they were designed to store trade goods for export, perhaps on behalf of a settlement or farms several miles inland. The hinterland was fertile; the hills of the Gebel Akhdar were also fertile, and Cyrene could not consume everything produced locally. Just the opposite: the wealth of Cyrene depended upon exports to Rome, Athens, and Ephesus, and the port for loading large cargo vessels could only be Apollonia. The chain of settlements from Ausigda to Phycus, Brak Nota, and Apollonia itself, including all the tanks on islands, such as Martin's Island first sketched in 1959, were probably assembly depots where produce was brought down to the shore for loading onto small vessels that sailed along from one depot to the next, transferring the oil, grain, nuts, and the drug silphium to warehouses in Apollonia. The Grid Building in the inner port at Apollonia makes sense in this scenario. On the other hand, the eastern harbour basin would have been capable of servicing and replenishing larger vessels, probably with loading quays on the east side of the great wall leading out to the Block Forts. Immense quantities of grain, oil, and wine must have been stored in warehouses in preparation for loading onto ships in the few months of late summer after the harvest.

The average spacing of the settlements that we visited west of Apollonia was less than 10 km (6 mi) along the shore, and there were several smaller inlets that might have served as additional mooring places for coastal boats. The hinterland of these settlements was the widest part of the fertile coastal plain. Further west is a gap of 25 km (15.5 mi) to the port of Ptolemais, which is comparable in scale to Apollonia, with a strong port construction providing shelter in all conditions. The coastal plain is narrow along this part of the coast. Study by diving teams over several decades, including meticulous work by Bob Yorke and David Davidson, shows that there are roads and foundations underwater at Ptolemais, indicating submergence of about 2 m (6 ft). In the opposite direction, to the east of Apollonia, there is a stretch of narrow coastal plain for 24 km (15 mi), with some small settlements but no harbours.

The rocky eroded shore throughout this area precluded beaching vessels as a method of loading and unloading safely. This limitation probably applied to the coastwise vessels as well as to the big grain carriers crossing to the northern Mediterranean cities. The average size for grain ships on the cross-Mediterranean voyage in about 200 AD was between 300 and 400 tons displacement. Many grain ships were smaller, and a few were as large as 1000 tons. The units of measure for grain were *modii*, with 150 *modii* making about 1 ton. Rome needed to receive 40 million *modii* of grain each year. But how the grain was stored on the ship is uncertain.

More than one thousand wrecks of cargo ships from the Graeco-Roman period have been found by divers in the Mediterranean, but not one of those that survives was obviously designed for carrying grain. They are mostly laden with hundreds or thousands of amphorae, presumably for wine, oil, garum, or other liquid produce, or nuts, or spices. In contrast, we know from literature and archaeological evidence in the big cities of the northern Mediterranean that grain was the principle south-to-north trade. How was it carried? Why have no grain wrecks been found?

The second question is easier to answer than the first. A ship full of hundreds of tons of grain, however it is stored (other than in amphorae, which have never been found to contain grain), will sink, if wrecked, and be carried to the seafloor by its weight of ballast. On the seafloor the grain will start to absorb water, it will suffer biological decay and microbial attack, while fish and other creatures will soon join the feast. As the grain is consumed or washed away, with no overburden of amphorae, the timbers of the ship will be exposed to bacteria, boring molluscs, and biological encrustation and algae, so that nothing will survive except possibly some bottom planks covered with ballast stones. As an archaeological target this is unlikely to attract attention, though the anchors and the contents of the crews' living quarters might provide some artefacts of interest.

Now for the first question. Anyone who has worked on a farm in a temperate climate will know that the need to protect harvested grain from moisture generates anxiety close to hysteria. Grain must not get wet, and it must not be moist when it is stored in bulk. Wooden-hulled ships, however well-constructed, and however well sailed at sea, will always ship some water onto the decks, and will always weep through the hull planking internally, however well caulked, so that some seawater accumulates in the bilges. The bilge-pumps fabricated from lead pipes and leather that divers have found on many ancient wrecks show that the ships were well-equipped to cope with this water, and the crews were presumably trained to use the pumps.

The problem is, how did they keep the grain dry while it was in the ship's cargo holds? Pumping out the bilges would prevent gross contamination of the grain with seawater, but a ship that rolls in a rough sea, or simply heels in the wind in a gentle breeze, would inevitably allow some spray or drops of bilge water to splash onto the grain if it were stored loose in bulk, or even in sacks. In sunny North Africa, storage in sacks would be cheap, logical, and safe. I imagine that the grain cultivated in Cyrenaica was, on calm days, loaded in sacks onto small coastal vessels at places like Phycus, Ausigda, or Brak Nota, for transport to Apollonia, where it was stored in the warehouses.

When the big cargo ships were ready for loading, slaves probably carried the grain in sacks onto the ship. Whereas photographs of amphora wrecks show the amphorae lying in direct contact with the planks and ribs of the hull, a grain carrier must have had an inner lining of small timbers tacked across the ribs, known as a spar ceiling, and then a layer of light planks to keep the grain away from the hull. By analogy with grain-carrying wooden ships in the 19th and even in the 20th centuries, there may also have been a further lining of canvas inside the spar ceiling, or more probably skins. Such inner timbers or planking have not been found on ancient shipwrecks, but, as explained above, a ship carrying grain, whether in bulk or in sacks, would rot away almost completely, so that is no surprise.

The spar ceiling and internal planking could be fitted temporarily to the inner surface of the ribs in the season when grain was being carried northwards. On the return journeys, the grain carriers often transported bricks or tiles as ballast, since these could easily be sold in North Africa. A shipwreck carrying bricks is not attractive to archaeologists, and although several have been found, they have not been excavated. It might be worth excavating one of these wrecks to see if there is an inner lining of planking to the hull that has been preserved by the overlay of bricks.

Marine archaeologist Stella Demesticha alerted me to a phrase in the Rhodian Sea Law, written about 600 AD but based on much earlier texts, that describes precautions to protect bulk grain:

> If a ship loaded with corn is caught in a gale, let the captain provide skins and the sailors work the pumps. If they are negligent and the cargo is wetted by the bilge, let the sailors pay the penalty. But if it is from the gale that the cargo is injured, let the captain and the sailors together with the merchant bear the loss....

This shows that sailors routinely used skins as a waterproof protection for grain, but applying them after a storm already threatens to spoil the cargo seems rather too late. The skins could only be applied as a top covering, while water accumulating and sloshing about in the bilge will inevitably damage the grain unless pumped out quickly. It would have been much safer to line the spar ceiling with skins before loading the grain.

This problem is not solved, but it worries me. There must be a simple answer. In the meantime, the largest trade cargo by volume in the ancient world has left no trace on the seafloor.

* * *

Moktar, whose family lived in Shahat, invited Jay and Kirsten to see their house and to learn about the design of houses where the men and women lived in separate quarters. One evening he also invited them to join a party for one of his nieces, who was preparing for a wedding. This party was not the wedding itself, but the bride-to-be and her female friends celebrating the forthcoming event, like a wedding shower. Jay and Kirsten returned with their hands beautifully decorated with designs in henna.

Next morning, the wind dropped at Apollonia so that our family could swim across the main area of ruins. The incoming swell was still heaving and surging with white surf over the Grotto Reef, so that we could not investigate it closely, but otherwise I could swim over all the buildings that I remembered so well. We swam as a family, zig-zagging back and forth over the Nine Quays, then along the major wall out to the Block Forts. The masonry of the Block Forts reminded me that so much of the ancient city must have been built with a defensive purpose. Ashlar blocks amongst the rubble of the eastern breakwater showed that the Hellenic city wall had continued along the breakwater to provide defence to the east. Even at the height of security of the Roman Empire, piracy was always a problem, so that military patrols were needed, with constant vigilance, lookouts, and watchmen.

I swam across the channel between the towers. Peter and Jay had already swum ahead to the slipways, but Kirsten and I were crossing between the towers when Jay turned back and photographed us (Fig. 11.13). We all swam over the gentle gradient of the slipways, but only examined the easternmost slips. If I had swum to the westernmost,

Figure 11.13. Looking across the channel between the Block Forts at Apollonia. The massive blocks in the foreground are the northern tower, with the southern tower visible beyond the divers. I am the diver without fins, and the other one is Kirsten (Photo Jay Kleinberg).

slipways 1 and 2 on the map (Chapter 1, Fig. 1.13), I might have discovered some surprises that would, in retrospect, have answered questions that arose later.

With sorrow, I turned back towards the city on land. Glancing through my mask splashed with water I scanned the columns of the basilicas standing so proudly on the shore. Those columns represented the last phase of Apollonia's active life, the Christian Byzantine stage, built high on the soil and ruins of earlier variants of the city's plan. Here, in the water, I was swimming over the earliest structures where they rested on the bedrock. Then back to the beach, and a tour west to look at the minor settlements under the wise guidance of Mr Hamid.

On the fourth day at Apollonia there was a bit of a breeze blowing, so that suspended sediment would make it difficult to see under water. I wondered if we could hire a boat to go out to the islands instead of swimming. I turned to Moktar, where we all sat under the shade between the Toyotas.

"Do you think we could hire one of the fishing boats, to go out to the islands?" I gestured to the little modern harbour close to the village where there were a dozen boats tied up to the jetty.

Moktar looked hesitant.

"I wish I could help you. There is nothing in the plan about boats. It would cost money, and I have no approval for that."

Moktar had been handling all the money throughout the trip, paying hotels, restaurants, purchasing fuel, and buying food in the markets.

"I have brought some dollars as a reserve, just in case. I am sure that the fishermen would accept dollars."

"But I don't know what a boat would cost. I am happy to do that, but I could not bargain properly with the boatman."

I handed Moktar a ten-dollar note.

"I am sure that if you offer the fishermen ten dollars, they will agree the time that they can provide us with the boat. I think an hour or two would be reasonable. I don't know the prices either, but it would be worth trying. I would be very grateful if we can land some people on the island today."

Moktar looked uncertain, but walked the few hundred metres to the jetty, and returned 20 minutes later smiling happily.

"One of the fishermen can take you out in his boat for two hours for ten dollars. He is very keen."

We drove to the jetty, where the drivers lifted me down into the small boat, in which I sat on my cushion without my wheelchair. Mr Hamid and the family followed. The outboard looked very battered and unreliable, but the fisherman was smiling, and off we went. The others brought their fins just in case.

In the shelter of the West Island, the fisherman brought the bow gently up to the rocks, so that everybody could jump ashore except me. We stood off in the boat, just stemming the waves, and watched the others clambering about. Peter soon emerged on the top of the island, having found the cave and storage rooms inside, with the rock-cut steps up to a watchtower on the summit. I called to Mr Hamid to come back into the boat and gestured to the boatman to go around the island. Mr Hamid explained what I wanted. The sea was choppy between the islands, but our little boat rode the waves smoothly, and we turned to cruise along the outside of the wave-trap, looking up at the

Figure 11.14. North face of the West Island. The wave in the foreground is breaking into the ancient wave-trap. The cliff face is completely artificial, being cut vertical by the same process of excavation and removal of rock while excavating the trough of the wave-trap. This vertical face made it difficult to attack the island from the sea (Photo Jay Kleinberg).

carved cliff on the north side of the island. It was splashed by the spray and mottled with encrusting algae (Fig. 11.14).

I tried to envisage the shape of the island before the excavation of the wave-trap. Suddenly I was staring straight through the cliff at the misty hills of the Gebel Akhdar on the mainland. My mind grappled unsteadily with the shock. I felt disoriented, like Alice tumbling down the rabbit hole. The cliff had become transparent, like a window. That made no sense. The cliff had disappeared. Not much more sense. I seemed to be looking through a window in the cliff at the distant hills. Still absurd.

I gestured to the boatman to stop and tried to explain my problem to Mr Hamid. When we looked carefully at the island we saw that a freshly broken rock surface gleamed palely on either side of a gap 20 m (85 ft) long, where the rock wall that I described at the start of Chapter 1 had been. The solid rock that I had perched on 45 years ago had vanished. The exposed surfaces at each end and across the base of the wall showed that it had been about 2 m (6 ft) thick. But where had it gone? And why? When?

Even stranger, I, who thought of myself as a trained observer, had been sitting on the mainland shore for several hours on each of the past three days looking towards the islands, and I had not noticed anything wrong. One does not expect hundreds of tons of solid bedrock to disappear, and, like the detectives in Edgar Allan Poe's *The Purloined Letter*, the missing element was so striking, so obvious, that I could not see it. Since it could not be true, I did not register the change in the profile of the island.

I did not feel humiliated so much as astonished and shocked. The human mind is very strange. We drove quickly around the west end of the island, turning in towards the submerged slipways. I called to Jay and Kirsten, both of whom had cameras.

"The wall has broken at the back of the island. Try to get pictures of the waves breaking across the island and the broken wall."

Jay waved that she had understood, so the boat backed off again while Jay and Kirsten both photographed the damage (Fig. 11.15). At dinner that evening, while consuming my second portion of Libyan soup, I pondered the strange occurrence of the missing wall. How rapidly was Apollonia collapsing under the attack of storms? Some years later, I was sent a picture of Apollonia in a winter storm, and it is astonishing that anything survives (Fig. 11.16).

* * *

The drive back to Benghazi next day was truly nostalgic. I had been privileged to enjoy a revisit to Apollonia, but would I ever return again? Once more I thought of finding archaeologists who could give the site the serious attention it needed under water. We stayed again at the plush, glitzy hotel in Benghazi, where we were the only guests. In the echoing, vaulted dining room, we sat at a large table, from which, halfway through dinner, the Man from the Ministry departed, perhaps to make a telephone call. He returned ten minutes later, and, seeing that Kirsten was sitting with her back to him, determined to creep up behind her and give her a surprise, perhaps by clapping his hands. With exaggerated contortions and facial expression, he tip-toed *à la* grandmother's footsteps towards Kirsten's chair. He failed to notice that in a hall surrounded by mirrors, his manoeuvres were not secret at all. Kirsten had been watching him carefully in the mirror and, when he was still a metre back from her chair, she whipped around and clapped her hands in his face. He jumped with surprise, before realizing the impossibility of his attempt. We all laughed at him, but he was not at all upset and re-joined us for the meal with good humour. As security officers go, he was pretty relaxed, at least in our company.

We covered the 900 km (570 mi) from Benghazi to Tripoli in one high-speed drive at full throttle, stopping only for security checkpoints and occasional food and drink. The road surface on the coast road is good, and traffic is sparse, although we did get stuck once behind a small pickup truck carrying two camels, whose long necks extending on either side blocked most of the road. As we approached Sirte, the wind began to drive the sand in swivelling streamers snaking across the road, and then the finer particles and dust were whipped up into the air in a small dust storm. The sun gleamed weakly through the clouds of dust, but the drivers assured us that this could not really be called a sandstorm. The storm lasted for several hours, but the density of dust was not so bad as to make breathing difficult, and the drivers were confident that the air filters on the engines could cope with the abrasion. I was reminded of the extraordinary work by Brigadier Ralph Bagnold who, between the wars, studied the movement of sand in the Libyan desert, writing his original and fundamental 1941 book, *The Physics of Blown Sand and Desert Dunes*. It is still a reliable reference work. Bagnold commanded the Long Range Desert Group in Libya during World War II, from which were born both the SAS and the SBS.

Figure 11.15. Jay Kleinberg photographs the broken wall that originally protected the West Island, both against military attack, and to prevent waves and spray washing onto the island. The ridge of the wave-trap is visible on the outside of the trough. The entrance to the storage caverns has half collapsed. Waves now wash right across the centre of the island.

Figure 11.16. Apollonia West Island in a winter storm, with waves and wind from the northwest. The effect of the broken rock wall is to allow the breaking waves to flow over the centre of the island (Photo Jean-Pierre Misson).

We stayed in the same grand high-rise hotel in Tripoli but enjoyed one last magnificent dinner at an excellent fish restaurant by the sea. The contemplation of the end of our adventure was surprisingly intense for all of us. Our support team of five Libyans were amongst the best people I have ever worked with. If I had tried to pick a team for the project, I could hardly have found people who would approach their efficiency, kindness, thoughtfulness, and quick responses to problems and emergencies. We exchanged many commitments and assurances that we would return to Libya soon, to conduct more research together at Apollonia. Many toasts of Libyan non-alcoholic beer were consumed emotionally.

On the final morning, as we prepared to drive to the airport in the six-seater van, Moktar quizzed me while the others did some last-minute shopping. "Please promise you will come back. This trip has been so interesting for all of us, and we have enjoyed your company. We must work together again. Your children are very unusual. They are so ..." He struggled to find the right word. Finally he said: "They are not complicated." I knew that he did not mean that they were simple. Perhaps he was looking for words like not moody, or not temperamental. I think he had sensed, correctly, that in a stressful or demanding situation, both Kirsten and Peter just go for the practical solution first, and postpone any quibbles or complaints for afterwards, if they are necessary. They do not make problems for their own sake and always try to simplify and ease the situation in a conflict. Moktar certainly intended it as a compliment. Once more I wondered if he had had to cope with tourist families where the children indulged in mild, or not so mild, fits of complaint or tantrums. I owe many thanks to Moktar and the whole Libyan team.

* * *

Back in England, I wondered how best to protect Apollonia from further damage, and whether the original cause of the smashed wall had been storm waves or fish-bombing. Cyrene is a designated UNESCO World Heritage Site, which technically includes Apollonia as its port. But no effort has ever been made to use the UNESCO designation as a reason to protect Apollonia. I consulted Dutch engineers during one of my visits to the Koninklijk Nederlands Meteorologisch Instituut in Den Haag about the costs of a breakwater encircling the whole city to protect it from direct wave attack. The best approach was probably a submerged breakwater that, rather like the present function of the ancient wave-trap, would cause the waves to dissipate most of their energy several hundred metres outside the city limits. This could be a crude rubble construction, using blocks large enough to stay in position, which could be built for about €8-9M. At that time, there was a huge publicity campaign supported by Colonel Gaddafi for a project called "Green Mountain", which implied an investment of many millions of dollars into development of the Gebel Akhdar area, so that I hoped that protection of Apollonia could be added to the agenda.

During this time also, the BBC contacted me to consider making a film about triremes, and we discussed using Apollonia as the site, with the perfect slipways, and using the reconstructed ancient trireme *Olympia* as part of an elaborate documentary. It was attractive but too ambitious and came to nothing.

In 2008-09, Carlo Beltrame, representing an environmental research group in Venice, dived on Apollonia to assess the possibility of developing an archaeological marine park, providing tourists with glass-bottomed boats that could travel over the city, allowing them to keep their feet dry while enjoying the ancient ruins. Such tours could have been combined with dive trails supervised by diving guides. The project was part of a general remit by the Libyan government for Evaluation and Conservation of the Cultural Libyan Heritage. I corresponded with Carlo about these ideas, which I supported, because I felt that the justification for protecting the ancient port properly would be improved if there was strong public interest, even providing a small financial return with the stimulation of local employment. Carlo's report assessed the visual appeal of each feature and target on the seafloor, showing also that, as the scheme was developed, there should be scope for research. I sent Carlo a digitized version on CD of the 16 mm film that we shot in 1958-59, and he has sent me some pictures which are used in this book. Again, we abandoned such plans as the political situation in Libya deteriorated in 2010-11.

Professor Claude Sintes, Director of the Museum of Ancient Arles, France, excavated at Apollonia between 1986 and 2003 as a member of the Mission Archéologique Française (MAF), in a series of projects some of which included diving. His work focused on the channel between the inner harbour and the east harbour and on the piscina, where his divers found part of a statue that matched the marble faun that we raised in 1959. French divers also found remains of a wreck in the eastern harbour and raised part of it. I sent Sintes the same CD I had sent to Carlo, and he also responded with several pictures of his work, but also with the comment that the Grotto Reef was, he thought, completely destroyed before he began work there. He also confirmed that the back wall of the West Island was broken already. This pushed back the date for the great wall breaking to a much earlier date than I had expected.

When Jon Henderson conducted the sophisticated, high-tech survey of Pavlopetri in 2008-09 (Chapter 6), we discussed the possibility of applying the same methods to Apollonia, which would, in the absence of more substantive protection, at least have documented the exact state of the remains at that date. Unfortunately, the civil war to overthrow Gaddafi started in 2011, which brought all such speculative ambitions to a stop.

* * *

I presented an illustrated report on our 2003 tour of the Libyan coast to the Society for Libyan Studies in the British Academy lecture hall in 2004. I showed a section of the 1959 movie, now on DVD, then followed with the original measurements that we had made on the sea levels all along the coast in the Gulf of Sirte. I stressed the need for new efforts to map and protect Apollonia from damage. In the following years, I kept in touch with Paul Bennett, the chairman of LibSoc, and with Claudio Vita Finzi, an old friend from Cambridge days, who also participated in the Farasan Islands project (Chapter 10). Vita Finzi has a strong interest in the climate change and sea level on the coast of Libya.

In 2007, LibSoc invited André Laronde to give a talk on his excavation of trenches on the acropolis at Apollonia, which I attended with interest and curiosity. The excavation revealed details of ceramics and coins in stratigraphic context, but Laronde made almost no mention of the underwater studies, although he did refer to my early mapping. He mentioned very briefly the MAF diving research in the 1980s.

As we were putting our coats on after dinner, Paul Bennett approached me in a chatty way.

"Nic, we have decided to return the Fitzwilliam Bronze Ram to Libya as soon as possible. All part of trying to re-establish friendly political relations, you know."

I gulped with astonishment.

"Ram? Bronze ram? I'm sorry, I did not know that there was any bronze warship's ram in England. Surely the Atlit ram in Haifa is the only one that has been found and published?"

"Oh no," said Paul. "This ram was found near Tobruk, off Wadi Belgammel, by British military divers in the 1960s. They brought it back to England and loaned it to the Fitzwilliam Museum in Cambridge. It's been stored in a basement archive there. The owner has decided that it would now be good to give it back to the Libyans, and I have agreed to arrange that at the diplomatic level."

"Wait a minute, Paul, please. Please. I must know more. I am not a "ship man", but I am sure that no nautical archaeologist or expert on ancient ships in the Nautical Archaeology Society even knows that this ram exists. The only warship ram ever published is in Haifa. If we send the Fitzwilliam Ram back to Libya, we will never be able to study it in detail again. We need several months to make proper scientific analysis of how it was cast, what it is made of, the alloys, how strong it is, how it worked, how it fitted onto the ship's bow, where the metal ores came from and so on. It will take time. We must have time. Please."

Paul hesitated.

"It's a bit sensitive. Of the people who found it, raised it to the surface, only one is still alive. The Fitzwilliam Museum has never been told exactly where it was found. It is now back in private hands, and I will have to ask the owner if he will agree to the sort of examination that you are suggesting. I'll try."

"Thank you, Paul. Thank you. Honestly, I think this is a unique chance to study an ancient bronze fighting ram in detail. It will never happen again."

Several months had passed when Paul told me that the owner of the bronze ram had agreed to it being examined scientifically before sending it back to Libya. I asked if I could see it before I started to contact the scientists who I thought needed to work on it. A few weeks later. Paul told me that the owner, Mr Ken Oliver, would be happy to show me the ram, and that he lived in a suburb of Southampton. This was an extraordinarily convenient coincidence, since the NOC is in Southampton, and I could request scientists from different groups at Southampton University to conduct the specialist studies.

In January 2008, I drove to Ken Oliver's house. Ken was a lean, active man, a retired teacher, older than me, and keen to ensure that the bronze ram was given a good home. He escorted me through to his kitchen where, on a metal display frame, the exquisite shape of the three-bladed ram dominated the room like a prima donna. It was impossible not to stare at it: 20 kg (44 lb) of solid bronze, cast in one piece, showing complexities of shape and decoration that demonstrated the skill of its makers (Fig. 11.17).

"The man who had it in his home in Cornwall died recently," said Ken, "and he left it to me in his will. The other two divers who found it died many years ago. Technically it was on loan to the Fitzwilliam Museum, but it really ought to go back to Libya."

"It is beautiful," I said. "Has Paul Bennett told you what we want to do before it goes home?"

Figure 11.17. The Belgammel ram: the bronze ram weighs almost 20 kg. The trident blades at the front would have been used to damage the oars on an enemy ship (Crown Copyright).

"Yes, and I think all the original finders would agree. But I think we need also to obtain permission from the Fitzwilliam."

Ken told me how the Belgammel ram was found by the group of RAF divers to which he belonged in the early 1960s. They found it in 24 m (80 ft) of water to the west of Tobruk, off Wadi Belgammel, 200 kilometres (124 mi) to the east of Apollonia. The discovery and lifting of the ram had never been published, and they did not reveal to the Fitzwilliam the position where it was found, not even the country. The find site, fairly close to Apollonia, provided a richness to the event that added to what we already knew about the use of warships in the area, both Greek and Roman. The Belgammel ram was smaller than the Haifa ram, and was clearly designed to fit above the waterline, not a massive battering ram to stave in the enemy timbers, but more probably to smash the oars. I imagined a small, armed patrol boat that assisted the Roman warships in the suppression of piracy along that important coast, where trade was so essential. Perhaps it had sometimes called at the port of Apollonia.

Ken lent me the files on the discovery, photographs from the date of discovery, and correspondence about the possible function of the small ram that he had in his records. I thanked Ken for the papers, after which I informed Paul Bennett and the Fitzwilliam what we intended to do. The museum agreed both to the return of the ram to Libya, and the study and sampling of the bronze, given ethical assessment by archaeologists.

I recruited a range of experts who could subject the ram to several tests and sampling: surface non-contact digitizing using a laser scanner; reflectance transformation imaging using polynomial texture mapping and hemispherical harmonics; digital photogrammetry with dense surface modelling; structured light optical scanning; and X-ray fluorescence analysis. The experts examined its internal structure by X-radiography and 3-D X-ray tomography. Metallurgical composition was studied by micro-drilling and subjecting the nine samples to scanning electron microscope X-ray micro-analysis, micro X-ray fluorescence, and X-ray backscatter. The organic materials and fragments of wood inside the ram were dated by radiometric Carbon-14, and an expert on ancient shipping estimated its age independently based on its design and decoration.

Paul Bennett and Ken Oliver joined our first meeting at the NOC Southampton. People responded dramatically to the sight of the mass of bronze supported on its frame, standing on the table. The alien shape with its aggressive trident blades forced everyone to stare at it. Eyes widened, gazes were fixed, jaws dropped, so that metaphors are not needed. We had to make a decision on the ethics of drilling for metallurgical samples. Bronze is an alloy of copper and tin, with possible additions of lead, and other impurities. From my days of undergraduate metallurgy at Cambridge, I knew that, when a complex alloy of two or three or more metals cools and the first crystals solidify, the residual liquid has a changed composition, so that the next crystals change to a different

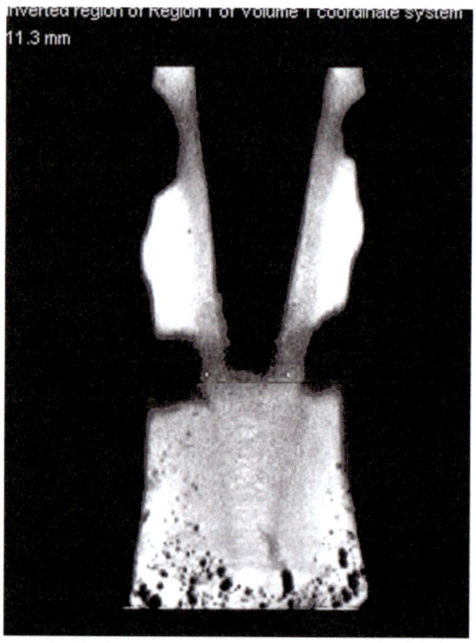

Figure 11.18 a) This section through an X-ray CAT-scan of the uppermost blade of the trident on the bronze Belgammel ram shows intense porosity concentrated into the up-curved tip. This shows that the ram was cast on an upright position with this blade at the top (Crown Copyright).

Figure 11.18 b) Section through a 3-D X-ray CAT-scan reconstruction of the Belgammel ram. The thin curved material would fasten round the stem-post of the ship, while the wales were inserted inside the main cavity, terminating at the solid bronze at the front of the Ram. Note extreme variation of the thickness of the metal, which was cast in one piece (Crown Copyright).

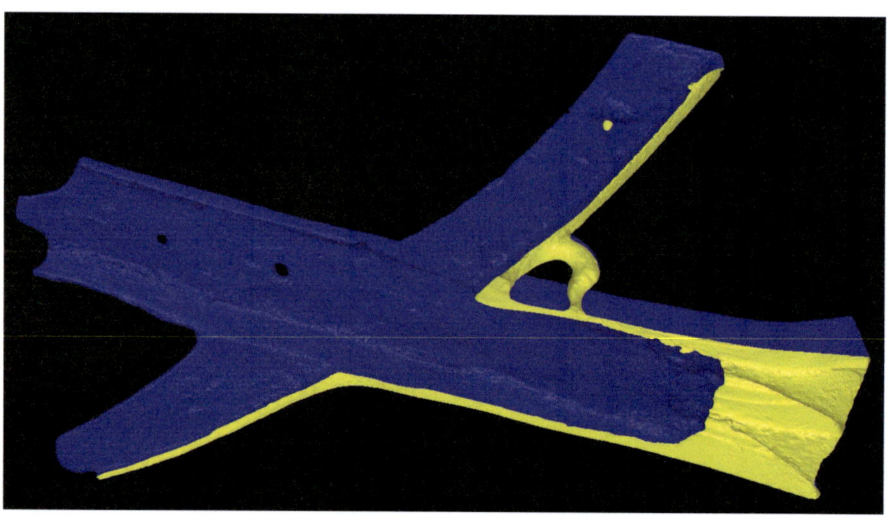

mix. Since some parts of the ram were only a few millimetres (about 0.1 in) thick, and others 20-25 cm (8-10 in) thick (Fig. 11.18) the rate of cooling would vary, and thus the composition. Also, bubbles probably formed in the molten metal when it was poured into the mould, and the bubbles would rise to the top of the mould, depending upon the location of riser vents, revealing the position and orientation of the ram when it was cast. Ian Croudace, the geochemical expert, reported that he needed to take about eight or nine micro-drilled samples, each of only 50 milligrams (one five-hundredth of an ounce). After the position for the samples was agreed, we recorded that this was an ethically allowable programme, in view of the knowledge to be gained.

We started work on the Belgammel ram with the assumption that we had only a few months to complete the examination, but the diplomacy with the Libyan Embassy proceeded slowly for two years, so that we were never told to hurry up. Analysis of the micro-drill samples showed quickly that the alloy was high in lead, with an overall composition of 86.9% copper, 6.3% tin, and 6.6% lead. Isotopic analysis of the lead showed that it could have come from mines near Athens, but this is uncertain, because of the habit of re-melting and re-casting multiple lead or bronze objects. The most difficult part of the study was to shine X-rays through the thick 20-25 cm (8-10 in) parts of the bronze, to reveal the internal integrity, cracks, porosity, and concentrations of bubbles. Ian Croudace discovered that no hospital or laboratory in the Southampton region had an X-ray machine of sufficient power to even consider such a project. His contacts with scientists at MoD Aldermaston provided us with the solution we needed. In a concrete bunker with walls of unbelievable thickness, the experts there had an X-ray generating machine that fired beams of 6-10 million electron volts at targets on a rotating table. Using this device, we could shoot X-rays through the ram at many different orientations, building up a complete picture of its internal structure, somewhat like a medical Computer Aided Tomography scan.

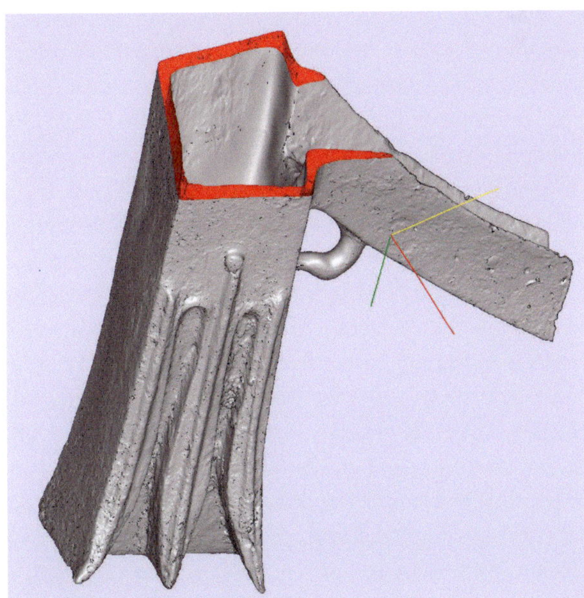

Figure 11.19. A cross-section cut through the 3-D reconstruction of the ram based on X-ray CAT-scan (Crown Copyright).

The security precautions were laborious but necessary. The outcome of the X-ray study was wonderful (Fig. 11.19), showing a porosity that would be regarded as unacceptable by modern standards, but which, by all reports, was quite common in ancient castings. The bubbles are concentrated in the tips of one of the up-curved blades, showing that this part of the ram was uppermost during casting. Consideration of the shape of the ram itself, the lack of seams, and the multiple curves, showed that it was cast in one piece, in a single pour, using the lost-wax process. This is the same process that was used by the ancient sculptors to make beautiful bronze statues like the Charioteer at Delphi. The technique was very highly developed in the ancient world and was apparently just as important in the realm of military technology as in art and aesthetics.

In May 2010, Paul Bennett and Ken Oliver travelled to Libya with the ram, where they were guests of honour in a ceremony at the National Museum that was so grand that Tripoli was bedecked with banners. We published the report on the ram first in an illustrated book which accompanied the donation in Tripoli, and then in the *International Journal of Nautical Archaeology*.

* * *

Mohamed Bouazizi, a Tunisian, burned himself to death in December 2010 out of despair at unemployment and a sense of the impossibility of change. This sad event started the spread of imitations and civil unrest in many North African and Middle Eastern Arab countries. The government of Tunisia fell a few months later, followed by elections, encouraging many observers and commentators to hope for a decline in dictatorships and a rise in responsive democratic governments throughout the region. The initial phase of optimism was dubbed the "Arab Spring", but it was soon followed by counter-attacks from governments, skirmishes, vendettas, suppression, and increased use of force in many countries. Only Tunisia seems to have survived without lapsing back into authoritarian or religious dictatorship.

Muammar Gaddafi's army marched on rebellious Benghazi in February 2011, prompting the expectation of an unrestricted massacre of civilians. The United Nations Security Council decided on 26th February to freeze the assets of Gaddafi and his inner circle and restrict their travel. On 19th March, a NATO force led by British and French intervened with aircraft, bombs, and rockets in support of the rebels. Various factions of rebels increased their attacks on Gaddafi's army, and civil war raged until October, when Gaddafi was captured and killed on the 20th, with Tripoli falling into the hands of the alternative government council on the 23rd October.

The destruction and suffering caused during the intense months of August to October were all the more agonizing to watch on TV, and in the press reports, because most of the battles were surging to and fro along the coast road that we had travelled in 2003. Every event, every picture of a shell-shattered town or village, brought back memories of the good people we had met in Libya, their kindness, generosity, and open-heartedness. Ironically, we returned the Belgammel bronze ram to Tripoli only a year before the city was bombed by the RAF. The survival or destruction of the ram was a trivial matter compared to the sufferings of tens of thousands of innocent people, but, as a matter of fact, the museum was not destroyed. I have enquired of the survival of the people we worked with, but have been unable to find out if most are alive or dead. Mr Hamid died of old age in 2017.

* * *

The academic publication of the uniquely detailed metallurgical and geochemical study of the Belgammel ram created some correspondence, including an e-mail from a Monsieur Jean-Pierre Misson. This began a long, friendly correspondence and, so far, one meeting, which has been surprisingly fruitful and constructive.

Jean-Pierre (JP), who was born in Alexandria, Egypt, had Belgian and English parents who stayed in Alexandria after World War II, so that his education was completed there. He speaks excellent Arabic, plus English, French, Italian, and possibly other languages. He trained as a telecommunications engineer, working from 1961-67 as a radio engineer on the microwave network in Libya, where his accommodation was in El Beda, not far from Cyrene. His spare time was often spent in the sea, swimming at many archaeological sites on the coast, frequently at Apollonia. He first dived with scuba tanks in 1963, getting his tanks filled with compressed air at the RAF base in Tobruk by the same people who later found the Belgammel ram. To record the ruins that he saw under water, he built a waterproof housing for a reflex 6" x 6" camera, which worked in shallow water, but also purchased a Calypso Phot 35 mm camera. He became completely obsessed with the details of the submerged buildings at Apollonia. Richard Goodchild showed JP a copy of the map that Nick Wood and I had prepared a few years earlier, which had been published in the *Geographical Magazine*, and in a book edited by Joan du Plat Taylor in 1965. JP annotated a copy for himself to use as guidance while he was diving regularly on the city.

JP's sharp eyes detected many small details of the construction of the Nine Quays and the slipways (Fig. 11.20). There were grooves and slots for dowels of lead or iron in several places. As our exchange of e-mails expanded, I noticed that some of the pictures

Figure 11.20. View along the Quay Number 2 of the Nine Quays. Each quay, which is 2 m wide, has an outer face of long blocks, while the internal space is filled with rubble and topped with rectangular blocks. There are no footings for a wall or pillars that could have supported a roof (Photo J-P Misson).

included the West Island in the background. To my surprise, the great wall at the back of the West Island was already broken away and missing in pictures from the early 1960s. When I discussed this with JP, he found pictures of colossal blocks of broken stone lying on the slipways (Fig. 11.21). They were lying on slipways 1 and 2, which I did not inspect in 2003. It appeared that the wall had already been smashed by 1963. The light-coloured fresh surfaces of the broken rock where the wall had collapsed were only relative to the heavily weathered rock around them. They had been exposed for at least 40 years.

One day, JP sent me a slightly blurred monochrome image of an aerial photograph of the underwater city which, he said, had been part of a survey started in 1959. Careful scrutiny of the image of the slipways 1 and 2 shows that there are large shadowy blocks lying on the seafloor. The photograph is too blurred to check for the existence of the wall itself, but this seems to show that the great wall was probably smashed in the winter of 1959, very soon after we left. Such an early date would also explain why nobody I spoke to in the village, and no archaeologists, recalled the event of the wall being broken. While this occurrence has now been more or less dated, the destruction of the Grotto Reef is less well documented, since nobody has examined it for many decades. Sadly, we are losing information as the city surrenders to the waves.

Figure 11.21. This huge block of stone used to be part of the rock wall that protected the West Island. The storms broke the wall and pushed hundreds of tons of rock onto the slipways. This block is about 3 m across. Smaller debris from the wall are visible in the foreground. These rocks were not on the slipway in 1958-59 (Photo J-P Misson).

* * *

One final postscript on factual research from Apollonia: one day in early 2016, I received an e-mail from my colleague Udi Galili in Israel, attaching a recent academic paper of which he was a co-author, describing geochemical analysis of the production of industrial quantities of raw glass in various ancient cities along the coast of Israel and Egypt. I was about to classify it as interesting but not important, when I glanced up at the bookshelf to the right of my desk, on which there was a chunk of raw bluish-green glass that the CUUEG divers had found in Apollonia in 1958.

In the first few years after the Apollonia expeditions, I did try to have the glass examined or dated by experts, but nobody could help. It might be any date from early Roman to late Victorian. Any ship carrying bulk glass to be used somewhere else could have dropped it.

Udi's paper persuaded me that it was worth trying again. I sent him photographs of the chunk which he copied to his co-author in Haifa, Yael Goren Rosen. Yael suggested that there was an expert on ancient glass at the Institute of Archaeology in London named Iain Freestone. I immediately wrote to Professor Freestone and took him the sample, which he agreed to date and analyse chemically. He was puzzled by the shape, which looked as if it had been broken out of shallow mould, but assumed that everything would work out when they knew the chemistry.

The dating of the glass places it in the in the 4th to 7th century AD, most probably 5th century, with a chemical composition that ironically, suggests that it was manufactured in another city called Apollonia, the modern Arsuf, in Israel (Chapter 6). I have dived at Arsuf several times with my friend Dr Yehuda Melamed. The late date surprised me, but it confirms the observations of Philip Kenrick that the port of Apollonia in Libya was still active and commercially viable in the Byzantine era. Any model of the earthquakes and earth movements that resulted in Apollonia being under water, as it is now, must include a stage of the harbour in the Byzantine period which was still usable (Fig. 11.22).

Figure 11.22. This chunk of raw glass was found by the divers in the inner port at Apollonia in 1958. It has been dated to the 4th-7th century AD, in the Byzantine period. The chemical composition of the glass, and the isotopes of contaminants, are typical of glass manufactured at the city of Apollonia in Israel, the modern Arsuf (Apollonia was a popular city name) (Photo Mike Tibbots).

* * *

Sixty years after the Cambridge team of divers made the first complete map of the underwater city of Apollonia, do we fully understand how the city evolved after its foundation, how it adapted to changes during the 1300 years of active life as a commercial and military port? How ships were managed? How warships were stored and launched? How bulk cargo was loaded? How valuable produce was stored and protected? And what happened during the further 1300 years after the port was abandoned? The quick answer to all these questions is still no, certainly not in a precise and reliable way.

As I have stressed many times, I am not an archaeologist. During the decades after 1959, I did not have any reason to observe or record the subsequent short visits to Cyrenaica by small research teams from other countries, nor to try to understand what was happening locally on the site. But I shall try here to fill in some of the gaps, and note what happened during those years.

There are two main factors that drove the changes at Apollonia: the effect of tectonic earth movements on the coast causing changes of relative sea level, and the cultural, political, and military changes in North Africa and the wider Mediterranean empires.

The people of the city responded to both. They could not avoid or ignore the changes of relative sea level that caused the harbour to contract, or flooded it and drowned other buildings. After such changes of sea level, the relative importance of certain activities in different parts of the town and harbour would change overnight. That is in addition to the destructive effects on buildings that may have occurred from the earthquakes. On a different timescale, the prosperity of Cyrene, the demand for trade from the great cities of the northern Mediterranean, shifting political alliances in the region, military and naval security, and raiding parties from the Arab expansion combined to force changes in priorities, increased defences, or required more warehouses, docks for larger vessels, and so on.

Modern maps of seismicity, combined with historical records of ancient earthquakes over the past 2000 years, confirm that there are two regions of intense activity on the Libyan coast, from Sabratha to Tripoli in the west, and from Ptolemais to Apollonia in the east. This fits exactly with observations of uplift and submergence, but leaves a question mark over the timing, and the possibility of multiple earthquake events.

In 1959, I noted that the walkway around the piscina is at present under 2.5 m (8 ft) of water. If the land uplifted so that the relative sea level was lowered sufficiently to make the walkway dry, with a bit of clearance, then the slipways on the West Island would have been entirely out of the water. All relevant evidence indicates that they should be submerged by about one-quarter of their length. Since the piscina is probably Roman, and the slipways are much earlier, this suggests that there was an earthquake or seismic event shortly before the year zero, that uplifted the city by about 50 cm (20 in). This is compatible with the circular line of buildings showing a contraction of the radius of the inner harbour, and existence of the buildings constructed on top of the slipways. It also explains a wall in square B9 (Fig. 1.7) that runs across the seafloor about 20 m (66 ft) beyond the bottom of the slipways, which would be nonsense if they were still in use.

On this evidence alone, we can speculate that there was a tectonic uplift on the order of 50 cm (20 in) at the time of the Early Roman Empire, which was followed by tectonic submergence of the order of 3 m (10 ft) at an unspecified date, probably in the late Byzantine period or shortly afterwards. Such multiple stick-slip phenomena are quite common in the way that large faults yield to the stresses caused by plate movements. There is no direct evidence for this double movement at the other ports in the region, nor would one expect the displacement to be identical at all ports, but the final net subsidence is confirmed, or at least compatible, with the evidence at all the settlements between Apollonia and Ptolemais.

The post-glacial GIA (Chapter 6) component for Cyrenaica in the past 2000 years is close to zero, and the most recent calculation by Giorgio Spada from Urbino University shows that it is probably about 10 cm (4 in) at Apollonia.

This simple model therefore proposes three phases bounded by tectonic change: before the uplift; after the uplift; and then after the submergence which is both tectonic plus a small GIA component.

* * *

Almost 30 years after the CUUEG expeditions, the Mission Archéologique Français en Libye, led by Professor Claude Sintes, from the University of Arles, conducted a series of land-based and diving research projects, with some excavation by air lift suction hose between 1985 and 1998. Work on land at Apollonia continued under the direction of André Laronde until 2006. A diving team from the Department for Underwater Archaeological Research (DRASSM) of Marseille took part in the early years of the mission. They discovered two Roman wrecks in the eastern basin and excavated one of them, which they dated to the 2nd century BC. This pair of wrecks with amphorae dating to the later Republican period confirms the hypothesis that the external basin to the east, bounded by a massive rubble breakwater, was the main commercial basin for large cargoes. But the wrecks were not grain-carrying ships.

The DRASSM team also excavated a trench between the Block Forts at the entrance of the western harbour. The trench uncovered evidence that, between the 6th and 7th centuries AD, the entrance had been closed quickly with reused building material. The French divers' investigations at the Block Forts estimated a total change in sea level of 3.50 m (11 ft), with a slight variation for the small tides. They based this on indications of lithophaga, small molluscs that bore into stone, on the sides of the fortified towers facing the channel. The French concluded that there is additional support for the suggestion of a change of level of 3.70-3.80 m (12 ft 1 in – 12 ft 6 in) since the beginning of the Christian era. If the relative sea level has risen that much, it must be pointed out that the western inner harbour would have had a very small radius indeed, with a depth of less than 1 m (3 ft), while the slipways and Nine Quays would have been 1 m (3 ft) above sea level 2000 years ago. Nevertheless, the suggestion that the most recent relative sea level rise is on the order of 3.0 m (10 ft) or more, is consistent with my previous observations concerning the tectonics.

The CUUEG discovery of the 4th-7th century glass ingot in the western inner harbour suggests that ships were still using it as late as that. There is literary evidence that the harbour was active in the Byzantine period, and Apollonia was the capital of the Pentapolis, showing that it was not in a state of terminal decline.

A comprehensive and insightful publication in 2010 by Sintes proposes four phases in the evolution of the city plan and function, based on his analysis of changes in the architecture of buildings and structures of the sea defences. These successive plans are based on the excavation at the Block Forts and estimates of the depth of each channel and structure at each date (though, as far as I can see, they have not included the possibility of two phases of earth movements). Sintes notes that, during his work at Apollonia, a thick layer of sand covered the lower half of all ten slipways, concealing the observations that we made in 1958-59. It was impossible for him even to consider excavating through hundreds of tons of sand. This accumulation may be at least partly the result of waves washing over the West Island since the rock wall collapsed.

According to Sintes' model, in the first phase of the port's development, the West Island was connected to the mainland by a natural isthmus that separated what we now regard as the east and west harbour basins, while the entrance to the west harbour was over the ridge to the west of the West Island (see Fig. 1.7). The slipways were built on the West Island during this stage, but, since the basin was exposed to the worst northwest storms, the citizens eventually decided to fill in the gap between the island and the Grotto Reef with a rubble mole, and cut a channel through the isthmus by the Block Forts to create a port with two basins.

My rough calculations show that construction of the slipways by cutting away the overburden of rock, the creation of the wave-trap, the hollowing out of the West Island with the open work area and hollowed caves, plus the deep quarrying on the eastern side of the island, removed in the order of 100,000 m^3 (3.5 million cu ft) of rock. This could have been used to build rubble breakwaters and to provide squared ashlar blocks for buildings in the city itself. The sculpting of the East Island additionally produced about half as much re-usable stone as the West Island.

During the second phase, the Hellenistic period, the citizens built massive defensive walls with towers to protect the whole city, both on the seaward defences and inland, around to the theatre, and out along the eastern rubble breakwater. Sintes judges the inner western basin to be 4 m (13 ft) deep at that time, while boats or galleys not stationed on the slipways could be pulled up on the sandy beach on the southern side of the basin. This is the area where there are the parallel buildings that we proposed in 1958-59 as quays, which would have served the same purpose.

The third phase in Sintes' synthesis is the Roman epoch, which shows the reduced perimeter of the harbour clearly for the first time. There is no suggestion that this was caused by uplift, but the ten slipways are now covered over and defined as an area of dry land. The Nine Quays appear on the map for the first time, and are defined, tactfully, as warehouses, with a question mark. The channel between the West and East Islands is shown as filled in, as if there was only one island during all phases, which is curious, since the channel is now 7 m (23 ft) deep, and would have been open at all times in the past, unless blocked deliberately with rubble.

In the fourth phase, described simply as "late", the channel between the basins has been filled in, and the gap between the islands has been partly filled with rubble.

My primary reaction to Sintes' proposals is one of a total welcome for seizing the big ideas and the big questions about Apollonia. My own suggestion is that the channel between the ports probably existed as a natural connection quite early on, and that the blocking of the gap to the west of the West Island had to be early, because without it the basin was exposed to almost constant northwest winds and would have been useless. There are plenty of rubble breakwaters in ancient Greek ports from this date. The waste rock from excavating around the islands provided the material early on. When the tectonic uplift occurred at the time of the late Republic or Early Empire, the channel between the basins became impassable, and had to be deepened. This would explain the down-cutting that the French reported.

Apollonia continued to thrive in one way or another until the 7th century AD, and it is probable that at some stage an attempt was made to block the channel between the forts to protect against pirate attacks or Arab raids.

Finally, perhaps as late as 700 AD, the harbour sank by more than 3 m (10 ft).

Many of the details of these models, which I have described as plausible, including mine, still need checking more carefully. We need definitive confirmation, not plausible proposals. But the delightful consequence of the synthesis by Sintes is that dozens of factors can now be seen to be important in context, bringing the key issues sharply into focus. By further precise work, we can arrive at a complete description of what happened to Apollonia, how it grew, and how it died.

Sintes says that Apollonia is one of the most important ancient coastal cities in the Mediterranean, and he adds that neglect and relative lack of study of the city is scandalous, especially since the ruins are eroding so rapidly. He quotes the work of myself and the Cambridge group, crediting us with the first complete survey on which subsequent serious research should be based. He is generous with his comments on what I always presented as a non-archaeologist's common-sense attempt. Experts had to follow to do the real archaeology.

Incidentally, there are many dubious copies of the original Apollonia map published by rather careless scholars who, for reasons that I do not understand, have altered and simplified the plans of the city and individual buildings that were plotted so carefully by Nick Wood. The distorted and oversimplified maps conceal from modern scholars the true extent of our knowledge of this unique early Greek harbour. The best source for information is the map and plans in this book, my own publications on Apollonia, and detailed plans by Sintes.

* * *

The efforts that I have made over 60 years to map, excavate, protect, and understand Apollonia in all its glory have, in the end, produced useful but modest results. The sense of futility in attempting to combat and defeat the forces of the waves that I experienced on my first day there in 1958 has returned. The evocative Grotto Reef passageway has collapsed; storms destroyed the wall on the West Island in the early 1960s; the plans to map and excavate the city to modern standards and details were overtaken by the Libyan civil war; the Dutch proposal for a protecting breakwater was a non-starter in the political chaos. The building of a multi-storey hotel in Marsa Susa, almost within the area of ruins, has damaged the appearance of the ancient city and ruined its appeal to tourists. There are constant rumours that the tiny fishing harbour will, one day, be developed as a major port, though why anyone would think that such a shallow bit of the coast was suitable for large modern ships is inexplicable. The work under water by Sintes has been excellent, but serves rather to emphasize how much still needs to be done. I thoroughly approve of his speculative reconstructions of the sequence of inundation and adaptation of the city to changes of sea level but, in effect, this work produces a set of hypotheses that need proving or checking by many weeks of diving on the site. We need to understand the original function of many unusual structures. That work will, on the present evidence, not be attempted in the near future.

Cyrenaica, the eastern part of Libya, has for several years been within the area dominated by the secessionist army that is not recognized by the government in Tripoli, nor by the United Nations. A state of lawlessness exists there. Paul Bennett has reported that, on his visit to Cyrene and Apollonia in 2020, he saw the ancient ruins being flattened for modern buildings, and open ground with buried ruins being ploughed for farming. Dense groups of houses have been built at Apollonia with untreated sewage discharged over the ruins. The water is cloudy with algae and filthy with pollution.

Like Jean-Pierre Misson, I find that I have come to adopt Apollonia as a friend of the family. I feel personally about its fate. I know that I am watching its death, with the loss of the detailed and unique information that I had hoped to decipher and preserve in 1958. The sea is claiming everything: bit by bit, blocks are corroding, stones being shifted in storms, roofs collapsing, ceramics being removed by swimmers as souvenirs. Apollonia, I tried to save you. You have been neglected, ignored, bombed, smashed by waves, traduced, misinterpreted, distorted, covered in sand, polluted, exploited, and abandoned.

And yet ...

What can be salvaged from this tragic mess? Is Apollonia itself really so important as an ancient port? Are there others where we could study and preserve the same features? Sintes considers Apollonia to be exceptionally important, and as I have personally visited hundreds of other ancient ports (Chapter 6), I can confirm that there is no other early Greek port preserved to anything approaching its completeness. The great harbours, such as Leptis Magna near Tripoli, or Caesarea in Israel, are Late Roman constructions, built with all the resources available through the wealth and infrastructure of the Empire and the extensive use of hydraulic concrete. Apollonia displays intact the Greek or Phoenician construction techniques and the spatial plan from 600 years earlier.

Arising from a strange sequence of visits by interested parties from several countries, starting with Captain Beechey in 1822, we have gradually built a broadly reliable plan of the submerged city, certain buildings and features have been studied in detail (the Block Forts, the harbour entrance channel, the apsidal building, the fish-tank), and a small number of items have been raised and preserved in museums. That constitutes a fair academic outcome from a difficult geographical, environmental, and political situation. But the city itself is now in a very unfortunate state, and further work, either to study and understand it, or to preserve it, is scarcely possible for decades to come.

What will happen to the ruins in those years?

Let us first consider the worst case. If building and development at Marsa Susa continue unchecked and with no respect at all for the ancient city on land, it is just conceivable that the Byzantine basilicas could be bulldozed flat and built over. If that happened, the continuing erosion of the low cliff would be a threat to the new houses, and people would be tempted to reinforce the beach and the cliff to repel the waves. The obvious source of stone to do this would be the ruins just nearby in shallow water. If the stone blocks were dragged onto the beach, the last vestiges of the underwater city would be minimal indeed, and future study almost pointless. This scenario is very unlikely, so let us look at more optimistic possibilities.

On a scale of decades, it is possible that the modern buildings will be fitted with proper drainage and sewage treatment. If that were to happen, the seabed ruins as they are now would be more or less accessible, and only partly damaged – similar to the condition of some of the Roman buildings in the sea close to Naples. The attachments of algae to the stones will do some damage, but not much. Those areas that have collapsed, such as the Grotto Reef and the broken wall on the West Island, could be studied by divers, the fragments measured, and the original features graphically reconstructed. The very bottom layer of foundations, where cut blocks are in contact with bedrock, is unlikely to be disturbed. Thus, a survey by an acoustic or optically sensing 3D integrating robotic autonomous underwater vehicle (AUV) might still detect many accurately positioned remains. Cuttings in bedrock, such as slots for dowels of lead and iron, or sockets for wooden beams, will not be destroyed. Technology developed in the next decades may revolutionize the accuracy of underwater surveys, with Artificial Intelligence to create reconstructions.

Taking this long view, the situation is not so depressing. On a scale of 2600 years, another decade or so is not long to wait, even if some of the detailed trimmings and finer structures of the city are destroyed. The deep-seated foundations will remain for our descendants to visit. They will have the maps, photographs, and many publications to guide them.

On this more cheerful note, I am happy to hand over responsibility for understanding and nurturing Apollonia to a much younger generation.

Further Reading

Chapter 1

Bass G.F. (Ed.), 1972. *A History of Seafaring*. London: Thames and Hudson. ISBN 0 500 01077 3.

Beechey, F.W., 1828. *Proceedings of the Expedition to Explore the North African Coast*. John Murray. London.

Blackman D.J. and Rankov, B. *et al.*, 2014. *Shipsheds of the Ancient Mediterranean*. Cambridge University Press. ISBN-13: 978-1107001336.

Flemming N.C., 1959. Report of talk to RGS on Apollonia. *Triton*. v.4, no.2.

Flemming, N.C., 1959. Underwater adventure in Apollonia. *Geographical Magazine*, v.31, pp. 497-508.

Flemming, N.C., 1960. Apollonia revisited. *Geographical Magazine*. v.32. pp. 522-530.

Flemming, N.C., 1962. Sunken cities and forgotten wrecks, in: Deacon, G.E.R. (Ed.), *Oceans*, pp. 123-171. London: Hamlyn.

Flemming, N.C., 1964. Underwater survey of Apollonia; preliminary report, in: du Plat Taylor, J. (Ed.), *Marine Archaeology*, pp. 168-178. London: Hutchinson.

Flemming, N.C., 1971. *Cities in the Sea*. New York: Doubleday.

Flemming, N.C., 1971. *Cities in the Sea*. London:. New English Library.

Flemming, N.C., 1974. *Cities in the Sea*. Japanese Edition, Tokyo. Gukusei-sha Publishing Co.

Goodchild, R.G., Pedley, J.G., and White, D., 1976. *Apollonia, the Port of Cyrene: Excavations by the University of Michigan,1965-1967*. Tripoli. *Libya Antiqua*, suppl. 4.

Higginbotham, J.A., 1997. *Piscinae: Artificial Fishponds in Roman Italy*. Chapel Hill, NC: University of North Carolina Press.

Muckelroy, K., 1980. *Archaeology Underwater, an atlas of the world's submerged sites* (see pages 174-175). New York, London: McGraw-Hill Book Company. ISBN 0-07-043951-6.

Lamboglia, N. and Roghi, G., 1965. Wreck excavation 1958-59. in: du Plat Taylor, J. (Ed.), *Marine Archaeology*, pp.103-107. London: Hutchinson.

Pizzinato C. and Beltrame, C., 2012. A project for the creation of an underwater archaeological park at Apollonia, Libya. *International Journal of the Society for Underwater Technology*, v.30, no.4, pp. 217-224.

Schmiedt, G., 1972. *Il livello antico del Mar Tirreno*. Firenze: Leo Olschki. (Includes extensive details of ancient fish tanks on the Italian coast).

Sintes, C., 2010. Les Neosoikoi d'Apollonia de Cyrénaïque. in: Blackman, D.J., and Lentini, M.C. (Eds), *Ricoveri por navi militari nei porti del Mediterraneo antico e medieval,l* pp. 83-96. Atti del Laboratorio di Ricerca. Ravello, 4-5 Nov. 2005.

Throckmorton, P. (Ed.), 1987. *History from the Sea.* London: Mitchell Beazley. ISBN 0-85533-614-5.

Woods, J.D., 2011. *Xlendi Reef and its Ancient Shipwrecks.* Fondazioni Patrimonju Malti, Palazzo Falson Historic House Museum Publications, 215 pages. ISBN 978-99909-3-190-7.

Websites

https://lsa.umich.edu/kelsey/research/past-field-projects/apollonia-libya.html

Chapter 2

Baddeley, A.D. and Flemming, N.C., 1967. The efficiency of divers breathing oxy-helium. *Ergonomics*, v.10, no.3, pp. 311-319.

Binns, P.E., Flemming, N.C., and Roberts, D.G., 1973. Helen's Reef, geology and bathymetry. *Institute of Geological Sciences Report,* 75/1. pp. 53-60. HMSO.

David, S., 2021. SBS – Silent Warriors: The Authorised Wartime History. Harper Collins, ISBN 0008394547; 9780008394547.

Diving on Rockall. *Scottish Field*, no.120, p. 63.

Express newspaper, July 2, 1972.

Flemming, N.C., 1966. Operational diving with oxy-helium self-contained diving apparatus. in: Lythgoe, J.N. and Woods, J.D. (Eds), *Underwater Association, 1st Symposium Report*, pp. 3-12.

Flemming, N.C., 1973. Diving on Rockall. *Triton*, Jan. 1973, pp. 30-31.

Guardian newspaper, July 3, 1972,p. 10. "Wider yet and wider...". London.

Owen, J., 1972. Rockall Oil Hope, in: *Sunday Telegraph*, 18 June 1972, p. 3.

Parker, J., 2004. *SBS: the Inside Story of the Special Boat Service.* Headline Press. ISBN 9-7807- 5531-2252.

"Peter Simple" (Michael Wharton), 1972. Way of the World, in: *The Telegraph*, 5 July, 1972, p. 12.

Roberts, D.G., Flemming, N.C., Harrison, R.K., Binns, P.E.,1974. Helen's Reef: A Microgabroic Intrusion. *Marine Geology*, v.16, pp. 21-30.

Thorne, T., 1998. *Brasso, Blanco and Bull.* London. Constable Robinson Ltd. ISBN 1-84119-178-7.

Warner, P., *1983. The SBS. Sphere Books. ISBN 0-7221-8911-7.*

Chapter 3

Coit, S., 1891. *Neighbourhood Guilds: an instrument for Social Reform.* London. Swan Sonnenschein and Co. (Many of Stanton Coit's books are available as modern facsimile reproductions.)

Walter, W.G., 1953. *The Living Brain.* W.W.Norton. ISBN 978-0393-00153-2.

Websites

Chambers, P. A History of Rugby Radio. https://web.archive.org/web/20061015113611/http://www.alan.melia.btinternet.co.uk/rugbyrs.htm

Emmerson Andrew on 28 May, 2003. https://www.subbrit.org.uk/sites/rugby-radio/ http://rugbyradiostation.co.uk/articles/1927-article/

Chapter 4

Billaud, Y., 2017. Submerged Karst Structures of the French Mediterranean Coast, in: Flemming, N.C., Harff, J., Moura, D., Burgess, A., and Bailey, G. (Eds). *Submerged Landscapes of the European Continental Shelf: Quaternary Geology and Prehistoric Paleoenvironments*, pp. 333-340. Chichester, UK: Wiley-Blackwell. ISBN 978-1-118-92213-2.

Busk, G., 1865. On a very ancient human cranium from Gibraltar, *Report of the 34th meeting of the British Association for the Advancement of Science, Bath, 1864*. pp. 91-92.

Clottes, J. and Courtin, J., 1994. *The Cave beneath the sea*. New York: Harry N. Abrams.

Clottes, J., Courtin J., and Vanrell, L. (Eds), 2005. *Cosquer redécouvert*. Paris: Seuil.

Flemming, N.C., 1963. Underwater caves of Gibraltar, in: *Report of Second World Congress of Underwater Activities, British Sub-Aqua Club*, pp. 96-106. London: Palantype.

Flemming, N.C., 1965. Form and Relation to Present Sea Level of Pleistocene Marine Erosion Features. *The Journal of Geology*, v.73, no.5. pp. 799-811. University of Chicago.

Flemming, N.C., 1968. Derivation of Pleistocene Marine Chronology from Morphometry of Erosion Profiles. *The Journal of Geology*, v.76, no.3. pp. 280-296. University of Chicago.

Flemming, N.C. and Antonioli, F., 2017. Prehistoric archaeology, palaeontology, and climate change indicators from caves submerged by change of sea level. in: Campbell, P.B. (Ed.), *The Archaeology of Underwater Caves*, pp. 23-38. Southampton, Highfield Press. ISBN 978-0992633677.

Finlayson, C., Fa, D.A., Jimenez Espejo, F., Carrion, J.S., Finlayson, G., Pacheco, F.G., Rodrıguez Vidal, J., Stringer, C., and Martınez Ruiz, F., 2008. Gorham's Cave, Gibraltar – The persistence of a Neanderthal population. *Quaternary International*, v.181, pp. 64-71.

Martineau, M.P., 1966. Marine terraces in Malta, in: Lythgoe, J.N., and Woods, J.D. (Eds), *Symposium of the Underwater Association 1965*, p.68-71. Underwater Association, UK.

Waechter, J. d'A., 1951. Excavations at Gorham's Cave, Gibraltar. *Proceedings of the Prehistoric Society*. v.17, pp. 83-92.

Website

CUUEG report for 1969 and web-reference: http://cuueg.org.uk/about-us/history/#apollonia; http://cuueg.org.uk/downloads/Elaphonisos_1968.pdf

Chapter 5

Cloud, W., 1968. Science Cities Under the Sea. *Popular Mechanics*, Feb. 1968. pp. 123-127.

Cousteau, J-Y, and Dumas, F., 1953. *The Silent World*. Harper.

Craven, J.P., 2001. *The Silent War*. New York: Simon and Schuster. ISBN 10:0684872137.

Flemming, N.C. (Ed.), 1977. *The Undersea*. Rainbird books. London, Cassell and Co. ISBN 0-02-538740-5; New York, McMillan; Rigby, Australia. ISBN 0727 0032 275.

Flemming, N.C., 1968. Functional requirements for research work submersibles. *Journal of the Royal Aeronautical Society*, v.72, pp. 123-131.

Flemming, N.C., 1964. Tank experiments on the sorting of beach material during cusp formation. *Journal of Sedimentary Petrology*, v.34, pp. 112-122.

High, L., Ellis, L.E., Schroeder W.W., and Loverich, G., 1973. Evaluation of the undersea habitats – Tektite II, Hydro-Lab, and Edalhab – for scientific saturation diving programs. *Helgoländer wissen Meeresunter* v.24, pp. 16-44.

Hill, M.N. (Ed.), 1962. *The Sea. Ideas and Observations on Progress in the Study of the Seas. Vol. 1: Physical Oceanography*. New York, London: Interscience Publishers, John Wiley & Sons Ltd.

Hill, M.N. (Ed.), 1963. *The Sea. Ideas and Observations on Progress in the Study of the Seas. Vol. 2: Composition of Sea-Water. Comparative and Descriptive Oceanography*. New York, London: Interscience Publishers, John Wiley & Sons Ltd.

Mero, J., 1965. *The mineral resources of the sea*. Amsterdam: Elsevier Publishing Company.

Pauli, D.C. and Coles, H.A. (Eds), 1970. Project Tektite I: A Multi-agency 60-day saturated dive conducted by the US Navy, NASA, the Dept of Interior and General Electric Company. ONR Report DR 153. Washington DC. ONR. Synopsis available at: https://play.google.com/books/reader?id=JtVHAQAAIAAJ&hl=en_GB&pg=GBS.PA8

Piccard, J., 1960. Man's Deepest Dive, *National Geographic Magazine*, August 1960. Available at: http://www.deepseachallenge.com/the-expedition/1960-dive/

Robinson, S.A., 2018. *Ocean Science and the British Cold War State* (Palgrave Studies in the History of Science and Technology, XIV). Macmillan.

Sontag, S., Drew, C. with Drew, A.L., 1998. *Blind Man's Bluff. The Untold Story of American Submarine Espionage*. London. Hutchinson. ISBN 009-180075-7.

US Commission on Marine Science, Engineering and Resources, 1969. *Our Nation and the Sea: a plan for national action*: report of the Commission on Marine Science Engineering and Resources (the "Stratton Commission"). Washington DC.

Vadus, J.R., (Manager, Technology Manned Undersea Science and Technology), 1975. International Review of Manned Submersibles and Habitats, U.S. Department of Commerce.

Wenk, E., 1972. *The Politics of the Ocean*. Seattle and London. University of Washington Press.

Chapter 6

Abulafia, D., 2011. *The Great Sea A human history of the Mediterranean*. London. Penguin Books. SBN 978 0 141 97716 4.

Bean, G.E., 1968. *Turkey's Southern Shore: Archaeological Guide*. Ernest Benn Ltd. ISBN 10: 0510032516.

Blue, L., 2007. Locating the harbour: Myos Hormos/ Quseir al-Qadim: A Roman and Islamic port on the Red Sea coast of Egypt. *International Journal of Nautical Archaeology*, v.36, no.2, pp. 265-281.

Blue, L., 2011. The Red Sea, in: Catsambis, A., Ford, B., and Hamilton, D. (Eds), Oxford Handbook of Maritime Archaeology, pp. 295-512. Oxford: Oxford University Press.

Brandon, C., 1996. Cements, Concrete, and Settling Barges at Sebastos: Comparisons with Other Roman Harbor Examples and the Descriptions of Vitruvius. *Caesarea Maritima: A Retrospective after Two Millennia*, pp. 25-40. Leiden, Brill.

Diolé, P., 1954. *4000 Years under the sea: Excursions in Undersea Archaeology* (translated from French edition, 1953). London. Pan Books.

Dvorak, J.J., Mastrolorenzo, G., 1991. The mechanisms of recent vertical crustal movements in Campi Flegrei caldera, southern Italy. *GSA Special Papers*, Geological Society of America. DOI: 10.1130/SPE263.

Flemming, N.C., 1969. Archaeological evidence for eustatic change of sea level and earth movements in the Western Mediterranean in the last 2,000 years. Geological Society of America, Special paper 109.

Flemming, N.C., 1971. *Cities in the Sea*. London. New English Library.

Flemming, N.C., 1980. Submerged harbours of the Mediterranean, in Muckelroy, K. (Ed.), *Atlas of Underwater Archaeology*, pp. 162-177. London, Russel-Sharp.

Flemming, N C., and Webb, C.O., 1986. Tectonic and eustatic coastal changes during the last 10,000 years derived from archaeological data. *Zeitschrift für Geomorphologie*, December, Suppl – Bd62, pp. 129.

Flemming, N.C.,1968. Holocene earth movements and eustatic sea level change in the Peloponnese. *Nature*, v.217, pp. 1031-1032.

Flemming, N.C.,1971. *Cities in the Sea*. New York: Doubleday.

Flemming, N.C., 1973. Archaeological evidence for eustatic and tectonic components of relative sea level change in the South Aegean, in: Blackman, D.J. (Ed.), *Marine Archaeology* pp. 1-66. Colston Papers 23. Butterworth Press.

Flemming, N.C., 1978. Holocene eustatic changes and coastal tectonics in the north-east Mediterranean: implications for models of crustal consumption. *Philosophical Transactions of the Royal Society*, v.289, pp. 405-458.

Frost H., 1963. *Under the Mediterranean: Marine Antiquities*. Routledge & Kegan Paul PLC.

Garland, R., 1987. *The Piraeus*. London. Duckworth. ISBN 0-7156-2159-9.

Goddio, F., 1998. *Alexandria: The Submerged Royal Quarters*. London: Periplus Publishing. ISBN-10: 1902699009.

Goddio, F., 2007. *Underwater Archaeology in the Canopic Region in Egypt: The Topography and Excavation of Heracleion-Thonis and East Canopus (1996 – 2006)*. Oxford Centre for Maritime Archaeology, University of Oxford, Oxbow Books. ISBN: 978-0-9549627-3-9.

Harding, A.F., Cadogan, G. and Howell, R., 1969. Pavlopetri, an underwater Bronze Age town in Laconia. *Annual of the British School at Athens* v.64, pp. 112-142.

Henderson, J.C., Gallou, C., Flemming, N.C. and Spondylis, E., 2011. The Pavlopetri Underwater Archaeology Project: investigating an ancient submerged town. EAA volume., In: Benjamin, J., Bonsall, C., Pickard, C., and Fischer, A. (Eds), *Submerged Prehistory, papers presented at the 2009 Conference of the European Archaeologists' Association*, pp. 207-218. Oxbow Books, ISBN-13:978-1-84217-418-0.

Hohlfelder, R., 2007. Constructing the Harbour of Caesarea Palaestina, Israel: New Evidence from ROMACONS Field Campaign of October 2005. *International Journal of Nautical Archaeology*, v.36, pp. 409-415.

Holum, K., 1988. *King Herod's Dream: Caesarea on the Sea*. New York: Norton.

Keay, S. and Paroli, L., 2011. *Portus and Its Hinterland.* Recent Archaeological Research Archaeological Monographs of the British School at Rome 18. London: The British School at Rome.

Keay, S., Millett, M., Paroli, L., and Strutt, K., 2005. *Portus: An Archaeological Survey of the Port of Imperial Rome*. Archaeological Monographs of the British School at Rome 15, London: The British School at Rome.

Keay, S., 2012. *Rome, Portus and the Mediterranean*. London: British School at Rome.

Knapp, A.B., 2018. *Seafaring and Seafarers in the Bronze Age Eastern Mediterranean*. Sidestone Press. ISBN: 9789088905544

Marriner, N., Morhange, C., Kaniewski, D., and Carayon, N., 2014. Ancient harbour infrastructure in the Levant: Tracking the birth and rise of new forms of anthropogenic pressure. *Scientific Reports* 4, 5554 (2014). https://doi.org/10.1038/srep05554.

Raban, A., 1992. Sebastos: the royal harbour at Caesarea Maritima – a short-lived giant, *International Journal of Nautical Archaeology* v.21, pp. 111-124.

Raban, A., and Holum, K.G., 1996. *Caesarea Maritima: a retrospective after two millennia. Leiden: Brill.* ISBN 90-04-10378-3.

Reinhardt, E., Goodman, B., Boyce, J., Lopez, G., Hengstum, P., Rink, W., Mart, Y., Raban, A., 2006. The Tsunami of 13 December AD 115 and the Destruction of Herod the Great's Harbor at Caesarea Maritima, Israel. *Geology* v.34, pp. 1061-1064.

Spratt, T.A.B., 1865. *Travels and researches in Crete* (2 vols.). J. van Vorst.

Tallet, P., and Sukhna, A., 2012. Wadi el-Jarf: Two newly discovered pharaonic harbours on the Suez Gulf. *The British Museum Studies in Ancient Egypt and Sudan*, v.18, pp. 147-68.

Vacchi, M., Marriner, N., Morhange, C., Spada, G., Fontana, A. and Rovere, A., 2016. Multiproxy assessment of Holocene relative sea-level changes in the western Mediterranean: Sea-level variability and improvements in the definition of the isostatic signal. *Earth-Science Reviews*, v.155, pp. 172-197. https://doi.org/10.1016/j.earscirev.2016.02.002.

Website

Graauw, A.: http://www.ancientportsantiques.com/; http://www.ancientportsantiques.com/wp-content/uploads/Documents/AUTHORS/AdG/AncientPortsVol-III-Structures.pdf

Chapter 7

Croucher, N., 2000. *Legless But Smiling*. St Ives Printing and Publishing Co. ISBN: 9780948385346.

Flemming, N.C., 1991. pp. 341, 426, 504, 505, 506, 518, 519, 520, 524, 528, 536, 537, in: Walsh, A. (Ed.), *Nothing Ventured: Disabled People Travel The World.* London: Rough Guides. ISBN 9780747102083.

Fraser, D., 2017. *The Big Little Things*. Seven Dials: Orion Publishing Group. ISBN 978-1-4091-6778-5.

Gardner, Frank, 2006. *Blood and Sand*. Bantam Editions, Penguin Books. ISBN 0593055780.

Hockenberry, J., 1995. *Declarations of Independence: War Zones and Wheelchairs*. London: Viking, Penguin Group. ISBN 0670865044.

Websites

Disabled-diving organizations

Spinal Injuries Association, 2 Trueman Pl, Oldbrook, Milton Keynes MK6 2HH, phone 0800 980 0501. https://www.spinal.co.uk/about-us/

https://www.bsac.com/training/instructor-courses/diving-for-all-programme/

http://www.diveability.org/

https://www.2dive4.co.uk/scuba-diving-with-disibilities/

https://www.iahd.org/en/

https://www.scubatravel.com/extras/disability/disability.html

Sailing

https://www.rya.org.uk/programmes/rya-sailability/equipment-directory/Pages/hub.aspx

Magazines

Enable Magazine: http://enablemagazine.co.uk/

Able Magazine: https://ablemagazine.co.uk/

Posability Magazine: https://posabilitymagazine.co.uk/

Ucan2 Magazine: https://www.ucan2magazine.co.uk/

Ability Needs Magazine: http://www.abilityneeds.co.uk/

Disabled Motoring Magazine: https://www.disabledmotoring.org/?gclid=EAIaIQobChMI4tvSmuCk6QIV0e3tCh07WgpoEAAYASAAEgLC8fD_BwE

Chapter 8

Note: The bulk of the references cited for this chapter refer to the period 1970-2000, when diving equipment and safety procedures were different from today. Equipment has evolved dramatically in the past 20 years, so that the documents of early date cited below must not be taken as a guide to modern diving practice. They refer only to the practices at the date when they were published and are listed to demonstrate the degree of technical attention devoted to diving safety at that time, including the use of scuba in scientific research.

American Academy of Underwater Sciences, 1996. *Methods and Techniques of Underwater Research*. Washington DC: Smithsonian Institution.

Bellamy, N., 1996. A Quantitative Risk Assessment SCUBA DIVING, Health and Safety Executive Report 0090. Paras Limited, http://www.hse.gov.uk/research/crr_pdf/1997/crr97140.pdf

Benjamin, J., and MacKintosh, R., 2015. Regulating Scientific Diving and Underwater Archaeology: legal and historical considerations. *International Journal of Nautical Archaeology*, v.45, no.1, pp. 1-17. https://doi.org/10.1111/1095-9270.12141.

British Sub Aqua Club, 1979. *A Comprehensive Guide to the Techniques of Underwater Swimming*. Eyre and Spottiswoode Ltd.

Health and Safety Executive, 2014. Diving at Work Regulations (Part 1- Chamber sat and platforms and dynamic position; Part II – surface supply maximum depth 50m. Kirby Morgan mask and fins or boots; Parts III and IV are inshore engineering diving not deeper than 50m). http://www.hse.gov.uk/diving/qualifications/approved-list.pdf

Dean, M., Farrari, B., Oxley, I., Redknapp, M., and Watson K., 1992. *Archaeology Underwater, The NAS Guide to Principles and Practice.* Nautical Archaeology Society and Archetype Publications Ltd.

Directorate-General 'Social Affairs' and Directorate 'Health and Safety' Luxembourg, 1978. *Guidance Notes for Safe Diving compiled by the European Diving Technology Committee, Mines Safety and Health Commission, Industrial Health and Safety, Volume 1,* Commission of the European Communities.

Dugan, J., 1956. Professors with Fins, pp. 202-225 in Dugan, J., *Man explores the Sea: the Story of Undersea Exploration.* London: Hamish.

Eaton, B. (Ed.), 1963. *The Undersea Challenge: report of the proceedings of the Second World Congress of Underwater Activities.* London: British Sub Aqua Club, Palantype.

Ellerby, D., Holbrook, M., and Longworth, G. (Eds), 1990. *Advanced Sport Diving (BSAC).* Musselburgh, Scotland: Stanley Paul & Co Ltd, Scotprint Ltd.

Flemming, N.C., 1967. The Technology of Deep Diving, in: *Ministry of Technology Conference "The Technology of the sea and the sea bed"*, pps 300-321. HMSO, AERE.R, 5500.

Flemming, N.C., 1971. Underwater Geomorphology, in: Woods, J.D. and Lythgoe, J.N. (Eds), *Underwater Science: An introduction to experiments by divers*, pp. 265-290. Oxford: Clarendon Press.

Flemming, N.C., 1972 (Ed.). *Code of Practice for Scientific Diving.* Underwater Association.

Flemming, N.C., 1973 (Ed.). *Science Diving International.* London: British Sub Aqua Club.

Flemming, N.C., 1980. Safety Procedures and Training for the Scientific Diver: Technical and Human Aspects of Diving and Diving Safety. pp.137-176 in Document No. 3000/80E. Directorate of Health and Safety, Commission of the European Communities, Luxembourg.

Flemming, N.C., and Miles, D. (Eds), 1974. *Code of Practice for Scientific Diving, 2nd edition.* Swindon: Natural Environment Research Council.

Flemming, N.C. (Ed., with Michael Max), 1987. *International Code of Practice for Scientific Diving.* UNESCO Marine Sciences Division, Technical Series, No. 53. Paris: UNESCO.

Flemming, N.C. and Max, M.D. (Eds), 1990. *Scientific Diving: A General Code of Practice.* UNESCO/CMAS/Florida Sea Grant. ISBN 92-3-102641-0.

Flemming, N.C. and Max, M.D. (Eds), 1996. *Scientific diving: a general code of practice, 2nd edn* Flagstaff, Arizona: Best Publishing/Unesco Publishing.

Flemming, N.C. (with P. J. Lonsdale), 1994. Scientific Diving in Europe: Report from the UK Scientific Diving Supervisory Committee, in: *European Diving Technology Committee conference, Harmonisation of Diving Standards in Europe: 13-15 April 1994.*

Florida State University, 1989. Scientists in the Sea. Seminar Series, Scientific Diving Techniques.

Gamble, J.C. and George, J.D. (Eds), 1979. Progress in Underwater Science, Vol. 4 (New Series) of the Report of the Underwater Association. *Proceedings of the 12th Symposium of the Underwater Association, at the British Museum (Natural History). March 10-11th, 1978.* London: Pentech Press. ISBN 0 7273 1608 7.

Hamilton, R.W., Crosson, D.J., and Hulbert, A.W. (Eds), 1989. Workshop on Enriched Air Nitrox Diving. National Undersea Research Program, Research Report 89-1, US Government Printing Office.

Hempleman, V., 1971. Experimental Observations on Men at Pressures between 4 bars (100ft) and 47 bars (1500ft). Report No. 1-71 Copy No. 10.

Hicks, M., 1994. Findings of a Human Factors study of HSE Part-IV diver training. Health and Safety Executive, Offshore Technology Report OTH 94 447.

Limbrick, J., 2002. *North Sea Divers: A Requiem.* New Generation Publishing. ISBN-10: 0755200365.

Marine Technology Support Unit. 1971. *Submersibles Underwater Vehicles and Habitats.* CMT Publication Number 7.

National Oceanic and Atmospheric Administration, 1975. *NOAA Diving Manual: Diving for Science and Technology 1975.* NOAA Manned Undersea Science and Technology Office, United States Department of Commerce.

Occupational Safety and Health Standards Board, 1979. *Draft Report of the Advisory Committee on Scientific and Technical Diving Operations.* Department of Industrial Relations, State of California.

Schenck, H.V., McAniff, J.J., 1973. United States Underwater Fatality Statistics 1973, NOAA Grant No. 4-3-158-31. Report No. URI-SSR-75-9.

Somers, L.H. (Ed.), 1972. Research Diver's Manual, Sea Grant Technical Report, No. 16. MICHU-SG-71-212. The University of Michigan Sea Grant Program.

South African Underwater Union, 1995. A Diving Force in Science, International Diving Science Symposium, Confédération Mondiale des Activitées Sub Aquatiques. University of Cape Town.

Statutory Instrument 399. 1981 diving regulations: Health and Safety: Diving Operations at Work Regulations. http://www.legislation.gov.uk/uksi/1981/399/pdfs/uksi_19810399_en.pdf.

Underwater Association of Malta 1966 Ltd., 1966. Memorandum and Articles of Association, No. 894247, Gilbert Samuel and Co.

University of British Columbia, 1986. Diving Safety Manual.

University of Delaware, 1972. Guide to Diving Safety, A Diving Manual. College of Marine Studies.

Vallintine, R., 2003. *The Club, 1953 – 2003. Fifty years of British Sub Aqua Club Diving.* Richmond, Surrey: Circle Books. ISBN 0-9538919-5-X.

Warner, J. and Park, F. 1990. Requiem for a Diver. Brown Son and Ferguson, Glasgow. ISBN 0 85174 578 4.

Chapter 9

Adams, R., Martin Brown, M., Colgan, C., Flemming, N., Kite-Powell, H. L., McCarl, B., Mjelde, J., Solow, A., Teisberg, T., and Weiher, R., 2000. The Economics of ISOOS: Benefits and the Rationale for Public Funding. US Dept. of Commerce, NOAA Office of Policy and Strategic Planning. NOOA and Office of Naval Research.

Dahlin, H., Flemming, N.C., and Petersson, S.E., 2013. Sustainable Operational Oceanography. *Proceedings of the Sixth International Conference on EuroGOOS. Sopot, Poland.* EuroGOOS Office, SMHI, 601 76 Norrköping, Sweden. ISBN 978-91-974828.

Flemming, N.C., 1994. Analytical report (The role and objectives of megascience In oceanography, developing GOOS), in: *Megascience: the OECD Forum. Oceanography.* Paris, pp. 71-126. Organisation for Economic Cooperation and Development.

Flemming, N.C., 2007. The SEPRISE (Sustained, Efficient Production of Required Information and Services within Europe) Cost Benefit Analysis, Scoping Report. EuroGOOS document for the European Commission, DG XII Framework 6. Contract 009876. 131 pages. EuroGOOS Website, SEPRISE final documents: www.copernicus.eu/projects/seprise.

Flemming, N.C. (Ed.), 1977. *The Undersea.* London: Rainbird books, Cassell and Co. ISBN 0-02-538740-5; New York: McMillan; Australia: Rigby. ISBN 0727 0032 275.

Flemming, N.C. (Ed.), 1990. Marine Science and Technology in the United Kingdom, A Strategic Framework. Report to Government by the Coordinating Committee on Marine Science and Technology (CCMST). HMSO. ISBN 011701923.

Guymer, T.H., Flemming, N.C., Font, J., Gaspar, P., Johannessen, J., van der Kolff, G.J., le Provost, C., Ratier, A., and Williams, D., 2001. *EuroGOOS Conference on Operational Ocean Observations from Space.* EuroGOOS Publication No. 16. Southampton. Southampton Oceanography Centre. ISBN 0-904175-44-8.

Intergovernmental Oceanographic Commission, 1998. GOOS Prospectus 1998. ISBN 0-904175-39-1.

Laughton, A., Gould, J., Tucker, M.J. and Roe, H. (Eds), 2010. *Of Seas and Ships and Scientists: The Remarkable History of the UK's National Institute of Oceanography, 1949-1973.* Lutterworth Press. ISBN: 9780718892302.

Robinson, S.A., 2018. *Ocean Science and the British Cold War State.* Palgrave Studies in the History of Science and Technology XIV. Macmillan. http://10.1007/978-3-319-73096-7.

Stern, N, 2007. Economics of Climate Change. Government of the United Kingdom. Cambridge University Press. Online ISBN: 9780511817434. https://doi.org/10.1017/CBO9780511817434.

Stratton, J.A. (Ed.), 1969. *Our Nation and the Sea: a plan for national action: report of the commission on marine science, engineering and resources.* United States Commission on Marine Science, Engineering, and Resources. Washington DC. Government Printing Office.

UNCLOS 1982. United Nations Convention on the Law of the Sea. https://www.un.org/depts/los/convention_agreements/texts/unclos/unclos_e.pdf

Wenk, E., 1972. *The Politics of the Ocean.* Seattle and London: University of Washington Press. ISBN 0295952407.

Websites

https://www.oceandecade.org/

http://eurogoos.eu/

http://www.goosocean.org/

http://www.sut.org

http://www.mtsociety.org

http://www.argo.ucsd.edu/

TAO buoy chain array design, and data sources: https://www.pmel.noaa.gov/gtmba/pmel-theme/pacific-ocean-tao

Chapter 10

Ashton, N., Lewis, S.G., De Groote, I., Duffy, S.M., Bates, M., Bates, R., Hoare, P., Lewis, M., Parfitt, S.A., Peglar, S., Williams, C., and Stringer, C., 2014. Hominin footprints from early Pleistocene deposits at Happisburgh, UK. *PLoS One*, v.9 no. 2: e88329. https://doi.org/10.1371/journal.pone.0088329.

Bailey, G.N., King, G.C.P., Lambeck, K., Momber, G., Moran, L.J., Al-Sharekh, A., and Vita-Finzi, C. 2007. Coastlines, Submerged landscapes and Human Evolution: the Red Sea Basin and the Farasan Islands. *Journal of Island and Coastal Archaeology*. v.2, pp. 127-160.

Bailey G.N., Devès M.H., Inglis, R.H., Meredith-Williams, M.G., Momber, G., Sakellariou, D., Sinclair, A.G.M., Rousakis, G., Al Ghamdi, S., and Alsharekh, A.M., 2015. Blue Arabia: Palaeolithic and underwater survey in SW Saudi Arabia and the role of coasts in Pleistocene dispersal. *Quaternary International* 382, pp. 42-57. https://doi.org/10.1016/j.quaint. 2015.01.002.

Bailey, G.N. and Flemming, N.C., 2008. Archaeology of the continental shelf: marine resources, submerged landscapes and underwater archaeology. *Quaternary Science Reviews*. v.3, no.24, pp. 2153-65.

Bailey, G.N., Galanidou, N., Peeters, H., Jöns, H. and Mennenga, M., 2020. *The Archaeology of Europe's Drowned Landscapes*. Cham, Switzerland. Springer Open. ISBN 978-3-030-37366-5. https://link.springer.com/book/10.1007/978-3-030-37367-2.

Benjamin, J., Bonsall, C., Pickard, C., and Fischer, A., 2011. *Submerged Prehistory, Submerged Prehistory*. Oxford. Oxbow Books. doi: 10.1029/2004PA001071.

Blanc, A.C., 1940. Industrie musteriane e paleolitiche superior nelle dune fossili e nelle grotte litoranee del Capo Palinuro. *Rendiconti della Reale Academia d'Italia*, Series VIII v.10, no.1.

Clausen, C.J., Cohen, A.D., Emiliani, C., Holman, J.A., and Stipp, J., 1979. Little Salt Spring, Florida: a unique underwater site. *Science*, v.203, pp. 609-614. DOI: 10.1126/science.203.4381.609

Cliquet, D., Coutard, S., Clet, M., Allix, J., Tessier, B., Lelong, F., Baltzer, A., Mear, Y., Poizot, E., Auguste, P., Alix, P., Olive, J., and Guesnon, J., 2011. The middle palaeolithic underwater site of La Mondrée, Normandy, France. in: Benjamin, J., Bonsall, C., Pickard, C., and Fischer, A. (Eds), *Submerged Prehistory*, pp. 111-128. Oxford: Oxbow Books.

Clottes, J. and Courtin, J., 1996. *The Cave Beneath the Sea: Paleolithic Images at Cosquer*. New York: Harry N. Abrams.

Clottes, J., Courtin, J., and Vanrell, L., 2005. *Cosquer redécouvert*. Paris. Seuil.

Dixon, J.E., Monteleone, K., 2014. Gateway to the Americas: Underwater Archeological Survey in Beringia and the North Pacific. in: Evans, A., Flatman, J.C., and Flemming, N.C. (Eds), *Prehistoric Archaeology on the Continental Shelf: A Global Review*, pp. 95-114. New York: Springer.

Efstratiou, N., 1985. *Aghios Petros: A Neolithic site in the northern Sporades: relationships during the Neolithic of the 5th millennium*. British Archaeological Reports International Series 241.

Evans, A., Flatman, J., and Flemming, N.C. (Eds), 2014. *Prehistoric Archaeology on the Continental Shelf: a Global Review*. New York: Springer.

Flemming, N.C., 1983a. Survival of submerged lithic and Bronze Age artifact sites: a review of case histories, in: Masters and Flemming, pp. 135-174.

Flemming, N.C., 1983b. Preliminary geomorphological survey of an early Neolithic submerged site in the Sporadhes, N. Aegean, in: Masters and Flemming, pp. 233-268.

Flemming, N.C., 1986. A survey of the late Quaternary landscape of the Cootamundra Shoals, North Australia: a preliminary report. in: Flemming, N.C., Marchetti, F., and Stefanon, A. (Eds), *Proceedings of the 7th International Diving Science Symposium of CMAS 1983, Padova, Italy*. pp. 149-180.

Flemming, N.C., 2002. The scope of Strategic Environmental Assessment of North Sea areas SEA3 and SEA2 in regard to prehistoric archaeological remains. Department of Trade and Industry, Strategic Assessment of Parts of the Central and Southern North Sea: SEA3. Consultation Document. Technical Report. 52 pages. CD ROM and website.

Flemming, N.C., 2004 (Ed.). *Submarine Prehistoric Archaeology of the North Sea, Council for British Archaeology*. York: English Heritage. ISBN 1-902771-46-X.

Flemming, N.C., Çağatay, M.N., Chiocci, F.L., Galanidou, N., Jöns, H., Lericolais, G., Missiaen, T., Moore, F., Rosentau, A., Sakellariou, D., Skar, B., Stevenson, A., Weerts, H., 2014. Land Beneath the Waves: Research strategies in submerged landscapes and sea level change: A joint geoscience-humanities research strategy for European Continental Shelf Prehistoric Research, in: Chu, N.-C. and McDonough, N. (Eds), Position paper 21 of the European Marine Board, Ostend, Belgium.

Flemming, N.C., Harff, J., Moura, D., Burgess, A., and Bailey, G. (Eds), 2017. *Submerged Landscapes of the European Continental Shelf: Quaternary Geology and Prehistoric Paleoenvironments*. Chichester, UK: Wiley-Blackwell. ISBN 978-1-118-92213-2.

Galili, E., Weinstein-Evron, M., Hershkovitz, I., Gopher, A., Kislev, M., Lernau, O., Lolska-Horwitz, L., and Lernau, H., 1993. Atlit-Yam: A prehistoric site on the sea floor off the Israeli coast. *Journal of Field Archaeology*, v.20, pp.133-157.

Harding, A. F., Cadogan, G., and Howell, R., 1969. Pavlopetri, an underwater Bronze Age town in Laconia. *Annual of the British School at Athens*, v.64, pp. 113-142.

Harff, J., Bailey, G.N., and Lüth, F. (Eds), 2016. Geology and Archaeology: Submerged Landscapes of the Continental Shelf. *Geological Society, London, Special Publications* 411. http://doi.org/10.1144/SP411.10.

Jöns, H., 2011. Settlement Development in the Shadow of Coastal Changes – Case Studies from the Baltic Rim, in: Harff, J., Björck, S., and Hoth, P. (Eds), *The Baltic Sea Basin*, pp. 301-336. New York: Springer.

Lambeck, K., Purcell, A., Flemming, N.C., Bailey, G.N., and Vita-Finzi, C., 2011. Sea level and shoreline reconstructions for the Red sea during the last glacial cycle and implications for hominin migration. *Quaternary Science Reviews*. v.30 nos.25-26, pp. 3542-3574.

Marriner, N., Morhange, C., Kaniewski, D., and Carayon, N., 2014. Ancient harbour infrastructure in the Levant: Tracking the birth and rise of new forms of anthropogenic pressure. *Scientific Reports* 4, 5554 (2014). https://doi.org/10.1038/srep05554.

Masters, P.M. and Flemming, N.C., 1983. *Quaternary Coastlines and Marine Archaeology*. London and New York: Academic Press.

Murphy, L. 1978. 8-SLO-19: Specialised methodological, technological and physiological approaches to deep water excavation of a prehistoric site at Warm Mineral Springs, Florida. in: Barto Arnold, J. (Ed.) *Beneath the Waters of Time. 9th Conference of Underwater Archaeology*, pp.123-128. Texas Antiquities Committee, No.6.

Oppenheimer, S., 2003. *Out of Eden: The peopling of the World*. London: Constable. ISBN 1-84119-697-5.

Pedersen, L., Fischer, A., and Aaby, B., 1997. The Danish Storebælt since the Ice Age – man, sea and forest. A/S Storebæltsforbindelsen, København.

Peeters, H., Murphy, P., and Flemming, N.C. (Eds), 2009. *North Sea Prehistory research and management framework (NSPRMF).* Amersfoort: English Heritage and Rijksdienst voor het Cultureel Erfgoed.

Reid, C., 1913. *Submerged Forests*. Cambridge: Cambridge University Press.

Scuvée, F., and Verague, J., 1978. *Le gisement sous-marine du Paléolithique Moyen de l'anse de la Mondrée à Fermanville (Manche).* Ministère des Affaires Culturelles, Autorisation No.001740, 1971, CEHP- Littus, BP.306, Cherbourg 50104.

Skaarup, J., 1980. Undersoisk stenhalder. *Tiddskriftet Skalk*, v.1, pp. 3-8.

Stringer, C., 2006. *Homo Britannicus: The incredible story of human life in Britain*. London: Allen Lane. ISBN 978-0-713-99795-8.

Stringer, C., 2000. Coasting out of Africa. *Nature* v.405, pp. 24-25. doi:10.1038/35011166.

Tizzard, L., Bicket, A.R., Benjamin, J.,and Loecker, D. De, 2014. A Middle Palaeolithic site in the southern North Sea: Investigating the archaeology and palaeogeography of Area 240. *Journal of Quaternary Science*, v.29, pp. 698-710. doi:10.1002/jqs.2743.

Werz, B. and Flemming, N.C., 2001. Discovery in Table Bay of the oldest handaxes yet found underwater demonstrates preservation of hominid artefacts on the continental shelf. *South African Journal of Science*, v.97, pp. 183-185.

Wreschner, E.E., 1977. Newe Yam, a submerged late Neolithic settlement near Mount Carmel. in: Arensburg, B. and Bar-Yosef, O. (Eds), *Eretz-Israel* 13, pp. 260-271. Jerusalem. Israel Exploration Society, Institute of Archaeology of the Hebrew University.

Chapter 11

Abdulkariem, A., and Bennett, P., 2014. Heritage under threat: the case of Cyrene. *Libyan Studies* 45, pp. 155-161.The Society for Libyan Studies. doi:10.1017/lis.2013.1

Adams, J.R., Antoniadou, A., Hunt, C.O., Bennett, P., Croudace, I.W., Taylor, R.N., Pearce, R.B., Earl, G.P., Flemming, N.C., Moggeridge, J., Whiteside, T., Oliver, K. and Parker, A.J., 2012. The Belgammel Ram, a Hellenistic-Roman Bronze *Proembolion* Found off the Coast of Libya: test analysis of function, date and metallurgy, with a digital reference archive. *International Journal of Nautical Archaeology*, v.42, no.1, pp. 60-75. doi: 10.1111/1095-9270.12001.

Ashburner, W., [1909] 1976. *The Rhodian Sea Law*. Darmstadt: Scientia Verlag Aalen.

Bennett, P. and Barker, G., 2011. Protecting Libya's Archaeological Heritage. *The African Archaeological Review*, v.28, no.1, pp. 5-25. Africa's Fragile Heritages.

Goodchild, R.G., Pedley, J.G.. and White, D. (Eds), 1976. Apollonia, the port of Cyrene: excavations by the University of Michigan 1965-1967. *Supplements to Libya Antiqua, IV* Tripoli. Department of Antiquities.

Kenrick, P., 2013. *Cyrenaica* (Libya Archaeological Guides). London. Silphium Press. ISBN 978-1900971140.

Laronde A., 1985. Apollonia de Cyrénaique et son histoire: neuf ans de recherches de la mission archéologique francaise en Libye. *Comptes rendus des séances de l'Académie des inscriptions et belles-lettres* 85, pp. 102-115.

Laronde A., 1986. Les ports de la Cyrénaïque; Ptolémaïs et Apollonia. in: Mastino A. (a cura di). *L'Africa Romana*. Atti del III Convegno di studio, Sassari, 13-15 dicembre 1985, pp. 167-177.

Nicholls, R.V., 1970-71. The Trinity College Collection and other recent loans. Fitzwilliam Museum. Archaeological Reports, p. 85.

Pizzinato, C. and Beltrame, C., 2012. A project for the creation of an underwater archaeological park at Apollonia, Libya. *International Journal of the Society for Underwater Technology*, v.30, no.4, pp. 217-224. doi:10.3723/ut.30.217.

Pridemore, M.G., 1996. The form, function, and interrelationships of naval rams: a study of naval rams from antiquity. Thesis submitted to the Office of Graduate Studies, Texas A & M University. 124 pages. https://nautarch.tamu.edu/pdf-files/Pridemore-MA1996.pdf.

Sintes, C., 2010. Les Neosoikoi d'Apollonia de Cyrénaïque. in: Blackman, D.J., and Lentini, M.C. (Eds), *Ricoveri por navi militari nei porti del Mediterraneo antico e medieval*, pp. 83-96. Att del Laboratorio di Ricerca. Ravello, 4-5 Nov, 2005.

Yorke, R.A. and Davidson, D.P., 2017. The Harbour at Ptolemais: Hellenistic City of the Libyan Pentapolis. *International Journal of Nautical Archaeology*, v.46, no.1, pp. 48-71. https://doi.org/10.1111/1095-9270.12212.

Index

Numbers in *italic* denote pages with figures.
NF refers to Nic Flemming.